The Geographical Distribution of Animals

With a Study of the Relations of Living and Extinct Faunas

VOLUME 2

ALFRED RUSSEL WALLACE

CAMBRIDGE
UNIVERSITY PRESS

CAMBRIDGE UNIVERSITY PRESS

Cambridge, New York, Melbourne, Madrid, Cape Town,
Singapore, São Paolo, Delhi, Tokyo, Mexico City

Published in the United States of America by Cambridge University Press, New York

www.cambridge.org
Information on this title: www.cambridge.org/9781108037853

© in this compilation Cambridge University Press 2011

This edition first published 1876
This digitally printed version 2011

ISBN 978-1-108-03785-3 Paperback

THE GEOGRAPHICAL

DISTRIBUTION OF ANIMALS.

THE GEOGRAPHICAL

DISTRIBUTION OF ANIMALS

WITH A STUDY OF
THE RELATIONS OF LIVING AND EXTINCT FAUNAS
AS ELUCIDATING THE
PAST CHANGES OF THE EARTH'S SURFACE.

BY

ALFRED RUSSEL WALLACE,

AUTHOR OF "THE MALAY ARCHIPELAGO," ETC.

WITH MAPS AND ILLUSTRATIONS.

IN TWO VOLUMES.—VOLUME II.

London:
MACMILLAN AND CO.
1876.

LONDON :
R. CLAY, SONS, AND TAYLOR, PRINTERS,
BREAD STREET HILL.

CONTENTS OF THE SECOND VOLUME.

PART III. (*continued*).

ZOOLOGICAL GEOGRAPHY: A REVIEW OF THE CHIEF FORMS OF
ANIMAL LIFE IN THE SEVERAL REGIONS AND SUB-REGIONS,
WITH THE INDICATIONS THEY AFFORD OF GEOGRAPHICAL MU-
TATIONS,

CHAPTER XIV.

THE NEOTROPICAL REGION.

CHAPTER XVIII.

THE DISTRIBUTION OF THE FAMILIES AND GENERA OF BIRDS.

CHAPTER XIX.

THE DISTRIBUTION OF THE FAMILIES AND GENERA OF REPTILES AND
AMPHIBIA.

CHAPTER XX.

THE DISTRIBUTION OF THE FAMILIES OF FISHES, WITH THE RANGE OF SUCH
GENERA AS INHABIT FRESH WATER.

MAPS AND ILLUSTRATIONS IN VOL. II.

ERRATA IN VOL. II

As in Vol. I. mis-spellings are not given here, being mostly corrected in the Index.

Page 111, No. 642, *for* 1 *read* 2.

,, 111, No. 643, *for* 15 *read* 9.

,, 267, line 7, *add* Borneo.

,, 276, line 10, *for* 16 Genera *read* 11 Genera.

,, ,, 8 lines from foot, for *Drepanornis* read *Neodrepanis*.

,, 291, 5 lines from foot, for *Sayornis* read *Empidias*.

THE

GEOGRAPHICAL DISTRIBUTION

OF ANIMALS.

PART III. (*continued.*)

ZOOLOGICAL GEOGRAPHY:

*A REVIEW OF THE CHIEF FORMS OF ANIMAL LIFE IN THE
SEVERAL REGIONS AND SUB-REGIONS, WITH THE INDICA-
TIONS THEY AFFORD OF GEOGRAPHICAL MUTATIONS.*

NEOTROPICAL REGION

Scale 1 inch=1,000 miles

CHAPTER XIV.

THE NEOTROPICAL REGION.

THIS region, comprehending not only South America but Tropical North America and the Antilles, may be compared as to extent with the Ethiopian region ; but it is distinguished from all the other great zoological divisions of the globe, by the small proportion of its surface occupied by deserts, by the large proportion of its lowlands, and by the altogether unequalled extent and luxuriance of its tropical forests. It further possesses a grand mountain range, rivalling the Himalayas in altitude and far surpassing them in extent, and which, being wholly situated within the region and running through eighty degrees of latitude, offers a variety of conditions and an extent of mountain slopes, of lofty plateaus and of deep valleys, which no other tropical region can approach. It has a further advantage in a southward prolongation far into the temperate zone, equivalent to a still greater extension of its lofty plateaus; and this has, no doubt, aided the development of the peculiar alpine forms of life which abound in the southern Andes. The climate of this region is exceptionally favourable. Owing to the lofty mountain range situated along its western margin, the moisture-laden trade winds from the Atlantic have free access to the interior. A sufficient proportion of this moisture reaches the higher slopes of the Andes, where its condensation gives rise to innumerable streams, which cut deep ravines and carry down such an amount of sediment, that they have formed the vast plains of the Amazon, of Para-

B 2

guay, and of the Orinooko out of what were once, no doubt, arms of the sea, separating the large islands of Guiana, Brazil, and the Andes. From these concurrent favourable conditions, there has resulted that inexhaustible variety of generic and specific forms with a somewhat limited range of family and ordinal types, which characterise neotropical zoology to a degree nowhere else to be met with.

Together with this variety and richness, there is a remarkable uniformity of animal life over all the tropical continental portions of the region, so that its division into sub-regions is a matter of some difficulty. There is, however, no doubt about separating the West Indian islands as forming a well-marked subdivision; characterised, not only by that poverty of forms which is a general feature of ancient insular groups, but also by a number of peculiar generic types, some of which are quite foreign to the remainder of the region. We must exclude, however, the islands of Trinidad, Tobago, and a few other small islands near the coast, which zoologically form a part of the main land. Again, the South Temperate portion of the continent, together with the high plateaus of the Andes to near the equator, form a well-marked subdivision, characterised by a peculiar fauna, very distinct both positively and negatively from that of the tropical lowland districts. The rest of Tropical South America is so homogeneous in its forms of life that it cannot be conveniently subdivided for the purposes of a work like the present. There are, no doubt, considerable differences in various parts of its vast area, due partly to its having been once separated into three or more islands, in part to existing diversities of physical conditions; and more exact knowledge may enable us to form several provinces or perhaps additional sub-regions. A large proportion of the genera, however, when sufficiently numerous in species, range over almost the whole extent of this sub-region wherever the conditions are favourable. Even the Andes do not seem to form such a barrier as has been supposed. North of the equator, where its western slopes are moist and forest-clad, most of the genera are found on both sides. To the south of this line its western valleys are arid and its lower plains almost deserts; and thus the absence of a number of groups to which verdant forests are essential, can be traced to the unsuitable conditions rather than to the existence of the mountain barrier. All Tropical South America, therefore, is here considered to form but one sub-region.

The portion of North America that lies within the tropics, closely resembles the last sub-region in general zoological features. It possesses hardly any positive distinctions; but there are several of a negative character, many important groups being wholly confined to South America. On the other hand many genera range into Mexico and Guatemala from the north, which never reach South America; so that it is convenient to separate this district as a sub-region, which forms, to some extent, a transition to the Nearctic region.

General Zoological Features of the Neotropical Region.—Richness combined with isolation is the predominant feature of Neotropical zoology, and no other region can approach it in the number of its peculiar family and generic types. It has eight families of Mammalia absolutely confined to it, besides several others which are rare elsewhere. These consist of two families of monkeys, Cebidæ and Hapalidæ, both abounding in genera and species; the Phyllostomidæ, or blood-sucking bats; Chinchillidæ and Caviidæ among rodents; besides the greater part of the Octodontidæ, Echimyidæ and Cercolabidæ. Among edentata, it has Bradypodidæ, or sloths, Dasypodidæ, or armadillos, and Myrmecophagidæ, or anteaters, constituting nearly the entire order; while Procyonidæ, belonging to the carnivora, and Didelphyidæ, a family of marsupials, only extend into the Nearctic region. It has also many peculiar groups of carnivora and of Muridæ, making a total of full a hundred genera confined to the region. Hardly less remarkable is the absence of many widespread groups. With the exception of one genus in the West Indian islands and a *Sorex* which reaches Guatemala and Costa Rica, the Insectivora are wholly wanting; as is also the extensive and wide-spread family of the Viverridæ. It has no oxen or sheep, and indeed no form of ruminant except deer and llamas; neither do its vast forests and grassy plains support a single form of non-ruminant ungulate, except the tapir and the peccary.

Birds.—In birds, the Neotropical region is even richer and more isolated. It possesses no less than 23 families wholly confined within its limits, with 7 others which only extend into the Nearctic region. The names of the peculiar families are : Cærebidæ, or sugar-birds; Phytotomidæ, or plant-cutters; Pipridæ, or manakins; Cotingidæ, or chatterers; Formicariidæ, or ant-thrushes; Dendrocolaptidæ, or tree-creepers ; Pteroptochidæ ; Rhamphastidæ, or toucans; Bucconidæ, or puff-birds; Galbulidæ, or jacamas; Todidæ, or todies; Momotidæ, or motmots; Steatornithidæ. the guacharo, or oil-bird; Cracidæ, or curassows; Tinamidæ, or tinamous ; Opisthocomidæ, the hoazin ; Thinocoridæ ; Cariamidæ Aramidæ; Psophiidæ, or trumpeters ; Eurypygidæ, or sun-bitterns ; and Palamedeidæ, or horned-screamers. The seven which it possesses in common with North America are : Vireonidæ, or greenlets ; Mniotiltidæ, or wood-warblers ; Tanagridæ, or tanagers ; Icteridæ, or hang-nests ; Tyrannidæ, or tyrant-shrikes ; Trochilidæ, or humming-birds ; and Conuridæ, or macaws. Most of these families abound in genera and species, and many are of immense extent ; such as Trochilidæ, with 115 genera, and nearly 400 species ; Tyrannidæ, with more than 60 genera and nearly 300 species ; Tanagridæ, with 43 genera and 300 species ; Dendrocolaptidæ with 43 genera and more than 200 species ; and many other very large groups. There are nearly 600 genera peculiar to the Neotropical region ; but in using this number as a basis of comparison with other regions we must remember, that owing to several ornithologists having made the birds of South America a special study, they have perhaps been more minutely subdivided than in the case of other entire tropical regions.

Distinctive Characters of Neotropical Mammalia.—It is important also to consider the kind and amount of difference between the various animal forms of this region and of the Old World. To begin with the Quadrumana, all the larger American monkeys (Cebidæ) differ from every Old World group in the possession of an additional molar tooth in each jaw ; and it is in this group alone that the tail is developed into a prehensile organ of wonderful power, adapting the animals to a purely arboreal life. Four of the genera, comprising more than half the species, have the prehensile tail, the remainder having this organ either short, or lax as in the Old World monkeys. Other differences from Old World apes, are the possession of a broad nasal septum, and a less opposable thumb; and the absence of cheek-pouches, ischial callosities, and a bony ear-tube. The Hapalidæ, or marmozets, agree with the Cebidæ in all these characters, but have others in addition which still more widely separate them from the Simiidæ; such as an additional premolar tooth, acute claws, and thumb not at all opposable ; so that the whole group of American monkeys are radically different from the remainder of the order.

The Procyonidæ are a distinct family of Carnivora, which make up for the scarcity of Mustelidæ in South America. The Suidæ are represented by the very distinct genus *Dicotyles*(Peccary) forming a separate sub-family, and differing from all other genera in their dentition, the absence of tail and of one of the toes of the hind feet, the possession of a dorsal gland, and only two mammæ. The rodents are represented by the Chinchillidæ and Caviidæ, the latter comprising the largest animals in the order. The Edentata are almost wholly confined to this region ; and the three families of the sloths (Bradypodidæ), armadillos (Dasypodidæ), and ant-eaters (Myrmecophagidæ), are widely separated in structure from any Old World animals. Lastly, we have the opossums (Didelphyidæ), a family of marsupials, but having no close affinity to any of the numerous Australian forms of that order. We have already arrived at the conclusion that the presence of marsupials in South America is not due to any direct transference from Australia, but that their introduction is comparatively recent, and that they came from the Old World by way of North America (vol. i., p. 155). But the numerous and deep-seated peculiarities of many other of its mammalia, would indicate a very remote origin ; and a long-continued isolation of South America from the rest of the world is required, in order to account for the preservation and development of so many distinct groups of comparatively low-type quadrupeds.

Distinctive Characters of Neotropical Birds.—The birds which are especially characteristic of this region, present similar distinctive features. In the enormous group of Passerine

birds which, though comprising nearly three-fourths of the
entire class, yet presents hardly any well-marked differences
of structure by which it can be subdivided—the families confined
to America are, for the most part, more closely related to each
other than to the Old World groups.   The ten families forming
the group of "Formicaroid Passeres," in our arrangement (vol.
i., p. 94), are characterised by the absence of singing muscles in
the larynx, and also by an unusual development of the first primary
quill; and seven of this series of families (which are considered
to be less perfectly developed than the great mass of Old World
passeres) are exclusively American, the three belonging to the
Eastern hemisphere being of small extent.   Another group of
ten families—our "Tanagroid Passeres," are characterised by the
abortion or very rudimentary condition of the first quill; and of
these, five are exclusively American, and have numerous genera
and species, while only two are non-American, and these are of
small extent.   On the other hand the "Turdoid Passeres," consisting of 23 families and comprising all the true "singing-birds,"
is poorly represented in America; no family being exclusively
Neotropical, and only three being at all fully represented in South
America, though they comprise the great mass of the Old World
passeres.   These peculiarities, which group together whole series
of families of American birds, point to early separation and long
isolation, no less surely than the more remarkable structural
divergences presented by the Neotropical mammalia.

In the Picariæ, we have first, the toucans (Rhamphastidæ);
an extraordinary and beautiful family, whose enormous gaily-
coloured bills and long feathered tongues, separate them widely
from all other birds.   The Galbulidæ or jacamars, the motmots
(Momotidæ), and the curious little todies (Todidæ) of the
Antilles, are also isolated groups.   But most remarkable of all
is the wonderful family of the humming-birds, which ranges
over all America from Tierra del Fuego to Sitka, and from the
level plains of the Amazon to above the snow-line on the Andes;
which abounds both in genera, species, and individuals, and is
yet strictly confined to this continent alone!   How vast must
have been the time required to develop those beautiful and

highly specialized forms out of some ancestral swift-like type;
how complete and long continued the isolation of their birth-
place to have allowed of their modification and adaptation to
such divergent climates and conditions, yet never to have per-
mitted them to establish themselves in the other continents.
No naturalist can study in detail this single family of birds,
without being profoundly impressed with the vast antiquity of
the South American continent, its long isolation from the rest of
the land surface of the globe, and the persistence through countless
ages of all the conditions requisite for the development and
increase of varied forms of animal life.

Passing on to the parrot tribe, we find the peculiar family of the
Conuridæ, of which the macaws are the highest development, very
largely represented. It is in the gallinaceous birds however that
we again meet with wholly isolated groups. The Cracidæ, including the curassows and guans, have no immediate relations
with any of the Old World families. Professor Huxley considers
them to approach nearest to (though still very remote from) the
Australian megapodes; and here, as in the case of the marsupials, we probably have divergent modifications of an ancient
type once widely distributed, not a direct communication between
the southern continents. The Tinamidæ or tinamous, point to a
still more remote antiquity, since their nearest allies are believed
to be the Struthiones or ostrich tribe, of which a few representatives are scattered widely over the globe. The hoazin of
Guiana (Opisthocomus) is another isolated form, not only the
type of a family, but perhaps of an extinct order of birds. Passing on to the waders, we have a number of peculiar family types,
all indicative of antiquity and isolation. The *Cariama* of the
plains of Brazil, a bird somewhat intermediate between a bustard
and a hawk, is one of these; the elegant *Psophia* or trumpeter of
the Amazonian forests; the beautiful little sun-bittern of the
river banks (*Eurypyga*) ; and the horned screamers (*Palamedea*),
all form distinct and isolated families of birds, to which the Old
World offers nothing directly comparable.

Reptiles.—The Neotropical region is very rich in varied forms
of reptile life, and the species are very abundant. It has six

altogether peculiar families, and several others which only range into the Nearctic region, as well as a very large number of peculiar or characteristic genera.   As the orders of reptiles differ considerably in their distributional features, they must be considered separately.

The snakes (Ophidia) differ from all other reptiles, and from most other orders of vertebrates, in the wide average distribution of the families; so that such an isolated region as the Neotropical possesses no peculiar family, nor even one confined to the American continent. The families of most restricted range are— the Scytalidæ, only found elsewhere in the Philippine islands; the Amblycephalidæ, common to the Oriental and Neotropical regions; and the Tortricidæ, most abundant in the Oriental region, but found also in the·Austro-Malay islands and Tropical South America.   Sixteen of the families of snakes occur in the region, the Colubridæ, Amblycephalidæ, and Pythonidæ, being those which are best represented by peculiar forms.   There are 25 peculiar or characteristic genera, the most important being *Dromicus* (Colubridæ) ; *Boa*, *Epicrates*, and *Ungalia* (Pythonidæ) ; *Elaps* (Elapidæ) ; and *Craspedocephalus* (Crotalidæ).

The lizards (Lacertilia) are generally more restricted in their range; hence we find that out of 15 families which inhabit the region, 5 are altogether peculiar, and 4 more extend only to N. America.   The peculiar families are Helodermidæ, Anadiadæ, Chirocolidæ, Iphisiadæ, and Cercosauridæ ; but it must be noted that these all possess but a single genus each, and only two of them (Chirocolidæ and Cercosauridæ) have more than a single species.   The families which range over both South and North America are Chirotidæ, Chalcidæ, Teidæ, and Iguanida., the first and second are of small extent, but the other two are very large groups, the Teidæ possessing 12 genera and near 80 species; the Iguanidæ 40 genera and near 150 species ; the greater part of which are Neotropical. There are more than 50 peculiar or highly characteristic genera of lizards, about 40 of which belong to the Teidæ and Iguanidæ, which thus especially characterize the region.   The most important and characteristic genera are the following : *Ameiva* (Teidæ) ; *Gymnopthalmus* (Gymnopthalmidæ) ;

Celestus and *Diploglossus* (Scincidæ); *Sphærodactylus* (Geckotidæ); *Liocephalus, Liolænus, Proctotretus,* and many smaller
genera (Iguanidæ). The three extensive Old World families
Varanidæ, Lacertidæ, and Agamidæ, are absent from the entire
American continent.

In the order Crocodilia, America has the peculiar family of
the alligators (Alligatoridæ), as well as several species of true
crocodiles (Crocodilidæ). The Chelonia (tortoises) are represented by the families Testudinidæ and Chelydidæ, both of wide
range; but there are six peculiar genera,—*Dermatemys* and *Staurotypus* belonging to the former family,—*Peltocephalus, Podocnemis, Hydromedusa,* and *Chelys,* to the latter. Some of the
Amazon river-turtles of the genus *Podocnemys* rival in size the
largest species of true marine turtles (Cheloniidæ), and are equally
good for food.

Amphibia.—The Neotropical region possesses representatives
of sixteen families of Amphibia of which four are peculiar; all
belonging to Anoura or tail-less Batrachians. The Cæciliadæ
or snake-like amphibia, are represented by two peculiar genera,
Siphonopsis and *Rhinatrema.* Tailed Batrachians are almost
unknown, only a few species of *Spelerpes* (Salamandridæ) entering Central America, and one extending as far south as the
Andes of Bogota in South America. Tail-less Batrachians on
the other hand, are abundant; there being 14 families represented, of which 4,—Rhinophryndæ, Hylaplesidæ, Plectromantidæ, and Pipidæ are peculiar. None of these families contain
more than a single genus, and only the second more than a
single species; so that it is not these which give a character to
the South American Amphibia-fauna. The most important and
best represented families are, Ranidæ (true frogs), with eleven
genera and more than 50 species; Polypedatidæ (tree-frogs)
with seven genera and about 40 species; Hylidæ (tree-frogs)
with eight genera and nearly 30 species; Engystomidæ (toads)
(5 genera), Bombinatoridæ (frogs), (4 genera), Phryniscidæ and
Bufonidæ (toads), (each with 2 genera), are also fairly represented. All these families are widely distributed, but the Neotropical genera are, in almost every case, peculiar.

Fresh-water fishes.—The great rivers of Tropical America abound in fish of many strange forms and peculiar types. Three families, and three sub-family groups are peculiar, while the number of peculiar genera is about 120. The peculiar families are Polycentridæ, with two genera; Gymnotidæ, a family which includes the electric eels, (5 genera); and Trygonidæ, the rays, which are everywhere marine except in the great rivers of South America, where many species are found, belonging to two genera. Of the extensive family Siluridæ, three sub-families Siluridæ anomalopteræ, S. olisthopteræ, and S. branchiolæ, are confined to this region. The larger and more important of the peculiar genera are the following: *Percilia*, inhabiting Chilian and *Percichthys* South Temperate rivers, belong to the Perch family (Percidæ); *Acharnes*, found only in Guiana, belongs to the Nandidæ, a family of wide range in the tropics; the Chromidæ, a family of exclusively fresh-water fishes found in the tropics of the Ethiopian, Oriental and Neotropical regions, are here represented by 15 genera, the more important being *Acara* (17 sp.), *Heros* (26 sp.), *Crenicichla* (9 sp.), *Satanoperca* (7 sp.). Many of these fishes are beautifully marked and coloured. The Siluridæ proteropteræ are represented by 14 genera, of which *Pimelodus* (42 sp.), and *Platystoma* (11 sp.), are the most important; the Siluridæ stenobranchiæ by 11 genera, the chief being *Doras* (13 sp.), *Auchenipterus* (9 sp.), and *Oxydoras* (7 sp.). The Siluridæ proteropodes are represented by 16 genera, many of them being among the most singular of fresh-water fishes, clothed in coats of mail, and armed with hooks and serrated spines. The following are the most important,—*Chætostomus* (25 sp.), *Loricaria* (17 sp.), *Plecostonus* (15 sp.) and *Callichthys* (11 sp.). The Characinidæ are divided between Tropical America and Tropical Africa, the former possessing about 40 genera and 200 species. The Haplochitonidæ are confined to South America and Australia; the American genus being *Haplochiton*. The Cyprinodontidæ are represented by 18 genera, the most important being, *Pœcilia* (16 sp.), *Girardinus* (10 sp.), and *Gambusia* (8 sp.) The Osteoglossidæ, found in Australian and African rivers, are represented in South America by the peculiar *Arapaima*, the "pirarucu" of the

Amazon. The ancient Sirenoidei, also found in Australia and Africa, have the *Lepidosiren* as their American representative. Lastly, *Ellipisurus* is a genus of rays peculiar to the fresh waters of South America. We may expect these numbers to be largely increased and many new genera to be added, when the extensive collections made by Agassiz in Brazil are described.

Summary of Neotropical Vertebrates.—Summarizing the preceding facts, we find that the Neotropical region possesses no less than 45 families and more than 900 genera of Vertebrata which are altogether peculiar to it; while it has representatives of 168 families out of a total of 330, showing that 162 families are altogether absent. It has also representatives of 131 genera of Mammalia of which 103 are peculiar to it, a proportion of $\frac{4}{5}$; while of 683 genera of land-birds no less than 576 are peculiar, being almost exactly $\frac{5}{6}$ of the whole. These numbers and proportions are far higher than in the case of any other region.

Insects.

The Neotropical region is so excessively rich in insect life, it so abounds in peculiar groups, in forms of exquisite beauty, and in an endless profusion of species, that no adequate idea of this branch of its fauna can be conveyed by the mere enumeration of peculiar and characteristic groups, to which we are here compelled to limit ourselves. Our facts and figures will, however, furnish data for comparison; and will thus enable those who have some knowledge of the entomology of any other country, to form a better notion of the vast wealth of insect life in this region, than a more general and picturesque description could afford them.

Lepidoptera.—The Butterflies of South America surpass those of all other regions in numbers, variety and beauty; and we find here, not only more peculiar genera and families than elsewhere, but, what is very remarkable, a fuller representation of the whole series of families. Out of the 16 families of butterflies in all parts of the world, 13 are found here, and 3 of these are wholly peculiar—Brassolidæ, Heliconidæ, and Eurygonidæ, with a fourth, Erycinidæ, which only extends into the Nearctic region; so that there are 4 families peculiar to America. These four families comprise 68 genera and more than 800 species; alone constituting a very important feature in the entomology of the region. But in almost all the other families there are numbers of peculiar genera, amounting in all to about 200, or not far short of half the total number of genera in the world— (431). We must briefly notice some of the peculiarities of the several families, as represented in this region. The Danaidæ consist of 15 genera, all peculiar, and differing widely from the generally sombre-tinted forms of the rest of the world. The delicate transparent-winged Ithomias of which 160 species are described, are the most remarkable. *Melinœa, Napeogenes, Ceratina* and *Dircenna* are more gaily coloured, and are among the chief ornaments of the forests. The Satyridæ are represented by 25 peculiar genera, many of great beauty; the most remarkable and elegant being the genus *Hœtera* and its allies, whose transparent wings are delicately marked with patches of orange, pink, or violet. The genus *Morpho* is perhaps the grandest development of the butterfly type, being of immense size and adorned with the most brilliant azure tints, which in some species attain a splendour of metallic lustre unsurpassed in nature. The Brassolidæ are even larger, but are crepuscular insects, with rich though sober colouring. The true Heliconii are magnificent insects, most elegantly marked with brilliant and strongly contrasted tints. The Nymphalidæ are represented by such a variety of gorgeous insects that it is difficult to select examples. Prominent are the genera *Catagramma* and *Callithea,* whose exquisite colours and symmetrical markings are unique and indescribable; and these are in some cases rivalled by *Agrias* and *Prepona,* which reproduce their style of coloration although not closely allied to them. The Erycinidæ, consisting of 59 genera and 560 species, comprise the most varied and beautiful of small butterflies; and it would be useless to attempt to indicate the unimaginable combinations of form and colour they present. It must be sufficient to say that nothing elsewhere on the globe at all resembles them. In Lycænidæ the world-wide genus *Thecla* is wonderfully developed, and the South

American species not only surpass all others in size and beauty, but some of them are so gorgeous on the under surface of their wings, as to exceed almost all the combinations of metallic tints we meet with in nature. The last family, Hesperidæ, is also wonderfully developed here, the species being excessively numerous, while some of them redeem the character of this generally sober family, by their rich and elegant coloration.

In the only other group of Lepidoptera we can here notice, the Sphingina, the Neotropical region possesses some peculiar forms. The magnificent diurnal butterfly-like moths, *Urania*, are the most remarkable ; and they are rendered more interesting by the occurrence of a species closely resembling them in Madagascar. Another family of day-flying moths, the Castniidæ, is almost equally divided between the Neotropical and Australian regions, although the genera are more numerous in the latter. The American Castnias are large, thick-bodied insects, with a coarse scaly surface and rich dull colours ; differing widely from the glossy and gaily coloured Agaristas, which are typical of the family in the East.

Coleoptera.—This is so vast a subject that, as in the case of the regions already treated, we must confine our attention to a few of the more important and best known families as representatives of the entire order.

Cicindelidæ.—We find here examples of 15 out of the 35 genera of these insects ; and 10 of these genera are peculiar. The most important are *Oxychila* (11 sp.), *Hiresia* (14 sp.), and *Ctenostoma* (26 sp.). *Odontochila* (57 sp.) is the most abundant and characteristic of all, but is not wholly peculiar, there being a species in the Malay archipelago. *Tetracha*, another large genus, has species in Australia and a few in North America and Europe. The small genus *Peridexia* is divided between Brazil and Madagascar,—a somewhat similar distribution to that of *Urania* noticed above. One genus, *Agrius*, is confined to the southern extremity of the continent.

Carabidæ.—Besides a considerable number of cosmopolitan or wide-spread genera, this family is represented by more than 100 genera which are peculiar to the Neotropical region. The most important of these are *Agra* (150 sp.), *Ardistomus* (44 sp.), *Schizogenius* (25 sp.), *Pelecium* (24 sp.), *Calophena* (22 sp.), *Aspidoglossa* (21 sp.), and *Lia, Camptodonotus, Stenocrepis*, and *Lachnophorus*, with each more than 12 species. These are all tropical; but there are also a number of genera (26) peculiar to Chili and South Temperate America. The most important of these are *Antarctia* (29 sp.), all except two or three confined to South Temperate America; *Scelodontis* (10 sp.), mostly Chilian; *Feronomorpha* (6 sp.) all Chilian; and *Tropidopterus* (4 sp.), all Chilian. *Helluomorpha* (18 sp.), is confined to North and South America; *Galerita, Callida,* and *Tetragonoderus*, are large genera which are chiefly South American but with a few species scattered over the other tropical regions. *Casnonia* and *Lebia* are cosmopolite, but most abundant in South America. *Pachyteles* is mostly South American but with a few species in West Africa; while *Lobodonotus* has one species in South America and two in Africa.

Lucanidæ.—The Neotropical species of this family almost all belong to peculiar genera. Those common to other regions are *Syndesus*, confined to Tropical South America and Australia, and *Platycerus* which is Palæarctic and Nearctic, with one species in Brazil. The most remarkable genus is undoubtedly *Chiasognathus*, confined to Chili. These are large insects of metallic green colours, and armed with enormous serrated mandibles. The allied genera, *Pholidotus* and *Sphenognathus*, inhabit Tropical South America. *Streptocerus* confined to Chili, is interesting, as being allied to the Australian *Lamprima*. The other genera present no remarkable features; but *Sclerognathus* and *Leptinoptera* are the most extensive.

Cetoniidæ.—These magnificent insects are but poorly represented in America; the species being mostly of sombre colours. There are 14 genera, 12 of which are peculiar. The most extensive genus is *Gymnetis*, which, with its allies *Cotinis* and *Allorhina*, form a group which comprehends two-thirds of the Neotropical species of the family. The only other genera of importance are, *Inca* (7 sp.), remarkable for their large size, and being the only American group in which horns are developed on the head;

and *Trigonopeltastes* (6 sp.), allied to the European *Trichius*. The non-peculiar genera are, *Stethodesma*, of which half the species are African and half tropical American; and *Euphoria*, confined to America both North and South.

Buprestidæ.—In this fine group the Neotropical region is tolerably rich, having examples of 39 genera, 18 of which are peculiar to it. Of these, the most extensive are *Conognatha* and *Halecia*, which have a wide range over most parts of the region ; and *Dactylozodes*, confined to the south temperate zone. Of important genera which range beyond the region, *Dicerca* is mainly Nearctic and Palæarctic ; *Cinyra* has a species in North America and one in Australia ; *Curis* is divided between Chili and Australia ; the Australian genus *Stigmodera* has a species in Chili ; *Polycesta* has a species in Madagascar, two in the Mediterranean region, and a few in North America ; *Acherusia* is divided between Australia and Brazil ; *Ptosima* has one species in south temperate America, the rest widely scattered from North America to the Philippines ; *Actenodes* has a single species in North America and another in West Africa ; *Colobogaster* has two in West Africa, one in Java and one in the Moluccas. The relations of South America and Australia as indicated by these insects has already been sufficiently noticed under the latter region.

Longicornia.—The Neotropical Longicorn Coleoptera are overwhelming in their numbers and variety, their singularity and their beauty. In the recent Catalogue of Gemminger and Harold, it is credited with 516 genera, 489 of which are peculiar to it ; while it has only 5 genera in common (exclusively) with the Nearctic, and 4 (in the same way) with the Australian region. Only the more important genera can be here referred to, under the three great families into which these insects are divided.

The Prionidæ are excessively numerous, being grouped in 64 genera, more than double the number possessed by any other region ; and 61 of these are peculiar. The three, common to other regions, are, *Parandra* and *Mallodon*, which are widely distributed ; and *Ergates*, found also in California and Europe. The most remarkable genera are, the magnificently-coloured *Psalidognathus* and *Pyrodes;* the large and strangely marked

Macrodontia; and *Titanus,* the largest insect of the entire family.

Of the Cerambycidæ there are 233 genera, exceeding by one-half, the number in any other region; and 225 of these are peculiar. Only 2 are common to the Neotropical and Nearctic regions exclusively, and 3 to the Neotropical and Australian. The most extensive genera are the elegant *Ibidion* (80 sp.); the richly-coloured *Chrysoprasis* (47 sp.); the prettily-marked *Trachyderes* (53 sp.); with *Odontocera* (25 sp.); *Criodon* (22 sp.); and a host of others of less extent, but often of surpassing interest and beauty. The noteworthy genera of wide range are, *Oeme* and *Cyrtomerus,* which have each a species in West Africa, and *Hammatocerus,* which has one in Australia.

The Lamiidæ have 219 genera, and this is the only tropical region in which they do not exceed the Cerambycidæ. This number is almost exactly the same as that of the Oriental genera, but here there are more peculiar groups, 203 against 160 in the other region. The most extensive genera are *Hemilophus* (80 sp.), *Colobothea* (70 sp.), *Acanthoderes* (56 sp.), *Oncoderes* (48 sp.), *Lepturgus* (40 sp.), *Hypsioma* (32 sp.), and *Tæniotes* (20 sp.). *Macropus longimanus,* commonly called the harlequin beetle, is one of the largest and most singularly-marked insects in the whole family. *Leptostylus* has a single species in New Zealand; *Acanthoderes* has one species in Europe, W. Africa, and Australia, respectively; *Spalacopsis* has a species in W. Africa; *Pachypeza* is common to S. America and the Philippines; *Mesosa* is Oriental and Palæarctic, but has one species on the Amazon; *Apomecyna* ranges through the tropics of the Eastern Hemisphere, but has two species in S. America; *Acanthocinus* has one species in Tasmania, and the rest in South America, North America, and Europe; *Phœa* is wholly Neotropical, except two species in the Philippine Islands.

General Conclusions as to the Neotropical Insect-fauna.— Looking at the insects of the Neotropical region as a whole, we are struck with the vast amount of specialty they present; and, considering how many causes there are which must lead to the dispersal of insects, the number of its groups which are scattered over the globe is not nearly so great as we might expect. This points to a long period of isolation, during which the various forms of life have acted and reacted on each other, leading to such a complex yet harmoniously-balanced result as to defy the competition of the chance immigrants that from time to time must have arrived. This is quite in accordance with the very high antiquity we have shown most insect-forms to possess; and it is no doubt owing to this antiquity, that such a complete diversity of *generic* forms has been here brought about, without any important deviation from the great *family* types which prevail over the rest of the globe.

Land Shells.—The Neotropical region is probably the richest on the globe in Terrestrial Mollusca, but this is owing, not to any extreme productiveness of the equatorial parts of the continent, where almost all other forms of life are so largely developed, but to the altogether exceptional riches of the West India Islands. The most recent estimates show that the Antilles contain more species of land shells than all the rest of the region, and almost exactly as many as all continental America, north and south.

Mr. Thomas Bland, who has long studied American land shells, points out a remarkable difference in the distribution of the Operculated and Inoperculated groups, the former being predominant on the islands, the latter on the continent. The Antilles possess over 600 species of Operculata, to about 150 on the whole American continent, the genera being as 22 to 14. Of Inoperculata the Antilles have 740, the Continent 1,250, the genera being 18 and 22. The proportions of the two groups in each country are, therefore:

		West India Islands.		American Continent.	
Operculata ...	... Gen. 22	Sp. 608 ...	... 14	151
Inoperculata...	... „ 18	„ 737 ...	... 22	1251

The extensive family of the Helicidæ is represented by 22 genera, of which 6 are peculiar. *Spiraxis* is confined to Central America and the Antilles; *Stenopus* and *Sagda* are Antillean only; *Orthalicus, Macroceramus,* and *Bulimulus* have a wider range, the last two extending into the southern United

States. Important and characteristic genera are, *Glandina*, in all the tropical parts of the region; *Cylindrella*, in Central America and the Antilles; *Bulimus*, containing many large and handsome species in South America; *Stenogyra*, widely spread in the tropics; and *Streptaxis*, in Tropical South America.

Among the Operculata, the Aciculidæ are mostly Antillean, two genera being peculiar there, and one, *Truncatella*, of wide distribution, but most abundant in the West Indian Islands. The Cyclostomidæ are represented by 15 genera, 9 being peculiar to the region, and 5 of these (belonging to the subfamily Licinidæ) to the Antilles only. Of these peculiar genera *Cistula* and *Chondropoma* are the most important, ranging over all the tropical parts of the region. Other important genera are *Cyclotus* and *Megalomastoma*; while *Cyclophorus* also occurs all over the region. The Helicinidæ are mostly Neotropical, six out of the seven genera being found here, and four are peculiar. *Stoastoma*, is one of the largest genera; and, with *Trochatella* and *Alcadia*, is confined to the Antilles, while the wide-spread *Helicina* is most abundant there.

The Limacidæ, or Old World slugs, are absent from the region, their place being taken by the allied family, Oncidiadæ.

Marine Shells.—We go out of our usual course to say a few words about the marine shells of this region, because their distribution on the two sides of the continent is important, as an indication of the former separation of North and South America, and the connection of the Atlantic and Pacific Oceans. It was once thought that no species of shells were common to the two sides of the Central American Isthmus, and Dr. Mörch still holds that opinion; but Dr. Philip Carpenter, who has paid special attention to the subject, considers that there are at least 35 species absolutely identical, while as many others are so close that they may be only varieties. Nearly 70 others are distinct but representative species. The genera of marine mollusca are very largely common to the east and west coasts, more than 40 being so named in the lists published by Mr. Woodward. The West Indian Islands being a rich shell district, produce a number of peculiar forms, and the west coast of

South America is, to some extent, peopled by Oriental and Pacific genera of shells. On the west coast there is hardly any coral, while on the east it is abundant, showing a difference of physical conditions that must have greatly influenced the development of mollusca. When these various counteracting influences are taken into consideration, the identity or close affinity of about 140 species and 40 genera on the two sides of the Isthmus of Panama becomes very important; and, combined with the fact of 48 species of fish (or 30 per cent. of those known) being identical on the adjacent coasts of the two oceans (as determined by Dr. Günther), render it probable that Central America has been partially submerged up to comparatively recent geological times. Yet another proof of this former union of two oceans is to be found in the fossil corals of the Antilles of the Miocene age, which Dr. Duncan finds to be more allied to existing Pacific forms, than to those of the Atlantic or even of the Caribbean Sea.

NEOTROPICAL SUB-REGIONS.

In the concluding part of this work devoted to geographical zoology, the sub-regions are arranged in the order best adapted to exhibit them in a tabular form, and to show the affinities of the several regions; but for our present purpose it will be best to take first in order that which is the most important and most extensive, and which exhibits all the peculiar characteristics of the region in their fullest development. We begin therefore with our second division.

II. Tropical South-America, or the Brazilian Sub-region.

This extensive district may be defined as consisting of all the tropical forest-region of South America, including all the open plains and pasture lands, surrounded by, or intimately associated with, the forests. Its central mass consists of the great forest-plain of the Amazons, extending from Paranaiba on the north coast of Brazil (long. 42° W.) to Zamora, in the province of Loja (lat. 4° S., long. 79° W.), high up in the Andes, on the west;— a distance in a straight line of more than 2,500 English miles,

along the whole of which there is (almost certainly) one continuous virgin forest. Its greatest extent from north to south, is
from the mouths of the Orinooko to the eastern slopes of the
Andes near La Paz in Bolivia and a little north of Sta. Cruz de
la Sierra (lat. 18° S.), a distance of about 1,900 miles. Within this
area of continuous forests, are included some open " campos," or
patches of pasture lands, the most important being,—the Campos
of the Upper Rio Branco on the northern boundary of Brazil; a
tract in the interior of British Guiana; and another on the
northern bank of the Amazon near its mouth, and extending
some little distance on its south bank at Santarem. On the
northern bank of the Orinooko are the Llanos, or flat open plains,
partly flooded in the rainy season; but much of the interior of
Venezuela appears to be forest country. The forest again prevails from Panama to Maracaybo, and southwards in the Magda-
lena valley; and on all the western side of the Andes to about
100 miles south of Guayaquil. On the N.E. coast of Brazil is a
tract of open country, in some parts of which (as near Ceara)
rain does not fall for years together; but south of Cape St.
Roque the coast-forests of Brazil commence, extending to lat.
30° S., clothing all the valleys and hill sides as far inland as the
higher mountain ranges, and even penetrating up the great valleys
far into the interior. To the south-west the forest country reappears in Paraguay, and extends in patches and partially
wooded country, till it almost reaches the southern extension of
the Amazonian forests. The interior of Brazil is thus in the
position of a great island-plateau, rising out of, and surrounded
by, a lowland region of ever-verdant forest. The Brazilian sub-
region comprises all this forest-country and its included open
tracts, and so far beyond it as there exists sufficient woody
vegetation to support its peculiar forms of life. It thus extends considerably beyond the tropic in Paraguay and south
Brazil; while the great desert of Chaco, extending from 25° to
30° S., lat. between the Parana and the Andes, as well as the high
plateaus of the Andean range, with the strip of sandy desert on
the Pacific coast as far as to about 5° of south latitude, belong to
south temperate America, or the sub-region of the Andes.

Having already given a sketch of the zoological features of the Neotropical region as a whole, the greater part of which will apply to this sub-region, we must here confine ourselves to an indication of the more important groups which, on the one hand, are confined to it, and on the other are absent; together with a notice of its special relations to other regions.

Mammalia.—Many of the most remarkable of the American monkeys are limited to this sub-region; as *Lagothrix, Pithecia,* and *Brachyurus,* limited to the great Amazonian forests; *Eriodes* to south-east Brazil; and *Callithrix* to tropical South America. All the marmosets (Hapalidæ) are also confined to this sub-region, one only being found at Panama, and perhaps extending a little beyond it. Among other peculiar forms, are 8 genera of bats; 3 peculiar forms of wild dog; *Pteronura,* a genus of otters; *Inia,* a peculiar form of dolphin inhabiting the upper waters of the Amazon; tapirs of the genus *Tapirus* (a distinct genus being found north of Panama); 4 genera of Muridæ; *Ctenomys,* a genus of Octodontidæ; the whole family of Echimyidæ, or spiny rats, (as far as the American continent is concerned) consisting of 8 genera and 28 species; *Chætomys,* a genus of Cercolabidæ; the capybara (*Hydrochærus*) the largest known rodent, belonging to the Caviidæ; the larger ant-eaters (*Myrmecophaga*); sloths of the genus *Bradypus;* 2 genera of armadillos (Dasypodidæ); and two peculiar forms of the opossum family (Didelphyidæ). No group that is typically Neotropical is absent from this sub-region, except such as are peculiar to other single sub-regions and which will be noticed accordingly. The occurrence of a solitary species of hare (*Lepus braziliensis*) in central Brazil and the Andes, is remarkable, as it is cut off from all its allies, the genus not being known to occur elsewhere on the continent further south than Costa Rica. The only important external relation indicated by the Mammalia of this sub-region is towards the Ethiopian region, 2 genera of Echimyidæ, *Aulacodes* and *Petromys,* occurring in South and South-east Africa.

Plate IV. Characteristic Neotropical Mammalia.—Our illustration represents a mountainous forest in Brazil, the part of South America where the Neotropical Mammalia are perhaps best developed. The central and most conspicuous figure is the collared ant-eater, (*Tamandua tetradactyla*), one of the handsomest of the family, in its conspicuous livery of black and white. To the left are a pair of sloths (*Arctopithecus flaccidus*) showing the curious black spot on the back with which many of the species are marked, and which looks like a hole in the trunk of a tree ; but this mark seems to be only found on the male animal. The fur of many of the sloths has a greenish tinge, and Dr. Seemann remarked its resemblance to the *Tillandsia usneoides*, or " vegetable horsehair," which clothes many of the trees in Central America; and this probably conceals them from their enemies, the harpy-eagles. On the right are a pair of opossums (*Didelphys azarœ*), one of them swinging by its prehensile tail. Overhead in the foreground are a group of howling monkeys (*Mycetes ursinus*) the largest of the American Quadrumana, and the noisiest of monkeys. The large hollow vessel into which the hyoid bone is transformed, and which assists in producing their tremendous howling, is alto-gether unique in the animal kingdom. Below them, in the dis-tance, are a group of Sapajou monkeys (*Cebus* sp.) ; while gaudy screaming macaws complete the picture of Brazilian forest life.

Birds.—A very large number of genera of birds, and some entire families, are confined to this sub-region, as will be seen by looking over the list of genera at the end of this chapter. We can here only notice the more important, and summarize the results. More than 120 genera of Passeres are thus limited, belonging to the following 12 families: Sylviidæ (1), Troglo-dytidæ (2), Cœrebidæ (4), Tanagridæ (26), Fringillidæ (8), Ic-teridæ (5), Pteroptochidæ (3), Dendrocolaptidæ (12), Formi-cariidæ (16), Tyrannidæ (22), Cotingidæ (16), Pipridæ (10). Of the Picariæ there are 76 peculiar genera belonging to 9 families, viz., Picidæ (2), Rhamphastidæ (1), Cuculidæ (1), Bucconidæ (2), Galbulidæ (5), Momotidæ (2), Podargidæ (1), Caprimalgidæ (4) Trochilidæ (58). There are 3 peculiar genera of Psittaci, 8 of Gallinæ, the only genus of Opisthocomidæ, 3 of Accipitres, 1 of Rallidæ, *Psophia* and *Eurypyga* types of distinct families, and 1 genus of Ardeidæ, Palamedeidæ, and Anatidæ respectively. The preceding enumeration shows how very rich this sub-region

PLATE XIV.

A BRAZILIAN FOREST, WITH CHARACTERISTIC MAMMALIA.

is in peculiar types of all the most characteristic American families, such as the Tanagridæ, Tyrannidæ, Cotingidæ, Formicariidæ, Trochilidæ, and Galbulidæ.   A considerable proportion of the genera of the Chilian and Mexican sub-regions also occur here, so that out of about 680 genera of Neotropical land-birds more than 500 are represented in this sub-region.

Without entering minutely into the distribution of species it is difficult to sub-divide this extensive territory with any satisfactory result.[1]   The upland tract between the Amazon and Orinooko, which may be termed Guiana, was evidently once an island, yet it possesses few marked distinctive features.   Brazil, which must have formed another great island, has more speciality, but the intermediate Amazonian forests form a perfect transition between them.   The northern portion of the continent west of the Orinooko has more character; and there are indications that this has received many forms from Central and North America, and thus blended two faunas once more distinct than they are now.   The family of wood-warblers (Mniotiltidæ) seems to have belonged to this more northern fauna; for out of 18 genera only 5 extend south of the equator, while 6 range from Mexico or the Antilles into Columbia, some of these being only winter immigrants and no genus being exclusively South American.   The eastern slopes of the Andes constitute, however, the richest and best marked province of this sub-region.   At least 12 genera of tanagers (Tanagridæ) are found here only, with an immense number of Fringillidæ,—the former confined to the forests, the latter ranging to the upland plains.   The ant-thrushes (Formicariidæ) on the other hand seem more abundant in the lowlands, many genera being peculiar to the Amazonian forests.   The superb chatterers (Cotingidæ) also seem to have their head-quarters in the forests of Brazil and Guiana, and to have thence spread

[1] Messrs. Sclater and Salvin, and Professor Newton, divide the Neotropical Region into six sub-regions, of which our " Brazilian sub-region" comprises three—the " Brazilian," the " Amazonian," and the " Columbian ; " but, after due consideration, it does not seem advisable to adopt this subdivision in a general work which treats of all the classes of terrestrial animals.   (See p. 27.)

into the Amazonian valley. Guiana still boasts such remarkable forms as the cardinal chatterer (*Phœnicocercus*), the military chatterer (*Hœmatoderus*), as well as *Querula, Gymnoderus,* and *Gymnocephalus;* but the first three pass to the south side of the Lower Amazon. Here also belong the cock of the rock (*Rupicola*), which ranges from Guiana to the Andes, and the marvellous umbrella-birds of the Rio Nigro and Upper Amazon (*Cephalopterus*), which extends across the Ecuadorean Andes and into Costa Rica. Brazil has *Ptilochloris, Casiornis, Tijuca, Phibalura,* and *Calyptura;* while not a single genus of this family, except perhaps *Heliochœra,* is confined to the extensive range of the Andes. Almost the same phenomena are presented by the allied Pipridæ or manakins, the greater part of the genera and species occurring in Eastern South America, that is in Brazil, Guiana, and the surrounding lowlands rather than in the Andean valleys. The same may be said of the jacamars (Galbulidæ) and puff-birds (Bucconidæ); but the humming-birds (Trochilidæ) have their greatest development in the Andean district. Brazil and Guiana have each a peculiar genus of parrots; Guiana has three peculiar genera of Cracidæ, while the Andes north of the equator have two. The Tinamidæ on the other hand have their metropolis in Brazil, which has two or three peculiar genera, while two others seem confined to the Andes south of the equator. The elegant trumpeters (Psophiidæ) are almost restricted to the Amazonian valley.

Somewhat similar facts occur among the Mammalia. At least 3 genera of monkeys are confined to the great lowland equatorial forests and 1 to Brazil; *Icticyon* (Canidæ) and *Pteronura* (Mustelidæ) belong to Guiana and Brazil; and most of the Echimyidæ are found in the same districts. The sloths, anteaters, and armadillos all seem more characteristic of the eastern districts than of the Andean; while the opossums are perhaps equally plentiful in the Andes.

The preceding facts of distribution lead us to conclude that the highlands of Brazil and of Guiana represent very ancient lands, dating back to a period long anterior to the elevation of the Andean range (which is by no means of great geological antiquity) and perhaps even to the elevation of the continuous land which forms the base of the mountains. It was, no doubt, during their slow elevation and the consequent loosening of the surface, that the vast masses of debris were carried down which filled up the sea separating the Andean chain from the great islands of Brazil and Guiana, and formed that enormous extent of fertile lowland forest, which has created a great continent; given space for the free interaction of the distinct faunas which here met together, and thus greatly assisted in the marvellous development of animal and vegetable life, which no other continent can match. But this development, and the fusion of the various faunas into one homogeneous assemblage must have been a work of time; and it is probable that most of the existing continent was dry land before the Andes had acquired their present altitude. The blending of the originally distinct sub-faunas has been no doubt assisted by elevations and depressions of the land or of the ocean, which have alternately diminished and increased the land-area. This would lead to a crowding together at one time, and a dispersion at others, which would evidently afford opportunity for many previously restricted forms to enter fresh areas and become adapted to new modes of life.

From the preceding sketch it will appear, that the great sub-region of Tropical South America as here defined, is really formed of three originally distinct lands, fused together by the vast lowland Amazonian forests. In the class of birds sufficient materials exist for separating these districts; and that of the Andes contains a larger series of peculiar genera than either of the other sub-regions here adopted. But there are many objections to making such a sub-division here. It is absolutely impossible to define even approximate limits to these divisions—to say for example where the "Andes" ends and where "Brazil" or "Amazonia" or "Guiana" begins; and the unknown border lands separating these are so vast, that many groups, now apparently limited in their distribution, may prove to have a very much wider range. In mammalia, reptiles, and insects, it is even more difficult to maintain such divisions, so that on the whole it seems better to treat the entire area as one sub-region, although recognizing the fact of its zoological and geographical diversity, as well as its vast superiority over every other sub-region in the number and variety of its animal forms.

The reptiles, fishes, mollusca, and insects of this sub-region have been sufficiently discussed in treating of the entire region, as by far the larger proportion of them, except in the case of land-shells, are found here.

Plate XV. Characteristic Neotropical Birds.—To illustrate the ornithology of South America we place our scene on one of the tributaries of the Upper Amazon, a district where this class of animals is the most prominent zoological feature, and where a number of the most remarkable and interesting birds are to be found. On the left we have the umbrella-bird (*Cephalopterus ornatus*), so called from its wonderful crest, which, when expanded, completely overshadows its head like an umbrella. It is also adorned with a long tassel of plumes hanging from its breast, which is formed by a slender fleshy tube clothed with broad feathers. The bird is as large as a crow, of a glossy blue-black colour, and belongs to the same family as the exquisitely tinted blue-and-purple chatterers. Flying towards us are a pair of curl-crested toucans (*Pteroglossus beauharnaisii*), distinguished among all other toucans by a crest composed of small black and shining barbless plumes, resembling curled whalebone. The general plumage is green above, yellow and red beneath, like many of its allies. To the right are two of the exquisite little whiskered hummers, or " frill-necked coquettes," as they are called by Mr. Gould, (*Lophornis gouldi*). These diminutive birds are adorned with green-tipped plumes springing from each side of the throat, as well as with beautiful crests, and are among the most elegant of the great American family of humming-birds, now numbering about 400 known species. Overhead are perched a pair of curassows (*Crax globulosa*), which represent in America the pheasants of the Old World. There are about a dozen species of these fine birds, most of which are adorned with handsome curled crests. That figured, is distinguished by the yellow car-uncular swellings at the base of the bill. The tall crane-like bird near the water is one of the trumpeters, (*Psophia leucoptera*), elegant

PLATE XV.

A FOREST SCENE ON THE UPPER AMAZON, WITH SOME CHARACTERISTIC BIRDS.

birds with silky plumage peculiar to the Amazon valley. They are often kept in houses, where they get very tame and affectionate; and they are useful in catching flies and other house insects, which they do with great perseverance and dexterity.

Islands of Tropical South America.

These are few in number, and, with one exception, not of much interest. Such islands as Trinidad and Sta. Catherina form parts of South America, and have no peculiar groups of animals. The small islands of Fernando Noronha, Trinidad, and Martin Vaz, off the coast of Brazil, are the only Atlantic islands somewhat remote from land; while the Galapagos Archipelago in the Pacific is the only group whose productions have been carefully examined, or which present features of special interest.

Galapagos Islands.—These are situated on the equator, about 500 miles from the coast of Ecuador. They consist of the large Albemarle island, 70 miles long; four much smaller (18 to 25 miles long), named Narborough, James, Indefatigable, and Chatham Islands; four smaller still (9 to 12 miles long), named Abingdon, Bindloes, Hood's, and Charles Islands. All are volcanic, and consist of fields of black basaltic lava, with great numbers of extinct craters, a few which are still active. The islands vary in height from 1,700 to 5,000 feet, and they all rise sufficiently high to enter the region of moist currents of air, so that while the lower parts are parched and excessively sterile, above 800 or 1,000 feet there is a belt of comparatively green and fertile country.

These islands are known to support 58 species of Vertebrates, —1 quadruped, 52 birds and 5 reptiles, the greater part of which are found nowhere else, while a considerable number belong to peculiar and very remarkable genera. We must therefore notice them in some detail.

Mammalia.—This class is represented by a mouse belonging to the American genus *Hesperomys*, but slightly different from any found on the continent. A true rat (*Mus*), slightly differing from any European species, also occurs; and as there can be little doubt that this is an escape from a ship, somewhat changed under its new conditions of life (the genus *Mus* not being indigenous to the American continent), it is not improbable, as Mr. Darwin remarks, that the American mouse may also have been imported by man, and have become similarly changed.

Birds.[1]—Recent researches in the islands have increased the number of land-birds to thirty-two, and of wading and aquatic birds to twenty-three. All the land birds but two or three are peculiar to the islands, and eighteen, or considerably more than half, belong to peculiar genera. Of the waders 4 are peculiar, and of the swimmers 2. These are a rail (*Porzana spilonota*); two herons (*Butorides plumbea* and *Nycticorax pauper*); a flamingo (*Phœnicopterus glyphorhynchus*); while the new aquatics are a gull (*Larus fuliginosus*), and a penguin (*Spheniscus mendiculus*).

The land-birds are much more interesting. All except the birds of prey belong to American genera which abound on the opposite coast or on that of Chili a little further south, or to peculiar genera allied to South American forms. The only *species* not peculiar are, *Dolichonyx oryzivorus*, a bird of very wide range in America and of migratory habits, which often visits the Bermudas 600 miles from North America,—and *Asio accipitrinus*, an owl which is found almost all over the world. The only genera not exclusively American are *Buteo* and *Strix*, of each of which a peculiar species occurs in the Galapagos, although very closely allied to South American species. There remain 10 genera, all either American or peculiar to the Galapagos; and on these we will remark in systematic order.

1. *Mimus*, the group of American mocking-thrushes, is represented by three distinct and well-marked species. 2. *Dendrœca*, an extensive and wide-spread genus of the wood-warblers (Mniotiltidæ), is represented by one species, which ranges over the greater part of the archipelago. The genus is especially abundant in Mexico, the Antilles, and the northern parts of

---

[1] Mr. Salvin, who has critically examined the ornithological fauna of these islands, has kindly corrected my MS. List of the Birds, his valuable paper in the *Transactions of the Zoological Society* not having been published in time for me to make use of it.

tropical America, only one species extending south as far as Chili. 3. *Certhidea*, a peculiar genus originally classed among the finches, but which Mr. Sclater, who has made South American birds his special study, considers to belong to the *Cœrebidæ*, or sugar-birds, a family which is wholly tropical. Two species of this genus inhabit separate islands. 4. *Progne*, the American martins (Hirundinidæ), is represented by a peculiar species. 5. *Geospiza*, a peculiar genus of finches, of which no less than eight species occur in the archipelago, but not more than four in any one island. 6. *Camarhynchus* (6 sp.) and 7. *Cactornis* (4 sp.) are two other peculiar genera of finches; some of the species of which are confined to single islands, while others inhabit several. 8. *Pyrocephalus*, a genus of the American family of tyrant-flycatchers (Tyrannidæ), has one peculiar species closely allied to *T. rubineus*, which has a wide range in South America. 9. *Myiarchus*, another genus of the same family which does not range further south than western Ecuador, has also a representative species found in several of the islands. 10. *Zenaida*, an American genus of pigeons, has a species in James Island and probably in some of the others, closely allied to a species from the west coast of America.

It has been already stated that some of the islands possess peculiar species of birds distinct from the allied forms in other islands, but unfortunately our knowledge of the different islands is so unequal and of some so imperfect, that we can form no useful generalizations as to the distribution of birds among the islands themselves. The largest island is the least known; only one bird being recorded from it, one of the mocking-thrushes found nowhere else. Combining the observations of Mr. Darwin with those of Dr. Habel and Prof. Sundevall, we have species recorded as occurring in seven of the islands. Albemarle island has but one definitely known species; Chatham and Bindloe islands have 11 each; Abingdon and Charles islands 12 each; Indefatigable island and James island have each 18 species. This shows that birds are very fairly distributed over all the islands, one of the smallest and most remote (Abingdon) furnishing as many as the much larger Chatham Island, which is also the nearest

to the mainland. Taking the six islands which seem tolerably
explored, we find that two of the species (*Dendrœca aureola* and
Geospiza fortis) occur in all of them; two others (*Geospiza
strenua* and *Myiarchus magnirostris*) in five; four (*Mimus
melanotis, Geospiza fuliginosa, G. parvula,* and *Camarhynchus
prosthemelas*) in four islands; five (*Certhidea olivacea, Cactornis
scandens, Pyrocephalus nanus*), and two of the birds of prey, in
three islands; nine (*Certhidea fusca, Progne concolor, Geospiza
nebulosa, G. magnirostris, Camarhynchus psittaculus, C. variegatus,
C. habeli* and *Asio accipitrinus*) in two islands; while the remaining
ten species are confined to one island each. These peculiar
species are distributed among the islands as follows. James,
Charles and Abingdon islands, have 2 each; Bindloes, Chatham,
and Indefatigable, 1 each. The amount of speciality of James
Island is perhaps only apparent, owing to our ignorance of the
fauna of the adjacent large Albemarle island; the most remote
islands north and south, Abingdon and Charles, have no doubt
in reality most peculiar species, as they appear to have. The
scarcity of peculiar species in Chatham Island is remarkable, it
being large, very isolated, and the nearest to the mainland.
There is still room for exploration in these islands, especially in
Albemarle, Narborough, and Hood's islands of which we know
nothing.

Reptiles.—The few reptiles found in these islands are very
interesting. There are two snakes, a species of the American
genus *Herpetodryas,* and another which was at first thought to
be a Chilian species (*Psammophis Temminckii*), but which is
now considered to be distinct. Of lizards there are four at least,
belonging to as many genera. One is a species of *Phyllodactylus,*
a wide-spread genus of Geckotidæ; the rest belong to the
American family of the Iguanas, one being a species of the Neotropical genus *Leiocephalus,* the other two very remarkable forms,
Trachycephalus and *Oreocephalus* (formerly united in the genus
Amblyrhynchus). The first is a land, the second a marine, lizard;
both are of large size and very abundant on all the islands; and
they are quite distinct from any of the very numerous genera of
Iguanidæ, spread all over the American continent. The last

reptile is a land tortoise (*Testudo nigra*) of immense size, and also abundant in all the islands. Its nearest ally is the equally large species of the Mascarene Islands; an unusual development due, in both cases, to the absence of enemies permitting these slow but continually growing animals to attain an immense age. It is believed that each island has a distinct variety or species of tortoise.

Insects.—Almost the only insects known from these islands are some Coleoptera, chiefly collected by Mr. Darwin. They consist of a few peculiar species of American or wide ranging genera, the most important being, a *Calosoma*, *Pœcilus*, *Solenophorus*, and *Notaphus*, among the Carabidæ ; an *Oryctes* among the Lamellicornes; two new genera of obscure Heteromera ; two Curculionidæ of wide-spread genera ; a Longicorn of the South American genus *Eburia;* and two small Phytophaga,—a set of species highly suggestive of accidental immigrations at rare and distant intervals.

Land-Shells.—These consist of small and obscure species, forming two peculiar sub-genera of *Bulimulus*, a genus greatly developed on the whole West coast of America ; and a single species of *Buliminus*, a genus which ranges over all the world except America. As in the case of the birds, most of the islands have two or three peculiar species.

General Conclusions.—These islands are wholly volcanic and surrounded by very deep sea; and Mr. Darwin is of opinion, not only that the islands have never been more nearly connected with the mainland than at present, but that they have never been connected among themselves. They are situated on the Equator, in a sea where gales and storms are almost unknown. The main currents are from the south-west, an extension of the Peruvian drift along the west coast of South America. From their great extent, and their volcanoes being now almost extinct, we may assume that they are of considerable antiquity. These facts exactly harmonize with the theory, that they have been peopled by rare accidental immigrations at very remote intervals. The only peculiar *genera* consist of birds and lizards, which must therefore have been the earliest immigrants. We know that small Passerine birds annually reach the Bermudas from America, and the Azores from Europe, the former travelling over 600, the latter over 1000 miles of ocean. These groups of islands are both situated in stormy seas, and the immigrants are so numerous that hardly any specific change in the resident birds has taken place. The Galapagos receive no such annual visitants; hence, when by some rare accident a few individuals of a species did arrive, they remained isolated, probably for thousands of generations, and became gradually modified through natural selection under completely new conditions of existence. Less rare and violent storms would suffice to carry some of these to other islands, and thus the archipelago would in time become stocked. It would appear probable, that those which have undergone most change were the earliest to arrive; so that we might look upon the three peculiar genera of finches, and *Certhidea*, the peculiar form of Cœrebidæ, as among the most ancient inhabitants of the islands, since they have become so modified as to have apparently no near allies on the mainland. But other birds may have arrived nearly at the same time, and yet not have been much changed. A species of very wide range, already adapted to live under very varied conditions and to compete with varied forms of life, might not need to become modified so much as a bird of more restricted range, and more specialized constitution. And if, before any considerable change had been effected, a second immigration of the same species occurred, crossing the breed would tend to bring back the original type of form. While, therefore, we may be sure that birds like the finches, which are profoundly modified and adapted to the special conditions of the climate and vegetation, are among the most ancient of the colonists; we cannot be sure that the less modified form of tyrant-flycatcher or mocking-thrush, or even the unchanged but cosmopolitan owl, were not of coeval date; since even if the parent form on the continent has been changed, successive immigrations may have communicated the same change to the colonists.

The reptiles are somewhat more difficult to account for. We know, however, that lizards have some means of dispersal over the sea, because we find existing species with an enormous range. The ancestors of the *Amblyrhynchi* must have come as early, probably, as the earliest birds ; and the same powers of dispersal have spread them over every island. The two American genera of lizards, and the tortoises, are perhaps later immigrants. Latest of all were the snakes, which hardly differ from continental forms ; but it is not at all improbable that these latter, as well as the peculiar American mouse, have been early human importations. Snakes are continually found on board native canoes whose cabins are thatched with palm leaves ; and a few centuries would probably suffice to produce some modification of a species completely isolated, under conditions widely different from those of its native country. Land-shells, being so few and small, and almost all modifications of one type, are a clear indication of how rare are the conditions which lead to their dispersal over a wide extent of ocean ; since two or three individuals, arriving on two or three occasions only during the whole period of the existence of the islands, would suffice to account for the present fauna. Insects have arrived much more frequently ; and this is in accordance with their habits, their lower specific gravity, their power of flight, and their capacity for resisting for some time the effects of salt water.

We learn, then, from the fauna of these islands, some very important facts. We are taught that tropical land-birds, unless blown out of their usual course by storms, rarely or never venture out to sea, or if they do so, can seldom pass safely over a distance of 500 miles. The immigrants to the Galapagos can hardly have averaged a bird in a thousand years. We learn, that of all reptiles lizards alone have some tolerably effective mode of transmission across the sea ; and this is probably by means of currents, and in connection with floating vegetation. Yet their transmission is a far rarer event than that of land-birds ; for, whereas three female immigrants will account for the lizard population, at least eight or ten ancestors are required for the birds. Land serpents can pass over still more rarely, as two such transmissions would have sufficed to stock the islands with their snakes ; and it is not certain that either of these occurred without the aid of man.

It is doubtful whether mammals or batrachians have any means of passing, independently of man's assistance ; the former having but one doubtfully indigenous representative, the latter none at all. The remarkable absence of all gay or conspicuous flowers in these tropical islands, though possessing a zone of fairly luxuriant shrubby vegetation, and the dependence of this phenomenon on the extreme scarcity of insects, has been already noticed at Vol. I. p. 461, when treating of a somewhat similar peculiarity of the New Zealand fauna and flora.

I. South Temperate America, or the Chilian Sub-region.

This sub-region may be generally defined as the temperate portion of South America. On the south, it commences with the cold damp forests of Tierra del Fuego, and their continuation up the west coast to Chiloe and northward to near Santiago. To the east we have the barren plains of Patagonia, gradually changing towards the north into the more fertile, but still treeless, pampas of La Plata. Whether this sub-region should be continued across the Rio de la Plata into Uruguay and Entre-rios, is somewhat doubtful. To the west of the Parana it extends northward over the Chaco desert, till we approach the border of the great forests near St. Cruz de la Sierra. On the plateau of the Andes, however, it must be continued still further north, along the " paramos" or alpine pastures, till we reach 5° of South latitude. Beyond this the Andes are very narrow, having no double range with an intervening plateau; and although some of the peculiar forms of the temperate zone pass on to the equator or even beyond it, these are not sufficiently numerous to warrant our extending the sub-region to include them. Along with the high Andes it seems necessary to include the western strip of arid country, which is mostly peopled by forms derived from Chili and the south temperate regions.

Mammalia.—This sub-region is well characterised by the possession of an entire family of mammalia having Neotropical affinities—the Chinchillidæ. It consists of 3 genera—*Chinchilla* (2 sp.), inhabiting the Andes of Chili and Peru as far as 9° south latitude, and at from 8,000 to 12,000 feet altitude; *Lagidium* (3 sp.), ranging over the Andes of Chili, Peru, and South Ecuador, from 11,000 to 16,000 feet altitude; and *Lagostomus* (1 sp.), the "viscacha," confined to the pampas between the Uruguay and Rio Negro. Many important genera are also confined to this sub-region. *Auchenia* (4 sp.), including the domesticated llamas and alpacas, the vicugna which inhabits the Andes of Peru and Chili, and the guanaco which ranges over the plains of Patagonia and Tierra del Fuego. Although this genus is allied to the Old World camels, it is a very distinct form, and its introduction from North America, where the family appear to have originated, may date back to a remote epoch. *Ursus ornatus*, the "spectacled bear" of the Chilian Andes, is a remarkable form, supposed to be most allied to the Malay bear, and probably forming a distinct genus, which has been named *Tremarctos*. Four genera of Octodontidæ are also peculiar to this sub-region, or almost so; *Habrocomus* (1 sp.) is Chilian ; *Spalacopus* (2 sp.) is found in Chili and on the east side of the southern Andes; *Octodon* (3 sp.) ranges from Chili into Peru and Bolivia ; *Ctenomys* (6 sp.) from the Straits of Magellan to Bolivia, with one species in South Brazil. *Dolichotis*, one of the Cavies, ranges from Patagonia to Mendoza, and on the east coast to 37½° S. latitude. *Myopotamus* (1 sp.), the coypu (Echimyidæ), ranges from 33° to 48° S. latitude on the west side of the Andes, and from the frontiers of Peru to 42° S. on the east side. *Reithrodon* and *Acodon*, genera of Muridæ, are also confined to Temperate South America; *Tolypeutes* and *Chlamydophorus*, two genera of armadillos, the latter very peculiar in its organization and sometimes placed in a distinct family, are found only in La Plata and the highlands of Bolivia, and so belong to this sub-region. *Otaria*, one of the "eared seals" (Otariidæ), is confined to the coasts of this sub-region and the antarctic islands. Deer of American groups extend as far as Chiloe on the west, and the Straits of Magellan on the east coast. Mice of the South American genera *Hesperomys* and *Reithrodon*, are abundant down to the Straits of Magellan and into Tierra del Fuego, Mr. Darwin having collected more than 20 distinct species. The following are the genera of Mammalia which have been observed on the shores of the Straits of Magellan, those marked * extending into Tierra del Fuego :

Pseudalopex (two wolf-like foxes), *Felis* (the puma), *Mephitis* (skunks), *Cervus* (deer), *Auchenia* (guanaco), *Ctenomys* (tucutucu), *Reithrodon* and *Hesperomys* (American mice).

Birds.—Three families of Birds are confined to this sub-region, —Phytotomidæ (1 genus, 3 sp.), inhabiting Chili, La Plata, and Bolivia; Chionididæ (1 genus, 2 sp.) the "sheath-bills," found only at the southern extremity of the continent and in Kerguelen's Island, which with the other antarctic lands perhaps comes best here; Thinocoridæ (2 genera, 6 species) an isolated family of waders, ranging over the whole sub-region and extending northward to the equatorial Andes. Many genera are also peculiar: 3 of Fringillidæ, and 1 of Icteridæ; 9 of Dendrocolaptidæ, 6 of Tyrannidæ, 3 of Trochilidæ, and 4 of Pteroptochidæ,—the last four South American families. There is also a peculiar genus of parrots (*Henicognathus*) in Chili; two of pigeons (*Metriopelia* and *Gymnopelia*) confined to the Andes and west coast from Peru to Chili; two of Tinamous, *Tinamotes* in the Andes, and *Calodromus* in La Plata; three of Charadriidæ, *Phægornis, Pluvianellus,* and *Oreophilus*; and *Rhea*, the American ostriches, inhabiting all Patagonia and the pampas. Perhaps the Cariamidæ have almost as much right here as in the last sub-region, inhabiting as they do, the "pampas" of La Plata and the upland "campos" of Brazil; and even among the wide-ranging aquatic birds, we have a peculiar genus, *Merganetta,* one of the duck family, which is confined to the temperate plateau of the Andes.

Against this extensive series of characteristic groups, all either of American type or very distinct forms of Old World families, and therefore implying great antiquity, we find, in mammalia and birds, very scanty evidence of that direct affinity with the north temperate zone, on which some naturalists lay so much stress. We cannot point to a single terrestrial genus, which is characteristic of the north and reappears in this south temperate region without also occurring over much of the intervening land. *Mustela* seems only to have reached Peru; *Lepus* is isolated in Brazil; true *Ursus* does not pass south of Mexico. In birds, the northern groups rarely go further south than Mexico or the Columbian Andes; and the only case of discontinuous distribution we can find recorded is that of the genus of ducks, *Camptolœmus*, which has a species on the east side of North America and another in Chili and the Falkland Islands, but these, Professor Newton assures me, do not properly belong to the same genus. Out of 30 genera of land-birds collected on the Rio Negro in Patagonia, by Mr. Hudson, only four extend beyond the American continent, and the same exclusively American character applies equally to its southern extremity. No list appears to have been yet published of the land-birds of the Straits of Magellan and Tierra del Fuego. The following is compiled from the observations of Mr. Darwin, the recent voyage of Professor Cunningham, and other sources; and will be useful for comparison.

TURDIDÆ.
  1. Turdus falklandicus.

TROGLODYTIDÆ.
  2. Troglodytes magellanicus.

FRINGILLIDÆ.
  3. Chrysomitris barbata.
  *4. Phrygilus gayi.
  *5.　　,,　　aldunatii.
  6.　　,,　　fruticeti.
  *7.　　,,　　xanthogrammus.
  8. Zonotrichia pileata.

ICTERIDÆ.
  9. Sturnella militaris.
  10. Curæus aterrimus.

HIRUNDINIDÆ.
  11. Hirundo meyeni.

TYRANNIDÆ.
  12. Tænioptera pyrope.
  13. Myiotheretes rufiventris.
  14. Muscisaxicola mentalis.
  15. Centrites niger.
  16. Anæretes parulus.
  17. Elainea griseogularis.

DENDROCOLAPTIDÆ.
  18. Upucerthia dumetoria.
  *19. Cinclodes patagonicus.
  *20.　　,,　　fuscus.
  *21. Oxyurus spinicauda.

PTEROPTOCHIDÆ.
  *22. Scytalopus magellanicus.

PICIDÆ.
  *23. Campephilus magellanicus.
  24. Picus lignarius.

ALCEDINIDÆ.
  25. Ceryle stellata.

TROCHILIDÆ.
  26. Eustephanus galeritus.

CONURIDÆ.
  27. Conurus patagonus.

VULTURIDÆ.
  28. Cathartes aura.
  29. Sarcorhamphus gryphus.

FALCONIDÆ.
  30. Circus macropterus.
  31. Buteo erythronotus.
  32. Geranoaëtus melanolencus.
  33. Accipiter chilensis.
  34. Cerchneis sparverius.
  35. Milvago albogularis.
  36. Polyborus tharus.

STRIGIDÆ.
  37. Asio accipitrinus.
  38. Bubo magellanicus.
  39. Pholeoptynx cunicularia.
  40. Glaucidium nana.
  41. Syrnium rufipes.

STRUTHIONIDÆ.
  42. Rhea darwinii.

In the above list the species marked * extend to Tierra del Fuego. It is a remarkable fact that so many of the species belong to genera which are wholly Neotropical, and that the specially South American families of Icteridæ, Tyrannidæ, Dendrocolaptidæ, Pteroptochidæ, Trochilidæ, and Conuridæ, should supply more than one-third of the species; while the purely South American genus *Phrygilus*, should be represented by four species, three of which abound in Tierra del Fuego.

Plate XVI. A Scene in the Andes of Chili, with characteristic Animals.—The fauna of South Temperate America being most fully developed in Chili, we place the scene of our illustration in that country. In the foreground we have a pair of the beautiful little chinchillas (*Chinchilla lanigera*), belonging to a family of animals peculiar to the sub-region. There are only two species of this group, both confined to the higher Andes, at about 8000 feet elevation. Coming round a projecting ridge of the mountain, are a herd of vicunas (*Auchenia vicugna*), one of that peculiar form of the camel tribe found in South America and confined to its temperate and alpine regions. The upper bird is a plant-cutter (*Phytotoma rara*), of sober plumage but allied to the beautiful chatterers, though forming a separate family. Below, standing on a rock, is a plover-like bird, the *Thinocorus orbignianus*, which is considered to belong to a separate family, though allied to the plovers and sheath-bills. Its habits are, however, more those of the quails or partridges, living inland in dry and desert places, and feeding on plants, roots, and insects. Above is a condor, the most characteristic bird of the high Andes.

Reptiles and Amphibia.—These groups show, for the most part, similar modifications of American and Neotropical forms, as those we have seen to prevail among the birds. Snakes do not seem to go very far south, but several South American genera of Colubridæ and Dendrophidæ occur in Chili; while *Enophrys* is peculiar to La Plata, and *Callorhinus* to Patagonia, both belonging to the Colubridæ. The Elapidæ do not extend into the temperate zone; but *Craspedocephalus*, one of the Crotalidæ, occurs at Bahia Blanca in Patagonia (Lat. 40° S.)

PLATE XVI.

THE CHILIAN ANDES, WITH CHARACTERISTIC ANIMALS.

Lizards are much more numerous, and there are several peculiar and interesting forms. Three families are represented; Teidæ by two genera—*Callopistes* peculiar to Chili, and *Ameiva* which ranges over almost the whole American continent and is found in Patagonia; Geckotidæ by four genera, two of which,—*Caudiverbera* and *Homonota*—are peculiar to Chili, while *Sphærodactylus* and *Cubina* are Neotropical, the former ranging to Patagonia, the latter to Chili; and lastly the American family Iguanidæ represented by eight genera, no less than six being peculiar, (or almost so,) to the South temperate region. These are *Leiodera, Diplolæmus* and *Proctrotretus*, ranging from Chili to Patagonia; *Leiolæmus*, from Peru to Patagonia; *Phrymaturus*, confined to Chili, and *Ptygoderus* peculiar, to Patagonia and Tierra del Fuego. The other two genera, *Oplurus* and *Leiosaurus*, are common to Chili and tropical South America.

Tortoises appear to be scarce, a species of *Hydromedusa* only being recorded. Of the Amphibia, batrachia (frogs and toads) alone are represented, and appear to be tolerably abundant, seventeen species having been collected by Mr. Darwin in this sub-region. Species of the South American genera *Phryniscus, Hylaplesia, Telmatobius, Cacotus, Hylodes, Cyclorhamphus, Pleurodema, Cystignathus,* and *Leiuperus*, are found in various localities, some extending even to the Straits of Magellan,—the extreme southern limit of both Reptilia and Amphibia, except one lizard (*Ptygoderus*) found by Professor Cunningham in Tierra del Fuego. There are also four peculiar genera, *Rhinoderma* belonging to the Engystomidæ; *Alsodes* and *Nannophryne* to the Bombinatoridæ; *Opisthodelphys* to the Hylidæ; and *Calyptocephalus* to the Discoglossidæ.

It thus appears, that in the Reptiles all the groups are typically American, and that most of the peculiar genera belong to families which are exclusively American. The Amphibia, on the other hand, present some interesting external relations, but these are as much with Australia as with the North temperate regions. The Bombinatoridæ are indeed Palæarctic, but a larger proportion are Neotropical, and one genus inhabits New Zealand. The Chilian genus *Calyptocephalus* is allied to Australian tropical genera.

The Neotropical genera of Ranidæ, five of which extend to Chili and Patagonia, belong to a division which is Australian and Neotropical, and which has species in the Oriental and Ethiopian regions.

Fresh-water Fishes.—These present some peculiar forms, and some very interesting phenomena of distribution. The genus *Percilia* has been found only in the Rio de Maypu in Chili; and *Percichthys*, also belonging to the perch family, has five species confined to the fresh waters of South Temperate America, and one far away in Java. *Nematogenys* (1 sp.) is peculiar to Chili; *Trichomycterus* reaches 15,000 feet elevation in the Andes,—both belonging to the Siluridæ; *Chirodon* (2 sp.), belonging to the Characinidæ, is peculiar to Chili; and several other genera of the same family extend into this sub-region from Brazil. The family Haplochitonidæ has a remarkable distribution; one of its genera, *Haplochiton* (2 sp.), inhabiting Tierra del Fuego and the Falkland Islands, while the other, *Prototroctes*, is found only in South Australia and New Zealand. Still more remarkable is *Galaxias* (forming the family Galaxidæ), the species of which are divided between Temperate South America, and Australia, Tasmania, and New Zealand; and there is even one species (*Galaxias attenuatus*) which is found in the Chatham Islands, New Zealand, and Tasmania, as well as in the Falkland Islands and Patagonia. *Fitzroya* (1 sp.) is found only at Montevideo; *Orestias* (6 sp.) is peculiar to Lake Titicaca in the high Andes of Bolivia; *Jenynsia* (1 sp.) in the Rio de la Plata —all belonging to the characteristic South American family of the Cyprinodontidæ.

Insects.—It is in insects more than in any other class of animals, that we find clear indications of a not very remote migration of northern forms, along the great mountain range to South Temperate America, where they have established themselves as a prominent feature in the entomology of the country. The several orders and families, however, differ greatly in this respect; and there are some groups which are only represented by modifications of tropical forms, as we have seen to be almost entirely the case in birds and reptiles.

Lepidoptera.—The butterflies of the South Temperate Sub-region are not numerous, only about 29 genera and 80 species being recorded. Most of these are from Chili, which is sufficiently accounted for by the general absence of wood on the east side of the Andes from Buenos Ayres to South Patagonia. The families represented are as follows: Satyridæ, with 11 genera and 27 species, are the most abundant; Nymphalidæ, 2 genera and 8 species; Lemoniidæ, 1 genus, 1 species; Lycænidæ, 3 genera, 8 species; Pieridæ, 6 genera, 14 species; Papilionidæ, 2 genera, 8 species; Hesperidæ, 4 genera, 13 species. One genus of Satyridæ (*Elina*) and 2 of Pieridæ (*Eroessa* and *Phulia*) are peculiar to Chili. The following are the genera whose derivation must be traced to the north temperate zone:—*Tetraphlybia, Neosatyrus*, and 3 allied genera of 1 species each, were formerly included under *Erebia*, a northern and arctic form, yet having a few species in South Africa; *Argyrophorus*, allied to *Æneis*, a northern genus; *Hipparchia*, a northern genus yet having a species in Brazil;—all Satyridæ. The Nymphalidæ are represented by the typical north temperate genus *Argynnis*, with 7 species in Chili; *Colias*, among the Pieridæ, is usually considered to be a northern genus, but it possesses representatives in South Africa, the Sandwich Islands, Malabar, New Grenada, and Peru, as well as Chili, and must rather be classed as cosmopolitan. These form a sufficiently remarkable group of northern forms, but they are accompanied by others of a wholly Neotropical origin. Such are *Stibomorpha* with 6 species, ranging through South America to Guatemala, and *Eteona*, common to Chili and Brazil (Satyridæ); *Apodemia* (Lemoniidæ) confined to Tropical America and Chili. *Hesperocharis* and *Callidryas* (Pieridæ), both tropical; and *Thracides* (Hesperidæ) confined to Tropical America and Chili. Other genera are widely scattered; as, *Epinephile* found also in Mexico and Australia; *Cupido*, widely spread in the tropics; *Euryades*, found only in La Plata and Paraguay, allied to South American forms of *Papilio*, to the Australian *Eurycus*, and the northern *Parnassius;* and *Heteropterus*, scattered in Chili, North America, and Tropical Africa. We find then, among butterflies, a large north-temperate element, intermingled in nearly equal proportions with forms derived from Tropical America; and the varying degrees of resemblances of the Chilian to the northern species, seems to indicate successive immigrations at remote intervals.

Coleoptera.—It is among the beetles of South Temperate America that we find some of the most curious examples of remote affinities, and traces of ancient migrations. The Carabidæ are very well represented, and having been more extensively collected than most other families, offer us perhaps the most complete materials. Including the Cicindelidæ, about 50 genera are known from the South Temperate Sub-region, the greater part from Chili, but a good number also from Patagonia and the Straits of Magellan. Of these more than 30 are peculiar, and most of them are so isolated that it is impossible to determine with precision their nearest allies.

The only remarkable form of Cicindelidæ is *Agrius*, a genus allied to the *Amblycheila* and *Omus* of N.W. America. Two genera of Carabidæ, *Cascellius* and *Baripus*, are closely allied to *Promecoderus*, an Australian genus; and another, *Lecanomerus*, has one species in Chili and the other in Australia. Five or six of the peculiar genera are undoubtedly allied to characteristic Palæarctic forms; and such northern genera as *Carabus, Pristonychus, Anchomenus, Pterostichus, Percus, Bradycellus, Trechus,* and *Bembidium,* all absent from Tropical America, give great support to the view that there is a close relation between the insects of the northern regions and South Temperate America. A decided tropical element is, however, present. *Tropopterus* is near *Colpodes*, a Tropical and South American genus; *Mimodromius* and *Plagiotelium* are near *Calleida*, a South American genus; while *Pachyteles, Pericompsus, Variopalpus,* and *Calleida* are widely spread American groups. The preponderance of northern forms seems, however, to be undoubted.

Six Carabidæ are known from Juan Fernandez, 3 being identical with Chilian species and 3 peculiar. As the island is 350 miles from the mainland, we have here a proof of how readily insects may be transported great distances.

The Palæarctic affinity of the South Temperate Carabidæ may be readily understood, if we bear in mind the great antiquity of the group, and the known long persistence of generic and specific forms of Coleoptera; the facility with which they may be transported to great distances by gales and hurricanes, either on land or over the sea; and, therefore, the probability that suitable stations would be rapidly occupied by species already adapted to them, to the exclusion of those of the adjacent tracts which had been specialised under different conditions. If, for example, we carry ourselves back to the time when the Andes had only risen to half their present altitude, and Patagonia had not emerged from the ocean (an epoch not very remote geologically), we should find nearly all the Carabidæ of South America, adapted to a warm, and probably forest-covered country. If, then, a further considerable elevation of the land took place, a large temperate and cold area would be formed, without any suitable insect inhabitants. During the necessarily slow process of elevation, many of the tropical Carabidæ would spread upwards, and some would become adapted to the new conditions; while the majority would probably only maintain themselves by continued fresh immigrations. But, as the mountains rose, another set of organisms would make their way along the highest ridges. The abundance and variety of the North Temperate Carabidæ, and their complete adaptation to a life on barren plains and rock-strewn mountains, would enable them rapidly to extend into any newly-raised land suitable to them; and thus the whole range of the Rocky Mountains and Andes would obtain a population of northern forms, which would overflow into Patagonia, and there, finding no competitors, would develope into a variety of modified groups. This migration was no doubt effected mainly, during successive glacial epochs, when the mountain-range of the Isthmus of Panama, if moderately increased in height, might become adapted for the passage of northern forms, while storms would often carry insects from peak to peak over intervening forest lowlands or narrow straits of sea. If this is the true explanation, we ought to find no such preponderant northern element in groups which are proportionally less developed in cold and temperate climates. Our further examination will show how far this is the case.

Lucanidæ.—Only four genera are known in the sub-region. Two are peculiar, *Chiasognathus* and *Streptocerus*, the former allied to Tropical American, the latter to Australian genera; the other two genera are exclusively South American.

Cetoniidæ.—These seem very scarce, only a few species of the Neotropical genus *Gymnetis* reaching Patagonia.

Buprestidæ.—These are rather numerous, many very beautiful species being found in Chili. Nineteen genera are represented in South Temperate America, and 5 of these are peculiar to it; 3 others are South American genera; 2 are Australian, and the remainder are wide-spread, but all are found also in Tropical America. The only north-temperate genus is *Dicerca*, and even this occurs also in the Antilles, Brazil, and Peru. Of the peculiar genera, the largest, *Dactylozodes* (26 sp.), has one species in South Brazil, and is closely allied to *Hyperantha*, a genus of Tropical America; *Epistomentis* is allied to *Nascis*, an Australian genus; *Tyndaris* is close to *Acmæodera*, a genus of wide range and preferring desert or dry countries. The other two are single species of cosmopolitan affinities. On the whole, therefore, the Buprestidæ are unmistakeably Neotropical in character.

Longicorns.—Almost the whole of the South Temperate Longicorns inhabit Chili, which is very rich in this beautiful tribe. About 75 genera and 160 species are known, and nearly half of the genera are peculiar. Many of the species are large and handsome, rivalling in beauty those of the most favoured tropical lands. Of the 8 genera of Prionidæ 6 are peculiar, but all are allied to Tropical American forms except *Microplophorus*, which belongs to a group of genera spread over Australia, Europe, and Mexico. The Cerambycidæ are much more abundant, and their affinities more interesting. Two (*Syllitus* and *Pseudocephalus*) are common to Australia and Chili. Twenty-three are Neotropical; and among these *Ibidion*, *Compsocerus*, *Callideriphus*, *Trachyderes*, and *Xylocharis*, are best represented. Twenty are altogether peculiar, but most of them are more or less closely allied to genera inhabiting Tropical America. Some, as the handsome *Cheloderus* and *Oxypeltus*, have no close allies in any part of the world. *Holopterus*, though very peculiar, shows most resemblance to a New Zealand insect. *Sibylla*, *Adalbus*, and *Phantagoderus*, have Australian affinities ; while *Calydon* alone shows an affinity for north-temperate forms. One species of the northern genus, *Leptura*, is said to have been found at Buenos Ayres.

The Lamiidæ are less abundant. Nine of the genera are Neotropical. Two (*Apomecyna* and *Exocentrus*) are spread over all tropical regions. Ten genera are peculiar; and most of these are related to Neotropical groups or are of doubtful affinities. Only one, *Aconopterus*, is decidedly allied to a northern genus, *Pogonochœrus*. It thus appears, that none of the Lamiidæ exhibit Australian affinities, although these are a prominent feature in the relations of the Cerambycidæ.

It is evident, from the foregoing outline, that the insects of South Temperate America, more than any other class of animals, exhibit a connection with the north temperate regions, yet this connection is only seen in certain groups. In Diurnal Lepidoptera and in Carabidæ, the northern element is fully equal to the tropical, or even preponderates over it. We have already suggested an explanation of this fact in the case of the Carabidæ, and with the butterflies it is not more difficult. The great mass of Neotropical butterflies are forest species, and have been developed for countless ages in a forest-clad tropical country. The north temperate butterflies, on the other hand, are very largely open-country species, frequenting pastures, mountains, and open plains, and often wandering over an extensive area. These would find, on the higher slopes of mountains, a vegetation and conditions suited to them, and would occupy such stations in less time than would be required to adapt and modify the forest-haunting groups of the American lowlands. In those groups of insects, however, in which the conditions of life are nearly the same as regards both temperate and tropical species, the superior

number and variety of the tropical forms has given them the advantage. Thus we find that among the Lucanidæ, Buprestidæ, and
Longicorns, the northern element is hardly perceptible. Most of
these are either purely Neotropical, or allied to Neotropical genera,
with the admixture, however, of a decided Australian element.
As in the case of the Amphibia and fresh-water fishes, the Australian affinity, as shown by insects, is of two kinds, near and
remote. We have a few genera common to the two countries;
but more commonly the genera are very distinct, and the affinity
is shown by the genera of both countries belonging to a group
peculiar to them, but which may be of very great age. In the
former case, we must impute some of the resemblance of the two
faunas to an actual interchange of forms within the epoch of
existing genera—a period of vast and unknown duration in the
class of insects; while in the latter case, and perhaps also in
many of the former, it seems more in accordance with the whole
of the phenomena, to look upon most of the instances as
survivals, in the two southern temperate areas, of the relics of
groups which had once a much wider distribution. That this is
the true explanation, is suggested by the numerous cases of discontinuous and scattered distribution we have had to notice, in
which every part of the globe, without exception, is implicated;
and there is a reason why these survivals should be rather more
frequent in Australia and temperate South America, inasmuch
as these two areas agree in the absence of a considerable number
of otherwise cosmopolitan vertebrate types, and are also in many
respects very similar in climatic and other physical conditions.
The preponderating influence of the organic over the physical
environment, as taught by Mr. Darwin, leads us to give most
weight to the first of the above-mentioned causes; to which we
may also impute such undoubted cases of survival of ancient
types as the Centetidæ of the Antilles and Madagascar—both
areas strikingly deficient in the higher vertebrate forms. The
probable mode and time of the cross migration between Australia
and South America, has been sufficiently discussed in our chapter
on the Australian region, when treating of the origin and affinities
of the New Zealand fauna.

Islands of the South Temperate Sub-region.

These are few, and of not much zoological interest. Tierra del Fuego, although really an island, is divided from the mainland by so narrow a channel that it may be considered as forming part of the continent. The guanaco (*Auchenia huanaco*) ranges over it, and even to small islands further south.

The Falkland Islands.—These are more important, being situated about 350 miles to the east of Southern Patagonia; but the intervening sea is shallow, the 100 fathom line of soundings passing outside the islands. We have therefore reason to believe that they have been connected with South America at a not distant epoch; and in agreement with this view we find most of their productions identical, while the few that are peculiar are closely allied to the forms of the mainland.

The only indigenous Mammals are a wolf-like fox (*Pseudalopex antarcticus*) said to be found nowhere else, but allied to two other species inhabiting Southern Patagonia; and a species of mouse, probably one of the American genera *Hesperomys* or *Reithrodon*.

Sixty-seven species of Birds have been obtained in these islands, but only 18 are land-birds; and even of these 7 are birds of prey, leaving only 11 Passeres. The former are all common South American forms, but one species, *Milvago australis*, seems peculiar. The 11 Passeres belong to 9 genera, all found on the adjacent mainland. Three, or perhaps four, of the species are however peculiar. These are *Phrygilus melanoderus, P. xanthogrammus, Cinclodes antarcticus,* and *Muscisaxicola macloviana.* The wading and swimming birds are of little interest, except the penguins, which are greatly developed; no less than eight species being found, five as residents and three as accidental visitors.

No reptiles are known to inhabit these islands.

Juan Fernandez.—This island is situated in the Pacific Ocean, about 400 miles west of Valparaiso in Chili. It is only a few miles in extent, yet it possesses four land-birds, excluding the powerful Accipitres. These are *Turdus falklandicus; Anæretes fernandensis*, one of the Tyrannidæ; and two humming-birds, *Eustephanus fernandensis* and *E. galeritus*. The first is a wide-spread South Temperate species, the two next are peculiar to the island, while the last is a Chilian species which ranges south to Tierra del Fuego. But ninety miles beyond this island lies another, called "Mas-a-fuero," very much smaller; yet this, too, contains four species of similar birds; one, *Oxyurus mas-a-fueræ*, allied to the wide-spread South Temperate *O. spinicauda*, and *Cinclodes fusus*, a South Temperate species—both Dendrocolaptidæ; with a humming-bird, *Eustephanus leyboldi*, allied to the species in the larger island. The preceding facts are taken from papers by Mr. Sclater in the *Ibis* for 1871, and a later one in the same journal by Mr. Salvin (1875). The former author has some interesting remarks on the three species of humming-birds of the genus *Eustephanus*, above referred to. The Chilian species, *E. galeritus*, is green in both sexes. *E. fernandensis* has the male of a fine red colour and the female green, though differently marked from the female of *E. galeritus*. *E. leyboldi* (of Mas-a-fuera) has the male also red and the female green, but the female is more like that of *E. galeritus*, than it is like the female of its nearer ally in Juan Fernandez. Mr. Sclater supposes, that the ancient parent form of these three birds had the sexes alike, as in the present Chilian bird; that a pair (or a female having fertilised ova) reached Juan Fernandez and colonised it. Under the action of sexual selection (unchecked by some conditions which had impaired its efficacy on the continent) the male gradually assumed a brilliant plumage, and the female also slightly changed its markings. Before this change was completed the bird had established an isolated colony on Mas-a-fuera; and here the process of change was continued in the male, but from some unknown cause checked in the female, which thus remains nearer the parent form. Lastly the slightly modified Chilian bird again reached Juan Fernandez and exists there side by side with its strangely altered cousin.

All the phenomena can thus be accounted for by known laws, on the theory of very rare accidental immigrations from the mainland. The species are here so very few, that the greatest advocate for continental extensions would hardly call such vast causes into action, to account for the presence of these three birds on so small and so remote an island, especially as the union must have continued down to the time of existing species. But if accidental immigration has sufficed here, it will also assuredly have sufficed where the islands are larger, and the chances of reaching them proportionately greater; and it is because an important principle is here illustrated on so small a scale, and in so simple a manner as to be almost undeniable. that we have devoted a paragraph to its elucidation.

A few Coleoptera from Juan Fernandez present analogous phenomena. All belong to Chilian genera, while a portion of them constitute peculiar species.

Land-shells are rather plentiful, there being about twenty species belonging to seven genera, all found in the adjacent parts of South America; but all the species are peculiar, as well as four others found on the island of Mas-a-fuera.

III. Tropical North America, or the Mexican Sub-region.

This sub-region is of comparatively small extent, consisting of the irregular neck of land, about 1,800 miles long, which connects the North and South American continents. Almost the whole of its area is mountainous, being in fact a continuation of the great range of the Rocky Mountains. In Mexico it forms an extensive table-land, from 6,000 to 9,000 feet above the sea, with numerous volcanic peaks from 12,000 to 18,000 feet high; but in Yucatan and Honduras, the country is less elevated, though still mountainous. On the shores of the Caribbean Sea and Gulf of Mexico, there is a margin of low land from 50 to 100 miles wide, beyond which the mountains rise abruptly; but on the Pacific side this is almost entirely wanting, the mountains rising almost immediately from the sea shore. With the exception of the elevated plateaus of Mexico and Guatemala, and the extremity of the peninsula of Yucatan, the whole of Central America is clothed with forests; and as its surface is much broken up into hill and valley, and the volcanic

E 2

soil of a large portion of it is very fertile, it is altogether well adapted to support a varied fauna, as it does a most luxuriant vegetation. Although many peculiar Neotropical types are absent, it yet possesses an ample supply of generic and specific forms ; and, as far as concerns birds and insects, is not perhaps inferior to the richest portions of South America in the number of species to be found in equal areas.

Owing to the fact that the former Republic of Mexico comprised much territory that belongs to the Nearctic region, and that many Nearctic groups extend along the high-lands to the capital city of Mexico itself, and even considerably further south, there is much difficulty in determining what animals really belong to this sub-region. On the low-lands, tropical forms predominate as far as 28° N. latitude ; while on the cordilleras, temperate forms prevail down to 20°, and are found even much farther within the tropics.

Mammalia.—Very few peculiar forms of Mammalia are restricted to tropical North America ; which is not to be wondered at when we consider the small extent of the country, and the facility of communication with adjacent sub-regions. A peculiar form of tapir (*Elasmognathus bairdi*) inhabits Central America, from Panama to Guatemala, and, with *Myxomys*, a genus of Muridæ, are all at present discovered. *Bassaris*, a remarkable form of Procyonidæ, has been included in the Nearctic region, but it extends to the high-lands of Guatemala. *Heteromys*, a peculiar genus of Saccomyidæ or pouched rats, inhabits Mexico, Honduras, Costa Rica, and Trinidad. Five genera of monkeys extend here,—*Ateles, Mycetes, Cebus, Nyctipithecus,* and *Saimiris*; the two former alone reaching Mexico, the last only going as far as Costa Rica. Other typical Neotropical forms are *Galera,* the tayra, belonging to the weasel family ; *Nasua*, the coatimundi; *Dicotyles,* the peccary ; *Cercolabes,* the tree porcupine ; *Dasyprocta,* the agouti ; *Cœlogenys,* the paca ; *Cholœpus,* and *Arctopithecus,* sloths; *Cyclothurus,* an ant-eater; *Tatusia,* an armadillo ; and *Didelphys,* oppossum. Of Northern forms, *Sorex, Vulpes, Lepus,* and *Pteromys* reach Guatemala.

Birds.—The productiveness of this district in bird life, may be estimated from the fact, that Messrs. Salvin and Sclater have catalogued more than 600 species from the comparatively small territory of Guatemala, or the portion of Central America between Mexico and Honduras. The great mass of the birds of this sub-region are of Neotropical families and genera, but these are intermingled with a number of migrants from temperate North America, which pass the winter here; with some northern forms on the high-lands; and with a considerable number of peculiar genera, mostly of Neotropical affinities.

The genera of birds peculiar to this sub-region belong to the following families:—Turdidæ (2 genera) ; Troglodytidæ (1 gen.); Vireonidæ (1 gen.); Corvidæ (2 gen.); Ampelidæ (1 gen.); Tanagridæ (1 gen.) ; Fringillidæ (2 gen.) ; Icteridæ (1 gen.); Formicariidæ (2 gen.) ; Tyrannidæ (2 gen.) ; Cotingidæ (1 gen.); Momotidæ (1 gen.); Trogonidæ (1 gen.) ; Trochilidæ (14 gen.); Conuridæ (1 gen.); Cracidæ (2 gen.); Strigidæ (1 gen.) ; in all 37 genera of land-birds. The Neotropical families that do not extend into this sub-region are, Pteroptochidæ; the sub-family *Furnariinæ* of the Dendrocolaptidæ; the sub-family *Conophaginæ* of the Tyrannidæ; the sub-family *Rupicolinæ* of the Cotingidæ; Phytotomidæ; Todidæ; Opisthocomidæ; Chionididæ; Thinocoridæ; Cariamidæ; Psophiidæ; Eurypygidæ; Palamedeidæ; and Struthionidæ. On the other hand Paridæ, Certhiidæ, Ampelidæ, and Phasianidæ, are northern families represented here, but which do not reach South America; and there are also several northern genera and species, of Turdidæ, Troglodytidæ, Mniotiltidæ, Vireonidæ, Fringillidæ, Corvidæ, Tetraonidæ, and Strigidæ, which are similarly restricted. Some of the most remarkable of the Neotropical genera only extend as far as Costa Rica and Veragua,—countries which possess a rich and remarkable fauna. Here only are found an umbrella bird, (*Cephalopterus glabricollis*); a bell bird (*Chasmorhynchus tricarunculatus*); and species of *Dacnis* (Cerœbidæ), *Buthraupis*, *Eucometis, Tachyphonus* (Tanagridæ), *Xiphorhynchus* (Dendrocolaptidæ) ; *Hypocnemis* (Formicariidæ); *Euscarthmus* (Tyrannidæ); *Attila* (Cotingidæ); *Piprites* (Pipridæ); *Capito, Tetragonops* (Megalæmidæ) ; *Selenidera* (Rhamphastidæ); *Neomorphus*

(Cuculidæ); *Monasa* (Bucconidæ); many genera of Trochilidæ; and *Nothocercus* (Tinamidæ); none of which extend further north. A considerable number of the peculiar genera noted above, are also found in this restricted area, which is probably one of the richest ornithological districts on the globe.

Reptiles.—These are much less known than the preceding classes, but they afford several peculiar and interesting forms. Snakes are perhaps the least remarkable; yet there are recorded 4 peculiar genera of Calamariidæ, 1 of Colubridæ, 1 of Homalopsidæ, 3 of Dipsadidæ; while *Boa* and *Elaps* are in common with South America. Lizards are much more specially developed. *Chirotes*, one of the Amphisbænians, is confined to Mexico and the southern part of the Nearctic region; *Heloderma* forming a peculiar family, Helodermidæ, is Mexican only; *Abronia* and *Barissia* (Zonuridæ) are also Mexican, as is *Siderolampus* belonging to the Scincidæ, while *Blepharactitis* (same family) inhabits Nicaragua; *Brachydactylus*, one of the geckoes, is from Costa Rica; while *Phymatolepis*, *Lamanctus*, *Corytheolus*, *Cachrix*, *Corythophanes* and *Chamæleopsis*, all belonging to the Iguanidæ, are confined to various parts of the sub-region. In the same family we have also the Antillean, *Cyclura*, and the Nearctic *Phrynosoma* and *Tropidolepis*, as well as the wide-spread American genus *Anolius*.

Among the tortoises, *Staurotypus*, allied to *Chelydra*, is found in Mexico and Guatemala; and another genus, *Claudius*, has been lately described from Mexico.

Amphibia.—These are chiefly Batrachians; *Rhinophryna* (forming a peculiar family) being confined to Mexico; *Triprion*, a genus of Hylidæ, inhabiting Yucatan, with *Leyla* and *Strabomantis* (Polypedatidæ) found only in Costa Rica and Veragua, are peculiar genera. The Salamandridæ, so abundant in the Nearctic region, are represented by a few species of *Amblystoma* and *Spelerpes*.

Fresh-water fish.—Since the British Museum catalogue was published, a valuable paper by Dr. Günther, in the Transactions of the Zoological Society for 1868, furnishes much additional information on the fishes of Central America. In that part of the region south of Mexico, 106 species of fresh-water fishes are

enumerated; and 17 of these are found in streams flowing into
both the Atlantic and Pacific Oceans. On the whole, 11 families
are represented among the fresh-water fish, and about 38 genera.
Of these, 14 are specially Nearctic,—*Amiurus* (Siluridæ); *Fundulus* (Cyprinodontidæ); *Sclerognathus* (Cyprinidæ); and *Lepidosteus*
(Ganoidei). A much larger number are Neotropical; and several
Neotropical genera, as *Heros* and *Pœcilia*, are more largely
developed here than in any other part of the region. There are
also a considerable number of peculiar genera;—*Petenia, Theraps,*
and *Neotrophus* (Chromides); *Ælurichthys* (Siluridæ); *Chalcinopsis* (Characniidæ); *Characodon, Belonesox, Pseudoxiphophorus,
Platypœcilus, Mollienesia,* and *Xiphophorus* (Cyprinodontidæ).
A few peculiar Antillean forms are also present; as *Agonostoma*
(Mugilidæ); *Gambusia* and *Girardinus* (Cyprinodontidæ). The
other families represented are Percidæ (1 genus); Pristopomatidæ
(2 gen.); Gobiidæ (1 gen.); Clupeidæ (2 gen.); and Gymnotidæ
(1 genus).

On the whole the fish-fauna is typically Neotropical, but with
a small infusion of Nearctic forms. There are a considerable
proportion of peculiar genera, and almost all the species are
distinct from those of other countries. The predominant family
is that of the Cyprinodontidæ, represented by 12 genera; and
the genus *Heros* (Chromidæ) has here its maximum development,
containing between thirty and forty species. Dr. Günther considers that a number of sub-faunas can be distinguished, corresponding to some extent, with the islands into which the country
would be divided by a subsidence of about 2,000 feet. The
most important of these divisions is that separating Honduras from
Costa Rica, and as it also divides a very marked ornithological
fauna we have every reason to believe that such a division must
have existed during the latter portion of the tertiary epoch.
We shall find some farther evidence of this division in the
next class.

Insects.—The butterflies of various parts of Central America
and Mexico, having been largely collected, offer us some
valuable evidence as to the relations of this sub-region. Their
general character is wholly Neotropical, about one half of the

South American genera being found here. There are also a few peculiar genera, as, *Drucina* (Satyridæ); *Microtia* (Nymphalidæ); *Eumæus* (Lycænidæ) ; and *Eucheira* (Pieridæ). *Clothilda* (Nymphalidæ) is confined to this sub-region and the Antilles. The majority of the genera range over the whole sub-region from Panama to Mexico, but there are a considerable number, comprising many of the most characteristic South American forms, which do not pass north of Costa Rica or Nicaragua. Such are *Lycorea, Ituna, Thyridia, Callithomia, Oleria* and *Ceratina*, —all characteristic South American groups of Danaidæ; *Pronophila* and *Dynastor* (Satyridæ); *Protogonius, Pycina, Prepona, Nica, Ectima* and *Colænis* (Nymphalidæ); *Eurybia* and *Methonella* (Nemeobiidæ) ; *Hades*, and *Panthemos* (Erycinidæ).

Colcoptera.—These present some interesting features, but owing to their vast number only a few of the more important families can be noticed.

Cicindelidæ.—The only specially Neotropical genera recorded as occurring in this sub-region, are *Ctenostoma* and *Hiresia*, both reaching Mexico.

Carabidæ.—Several genera are peculiar. *Molobrus* is found in all parts of the sub-region, while *Onychopterygia, Phymatocephalus*, and *Anisotarsus* are Mexican only. There are about 20 South American genera, most of which extend to Mexico, and include such characteristic Neotropical forms as *Agra, Callida, Coptodera, Pachyteles, Ardistomus, Aspidoglossa, Stenocrepis*, and *Pelecium.*

Lucanidæ.—Of this important family there is, strange to say, not a single species recorded in Gemminger and Harold's catalogue up to 1868 ! It is almost impossible that they can be really absent; yet their place seems to be, to some extent, supplied by an unusual development of the allied Passalidæ, of which there are five South American and six peculiar genera.

Cetoniidæ.—All the larger South American genera extend to Mexico, which country possesses 3 peculiar forms, *Ischnoscelis, Psilocnemis*, and *Dialithus*; while *Trigonopeltastes* is characteristic, having 4 Mexican, 1 Brazilian, and 1 North American species.

Buprestidæ.—In this family there are no peculiar genera. All the large South American groups are absent, the only important and characteristic genus being *Stenogaster*.

Longicorns.—This important group is largely developed, the country being well adapted to them ; and their distribution presents some features of interest.

In the Prionidæ there are 6 peculiar genera, the largest being *Holonotus* with 3 species; two others, *Derotrachus* and *Mallaspis*, are characteristic; 3 more are common to South America, and 1 to Cuba. The Cerambycidæ are much more numerous, and there are 24 peculiar genera, the most important being *Sphenothecus*, *Entomosterna*, and *Cyphosterna ;* while *Crioprosopus* and *Metaleptus* are characteristic of the sub-region, although extending into South America; about 12 Neotropical genera extend to Mexico or Guatemala, while 12 more stop short, as far as yet known, at Nicaragua. Lamiidæ have a very similar distribution ; 13 genera are peculiar, the most important being *Monilema*, *Hamatoderus*, and *Carneades*, while *Phœa* and *Lagochirus* are characteristic. About sixteen typical Neotropical genera extend to Mexico, and 15 more only reach Nicaragua, among which are such important genera as *Anisopus*, *Lepturgus*, and *Callia*.

The land-shells are not sufficiently known to furnish any corresponding results. They are however mostly of South American genera, and have comparatively little affinity for those of the Antilles.

Relations of the Mexican sub-region to the North and South American Continents.—The sudden appearance of numerous South American forms of Edentata in temperate North America, in Post-Tertiary times, as narrated in Chapter VII., together with such facts as the occurrence of a considerable number of identical species of sea fish on the two sides of the Central American isthmus, render it almost certain that the union of North and South America is comparatively a recent occurrance, and that during the Miocene and Pliocene periods, they were separated by a wide arm of the sea. The low country of Nicaragua was probably the part submerged, leaving the highlands of Mexico and Guatemala still united with the North

American continent, and forming part of the Tertiary "Nearctic region." This is clearly indicated both by the many Nearctic forms which do not pass south of Nicaragua, of which the turkeys (*Meleagris*) are a striking example, and by the comparative poverty of this area in typical Neotropical groups. During the Miocene period there was not that marked diversity of climate between North and South America that now prevails; for when a luxuriant vegetation covered what are now the shores of the Arctic Ocean, the country south of the great lakes must have been almost or quite tropical. At an early Tertiary period, the zoological differences of the Nearctic and Neotropical regions were probably more radical than they are now, South America being a huge island, or group of islands—a kind of Australia of the New World, chiefly inhabited by the imperfectly organized Edentata; while North America abounded in Ungulata and Carnivora, and perhaps formed a part of the great Old World continent. There were also one or more very ancient unions (in Eocene or Miocene times) of the two continents, admitting of the entrance of the ancestral types of Quadrumana into South America, and, somewhat later, of the Camelidæ; while the isthmus south of Nicaragua was at one time united to the southern continent, at another made insular by subsidence near Panama, and thus obtained that rich variety of Neotropical types that still characterises it. When the final union of the two continents took place, the tropical climate of the lower portions of Guatemala and Mexico would invite rapid immigration from the south; while some northern forms would extend their range into and beyond the newly elevated territory. The Mexican sub-region has therefore a composite character, and we must not endeavour too rigidly to determine its northern limits, nor claim as exclusively Neotropical, forms which are perhaps comparatively recent immigrants; and it would perhaps be a more accurate representation of the facts, if we were to consider all the highlands of Mexico and Guatemala above the limits of the tropical forests, as still belonging to the Nearctic region, of which the whole country so recently formed a part.

The long-continued separation of North and South America

by one or more arms of the sea, as above indicated, is further rendered necessary by the character of the molluscan fauna of the Pacific shores of tropical America, which is much more closely allied to that of the Caribbean sea, and even of West Africa, than to that of the Pacific islands. The families and many of the genera are the same, and a certain proportion of very closely allied or identical species, shows that the union of the two oceans continued into late Tertiary times. When the evidence of both land and sea animals support each other as they do here, the conclusions arrived at are almost as certain as if we had (as we no doubt some day shall have) geological proof of these successive subsidences.

Islands of the Mexican Sub-region.—The only islands of interest belonging to this sub-region, are Tres Marias and Socorro, recently investigated by Col. Grayson for some of the American Natural History societies.

Tres Marias consist of four small islands lying off the coast of north-western Mexico, about 70 miles from San Blas. The largest is about 15 miles long by 10 wide. They are of horizontally stratified deposits, of moderate height and flat-topped, and everywhere covered with luxuriant virgin forests. They appear to lie within the 100 fathom line of soundings. Fifty-two species of birds, of which 45 were land-birds, were collected on these islands. They consisted of 19 Passeres; 11 Picariæ (7 being humming-birds); 10 Accipitres; 2 parrots, and 3 pigeons. All were Mexican species except 4, which were new, and presumably peculiar to the islands, and one tolerably marked variety. The new species belong to the following genera;—*Parula* and *Granatellus* (Mniotiltidæ); *Icterus* (Icteridæ); and *Amazilia* (Trochilidæ). A small *Psittacula* differs somewhat from the same species on the mainland.

There are a few mammalia on the islands; a rabbit (*Lepus*) supposed to be new; a very small opossum (*Didelphys*), and a racoon (*Procyon*). There are also several tree-snakes, a *Boa*, and many lizards. The occurrence of so many mammalia and snakes is a proof that these islands have been once joined to the mainland; but the fact that some of the species of both birds and mammals are peculiar, indicates that the separation is not a very recent one. At the same time, as all the species are very closely allied to those of the opposite coasts when not identical, we may be sure that the subsidence which isolated them is not geologically remote.

Socorro, the largest of the Revillagigedo Islands, is altogether different from the Tres Marias. It is situated a little further south (19 S. Latitude), and about 300 miles from the coast, in deep water. It is about 2,000 feet high, very rugged and bare, and wholly volcanic. No mammalia were observed, and no reptiles but a small lizard, a new species of a genus (*Uta*) characteristic of the deserts of N.-Western Mexico. The only observed land-shell (*Orthalicus undatus*) also inhabits N.-W. Mexico. Only 14 species of birds were obtained, of which 9 were land-birds; but of these 4 were new species, one a peculiar variety, and another (*Parula insularis*) a species first found in the Tres Marias. With the exception of this bird and a *Buteo*, all the land-birds belonged to different *genera* from any found on the Tres Marias, though all were Mexican forms. The peculiar species belonged to the genera *Harporhynchus* (Turdidæ); *Troglodytes* (Troglodytidæ); *Pipilo* (Fringillidæ); *Zenaidura* (Columbidæ); and a variety of *Conurus holochrous* (Psittacidæ).

The absence of mammals and snakes, the large proportion of peculiar species, the wholly volcanic nature of these islands, and their situation in deep water 300 miles from land,—all indicate that they have not formed part of the continent, but have been raised in the ocean; and the close relation of their peculiar species to those living in N.-Western Mexico, renders it probable that their antiquity is not geologically great.

The Cocos Islands, about 300 miles S.-W. of the Isthmus of Panama, are known to possess one peculiar bird, a cuckoo of the *Coccyzus* type, which is considered by some ornithologists to constitute a peculiar genus, *Nesococcyx*.

IV. The West Indian Islands, or Antillean Sub-region.

The West Indian islands are, in many respects, one of the most interesting of zoological sub-regions. In position they form an unbroken chain uniting North and South America, in a line parallel to the great Central American isthmus ; yet instead of exhibiting an intermixture of the productions of Florida and Venezuela, they differ widely from both these countries, possessing in some groups a degree of speciality only to be found elsewhere in islands far removed from any continent.   They consist of two very large islands, Cuba and Hayti ;[1] two of moderate size, Jamaica and Portorico ; and a chain of much smaller islands, St. Croix, Anguilla, Barbuda, Antigua, Guadeloupe, Dominica, Martinique, St. Lucia, St. Vincent, Barbadoes, and Grenada, with a host of intervening islets.   Tobago, Trinidad, Margarita, and Curaçao, are situated in shallow water near the coast of South America, of which they form part zoologically.   To the north of Cuba and Hayti are the Bahamas, an extensive group of coral reefs and islands, 700 miles long, and although very poor in animal life, belonging zoologically to the Antilles.   All the larger islands, and most of the smaller ones (except those of coral formation) are very mountainous and rocky, the chains rising to about 8,000 feet in Hayti and Jamaica, and to nearly the same height in Cuba. All, except where they have been cleared by man, are covered with a luxuriant forest vegetation ; the temperature is high and uniform ; the rains ample ; the soil, derived from granitic and limestone rocks, exceedingly fertile ; and as the four larger islands together are larger than Great Britain, we might expect an ample and luxuriant fauna.   The reverse is however the case ; and there are probably no land areas on the globe, so highly favoured by nature in all the essentials for supporting animal life, and at the same time so poor in all the more highly organised groups of animals.   Before entering upon our sketch of the main features of this peculiar but limited fauna, it will be well to note a few peculiarities in the physical structure of the islands, which have an important bearing on their past

---

[1] This name will be used for the whole island of St. Domingo, as being both shorter and more euphonious, and avoiding all confusion with Dominica, one of the Lesser Antilles. It is also better known than " Hispaniola," which is perhaps the most correct name.

history, and will enable us to account for much that is peculiar in the general character of their natural productions.

If we draw a line immediately south of St. Croix and St. Bartholomew, we shall divide the Archipelago into two very different groups. The southern range of islands, or the Lesser Antilles, are, almost without exception, volcanic ; beginning with the small detached volcanoes of Saba and St. Eustatius, and ending with the old volcano of Grenada. Barbuda and Antigua are low islands of Tertiary or recent formation, connected with the volcanic islands by a submerged bank at no great depth. The islands to the north and west are none of them volcanic; many are very large, and these have all a central nucleus of ancient or granitic rocks. We must also note, that the channels between these islands are not of excessive depth, and that their outlines, as well as the direction of their mountain ranges, point to a former union. Thus, the northern range of Hayti is continued westward in Cuba, and eastward in Portorico; while the south-western peninsula extends in a direct line towards Jamaica, the depth between them being 600 fathoms. Between Portorico and Hayti there is only 250 fathoms; while close to the south of all these islands the sea is enormously deep, from more than 1,000 fathoms south of Cuba and Jamaica, to 2,000 south of Hayti, and 2,600 fathoms near the south-east extremity of Portorico. The importance of the division here pointed out will be seen, when we state, that indigenous mammalia of peculiar genera are found on the western group of islands only ; and it is on these that all the chief peculiarities of Antillian zoology are developed.

Mammalia.—The mammals of the West Indian Islands are exceedingly few, but very interesting. Almost all the orders most characteristic of South America are absent. There are no monkeys, no carnivora, no edentata. Besides bats, which are abundant, only two orders are represented ; rodents, by peculiar forms of a South American family ; and insectivora (an order entirely wanting in South America) by a genus belonging to a family largely developed in Madagascar and found nowhere else. The early voyagers mention " Coatis " and " Agoutis " as being

found in Hayti and the other large islands, and it is not improbable that species allied to *Nasua* and *Dasyprocta* did
exist, and have been destroyed by the dogs of the invaders ;
though, on the other hand, these names may have been applied
to the existing species, which do bear some general resemblance
to these two forms.

The Chiroptera, or bats, are represented by a large number of
species and by several peculiar genera. The American family
of Phyllostomidæ or vampires, has six genera in the Antilles, of
which three, *Lonchorina, Brachyphylla,* and *Phyllonycteris,* are
peculiar, the latter being found only in Cuba. The Vespertilionidæ have four genera, of which one, *Nycticellus,* is confined to
Cuba. There are six genera of Noctilionidæ, of which one,
Phyllodia, is confined to Jamaica.

The Insectivora are represented by the genus *Solenodon,* of
which two species are known, one inhabiting Cuba the other
Hayti. These are small animals about the size of a cat, with
long shrew-like snout, bare rat-like tail, and long claws. Their
peculiar dentition and other points of their anatomy shows that
they belong to the family Centetidæ, of which five different genera
inhabit Madagascar ; while there is nothing closely allied to
them in any other part of the world but in these two islands.

Seals are said to be found on the shores of some of the islands,
but they are very imperfectly known.

The rodents belong to the family Octodontidæ, or, according
to some authors, to the Echimyidæ, both characteristic South
American groups. They consist of two genera, *Capromys,* containing three or four species inhabiting Cuba and Jamaica ;
while *Plagiodontia* (very closely allied) is confined to Hayti.
A peculiar mouse, a species of the American genus *Hesperomys,*
is said to inhabit Hayti and Martinique, and probably other
islands. A *Dasyprocta* or agouti, closely allied to, if not identical
with, a South American species, inhabits St. Vincent, St. Lucia,
and Grenada, and perhaps St. Thomas, and is the only mammal
of any size indigenous to the Lesser Antilles. All the islands
in which sugar is cultivated are, however, overrun with European
rats and mice, and it is not improbable that these may have

starved out and exterminated some of the smaller native rodents.

Birds.—The birds of the Antilles, although very inferior in number and variety to those of the mainland, are yet sufficiently abundant and remarkable, to offer us good materials for elucidating the past history of the country, when aided by such indications as geology and physical geography can afford.

The total number of land-birds which are permanent residents in the West India islands is, as nearly as can be ascertained from existing materials, 203. There are, in addition to this number, according to Prof. Baird, 88 migrants from North America, which either spend the winter in some of the islands or pass on to Central or South America. These migrants belong to 55 genera, and it is an interesting fact that so many as 40 of these genera have no resident representatives in the islands. This is important, as showing that this northern migration is probably a recent and superficial phenomenon, and has not produced any (or a very slight) permanent effect on the fauna. The migratory genera which have permanent residents, and almost always representative species, in the islands, are in most cases characteristic rather of the Neotropical than of the Nearctic fauna, as the following list will show; *Turdus, Dendrœca, Vireo, Polioptila, Agelœus, Icterus, Contopus, Myiarchus, Tyrannus, Antrostomus, Chordeiles, Coccyzus, Columba.* By far the larger part of these birds visit Cuba only; 81 species being recorded as occurring in that island, while only 31 have been found in Jamacia, 12 in Porto Rico and St. Croix, and 2 in Tobago and Trinidad. Setting aside these migratory birds, as having no bearing on the origin of the true Antillean fauna, we will discuss the residents somewhat in detail.

The resident land-birds (203 in number) belong to 95 genera and 26 families. Of these families 15 are cosmopolitan or nearly so—Turdidæ, Sylviidæ, Corvidæ, Hirundinidæ, Fringillidæ, Picidæ, Cuculidæ, Caprimulgidæ, Cypselidæ, Trogonidæ, Psittacidæ, Columbidæ, Tetraonidæ, Falconidæ, and Strigidæ; 5 are American only—Vireonidæ, Muiotiltidæ, Icteridæ, Tyrannidæ, Trochilidæ ; 4 are Netropical only or almost exclusively—

Cœrebidæ, Tanagridæ, Cotingidæ, Conuridæ ; 1 is Antillean only—Todidæ ; while 1—Ampelidæ—is confined (in the western hemisphere) to North America, and almost to the Nearctic region. Of the 95 genera, no less than 31, or almost exactly one-third, are peculiar ; while of the 203 resident species, 177 are peculiar, the other 26 being all inhabitants of South or Central America. Considering how closely the islands approach the continent in several places—Florida, Yucatan, and Venezuela—this amount of speciality in such locomotive creatures as birds, is probably unexampled in any other part of the globe. The most interesting of these peculiar genera are the following: 4 of Turdidæ, or thrushes—1 confined to the large islands, 1 to the whole archipelago, while 2 are limited to the Lesser Antilles ; 2 genera of Tanagridæ, confined to the larger islands ; 2 of Trogonidæ, also confined to the larger islands ; 5 of humming-birds, 3 confined to the Greater, 1 to the Lesser Antilles ; 2 of cuckoos, one represented in all the large islands, the other in Jamaica only ; 2 of owls, one peculiar to Jamaica, the other represented in St. Croix, St. Thomas, Portorico, and Cuba ; and lastly, *Todus*, constituting a peculiar family, and having representative species in each of the larger islands, is especially interesting because it belongs to a group of families which are wholly Neotropical—the Momotidæ, Galbulidæ, and Todidæ. The presence of this peculiar form, with 2 trogons ; 10 species of parrots, all but one peculiar ; 16 peculiar humming-birds belonging to 8 genera ; a genus of Cotingidæ ; 10 peculiar tanagers belonging to 3 genera ; 9 Cœrebidæ of 3 genera ; together with species of such exclusively Netropical genera as *Cœreba, Certhiola, Sycalis, Phonipara, Elainea, Pitangus, Campephilus, Chloronerpes, Nyctibius, Stenopsis, Lampornis, Calypte, Ara, Chrysotis, Zenaida, Leptoptila,* and *Geotrygon,* sufficiently demonstrate the predominant affinities of this fauna ; although there are many cases in which it is difficult to say, whether the ancestors of the peculiar genera or species may not have been derived from the Nearctic rather than from the Neotropical region.

The several islands differ considerably in their apparent productiveness, but this is, no doubt, partly due to our knowledge of Cuba and Jamaica being much more complete than of Hayti. The species of resident land-birds at present known are as follows :—

Cuba	68	species of which 40 are peculiar to it.
Hayti	40	„ ‚‚ 17 „ „
Jamaica	67	„ „ 41 „ „
Portorico	40	„ „ 15 „ „
Lesser Antilles	45	„ „ 24 „ „

If we count the peculiar genera of each island, and reckon as ($\frac{1}{2}$) when a genus is common to two islands only, the numbers are as follows:—Cuba $7\frac{1}{2}$, Hayti $3\frac{1}{2}$, Jamaica $8\frac{1}{2}$, Portorico 1, Lesser Antilles $3\frac{1}{2}$. These figures show us, that although Jamaica is one of the smaller and the most isolated of the four chief islands, it yet stands in the first rank, both for the number of its species and of its peculiar forms of birds,—and although this superiority may be in part due to its having been more investigated, it is probably not wholly so, since Cuba has also been well explored. This fact indicates, that the West Indian islands have undergone great changes, and that they were not peopled by immigration from surrounding countries while in the condition we now see them ; for in that case the smaller and more remote islands would be very much poorer, while Cuba, which is not only the largest, but nearest to the mainland in two directions, would be immensely richer, just as it really is in migratory birds.

The number of birds common to the four larger islands is very small—probably not more than half a dozen ; between 20 and 30 are common to some two of the islands (counting the Lesser Antilles as one island) and a few to three ; but the great mass of the species (at least 140) are confined each to some one of the five islands or groups we have indicated. This is an amount of isolation and speciality, probably not to be equalled else-where, and which must have required a remarkable series of physical changes to bring about. What those changes probably were, we shall be in a better position to consider when we have completed our survey of the various classes of land animals.

PLATE XVII.

A SCENE IN CUBA, WITH CHARACTERISTIC ANIMALS

In the preceding enumeration the Bahamas have been included with Cuba, as regards the birds they have in common; but they possess some half dozen species not found elsewhere, and even one central American genus of humming-birds (*Doricha*) not found in any other part of the Antilles. We have thus given Cuba rather more peculiar species than it really possesses, so that the proportionate richness of Jamaica is rather greater than shown by our figures.

The destruction of the forests and the increase of population, with, perhaps, the use of firearms, seem to have led to the extermination of some species of birds in the smaller islands. Professor Newton has called attention to the work of M. Ledru, who, in 1796, described the birds of St. Thomas. He mentions a parrot and a parroquet in the island, the latter only being now known, and very scarce; also a green pigeon and a tody, both now unknown. No less than six species of parrots are said to have been formerly found in Guadeloupe and Martinique, which are now extinct.

Plate XVII. Illustrating the peculiar Mammalia and Birds of the Antilles.—The scene of this illustration is Cuba, the largest of the West Indian islands, and one in which all its peculiar zoological features are well developed. In the foreground is the agouta (*Solenodon cubanus*), a remarkable insectivorous animal which, with another species inhabiting Hayti, has no allies on the American continent; nor anywhere in the world but in Madagascar, where a group of animals are found constituting the family Centetidæ, to which *Solenodon* is said undoubtedly to belong. Above it are a pair of hutias (*Capromys fournieri*), rat-like animals belonging to the South American family Octodontidæ. They live in the forests, and climb trees readily, eating all kinds of vegetable food. Three species of the genus are known, which are found only in Cuba and Jamaica. Just above these animals is a white-breasted trogon (*Prionoteles temnurus*), confined to Cuba, and the only species of the genus. Near the top of the picture are a pair of todies (*Todus multicolor*), singular little insectivorous birds allied to the motmots, but forming a very distinct family which is confined to the islands of the

Greater Antilles.　They are beautifully-coloured birds,—green above, red and white beneath, and are exceedingly active in their movements.　To the right are a pair of small humming-birds (*Sporadinus ricordi*), not very remarkable in this beautiful family, but introduced here because they belong to a genus which is confined to the Greater Antilles.

Table of distribution of West-Indian Birds.—As the birds of the West-Indian islands are particularly interesting and their peculiarities comparatively little known, we give here a table of the genera of land-birds, compiled from all available sources of information.　Owing to the numerous independent observations on which it is founded, the discrepancies of nomenclature, and uncertainty in some cases as to the locality of species, it can only be looked upon as an approximative summary of the existing materials on Antillean ornithology.

TABLE OF THE RESIDENT LAND-BIRDS OF THE ANTILLES.

NOTE.—Genera confined to the West Indies are in Italics.　An (*a*) after (1) indicates a species common to two islands : but where there are two or more species in an island, or the localities are doubtful, this indication cannot be given.　All species not otherwise noted are peculiar to the Antilles.

Family and Genus.	Number of Species in each Island.						Total resident species.	Remarks.
	Cuba.	Bahamas.	Hayti.	Jamaica.	Portorico & St. Croix.	Lesser Antilles.		
TURDIDÆ.								
Turdus	—	—	—	1	--	—	1	Five species migrate to Cuba
Mimocichla	2	1	1	1	—	—	5	
Margarops	—	—	1*a*	—	1*a*	3	4	Martinique, St. Lucia, Guada.
Rhamphocinclus ...	—	—	—	—	—	1	1	Martinique and St. Lucia
Cinclocerthia... ...	—	—	—	—	—	3	3	Nevis to St. Lucia
Mimus	1	1	—	1	(?)	—	3	Another species migrates to the Antilles
SYLVIIDÆ.								
Myiadestes	1	—	—	1	—	1	3	St. Lucia
Polioptila	1	—	—	—	—	—	1	

Family and Genus.	Cuba.	Bahamas.	Hayti.	Jamaica.	Portorico & St. Croix.	Lesser Antilles	Total resident species.	Remarks.
VIREONIDIE.								
Vireosylvia	1	—	1	1	1	1	2	One S. American species
Vireo	1	1	—	1	1	—	4	Five species migrate to Cuba
Laletes	—	—	—	1	—	—	1	
Phœnicomanes ...	—	—	—	1	—	—	1	
CORVIDÆ.								
Corvus...	1	—	1*a*	1	1*a*	—	3	
Cyanocorax	—	—	—	1	—	—	1	S. American species
MNIOTILTIDÆ.								
Perissoglossa... ...	1	—	1	—	1	—	1	N. American species
Dendrœca	2	2	1	3	1	1	7	Twelve sp. migrate to W. I.
Teretristis	2	·	—	—	—	—	2	
CŒREBIDÆ.								
Certhiola	—	1	1	1	2	2	7	Dominica and Martinique
Glossiptila	—	—	—	1	—	—	1	
Cœreba	1	—	—	—	—	—	1	S. American species
AMPELIDÆ.								
Dulus	(?)	—	1	(?)	(?)	(?)	2	One species locality unknown
HIRUNDINIDÆ.								
Progne	—	—	1	1	1	—	1	
Pterochelidon ...	1	—	—	1	1	—	1	
Hirundo	1	—	1*a*	1*a*	—	—	2	One S. American species
TANAGRIDÆ.								
Euphonia	1*a*	—	1*a*	1	1	1	4	St. Bartholom. & Martinique
Spindalis	2	1	1	1	1	—	5	
Phœnicophilus ...	—	—	1	—	—	—	1	
Saltator	—	—	—	—	—	1	1	Guadeloupe and St. Lucia
FRINGILLIDÆ.								
Loxigilla	—	—	1	1	—	1	3	Martinique and Dominica
Melopyrrha ·... ...	1	—	—	—	—	—	1	
Sycalis	—	—	—	1	—	—	1	S. American species
Phonipara	3	—	3	3	2	—	4	One S. American species
Chrysomitris	—	—	1	—	—	—	1	
ICTERIDÆ.								
Icterus	1	—	1	1	2	2	6	
Agelæus	2	—	·—	—	1	—	3	
Sturnella	1	·—	—	—	—	—	1	Mexican species
Nesopsar	—	—	—	1	—	—	1	
Scolecophagus ...	1	—	—	—	—	—	1	
Quiscalus ...	—	—	1	1	2	2	4	St. Lucia, Martinique and Barbadoes

Family and Genus.	Number of Species in each Island						Total resident species.	Remarks.
	Cuba.	Bahamas.	Hayti.	Jamaica.	Porto rico & St. Croix.	Lesser Antilles.		
TYRANNIDÆ.								
Elainea	—	—	—	2	—	1	3	
Pitangus	1a	—	—	1a	1	—	2	St. Lucia
Contopus	—	—	—	1	—	1	2	
Myiarchus	2	—	1	3	1	1b	7	One S. American species (b)
Blaccicus	1a	—	1a	1	—	—	2	
Tyrannus	2	—	—	1b	1b	2b	3	One sp. in Cen. America (b)
COTINGIDÆ.								
Hadrostomus	—	—	—	1	—	—	1	
PICIDÆ.								
Campephilus	1	—	—	—	—	—	1	
Xiphidiopicus.	1	—	—	—	—	—	1	
Melanerpes	—	. —	—	—	1	—	1	
Chloronerpes	—	—	1	—	—	—	1	
Centurus	1	—	1	1	—	—	3	
Colaptes	2	—	—	—	—	—	2	
Nesocelcus	1	—	—	—	—	—	1	
Picumnus	—	—	?1	—	—	—	1	
CUCULIDÆ.								
Saurothera	1	—	1	1	1	—	4	
Hyetornis	—	—	1	1	—	—	2	
Coccygus	1	—	2	1	1	1	3	Dominica, St. Lucia, all Neo-tropical species
Crotophaga	1	—	—	1	1	1	2	N. & Cen. American species
TODIDÆ.								
Todus	1	—	1	2	1	—	5	
TROGONIDÆ.								
Prionoteles	1	—	—	—	—	—	1	
Temnotrogon	—	—	1	—	—	—	1	
CAPRIMULGIDÆ.								
Nyctibius	—	—	—	1	—	—	1	Neotropical species
Chordeiles	—	—	—	1	—	—	1	
Antrostomus	2	—	—	1	—	1	2	One Neotropical species
Siphonorhis	—	—	—	1	—	—	1	
Stenopsis	—	—	—	—	—	1	1	Martinique (S. America sp.)
CYPSELIDÆ.								
Cypselus	1	—	—	1	—	—	1	
Panyptila	—	—	1	—	—	—	1	S. American species
Hemiprocne	—	—	1	—	—	—	1	Mexican species
Cypseloides	—	—	—	1	—	—	1	

Family and Genus.	Number of Species in each Island.						Total resident species	Remarks.
	Cuba.	Bahamas.	Hayti.	Jamaica.	Porto rico & St. Croix.	Lesser An- tilles.		
TROCHILIDÆ.								
Lampornis	—	—	1a	1	2a	1a	3	
Doricha	—	2	—	—	—	—	2	
Eulampis	—	—	—	—	1	2	2	St. Croix, Dominica, St. Lucia, Martinique
Aithurus	—	—	—	1	—	—	1	
Mellisuga	—	—	1	1	—	—	1	
Calypte	1	—	—	—	—	—	1	
Orthorhynchus	—	—	—	—	1	2	3	Domin., Martini., St. Lucia
Sporadinus	1	—	1	—	1	—	3	
CONURIDÆ.								
Ara	1	—	—	—	—	—	1	S. American species
Conurus	1	—	—	1	1	1	1	St. Thomas
PSITTACIDÆ.								
Chrysotis	1	—	1	2	1	3	8	
COLUMBIDÆ.								
Columba	1	—	1	2	2	1	3	One in Honduras
Chamæpelia	—	—	—	1	1	1	1	
Zenaida	1	—	1	1	1	2	2	
Leptoptila	—	—	—	1	—	—	1	
Geotrygon	2	—	1	2	1	2	5	St. Lucia, Martinique, one species Mexican
Starnœnas	1	—	—	—	—	—	1	
TETRAONIDÆ.								
Ortyx	1	—	—	—	—	—	1	
FALCONIDÆ.								
Accipiter	2	—	—	—	—	—	2	
Hypotriorchis	—	—	—	1	—	—	1	Mexican species
Cerchneis	2	—	1	—	—	1	2	
Cymindis	1	—	—	—	—	—	1	
Polyborus	1	—	—	—	—	—	1	Mexican species
STRIGIDÆ.								
Nyctalops	1	—	—	—	—	—	1	S. American species
Pseudoscops	—	—	—	1	—	—	1	
Gymnoglaux	1	—	—	—	1	—	2	St. Croix and St. Thomas
Glaucidium	1	—	—	—	—	—	1	

TOTALS { Number of families of resident land-birds in the Antilles 26
,, ,, genera ,, ,, ,, ... 95
,, ,, species ,, ,, ,, ... 203

Reptiles and Amphibia.—These classes not having been systematically collected, and the numerous described genera not having undergone careful revision, little trustworthy information can be derived from them. The following enumeration of the chief groups hitherto noticed or described, will, however, show very similar features to those presented by the birds—a general relation to Neotropical forms, a more special relation to those of Central America and Mexico, and a considerable number of peculiar types.

Snakes.—*Arrhyton* (Calamariidæ) from Cuba, *Hypsirhynchus* from Barbadoes, *Cryptodacus* from Cuba, *Ialtris* from Hayti, and *Coloragia* from Cuba (all Colubridæ), have been described as genera peculiar to the Antilles. *Phylodryas* and *Dromicus* (Colubridæ) are Antillean and Neotropical; *Ahætulla* (Dendrophidæ) has the same distribution but extends to tropical Africa; *Epicrates* and *Corallus* (Pythonidæ) are Neotropical and Antillean; while *Chilabothrus* from Jamaica and *Ungalia* from Cuba and Jamaica (both Pythonidæ) are found elsewhere only in Central America and Mexico. There appear to be no Crotalidæ except an introduced species of *Craspedocephalus* in St. Lucia.

Lizards are more numerous. *Amciva* (Teidæ) is found all over America. *Gerrhonotus* (Zonuridæ) is Neotropical and occurs in Cuba; *Gymnopthalmus* is South American and Antillean. Of Scincidæ seven genera are noted. *Celestus* (with 9 species) is peculiar to the Antilles; *Camilia* (1 species) to Jamaica, *Panoplus* (1 species) and *Embryopus* (1 species) to Hayti; *Diplogossus* is Antillean and South American; while *Plestiodcn* and *Mabouya* are cosmopolite. Of Geckotidæ there are four genera; *Phyllodactylus* and *Hemidactylus* which are cosmopolite; *Sphærodactylus* which is wholly American; and *Cubina* found only in Martinique and Brazil. Of Iguanidæ there are six genera; *Anolis*, which ranges all over America; *Polychrus*, which is Neotropical; *Iguana* and *Liocephalus* which are South American; *Tropedurus* found in Cuba and Brazil; and *Cyclura* only known from Jamaica, Cuba, and Central America.

Amphibia.—The genus *Trachycephalus*, belonging to the

Hylidæ or tropical tree-frogs, is almost peculiar to the Antilles ; Cuba, Hayti, and Jamaica possessing seven species, while only one is recorded from South America. Other genera are, *Pelta-phryne* (Bufonidæ) from Portorico ; *Phyllobates* (Polypedatidæ) from Cuba ; *Leiuperus* (Ranidæ) from Hayti,—all Neotropical. Of the Urodela, or tailed batrachians, no representative occurs, although they are so characteristic a feature of the Nearctic region.

Fresh-water fish.—The same general remarks apply to these as to the reptiles. Only one peculiar genus is noted—*Lebistes,* a form of Cyprinodontidæ from Barbadoes; other genera of the same family being, *Haplochilus, Rivulus,* and *Girardinus,* widely spread in the Neotropical region ; while *Gambusia* is confined to Central America, Mexico, and the Antilles. Four other families are represented; Siluridæ by *Chætostomus,* found in Portorico and South America; Chromidæ by the South American *Acara* ; Mugillidæ by the Central American *Agonostoma ;* and Percidæ by the North American *Centrarchus,* of which a species is recorded from Cuba.

Insects.—The various West Indian islands have not been well explored entomologically ; one reason no doubt being, that their comparative poverty renders them little attractive to the professional collector, while the abounding riches of Central and South America lie so near at hand. We can, therefore, hardly tell whether the comparative poverty, or even total absence of some families while others seem fairly represented, is a real phenomenon of distribution, or only dependent on imperfect knowledge. Bearing this in mind, we proceed to give a sketch of what is known of the chief groups of Lepidoptera and Coleoptera.

Lepidoptera.—The Neotropical butterfly-fauna is but poorly represented, the majority of the most remarkable types being entirely wanting ; yet there are a few peculiar and very characteristic forms which show great isolation, while the majority of the species are peculiar. Four genera are exclusively or characteristically Antillean,—*Calisto* belonging to the Satyridæ, with four species, of which one ranges to South Carolina; *Clothilda*

(Nymphalidæ) a fine genus which has 4 Antillean species and 2 in Central America; *Lucinia* (Nymphalidæ) 2 species, confined to Jamaica and Hayti ; and *Kricogonia* belonging to the Pieridæ, which has 2 West Indian species, while 1 inhabits Mexico and Florida.   Genera which show a special relation to Central America are *Euptoieta, Eumæus,* and *Nathalis.*   Almost all the other genera are South American, the total number recorded in each family as occurring in the West Indian islands, being, 3 of Danaidæ ; 1 of Heliconiidæ ; 2 of Satyridæ ; 18 of Nymphalidæ ; 1 of Erycinidæ ; 4 of Lycænidæ ; 6 of Pieridæ ; 1 of Papilionidæ, and 10 of Hesperidæ.   The genus *Papilio* is represented by about 20 species, 2 of which are North American, 4 South American, while the rest form little characteristic groups allied to those of Central America. The most marked feature seems to be the scarcity of Satyridæ and the almost total absence of Erycinidæ, with a great deficiency in characteristic Neotropical forms of Danaidæ and Nymphalidæ.

Coleoptera.—Cicindelidæ and Carabidæ are very poorly represented, by a few species of wide-spread groups, and hardly any peculiar genera.   No Lucanidæ are recorded.   Of Cetoniidæ, *Gymnetis* only appears to be represented.   Buprestidæ seem to be more numerous ; 15 genera being recorded, but almost all of wide distribution.   One only is peculiar—*Tetragonoschoma,* found in Hayti ; *Halecia* is the only exclusively South American genus ; *Chalcophora* is widely scattered over the tropical regions but is absent from South America, yet it occurs in the Nearctic region and extends to Jamaica and Guadeloupe. We now come to the Longicorns, the only group of Coleoptera which seems to be well represented, or which has been carefully collected.   No less than 40 genera are known from the West Indian islands, and 15 of these are peculiar.   Prionidæ are proportionately very numerous, there being 10 genera, 2 of which are widely distributed in both South and North America, 1 is North American, and 1 South American, while the following are peculiar,— *Stenodontes* (Hayti and Cuba); *Dendroblaptus* (Cuba); *Monodesmus* (Cuba and Jamaica) ; *Prosternodes* (Cuba); *Solenoptera* and *Elateropsis,* the two largest genera found in most of the islands. Of Cerambycidæ there are 16 genera, 2 of which range all over America, 4 are Neotropical, 1 South American only, while the following are confined to the islands,—*Merostenus, Pentomacrus,* and *Eburiola* (Jamaica) ; *Bromiades* (Cuba) ; *Trichrous, Heterops,* and *Pœciloderma* (Antilles). One genus, *Smodicum,* is widely spread, having a species in Carolina, 1 in South America, 1 in Hayti, and 1 in West Africa. Of Lamiidæ there are 14 genera, 8 of which are Neotropical, 1 common to Central America and Mexico, 1 to the United States and Cuba, while 2, *Proecha* and *Phidola,* are confined to Cuba. Several of the genera are curiously distributed ;—*Spalacopsis* is South American, with 4 species in Cuba and Tropical Africa ; *Lagocheirus* is Neotropical, with a species in Australia ; while *Leptostilus* is characteristic of the Antilles and North America, with a few species in South America, and one in New Zealand. These cases of erratic distribution, so opposed to the general series of phenomena among which they occur, must be held to be sufficiently explained by the great antiquity of these groups and their former wide distribution. They may be supposed to be the remnants of types, now dying out, which were once, like *Callichroma, Clytus,* and many others, almost universally distributed.

All the peculiar Antillean genera of Cerambycidæ and Lamiidæ are allied to Neotropical forms. The peculiar Prionidæ, however, are mostly allied to Mexican and North American groups, and one, *Monodesmus,* belongs to a group all the other genera of which inhabit the East Indies and South Africa.

Land-shells.—This subject has already been generally treated under the Region, of which, in this class of animals, the Antilles form so important a part. We must therefore now confine ourselves mainly to the internal distribution of the genera, and to a few remarks on the general bearing of the facts.

The excessive and altogether unexampled productiveness of the West Indian islands in land-shells, may be traced to two main sets of causes. The first and least known, consist of the peculiar influences and conditions which render islands always more productive than continents. Whatever these conditions are, they will be more effective where the islands have been long separated from the mainland, as is here undoubtedly the case. It seems most probable that the great development of land-shells in islands, is due to the absence or deficiency of the verte-brata, which on continents supply a variety of species adapted to prey upon these molluscs. This view is supported by the fact, that in such islands as have been united to a continent at no very distant epoch, and still maintain a continental variety of vertebrata, no such special development of land-shells has taken place. If we compare the Philippine islands with the Sunda group, we find the development of vertebrata and land-molluscs in inverse ratio to each other. The same thing occurs if we compare New Zealand and Tasmania ; and we have a still more striking example in the Antillean group itself, continental Trinidad having only 20 genera and 38 species, while the highly insular Jamaica has about 30 genera and more than 500 species.

The other causes favourable to the increase and development of land-shells are of a physical nature. A great extent of lime-stone-rock is one; and in the larger West Indian islands we have a considerable proportion of the surface consisting of this rock. But perhaps equally or more important, is the character of the land surface, and the texture of the exposed rock itself. A much broken surface, with numerous deep ravines, cutting up the whole country into isolated valleys and ridges, seems very favourable to the specialization of forms in this very sedentary class of animals. Equally favourable is a honeycombed and highly-fissured rock-surface, affording everywhere cracks and crannies for concealment. Now, taking Jamaica as an example of the archipelago, we find all these conditions in a wonderful degree. Over a large part of this island, a yard of level ground can hardly be found; but ridges, precipices, ravines, and rock-bound valleys, succeed each other over the whole country. At least five-sixths of the entire surface is limestone, and under the influence of tropical rains this rock is worn, fissured, and honey-combed, so as to afford ample shelter and concealment for land-shells.

It is probable that the three chief islands, Cuba, Jamaica and Hayti, are nearly equally rich in land-shells; but the last is very much less known, and therefore, perhaps, appears to be much poorer. Cuba has rather more species than Jamaica; but while the former has only 1 peculiar genus (*Diplopoma*), the latter has 3 (*Geomelania, Chittya,* and *Jamaicea*), as well as two others only represented in the other islands by single species. From Hayti, only about one-third as many species are known as from the two former islands. It has no peculiar genera, but it has some forms in common with Cuba and others with Jamaica, which show that those islands have more connection with it, than with each other; just as we found to be the case in birds. Portorico and the Virgin islands have still fewer species than Hayti; and, as many of the genera common to the other three islands are wanting, there is, no doubt, here a real deficiency. In the islands farther south (Barbuda to Martinique) more Antillean genera disappear or become very rare, while some continental forms take their place. The islands from St. Lucia to Trinidad have a still more continental character; the genus *Bulimus,* so largely developed on the continent, only reaching St. Lucia. The Bahamas contain about 80 species of land-shells, of which 25 are Antillean, the rest peculiar; all the genera being Antillean. The affinity is chiefly with Hayti and Cuba, but closest with the latter island.

In the West Indian islands as a whole, there are 11 peculiar genera; 9 operculate (*Geomelania, Chittya, Jamaicea, Licina, Choanopoma, Ctenopoma, Diplopoma, Stoastoma, Lucidella*); and 2 inoperculate (*Sagda* and *Stenopus*), besides *Cyclostomus,* which belongs to the Old World and is not found on the American continent. Mr. Bland considers, that many of the Antillean land-shells exhibit decided African and Asiatic, rather than South American affinities. A species of the Asiatic genus *Diplommatina* has been found in Trinidad, and an Indian species of *Ennea* occurs in Grenada and St. Thomas; a clear indication that land-shells are liable to be accidentally imported, and to become established in the less productive islands.

Although these islands are so wonderfully rich even now,

there is good reason to believe that many species have become extinct since the European occupation of them. When small islands are much cultivated, many of these molluscs which can only live under the shade of forests, are soon extirpated. In St. Croix many species have become extinct at a comparatively recent period, from the burning of forests; and as we know that in all the islands many of the species are excessively local, being often confined to single valleys or ridges, we may be sure that wherever the native forests have disappeared before the hand of man, numbers of land-shells have disappeared with them. As some of the smaller islands have been almost denuded of their wood, and in the larger ones extensive tracts have been cleared for sugar cultivation, a very considerable number of species have almost certainly been exterminated.

General Conclusions as to the Past History of the West Indian Islands.—The preceding sketch of the peculiarities of the animal life of these islands, enables us to state, that it represents the remains of an ancient fauna of decided Neotropical type, having on the whole most resemblance to that which now inhabits the Mexican sub-region. The number of peculiar genera in all classes of animals is so great in proportion to those in common with the adjacent mainland, as to lead us to conclude that, subsequent to the original separation from the Mexican area, a very large tract of land existed, calculated to support a rich and varied fauna, and, by the interaction of competing types, give rise to peculiar and specially modified organisms. We have already shown that the outline of the present islands and the depths of the surrounding seas, give indications of the position and extent of this ancient land; which not improbably occupied the space enclosed by uniting Western Cuba with Yucatan, and Jamaica with the Mosquito Coast. This land must have stretched eastward to include Anguilla, and probably northward to include the whole of the Bahamas. At one time it perhaps extended southward so as to unite Hayti with northern Venezuela, while Panama and Costa Rica were sunk beneath the Pacific. At this time the Lesser Antilles had no existence.

The only large island of whose geology we have any detailed

account, is Jamaica; and taking this as a type of what will probably be found in Cuba, and Hayti, we must place the continental period as having occurred after the close of the Miocene, or during some part of the Pliocene epoch, since a large portion of the surface of the former island consists of beds of marine limestone from 2,000 to 3,000 thick, believed to be of Pliocene age. After some time, the land between Hayti and South America subsided, and still later that between Central America and Cuba with Jamaica; but a large tract of land remained insulated, and no doubt supported a very much richer and more varied fauna than now. We have evidence of this in extinct Mammalia of large size, belonging to the peculiar South American family of the chinchillas, which have been found in caves in the small islands of Anguilla, and which, from the character of the land-shells associated with them, are believed to be of Pliocene or Post-pliocene age. This discovery is most interesting, and gives promise of very valuable results from the exploration of the numerous caverns that undoubtedly exist in the abundant limestone strata of the larger islands. This extensive Antillean land, after long continuing undivided, was at length broken up by subsidence into several islands; but as this alone would not account for the almost complete annihilation of the mammalian fauna, it seems probable that the subsidence was continued much farther, so as greatly to reduce the size and increase the number of the islands. This is indicated, by the extensive alluvial plains in Cuba and Hayti, and to a less extent in Jamaica; and by elevated beds of Post-pliocene marls in the latter island.

The series of changes now suggested, will account for all the main features of the Antillean fauna in its relations to that of the American continent. There remains the affinity with Madagascar, indicated by *Solenodon*, and a few cases of African and Asiatic affinity in insects and land-shells; but these are far too scanty to call for any attempt at special explanation. Such cases of remote affinity and discontinuous distribution, occur in all the regions, and in almost every group of animals; and we look upon them almost all, as cases of survival, under favourable

conditions, of once wide-spread groups. If no wild species
of the genus *Equus* were now to be found, except in South
Africa (where they are still most abundant), and in South
Temperate America, where their fossil remains show us they did
exist not very long ago, what a strong fact it would have
appeared for the advocates of continental extensions! Yet it
would have been due to no former union of the great southern
continents, but to the former extensive range of the family or
the genus to which the two isolated remnants belonged. And if
such an explanation will apply to the higher vertebrata, it is
still more likely to be applicable to similar cases occurring among
insects or mollusca, the genera of which we have every reason to
believe to be usually much older than those of vertebrates. It
is in these classes that examples of widely scattered allied
species most frequently occur; and the facility with which they
are diffused under favourable conditions, renders any other
explanation than that here given altogether superfluous.

The *Solenodon* is a member of an order of Mammalia of low
type (Insectivora) once very extensive and wide-spread, but
which has begun to die out, and which has left a number of
curious and isolated forms thinly scattered over three-fourths of
the globe. The occurrence, therefore, of an isolated remnant of
this order in the Antilles is not in itself remarkable; and the
fact that the remainder of the family to which the Antillean
species belong has found a refuge in Madagascar, where it has
developed into several distinct types, does not afford the least
shred of argument on which to found a supposed independent
land connection between these two sets of islands.

Summary of the Past History of the Neotropical Region.

We have already discussed this subject, both in our account
of extinct animals, and in various parts of the present chapter.
It is therefore only necessary here, briefly to review and summarise the conclusions we have arrived at.

The whole character of Neotropical zoology, whether as regards
its deficiencies or its specialities, points to a long continuance
of isolation from the rest of the world, with a few very distant

periods of union with the northern continent. The latest important separation took place by the submergence of parts of Nicaragua and Honduras, and this separation probably continued throughout much of the Miocene and Pliocene periods; but some time previous to the coming on of the glacial epoch, the union between the two continents took place which has continued to our day. Earlier submergences of the isthmus of Panama probably occurred, isolating Costa Rica and Veragua, which then may have had a greater extension, and have thus been able to develope their rich and peculiar fauna.

The isthmus of Tehuantepec, at the south of Mexico, may, probably, also have been submerged; thus isolating Guatemala and Yucatan, and leading to the specialization of some of the peculiar forms that now characterise those countries and Mexico.

The West Indian Islands have been long isolated and have varied much in extent. Originally, they probably formed part of Central America, and may have been united with Yucatan and Honduras in one extensive tropical land. But their separation from the continent took place at a remote period, and they have since been broken up into numerous islands, which have probably undergone much submergence in recent times. This has led to that poverty of the higher forms of life, combined with the remarkable speciality, which now characterises them; while their fauna still preserves a sufficient resemblance to that of Central America to indicate its origin.

The great continent of South America, as far as we can judge from the remarkable characteristics of its fauna and the vast depths of the oceans east and west of it, has not during Tertiary, and probably not even during Secondary times, been united with any other continent, except through the intervention of North America. During some part of the Secondary epoch it probably received the ancestral forms of its Edentates and Rodents, at a time when these were among the highest types of Mammalia on the globe. It appears to have remained long isolated, and to have already greatly developed these groups of animals, before it received, in early Tertiary times, the ancestors of its marmosets and monkeys, and, perhaps also, some of its peculiar forms of

Carnivora. Later, it received its Camelidæ, peccaries, mastodons, and large Carnivora; and later still, just before the Glacial epoch, its deer, tapir, opossums, antelopes, and horses, the two latter having since become extinct. All this time its surface was undergoing important physical changes. What its earlier condition was we cannot conjecture, but there are clear indications that it has been broken up into at least three large masses, and probably a number of smaller ones; and these have no doubt undergone successive elevations and subsidences, so as at one time to reduce their area and separate them still more widely from each other, and at another period to unite them into continental masses. The richness and varied development of the old fauna of South America, as still existing, proves, however, that the country has always maintained an extensive area; and there is reason to believe that the last great change has been a long continued and steady increase of its surface, resulting in the formation of the vast alluvial plains of the Amazon, Orinoko, and La Plata, and thus greatly favouring the production of that wealth of specific forms, which distinguishes South America above all other parts of our globe.

The southern temperate portion of the continent, has probably had a considerable southward extension in late Tertiary times; and this, as well as the comparatively recent elevation of the Andes, has given rise to some degree of intermixture of two distinct faunas, with that proper to South Temperate America itself. The most important of these, is the considerable Australian element that appears in the insects, and even in the reptiles and fresh-water fishes, of South Temperate America. These may be traced to several causes. Icebergs and icefloes, and even solid fields of ice, may, during the Glacial epoch, have afforded many opportunities for the passage of the more cold-enduring groups; while the greater extension of southern lands and islands during the warm periods—which there is reason to believe prevailed in the southern as well as in the northern regions in Miocene times —would afford facilities for the passage of the reptiles and insects of more temperate zones. That no actual land-connection occurred, is proved by the total absence of interchange of the mammals or land-birds of the two countries, no less than by the very fragmentary nature of the resemblances that do exist. The northern element consists almost wholly of insects ; and is evidently due to the migration of arctic and north temperate forms along the ridges and plateaus of the Andes; and most likely occurred when these organisms were driven southward at successive cold or Glacial periods.

A curious parallel exists between the past history and actual zoological condition of South America and Africa. In both we see a very ancient land-area extending into the South Temperate zone, isolated at a very early period, and developing only a low grade of Mammalian life ; chiefly Edentates and Rodents on the one, Lemurs and Insectivora in the other. Later we find an irruption into both of higher forms, including Quadrumana, which soon acquired a large and special development in the tropical portions of each country. Still later we have an irruption into both of northern forms, which spread widely over the two regions, and having become extinct in the land from whence they came, have been long held to be the original denizens of their adopted country. Such are the various forms of antelopes, the giraffe, the elephant, rhinoceros, and lion in Africa ; while in America we have deer and peccaries, the tapir, opossums, and the puma.

On the whole, we cannot but consider that the broad outlines of the zoological history of the Neotropical region can be traced with some degree of certainty; but, owing to the absence of information as to the most important of the geological periods —the Miocene and Eocene—we have no clue to the character of its early fauna, or to the land connections with other countries, which may possibly have occurred in early Tertiary times.

TABLES OF DISTRIBUTION.

In drawing up these tables, showing the distribution of the various classes of animals in the Neotropical region, the following sources of information have been relied on, in addition to the general treatises, monographs, and catalogues used in the compilation of the Fourth Part of this work.

Mammalia.—D'Orbigny, and Burmeister, for Brazil and La Plata; Darwin, and Cunningham, for Temperate S. America; Tschudi, for Peru; Frazer, for Ecuador; Salvin, for Guatemala; Frantzius, for Costa Rica; Sclater, for Quadrumana N. of Panama; Gundlach, for Cuba; and papers by Dr. J. E. Gray, and Mr. Tomes.

Birds.—Sclater and Salvin's Nomenclator; Notes by Darwin, and Cunningham; Gundlach, March, Bryant, Baird, Elliot, Newton, Semper, and Sundevall, for various islands of the Antilles; and papers by Hudson, Lawrence, Grayson, Abbott, Sclater, and Salvin.

TABLE I.

FAMILIES OF ANIMALS INHABITING THE NEOTROPICAL REGION.

EXPLANATION.

Names in *italics* show the families which are peculiar to the region.
Names enclosed thus (......) indicate families which barely enter the region, and are not considered properly to belong to it.
Numbers correspond with those of the series of families in Part IV.

Order and Family.	Chili.	Brazil.	Mexico.	Antilles.	Range beyond the Region.
MAMMALIA.					
PRIMATES.					
4. *Cebidæ*		—	—		
5. *Hapalidæ* ...		—	(?)		
CHIROPTERA.					
10. *Phyllostomidæ*	—	—	—	—	California
12. Vespertilionidæ	—	—	—	—	Cosmopolite
13. Noctilionidæ...	—	—	—	—	All tropical regions
INSECTIVORA.					
Centetidæ				—	Madagascar
CARNIVORA.					
23. Felidæ	—	—	—		All regions but Australian
28. Canidæ	—	—	—		All regions but Australian
29. Mustelidæ ...	—	—	—		All regions but Australian
30. Procyonidæ ...	—	—	—		N. America
32. Ursidæ	—				All regions but Ethiopian and Australian
33. Otariidæ... ...	--				S. temperate zone
35. Phocidæ... ...	—			(?)	N. and S. temperate zones
CETACEA.					
36 to 41		—			Oceanic
SIRENIA.					
42. Manatidæ ...		—	—	—	Tropical shores
UNGULATA.					
44. Tapiridæ ...		—	—		Indo-Malaya
47. Suidæ		—	—		Cosmopolite, excl. Australia
48. Camelidæ ...	—				Palæarctic
50. Cervidæ... ...	—	—	—		All regions but Ethiopian and Australian

Order and Family.	Sub-regions.				Range beyond the Region.
	Chili.	Brazil.	Mexico.	Antilles.	
RODENTIA.					
55. Muridæ	—	—	—	—	Cosmopolite
59. Saccomyidæ ...			—		Nearctic
61. Sciuridæ			—	—	All regions but Australian
63. *Chinchillidæ* ...	—				
64. Octodontidæ ...	—	—		—	Africa
65. Echimyidæ ...	—	—			Ethiopian
66. Cercolabidæ ...			—	—	Nearctic
68. *Caviidæ*	—	—	—	—	
70. Leporidæ ...			—	—	All regions but Australian
EDENTATA.					
71. *Bradypodidæ* ...		—	—		
73. *Dasypodidæ* ...	—	—	—		
75. *Myrmecophagidæ*		—	—		
MARSUPIALIA.					
76. Didelphyidæ....	—	—	—		Temperate N. America
BIRDS.					
PASSERES.					
1. Turdidæ	—	—	—	—	Almost cosmopolite
2. Sylviidæ		—	—	—	Almost cosmopolite
5. Cinclidæ			—	—	Nearctic, Palæarctic, Oriental
6. Troglodytidæ ...	—	—	—	—	Nearctic, Palæarctic, Oriental
8. Certhiidæ ...				—	Nearctic, Palæarctic, Oriental
9. Sittidæ ...				—	All regions, excl. Africa
10. Paridæ				—	Nearctic, Palæarctic, Oriental
20. Corvidæ	—	—	—	—	Cosmopolite
26. *Cœrebidæ* ...		—	—	—	
27. Mniotiltidæ ...		—	—	—	Nearctic
28. Vireonidæ ...		—	—	—	Nearctic
29. Ampelidæ ...			—	—	Nearctic, Palæarctic
30. Hirundinidæ ...	—	—	—	—	Cosmopolite
31. Icteridæ	—	—	—	—	Nearctic
32. Tanagridæ ..	—	—	—	—	Nearctic
33. Fringillidæ ...	—	—	—	—	All regions but Australian
38. Motacillidæ ...	—	—	—	—	Cosmopolite
38a. *Oxyrhamphidæ*		—	—	—	
39. Tyranuidæ ...	—	—	—	—	Nearctic
40. *Pipridæ* ...		—	—		
41. *Cotingidæ* ...	—	—	—	—	
42. *Phytotomidæ* ...	—				
44. *Dendrocolaptidæ*	—	—	—		
45. *Formicariidæ* ...	—	—	-		
46. *Pteroptochidæ* ...	—	—			
PICARIÆ.					
51. Picidæ	—	—	—	—	All regions but Australian
54. Megalæmidæ ...		—	-		Ethiopian, Oriental
55. *Rhamphastidæ*		—	—		

Order and Family.	Sub-regions.				Range beyond the Region.
	Chili.	Brazil.	Mexico.	Antilles.	
58. Cuculidæ ...	—	—	—	—	Cosmopolite
60. *Bucconidæ* ...		—	—		
61. *Galbulidæ* ...		—	—		
64. *Todidæ...* ...				—	
65. *Momotidæ* ...		—	—		
66. Trogonidæ ...		—	—	—	Ethiopian, Oriental
67. Alcedinidæ ...	—	—	—	—	Cosmopolite
72. *Steatornithidæ*		—			
73. Caprimulgidæ	—	—	—	—	Cosmopolite
74. Cypselidæ ...	—	—	—	—	Almost cosmopolite
75. Trochilidæ ...	—	—	—.	—	Nearctic
PSITTACI.					
80. Conuridæ ...	—	—	—	—	S. United States
81. Psittacidæ ...		—	—	—	Ethiopian
COLUMBÆ.					
84. Columbidæ ..	—	—	—	—	Cosmopolite
GALLINÆ.					
87. Tetraonidæ ...		—	—	—	Almost cosmopolite
88. Phasianidæ			—		All regions but Australian
91. *Cracidæ* ...		—	—		
92. *Tinamidæ* ...	—	—	—		
OPISTHOCOMI.					
93. *Opisthocomidæ*		—			
ACCIPITRES.					
94. Vulturidæ ..	—	—	—	—	All regions but Australian
96. Falconidæ ...	—	—	—	—	Cosmopolite
97. Pandionidæ ...		—	—	—	Cosmopolite
98. Strigidæ	—	—	—	—	Cosmopolite
GRALLÆ.					
99. Rallidæ ...	—	—	—	—	Cosmopolite
100. Scolopacidæ...	—	—	—	—	Cosmopolite
101. *Chionididæ* ...	—				
102. *Thinocoridæ*...	—				
103. Parridæ... ...		—	—		Tropical regions
105. Charadriidæ...	—	—	—	—	Cosmopolite
108. *Cariamidæ* ...	—	—			
109. *Aramidæ* ...		—	—		
110. *Psophiidæ* ...		—			
111. *Eurypygidæ*...		—	—		
113. Ardeidæ ...	—	—	—	—	Cosmopolite
114. Plataleidæ ...	—	—	—	—	Almost cosmopolite
115. Ciconiidæ ...	—	—	—		Nearly cosmopolite
116. *Palamedeidæ*	—	—			
117. Phœnicopteridæ	—		—	—	Ethiopian, Indian

Order and Family.	Sub-regions.				Range beyond the Region.
	Chili.	Brazil.	Mexico.	Antilles.	
ANSERES.					
118. Anatidæ ...	—	—	—	—	Cosmopolite
119. Laridæ	—	—	—	—	Cosmopolite
120. Procellariidæ	—	—	—	—	Cosmopolite
121. Pelecanidæ ...	—	—	—	—	Cosmopolite
122. Spheniscidæ...	—				S. temperate zone
124. Podicipidæ ...	—	—	—	—	Cosmopolite
STRUTHIONES.					
126. Struthionidæ	—				Ethiopian
REPTILIA.					
OPHIDIA.					
1. Typhlopidæ ...		—	—	—	Tropical regions and S. Palæarctic
2. Tortricidæ......			—		Oriental, N. W. America
5. Calamariidæ...	—	—	—	—	All warm countries
6. Oligodontidæ			—		Oriental, Japan
7. Colubridæ ...	—	—	—	—	Almost cosmopolite
8. Homalopsidæ	—			—	All the regions
11. Dendrophidæ	—	—	—	—	All tropical regions
12. Dryiophidæ ..		—	—		Oriental, Ethiopian
13. Dipsadidæ ...		—	—		All tropical regions
14. Scytalidæ ...		—	—		Philippine Islands
16.Amblycephalidæ		—	—		Oriental
17. Pythonidæ ...	—	—	—	—	All tropical regions, California
20. Elapidæ ...	—	—	—		Tropical regions, Japan, S. Carolina
23. Hydrophidæ...			—		Oriental, Australian, Madagascar
24. Crotalidæ ...	—	—	—	—	Nearctic, Palæarctic, Oriental
LACERTILIA.					
27. Chirotidæ ...			—		Missouri
28. Amphisbænidæ	—	—		—	Ethiopian, S. Palæarctic
29. Lepidosternidæ	—	—			Ethiopian
31. *Helodermidæ*				—	
32. Teidæ	—	—	—	—	Nearctic
34. Zonuridæ ...		—	—		Nearctic, Ethiopian, S. Europe, and N. India
35. *Chalcidæ* ...	—	—	—		Nearctic
36. *Anadiadæ* ...		—			
37. *Chirocolidæ* ...		—			
38. *Iphisadæ* ...		—			
39. *Cercosauridæ*		—			
41. Gymnopthal-} midæ ... }		—		—	Australian, Ethiopian, Palæarctic
45. Scincidæ ...	—	—	—	—	Almost cosmopolite
49. Geckotidæ ..	—	—	—	—	Almost cosmopolite
50. Iguanidæ ...	—	—	—	—	Nearctic
CROCODILIA.					
55. Crocodilidæ ...		—	—	—	Ethiopian, Oriental, N. Australian
56. Alligatoridæ...		—	—		Nearctic

Order and Family.	Chili.	Brazil.	Mexico.	Antilles.	Range beyond the Region.
CHELONIA.					
57. Testudinidæ ...	—	--	—	—	All continents but Australian
58. Chelydidæ ...		—			Ethiopian, Australian
60. Cheloniidæ ...					Marine
AMPHIBIA.					
PSEUDOPHIDIA.					
1. Ceciliadæ			—	—	Oriental, Ethiopian
URODELA.					
6. (Salamandridæ)			—	—	Nearctic, Palæarctic
ANOURA.					
7. *Rhinophrynidæ*			—		
8. Phryniscidæ ...	—	—	—		Ethiopian, Australian, Java
9. *Hylaplesidæ* ...	—	—		—	
10. Bufonidæ ..	—	—	—	—	All continents but Australia
12. Engystomidæ...	—	—	—		All regions but Palæarctic
13. Bombinatoridæ	—	—			Palæarctic, New Zealand
14. *Plectromantidæ*	—				
15. Alytidæ		—			All regions but Oriental
16. Pelodryadæ ...	—	—			Australia
17. Hylidæ	—			—	All regions but Ethiopian
18. Polypedatidæ...	—	—	—	--	All the regions
19. Ranidæ	—	—	—	—	Almost cosmopolite
20. Discoglossidæ	—	—			All regions but Nearctic
21. *Pipidæ*	—				
FISHES.					
(FRESHWATER).					
ACANTHOPTERYGII.					
3. Percidæ	—	—		—	All regions but Australian
11. (Trachinidæ) ...	—				Australia
12. Scienidæ ...	(?)	—	—	(?)	All regions but Australian
33. Nandidæ ...		—			Oriental
34. *Polycentridæ* ...		—	—		
38. Mugillidæ ...		(?)	--	—	Australian, Ethiopian
52. Chromidæ ...	—	—	—	—	Ethiopian, Oriental
PHYSOSTOMI.					
59. Siluridæ	—	—	—	—	All warm regions
60. Characinidæ ...	—	—	—	—	Ethiopian
61. Haplochitonidæ	--				S. Australia
67. Galaxidæ ...	—				Tasmania and New Zealand
73. Cyprinodontidæ	—	—	—	--	Absent from Australia
78. Osteoglossidæ...		—			All tropical regions
84. *Gymnotidæ* ...		--			
85. Symbranchidæ		—	—		Oriental, Australian, (? marine)

Order and Family.	Sub-regions.				Range beyond the Region.
	Chili.	Brazil.	Mexico.	Antilles.	
DIPNOI.					
92. Sirenoidei ...		—			Ethiopian, Australian
PLAGIOSTOMATA.					
112. *Trygonidæ* ...		—			
INSECTS.					
LEPIDOPTERA (PART).					
DIURNI (BUTTERFLIES).					
1. Danaidæ... ...	—	—	—	—	All warm regions, and to Canada
2. Satyridæ... ...	—	—	—	—	Cosmopolite
4. Morphidæ ...		—	—		Australian, Oriental
5. *Brassolidæ* ...		—	—		
6. Acræidæ... ...		—	—		All tropical regions
7. *Heliconiidæ* ...		—	—	—	
8. Nymphalidæ ...	—	—	—	—	Cosmopolite
9. Libytheidæ ...		—		—	Absent from Australia
10. Nemeobiidæ ..		—	—		Not in Australia or Nearctic regions
11. *Eurygonidæ* ...		—	—		
12. Erycinidæ ...		—	—	—	Nearctic
13. Lycænidæ ...	—	—	—	—	Cosmopolite
14. Pieridæ	—	—	—	—	Cosmopolite
15. Papilionidæ ...	—	—	—	—	Cosmopolite
16. Hesperidæ ...	—	—	—	—	Cosmopolite
SPHINGIDEA.					
17. Zygænidæ ...	—	—	—	—	Cosmopolite
18. Castniidæ ...		—	—	—	Australian
20. Uraniidæ ...		—	—	—	All tropical regions
21. Stygiidæ... .		—			Palæarctic
22. Ægeriidæ ...	—	—	—	—	Not in Australia
23. Sphingidæ ...	—	—	—	—	Cosmopolite

TABLE II.

GENERA OF TERRESTRIAL MAMMALIA AND BIRDS INHABITING THE NEOTROPICAL REGION.

EXPLANATION.

Names in *italics* show the genera peculiar to the region.
Names enclosed thus (......) indicate genera which barely enter the region, and are not considered properly to belong to it.
Genera undoubtedly belonging to the region are numbered consecutively.

MAMMALIA.

Order, Family, and Genus.	No. of Species	Range within the Region.	Range beyond the Region.
PRIMATES.			
CEBIDÆ.			
1. *Cebus* ...	18	Costa Rica to Paraguay	
2. *Lagothrix*	5	Upper Amazon and E. Andes	
3. *Eriodes*	3	East Brazil, S. of Equator	
4. *Ateles*	14	Almost all tropical America	
5. *Mycetes*	10	E. Guatemala to Paraguay	
6. *Pithecia*	7	Equatorial Forests	
7. *Brachiurus* ...	5	Equatorial Forests	
8. *Nyctipithecus* ...	5	Nicaragua to Amazonia	
9. *Saimiris*	3	Costa Rica to Brazil and Bolivia	
10. *Callithrix* ...	11	Panama to Paraguay	
HAPALIDÆ.			
11. *Hapale*	9	Brazil and Upper Amazon	
12. *Midas*	24	Equatorial America to Panama	
CHIROPTERA.			
PHYLLOSTOMIDÆ.			
13. *Lonchorina* ...	1	West Indian Islands	
14. *Macrophyllum* ...	1	Brazil	
15. *Vampyrus* ⎫			
16. *Lophostoma* ⎬	25	Tropical America and Chili	
17. *Phyllostoma* ⎭			
18. Macrotus	1	Antilles and Mexico	California
19. *Schizostoma* ...	5	South America	
20. *Brachyphylla* ...	1	Antilles	
21. *Glossophaga* ...	8	Tropical America	
22. *Phyllonycteris* ...	2	Cuba	
23. *Artibeus*	4	S. America & Antilles, Costa Rica	
24. *Stenoderma* ...	7	The whole region	
25. *Sturnira*	3	Chili to Guatemala	
26. *Desmodus* ...	3	Chili to Mexico	
27. *Saccopteryx* ...	1	Ecuador	

Order, Family, and Genus.	No. of Species	Range within the Region.	Range beyond the Region.
28. *Diphylla*	1	Brazil	
29. *Centurio*	3	Brazil to Mexico	
VESPERTILIONIDÆ.			
30. Lasiurus	2	Tropical America	Nearctic
31. Scotophilus ...	7	Antilles, Mexico to S. America	Nearc., Austral., Orien.
32. Vespertilio ..	12	The whole region	Cosmopolite
33. Nycticejus ...	3	S. Temperate America	Nearctic, India, Tropical Africa
34. *Natalus*	1	S. America and Antilles	
35. *Furipterus* ...	2	S. America	
36. *Thyroptera* ...	2	S. America	
37. *Nycticellus* ...	1	Cuba	
38. Taphozous ...	5	S. America	Ethiopian, Oriental, Austro-Malayan
39. *Diclidurus* ...	1	Brazil	
NOCTILIONIDÆ.			
40. *Noctilio*	2	Paraguay to W. Indies	
41. *Mormops*	1	Antilles and Mexico	
42. *Phyllodia*... ...	1	Jamaica	
43. *Chilonycteris* ...	5	Brazil and West Indies	
44. *Pteronotus* ...	1	Trinidad	
45. Nyctinomus ...	2	La Plata to Antilles & Costa Rica	S. Nearc., Orien.,Madag.
46. Molossus	16	Paraguay and Chili to Antilles	Ethiopian, S. Palæarc., Australian
INSECTIVORA.			
CENTETIDÆ.			
47. *Solenodon*... ..	2	Cuba and Hayti	
SORICIDÆ.			
(Sorex	1	Guatemala and Costa Rica)	All other reg. but Austrl.
CARNIVORA.			
FELIDÆ.			
48. Felis	13	The whole region, excl. Antilles	All regions but Austral.
CANIDÆ.			
49. *Icticyon*	1	Brazil	
50. *Chrysocyon* ...	1	S. America	
(Lupus	2	Mexico to Costa Rica)	Northern genus
51. *Lycalopex*... ...	2	S. America	
52. *Pseudalopex* ...	5	S. America, Falkland Islands, & Tierra del Fuego	
53. *Thous*	2	S. America to Chili	
MUSTELIDÆ.			
54. Mustela	2	Andes of Peru	All other reg. but Austrl.
55. *Galictis*	2	S. America to Chili & Patagonia	
56. *Lontra*	3	Central and S. America to Chonos Archipelago	
57. Nutria	1	W. coast of America to Chiloe	W. coast of N. America

Order, Family, and Genus.	No. of Species.	Range within the Region.	Range beyond the Region.
58. *Pteronura* ...	1	Surinam and Brazil	
59. Mephitis	3	Mexico to Sts. of Magellan	Nearctic to Canada
PROCYONIDÆ.			
60. Procyon	1	Tropical America	Nearctic to Canada
61. *Nasua*	5	Mexico to Paraguay & La Plata	
62. *Cercoleptes* ...	1	Mexico to Peru and N. Brazil	
63. Bassaris	2	Mexico and Guatemala	California and Texas
URSIDÆ.			
64. *Tremarctos* ...	1	Andes of Peru and Chili	
OTARIIDÆ.			
65. *Otaria*	1	Chili, La Plata, and Patagonia	
66. Arctocephalus ...	1	Falkland Islands & Cape Horn	New Zealand
PHOCIDÆ.			
67. Stenorhynchus	1	Falkland Islands	New Zealand
68. Lobodon	1	Antarctic shores	
69. Leptonyx	1	Antarctic shores, E. Patagonia	S. Australia
70. Ommatophoca...	1	Antarctic shores	
71. Morunga	1	Falkland Islands	California, S. temp. zone
72. Cystophora ..	1	Antilles	N. Atlantic
CETACEA.			
DELPHINIDÆ.			
73. *Inia*	1	Upper Amazon	
SIRENIA.			
MANATIDÆ.			
74. Manatus	1	Gulf of Mexico to N. Brazil, Amazon R.	W. Africa
UNGULATA.			
TAPIRIDÆ.			
75. Tapirus	2	Equatorial S. America	Indo-Malaya
76. *Elasmognathus*	1	Panama to Guatemala	
SUIDÆ.			
77. *Dicotyles*	2	Mexico to Paraguay	Texas
CAMELIDÆ.			
78. *Auchenia*	4	Temp. S. America, from Cape Horn to Andes of Peru	
CERVIDÆ.			
79. Cervus	12	Mexico to Patagonia and Tierra del Fuego	All regions but Ethiopian and Australian
RODENTIA.			
MURIDÆ.			
80. Reithrodon ...	4	South Temp. America to Tierra del Fuego	United States

Order, Family, and Genus.	No of Species.	Range within the Region.	Range beyond the Region.
81. *Acodon*	1	Peru, 14,000 ft. elevation	
82 *Myxomys*	1	Guatemala	
83. Hesperomys ...	76	The whole region	Nearctic
84. *Holochilus* ...	4	S. America	
85. *Oxymycterus* ..	3	Brazil and La Plata	
86. *Drymomys* ...	1	Peru	
87. *Neotomys*	2	S. America	
(Fiber	1	Mexico)	Nearctic genus
SACCOMYIDÆ.			
88. *Heteromys* ...	6	Mexico, Honduras, Costa Rica & Trinidad	
SCIURIDÆ.			
89. Sciurus	30	Mexico to Paraguay	All reg. but Australian
CHINCHILLIDÆ.			
90. *Chinchilla* ...	2	Andes of Chili and Peru	
91. *Lagidium* ...	3	Chili to Ecuador (11,000 to 16,000 ft.)	
92. *Lagostomus* ...	1	Uruguay to Rio Negro of Patagonia	
OCTODONTIDÆ.			
93. *Habrocomus* ...	2	Chili	
94. *Capromys*	3	Cuba and Jamaica	
95. *Plagiodont·a* ...	1	Hayti	
96. *Spalacopus* ..	2	Chili and E. of Andes	
97. *Octodon*	3	Chili, Peru, and Bolivia	
98. *Ctenomys*	6	S. Brazil to Tierra del Fuego	
ECHIMYIDÆ.			
99. *Dactylomys* ...	2	Guiana and Brazil	
100. *Cercoxys* ...	1	Central Brazil	
101. *Lasiuromys* ...	1	St. Paulo, Brazil	
102. *Myopotamus* ...	1	S. half of tropical S. America	
103. *Carterodon* ...	1	Central Brazil	
104. *Mesomys*	1	Upper Amazon	
105. *Echimys.*	11	Equatorial America to Paraguay	
106. *Loncheres* ...	10	New Granada to Brazil	
CERCOLABIDÆ.			
107. *Cercolabes* ...	12	Mexico to Paraguay	
108. *Chætomys* ...	1	N. Brazil	
CAVIIDÆ.			
109. *Dasyprocta* ...	9	Paraguay to Mexico and Lesser Antilles	
110. *Cælogenys* ..	2	Guatemala to Paraguay	
111. *Hydrochærus* ...	1	Guiana to La Plata	
112. *Cavia*	9	Brazil and Peru to Magellan Sts.	
113. *Kerodon*	6	Brazil and Peru to Magellan Sts.	
114. *Dolicholis* ...	1	The Pampas and Patagonia	

Order, Family, and Genus.	No. of pecies.	Range within the Region.	Range beyond the Region.
LEPORIDÆ.			
115. Lepus	1	Central Brazil and Andes, Costa Rica to Mexico	All regions but Austral.
EDENTATA.			
BRADYPODIDÆ.			
116. *Cholœpus*	2	Costa Rica to Brazil	
117. *Bradypus* ..	2	Amazon to Rio de Janeiro	
118. *Arctopithecus*...	8	Costa Rica to Brazil and Bolivia	
DASYPODIDÆ.			
119. *Tatusia*... ...	5	Rio Grande, Texas, to Patagonia	
120. *Prionodontes* ...	1	Surinam to Paraguay	
121. *Dasypus*... ...	4	Brazil to Chili and La Plata, Costa Rica ?	
122. *Xenurus* ...	3	Guiana to Paraguay, Costa Rica ?	
123. *Tolypeutes* ...	2	Bolivia and La Plata	
124. *Chlamydophorus*	2	La Plata and Bolivia	
¦ MYRMECOPHAGIDÆ.			
125. *Myrmecophaga*	1	Costa Rica ?, & N. Braz., to Parag.	
126. *Tamandua* ...	2	Guatemala to Paraguay	
127. *Cyclothurus* ...	2	Honduras and Costa Rica to Paraguay	
MARSUPIALIA.			
DIDELPHYIDÆ.			
128. Didelphys ...	20	Mexico to Uruguay and S. Chili	Temperate N. America
129. *Chironectes* ...	1	Guiana and Brazil, Costa Rica	
130. *Hyracodon* ...	1	Ecuador	

BIRDS.

PASSERES.			
TURDIDÆ.			
1. Turdus	32	The whole reg to Tierra del Fuego	Almost cosmopolite
2. *Rhodinocichla* ...	1	Mexico to Venezuela	
3. *Melanoptila* ...	1	Honduras	
4. *Catharus*	10	Mexico to Ecuador and Columbia	
5. *Margarops* ...	4	Hayti and Lesser Antilles	
6. Mimus	16	Nearly the whole region	Nearctic
7. *Melanotis*... ...	2	Mexico and Guatemala	
8. Galeoscoptes ...	1	Mexico to Panama	Nearctic
9. *Mimocichla* ...	4	Cuba to Porto Rico	
(Harporhynchus	3	Mexico)	Nearctic genus
10. *Cinclocerthia* ...	3	Lesser Antilles	
11. *Ramphocinclus*	1	Martinique and St. Lucia	
SYLVIIDÆ			
12. Myiadestes ...	8	Mexico and Antilles to Peru and Bolivia	N. & W. of N. America

Order, Family, and Genus.	No. of Species.	Range within the Region.	Range beyond the Region.
13. *Cichlopsis* ...	1	Brazil	
(Sialia	2	Mexico and Guatemala)	United States & Canada
14. Regulus	2	Mexico and Guatemala	Nearctic, Palæarctic
15. Polioptila ...	6	Mexico and Cuba to Bolivia and La Plata	Cen. and S. U. States
CINCLIDÆ.			
16. Cinclus	4	Mexico to Venezuela and Peru	Nearctic, Palæarctic
TROGLODYTIDÆ..			
17. Troglodytes ...	5	Mexico to Straits of Magellan	Nearctic, Palæarctic
18. Thryophilus ...	13	Mexico to Central Brazil	N. W. America
19. Thryothorus ..	12	Mexico to S. Brazil	N. America
20. Cistothorus ...	3	Mexico to Chili and Patagonia	N. America
21. *Donacobius* ...	2	Columbia to Brazil and Bolivia	
22. *Campylorhynchus*	18	Mexico to Brazil and Bolivia	New Mexico
23. *Cyphorhinus* ...	5	Costa Rica to Peru	
24. *Microcerculus* ...	5	Mexico to Peru	
25. *Henicorhina* ...	2	Mexico to Peru	
(Salpinctes ...	1	Mexico and Guatemala)	Nearctic genus
(Catherpes ...	1	Mexico)	Gila and Colorado
26. *Cinnicerthia* ...	2	Columbia and Ecuador	
27. *Uropsila*	1	Mexico	
CERTHIIDÆ			
(Certhia	1	Mexico and Guatemala)	North temperate genus
SITTIDÆ.			
(Sitta	2	Mexico)	North temperate genus
PARIDÆ.			
(Parus	1	Mexico)	Nearc., Palæarc. Orient.
(Lophophanes...	2	Mexico)	North temperate genus
(Psaltriparus ...	1	Mexico and Guatemala)	Nearctic
CORVIDÆ.			
28. Cyanocitta ...	16	Mexico to Peru and Bolivia	Nearctic
29. *Cyanocorax* ...	12	Mexico to Paraguay, Jamaica	
30. Calocitta... ..	2	Mexico to Guatemala	
31. *Psilorhinus* ...	3	Mexico to Costa Rica	
32. Corvus	4	Mexico to Guatemala, Cuba to Porto Rico	Cosmop., excl. S. Amer.
CŒREBIDÆ.			
33. *Diglossa*	14	Mexico to Guiana, Peru, and Bolivia	
34. *Diglossopis* ...	1	Venezuela to Ecuador	
35. *Oreomanes* ...	1	Ecuador	
36. *Conirostrum* ...	6	Columbia to Bolivia	
37. *Hemidacnis* ...	1	Columbia and Upper Amazon	
38. *Dacnis*	13	Costa Rica to Guiana & S. Brazil	
39. *Certhidea.*. ...	2	Galapagos Islands	

Order, Family, and Genus.	No. of Species.	Range within the Region.	Range beyond the Region.
40. *Chlorophanes* ...	2	Brazil to Central America, Cuba	
41. *Cœreba*	4	Mexico and Cuba to Guiana and Brazil	
42. *Certhiola*	10	Antilles to Ecuador and Brazil	Florida
43. *Glossiptila* ...	1	Jamaica	
MNIOTILTIDÆ.			
44. Siurus	3	Mexico to Columbia, Antilles	S. & E. States & Canada
45. Mniotilta	1	Columbia to Mexico and Antilles	Eastern United States
46. Parula	5	Brazil and Ecuador to Mexico	Eastern U. S. & Canada
47. Protonotaria ...	1	Venezuela to Central America and W. India	Florida to Ohio
48. Helminthophaga	5	Mexico to Columbia	North America
49. Helmintherus ...	1	Mexico to Veragua	U. States to Canada
50. Perissoglossa ...	1	Cuba, Hayti, and Porto Rico	E. United States
51. Dendrœca ...	25	Mexico & W. Indies to Ecuador and Chili	All N. America
52. Oporornis ...	1	Guatemala to Panama	
53. Geothlypis ...	10	Brazil to Mexico	All N. America
54. Setophaga ...	12	Mexico to Brazil	E. U. States & Canada
55. Cardellina ...	1	Gautemala and Mexico	
56. Ergaticus... ...	2	Guatemala and Mexico	
57. Myioodioctes ...	3	Columbia to Mexico	U. States and Canada
58. *Basileuterus* ...	22	Mexico to Brazil	
59. Icteria	1	Costa Rica to Mexico	E. and Central United States to Canada
60. *Granatellus* ...	3	Amazon to Mexico	
61. *Teretristis* ...	2	Cuba	
VIREONIDÆ.			
62. Vireosylvia ...	9	Venezuela to Mexico & Antilles	All N. America
63. Vireo	10	Mexico to Costa Rica & Antilles	All United States
64. *Neochloe*	1	Mexico	
65. *Hylophilus* ...	16	Brazil to Mexico	
66. *Laletes*	1	Jamaica	
67. *Phœnicomanes*...	1	Jamaica	
68. *Vireolanius* ...	4	Mexico to Amazon	
69. *Cychloris*	9	Mexico to Paraguay	
AMPELIDÆ.			
70. Dulus	2	Hayti	
(Ampelis... ...	1	Mexico and Guatemala)	N. temperate genus
71. Ptilogonys ...	2	Mexico to Costa Rica	
(Phainopepla ...	1	Mexico)	Gila and Lower Colorado
HIRUNDINIDÆ.			
72. Hirundo	9	Mexico and Antilles to Chili and La Plata	Almost cosmopolite
73. Petrochelidon ...	3	Mexico and Antilles to Paraguay	Nearctic
74. *Atticora*	6	Guatemala tc Peru and Brazil	
75. Cotyle	2	Central America to La Plata	All regions but Austral
76. Stelgidopteryx	4	Mexico to Brazil	S. United States
77. Progne	4	The whole region	Nearctic

Order, Family, and Genus.	No. of Species.	Range within the Region.	Range beyond the Region.
ICTERIDÆ.			
78. *Clypeicterus* ...	1	Upper Amazon	
79. *Ostinops*	8	Mexico to Guiana, Brazil, and Bolivia	
80. *Cassiculus*	1	Mexico	
81. *Cassicus*	10	Mexico to S. Brazil and Bolivia	
82. Icterus	33	Mexico to Antilles and La Plata	All U. States & Canada
83. Dolichonyx ...	1	Mexico to Paraguay, Galapagos	E. U. States and Canada
84. Molothrus ...	8	Mexico to La Plata and Bolivia	All U. States & Canada
85. Agelæus... ...	6	Mexico to Paraguay, Cuba, Porto Rico	All U. States & Canada
(Xanthocephalus	1	Mexico)	Nearctic genus
86. *Xanthosomus*...	4	Venezuela to La Plata	
87. *Amblyrhamphus*	1	Bolivia and La Plata	
88. *Gymnomystax*	1	Guiana and Amazonia	
89. *Pseudoleistes* ...	2	Brazil and La Plata	
90. *Leistes*	3	Venezuela to Paraguay & Bolivia	
91. Sturnella ...	4	Cuba and Mexico to Chili, Falkland Islands & Tierra del Fuego	All U. States & Canada
92. *Curæus*	1	Chili to Magellan Straits	
93. *Nesopsar* ...	1	Jamaica	
(Scolecophagus	1	Mexico, Cuba ?)	Nearctic genus
94. *Lampropsar* ...	4	Guatemala to Peru and Guiana	
95. Quiscalus... ...	9	Mexico to Antilles & Venezuela	S. and E. United States to Labrador
96. *Hypopyrrhus*...	1	Columbia	
97. *Aphobus*... ...	1	Brazil Paraguay and Bolivia	
98. *Cassidix*... ...	1	Mexico to Brazil and Guiana	
TANAGRIDÆ.			
99. *Procnias*... ...	2	Brazil and Peru to Columbia	
100. *Chlorophonia*...	7	Brazil to Mexico	
101. *Euphonia* ..	32	Mexico and W. Indies to Brazil and Bolivia	
102. *Tanagrella* ...	4	Columbia to Guiana and Brazil	
103. *Chlorochrysa* ...	2	Columbia to Peru	
104. *Pipridea* ...	2	Venezuela to Brazil and Bolivia	
105. *Diva*	1	Columbia and Ecuador	
106. *Calliste*	56	Guatemala to Bolivia & Paraguay	
107. *Iridornis* ...	4	Columbia to Peru	
108. *Pœcilothraupis*	4	Columbia to Bolivia	
109. *Stephanophorus*	1	Brazil and La Plata	
110. *Buthraupis* ...	5	Veragua to Bolivia	
111. *Compsocoma* ...	5	Columbia to Bolivia	
112. *Dubusia*... ...	2	Columbia and Ecuador	
113. *Tanagra*... ...	12	Mexico to Bolivia and La Plata	
114. *Spindalis* ...	5	Porto Rico to Bahamas	
115. *Rhamphocœlus*	11	Guatemala to Brazil and Bolivia	
116. *Phlogothraupis*	1	Mexico to Costa Rica	
117. *Euchœtes* ..	1	Eastern Ecuador	
118. *Pyranga*... ...	11	Mexico to Bolivia and Paraguay	U. States and Canada
119. *Orthogonys* ...	2	Brazil and Guiana	
120. *Lamprotes* ...	2	Brazil and Columbia	
121. *Phœnicothraupis*	7	Mexico to Paraguay and Bolivia	

Order, Family, and Genus.	No. of Species.	Range within the Region.	Range beyond the Region.
122. *Lanio*	4	Mexico to Bolivia	
123. *Eucometis* ...	5	Costa Rica to Bolivia	
124. *Trichothraupis*	1	S. Brazil and Paraguay	
125. *Creurgops* ...	1	West Ecuador	
126. *Tachyphonus*	11	Nicaragua to Paraguay	
127. *Cypsnagra* ...	1	S. Brazil and Bolivia	
128. *Nemosia*... ...	11	Venezuela, W. Ecuador, to Brazil and Bolivia	
129. *Pyrrhocoma* ...	1	S. Brazil and Paraguay	
130. *Chlorospingus*	18	Mexico to Peru and Bolivia	
131. *Buarremon* ...	20	Mexico to S. Brazil and Bolivia	
132. *Phænicophilus*	1	Hayti	
133. *Arremon* ...	12	Mexico to S. Brazil	
134. *Oreothraupis*...	1	East Ecuador	
135. *Cissopis*	3	Columbia to Peru and Bolivia	
136. *Lamprospiza*...	1	Guiana	
137. *Psittospiza* ...	2	Columbia to Peru	
138. *Saltator*	17	Mexico to La Plata and Bolivia	
139. *Diucopis*... ...	2	Upper Amazon and S. Brazil	
140. *Orchesticus* ...	3	Tropical S. America	
141. *Pitylus*	8	Mexico to Brazil and Ecuador	
FRINGILLIDÆ.			
142. Chrysomitris...	12	Mexico to Brazil, Chili and Patagonia	Nearctic, Palæarctic
143. *Sycalis*	9	Mexico to Chili and La Plata, Jamaica	
144. Coccothraustes	2	Mexico and Guatemala	Nearctic, Palæarctic
145. *Geospiza*... ...	7	Galapagos Islands	
146. *Camarhynchus*	5	Galapagos Islands	
147. *Cactornis* ...	4	Galapagos Islands	
148. *Phrygilus* ...	10	Columbia to Fuegia and Falkland Islands	
149. *Xenospingus* ...	1	Peru	
150. *Diuca*	3	Peru, Chili, and Patagonia	
151. *Emberizoides*...	3	Venezuela to Paraguay	
152. *Donacospiza* ...	1	S. Brazil and La Plata	
153. *Chamæospiza*	1	Mexico	
154. Embernagra ...	9	Mexico to La Plata	Rocky Mountains
155. *Hæmophila* ...	6	Mexico to Costa Rica	
156. Atlapetes ...	1	Mexico	Nearctic ?
157. *Pyrgisoma* ...	5	Mexico to Costa Rica	
158. Pipilo	4	Mexico to Guatemala	All Nearctic region
159. Junco	2	Mexico and Guatemala	United States
160. Zonotrichia ...	5	Mexico to Straits of Magellan	Nearctic
(Melospiza ...	2	Mexico and Guatemala)	Nearctic genus
(Spizella ...	3	Mexico and Guatemala)	Nearctic genus
(Passerculus ...	1	Mexico and Guatemala)	Nearctic genus
(Pocæcetes ...	1	Mexico)	Nearctic genus
161. Ammodramus	1	Guatemala	Nearctic
162. Coturniculus...	4	Mexico to Bolivia, Jamaica	E. & N. of N. America
163. Peucæa	4	Mexico	S. E. States & California
164. *Tiaris*	1	Brazil	
165. *Volatinia* ...	1	Mexico to Brazil	

Order, Family, and Genus.	No. of Species	Range within the Region.	Range beyond the Region.
(Cyanospiza ...	4	Mexico and Central America)	Nearctic
166. *Paroaria* ...	6	Trop. S. America, E. of Andes	
167. *Coryphospingus*	4	Tropical S. America	
168. *Porphyrospiza*	1	Brazil	
169. *Haplospiza* ...	2	Mexico and Brazil	
170. *Phonipara* ...	5	Mexico to Columbia, Greater Antilles	
171. Poospiza ...	12	Mexico to Bolivia and La Plata	W. & Central U. States
172. *Spodiornis* ...	1	Ecuador	
(Carpodacus ...	2	Mexico)	Nearctic, Palæarctic
173. Cardinalis ...	2	Mexico to Venezuela	S. & S. Cent. U. States
174. Guiraca	6	Mexico to Brazil and La Plata	Southern U. States
175. *Amaurospiza*	2	Costa Rica and Brazil	
176. Hedymeles ...	2	Mexico to Columbia	Nearctic
177. *Pheucticus* ...	5	Mexico to Peru and Bolivia	
178. *Oryzoborus* ...	6	Mexico to Ecuador and S. Brazil	
179. *Melopyrrha* ...	1	Cuba	
180. *Loxigilla* ...	4	Antilles	
181. *Spermophila* ...	44	Mexico to Bolivia and Uruguay	Texas
182. *Catamenia* ...	4	Columbia to Bolivia	
183. *Neorhynchus* ...	1	W. Peru	
184. *Catambly- rhynchus* } ...	1	Columbia	
(Loxia	1	Mexico)	North temperate genus
(Calamospiza	1	Mexico)	Arizona and Texas
(Chondestes ...	1	Mexico)	W. and Cent. U. States
(Euspiza	1	Mexico to Columbia)	S.-E. U. States, Palæarc.
185. *Gubernatrix* ...	1	Paraguay and La Plata	
(Plectrophanes	1	Mexico)	N. temp. & Arctic genus
ALAUDIDÆ.			
186. Otocorys ...	1	Mexico, Andes of Columbia	Nearc. & Palæarc. genus
MOTACILLIDÆ.			
187. Anthus ...	4	Mexico to Patagonia and Falkland Islands	Cosmopolite
OXYRHAMPHIDÆ.			
187*a*. *Oxyrhamphus*	2	Brazil to Costa Rica	
TYRANNIDÆ.			
188. *Conophaga* ...	11	Columbia to Bolivia and Brazil	
189. *Corythopis* ...	2	Brazil and Guiana	
190. *Agriornis* ...	5	Ecuador, Peru, and Chili	
191. *Myiotheretes* ...	3	Columbia to Ecuador, Patagonia	
192. *Tœnioptepa* ...	8	S. Brazil and Bolivia to Patago.	
193. *Ochthodiæta* ...	1	Columbian Andes	
194. *Ochthœca* ...	17	Andes, Bolivia to Columbia and Venezuela	
195. Sayornis ...	4	Mexico to Ecuador	E. United Sts. to Canada
196. *Fluvicola* ...	4	Guiana & W. Ecuador to Brazil, and Bolivia	
197. *Arundinicola*	1	Tropical S. America	
198. *Alectorurus* ...	2	S. Brazil and La Plata	

Order, Family, and Genus.	No. of Species	Range within the Region.	Range beyond the Region.
199. *Cybernetes* ...	1	Brazil	
200. *Sysopygis* ...	1	S. Brazil and La Plata	
201. *Cnipolegus* ...	9	Amazonia to Patagonia	
202. *Lichenops* ...	1	Brazil and La Plata	
203. *Muscipipra* ...	1	S. Brazil	
204. *Copurus...* ...	3	Costa Rica to S. Brazil	
205. *Machetornis* ...	1	Venezuela to Brazil	
206. *Muscisaxicola*	11	Andes of Ecuador to Chili and Patagonia	
207. *Centrites* ...	2	Bolivia to Patagonia	
208. *Muscigralla* ...	1	W. Ecuador	
209. *Platyrhynchus*	7	Mexico to Brazil	
210. *Todirostrum* ...	11	Tropical N. and S. America	
211. *Oncosotma* ...	2	Tropical N. America	
212. *Euscarthmus...*	12	Costa Rica to W. Ecuador, Brazil, and Bolivia	
213. *Orchilus...* ...	2	Costa Rica to Brazil and Bolivia	
214. *Colopterus* ...	2	Veragua to Columbia and Guiana	
215. *Hemitriccus* ...	1	Brazil	
216. *Phylloscartes* ...	1	Columbia to Brazil	
217. *Hapalocercus...*	3	Brazil to Chili and La Plata	
218. *Habrura* ...	1	Uruguay	
219. *Pogonotriccus...*	2	Brazil and Columbia	
220. *Leptotriccus* ...	2	Brazil and Veragua	
221. *Stigmatura* ...	2	Upper Amazon to La Plata	
222. *Serphophaga* ...	7	Columbia to Chili and La Plata	
223. *Anæretes* ...	4	Columbia to Chili and La Plata, Magell. Sts. & Juan Fernand.	
224. *Cyanotis* ...	1	W. Peru to La Plata	
225. *Mionectes* ...	4	Mexico to Brazil and Bolivia	
226. *Leptopogon* ...	6	Mexico to Peru and Brazil	
227. *Capsiempis* ...	1	Chiriqui to Brazil	
228. *Phyllomyias* ...	5	Columbia to Brazil	
229. *Ornithion* ...	4	Mexico to Brazil	
230. *Tyrannulus* ...	3	Guatemala to Amazonia	
231. *Tyranniscus* ...	9	Guatemala to E. Peru	
232. *Elainea*	18	Mexico to Tierra del Fuego, Antilles	
233. *Empidagra* ...	1	Bolivia and La Plata	
234. *Legatus*	2	Mexico to Brazil	
235. *Sublegatus* ...	2	Venezuela and Lower Amazon	
236. *Myiozetetes* ...	8	Mexico to W. Peru and Brazil	
237. *Rhynchocyclus*	10	Mexico to W. Ecuador & Brazil	
238. *Conopias* ...	3	Venezuela to Peru and Brazil	
239. *Pitangus* ...	7	Mexico to La Plata, Antilles	
240. *Sirystes*	2	Panama to Brazil	
241. *Myiodynastes* ..	6	Mexico to Bolivia and Paraguay	
242. *Megarhynchus*	1	Mexico to Brazil	
243. *Muscivora* ...	5	Mexico to W. Ecuador & Brazil	
244. *Hirundinea* ...	3	Columbia & Guiana to Paraguay	
245. *Cnipodectes* ...	1	Panama to W. Ecuador & Amazon	
246. *Myiobius* ...	13	Mexico to W. Peru, Bolivia, and La Plata	
247. *Pyrocephalus...*	3	Tropical N. and S. America and Galapagos Islands	Gila and Rio Grande

Order, Family, and Genus.	No. of Species.	Range within the Region.	Range beyond the Region.
248. *Empidochanes*	4	Venezuela to S. Brazil.	
249. *Mitrephorus* ...	2	Mexico to Costa Rica	
250. Empidonax ...	12	Mexico to Columbia & Ecuador	All N. America
251. Contopus ...	10	Mexico to Amazonia, Antilles	N. & E. of Rocky Mtns.
252. *Myiochanes* ...	1	Amazonia and Brazil	
253. Myiarchus ...	12	Mexico to W. Ecuador & Brazil, Galapagos and Antilles	East and West Coasts to Canada
254. *Blacicus*... ...	2	Cuba, Hayti, Jamaica	
(*Empidias* ...	1	Mexico)	Eastern United States
255. *Empidonomus*	1	Guiana and Brazil	
256. Tyrannus ...	11	All tropical sub-regions	All U. States to Canada
257. *Milvulus* ...	2	Tropical N. and S. America	Texas
PIPRIDÆ.			
258. *Piprites*	4	Costa Rica to Brazil	
259. *Masius*	2	Columbia and Ecuador	
260. *Chloropipo* ...	1	Columbia	
261. *Xenopipo* ...	1	Guiana and Columbia	
262. *Pipra*	19	Trop. N. and S. America	
263. *Neopipo*	1	Upper Amazon	
264. *Machæropterus*	4	Columbia to Brazil	
265. *Ilicura* ...	1	Brazil	
266. *Chiroxiphia* ...	5	Guatemala to Brazil	
267. *Metopia*	1	Brazil	
268. *Metopothrix* ...	1	Upper Amazon	
269. *Chiromachæris*	6	Mexico to Ecuador and Brazil	
270. *Heteropelma* ...	10	Mexico to Guiana and Brazil	
271. *Heterocercus* ...	2	Guiana and Upper Amazon	
272. *Schiffornis* ...	2	Upper Amazon and Brazil	
COTINGIDÆ.			
273. *Tityra*	6	Tropical N. and S. America	
274. *Hadrostomus*...	5	Mexico to W. Ecuador & Brazil, Jamaica	
275. *Pachyhamphus*	11	Mexico to W. Ecuador & Brazil	
276. *Lathria*	5	Mexico to Brazil	
277. *Aulia*	3	Veragua to Brazil	
278. *Lipaugus* ...	3	Guatemala to Brazil and Guiana	
279. *Ptilochloris* ...	2	Brazil	
280. *Attila*	8	Costa Rica to Brazil and Guiana	
281. *Casiornis* ...	2	S. Brazil to Paraguay	
282. *Rupicola* ...	3	Guiana to W. Ecuador & Bolivia	
283. *Phœnicocercus*	2	Guiana and Amazonia	
284. *Tijuca*	1	Brazil	
285. *Phibalura* ...	1	Brazil	
286. *Pipreola*... ...	7	Venezuela to Ecuador and Peru	
287. *Ampelio*... ...	4	Columbia to Peru and Brazil	
288. *Carpodectes* ...	1	Nicaragua and Costa Rica	
289. *Heliochæra* ...	2	Columbia to Peru and Bolivia	
290. *Cotinga*	6	Guatemala to Peru and Brazil	
291. *Xipholena* ...	3	Guiana to Brazil	
292. *Iodopleura* ...	3	Guiana to Brazil	
293. *Calyptura* ...	1	Brazil	
294. *Querula*... ...	1	Panama to Amazonia	

Order, Family, and Genus.	No. of Species.	Range within the Region.	Range beyond the Region.
295. *Hæmatoderus*	1	Guiana and Lower Amazon	
296. *Chasmorhynchus*	4	Costa Rica to Guiana and Brazil	
297. *Gymnocephalus*	1	Guiana and Rio Negro	
298. *Gymnoderus* ...	1	Guiana and Upper Amazon	
299. *Pyroderus* ...	3	Venezuela to Brazil	
300. *Cephalopterus*	3	Costa Rica to W. Ecuador & Upr. Amazon	
PHYTOTOMIDÆ.			
301. *Phytotoma* ...	3	Bolivia, Chili, and La Plata	
DENDROCOLAPTIDÆ.			
302. *Geobates*	1	South Brazil	
303. *Geositta*	6	Peru to Chili and Patagonia	
304. *Furnarius* ...	9	Guiana & W. Ecuador to La Plata	
305. *Clibanornis* ...	1	S. Brazil	
306. *Upucerthia* ...	4	Andes of Ecuador to Chili and Patagonia	
307. *Cinclodes* ...	5	Ecuador to Chili, Patagonia and Tierra del Fuego	
308. *Henicornis* ...	2	Patagonia	
309. *Lochmias* ...	2	Venezuela and Brazil	
310. *Sclerurus* ...	6	Mexico to Brazil	
311. *Oxyurus*	2	Chili to Tierra del Fuego, and Masafuera Islands	
312. *Sylviorthorhynchus* ...	1	Chili	
313. *Phlæocryptes* ...	1	W. Peru to La Plata	
314. *Leptasthenura*	5	Andes of Ecuador to Brazil and Patagonia	
315. *Synallaxis* ...	55	The whole region (excl. Antilles)	
316. *Coryphistera* ...	1	La Plata	
317. *Anumbius* ...	1	Paraguay and La Plata	
318. *Limnornis* ...	1	Uruguay and La Plata	
319. *Placellodomus*	4	Venezuela to Peru and La Plata	
320. *Thripophaga* ...	3	Brazil and Columbia	
321. *Pseudocolaptes*	1	Columbia to Peru	
322. *Homorus* ...	3	Brazil, Bolivia, and La Plata	
323. *Thripadectes* ...	1	Columbia	
324. *Ancistrops* ...	1	Upper Amazon	
325. *Automolus* ...	9	Mexico to Amazonia	
326. *Philydor* ...	14	Tropical South America	
327. *Heliobletus* ..	1	Brazil	
328. *Anabatoides* ...	1	Brazil	
329. *Anabazenops* ...	5	Mexico to Brazil	
330. *Xenops*	3	Trop. North and South America	
331. *Sittasomus* ...	3	Mexico to Ecuador and Brazil	
332. *Margarornis* ...	4	Costa Rica to Peru and Bolivia	
333. *Glyphorhynchus*	1	Trop. North and South America	
334. *Pygarrhicus* ...	1	Chili	
335. *Dendrocincla* ...	10	Mexico to Venezuela and Brazil	
336. *Dendrocolaptes*	7	Guatemala to Peru and Brazil	
337. *Nasica*	1	Guiana	
338. *Drymornis* ...	1	La Plata	
339. *Xiphocolaptes*	5	Mexico to Bolivia and Paraguay	

Order, Family, and Genus.	No. of Species.	Range within the Region.	Range beyond the Region.
340. *Dendrexetastes*	2	Guiana	
341. *Dendrornis* ...	14	Mexico, W. Ecuador and Brazil	
342. *Dendroplex* ...	2	Columbia & Venezuela to Brazil	
343. *Picolaptes* ...	14	Mexico to Bolivia and La Plata	
344. *Xiphorhynchus*	4	Veragua to Brazil	
FORMICARHDÆ.			
345. *Cymbilanius* ...	1	Amazonia and Guiana	
346. *Batara*	1	S. Brazil	
347. *Thamnophilus*	47	Trop. North and South America	
348. *Biatas*	1	Brazil	
349. *Thamnistes* ...	2	Central America and Ecuador	
350. *Pygoptila* ...	2	Amazonia	
351. *Neoctantes* ...	1	Amazonia	
352. *Clytoctantes* ...	1	Eastern Ecuador	
353. *Dysithamnus*...	12	Mexico to Bolivia and Brazil	
354. *Thamnomanes*	2	Ecuador, Guiana, and Brazil	
355. *Herpsilochmus*	4	Venezuela to Brazil and Bolivia	
356. *Myrmotherula*	21	Tropical S. America	
357. *Formicivora* ...	14	Trop. North and South America	
358. *Terenura* ...	3	Veragua to W. Ecuador & Brazil	
359. *Psilorhamphus*	1	Central Brazil	
360. *Microbates* ...	1	Cayenne	
361. *Rhamphocœnus*	4	Guatemala to Brazil	
362. *Cercomaera* ...	9	Cen. America to W. Equador & S. Brazil	
363. *Pyriglena* ...	4	Ecnador to Peru and Brazil	
364. *Gymnocichla* ...	2	Honduras to Panama	
365. *Percnostola* ...	3	Guiana and Upper Amazon	
366. *Hetcrocnemis*...	3	Guiana and Upper Amazon	
367. *Myrmeciza* ...	11	Veragua to W. Ecuador, Bolivia, and Brazil	
368. *Hypocnemis* ...	15	Costa Rica to W. Ecuador & Brazil	
369. *Pithys*	5	Nicaragua to Amazonia	
370. *Rhopoterpe* ...	1	Guiana	
371. *Phlogopsis* ...	4	Nicaragua to Guiana and Bolivia	
372. *Formicarius* ...	9	Mexico to Brazil and Bolivia	
373. *Pittasoma* ...	1	Panama and Veragua	
374. *Chamœza* ..	4	Columbia to Brazil	
375. *Grallaria* ...	20	Mexico to W. Ecuador & Brazil	
376. *Grallaricula* ...	5	Costa Rica to Ecuador	
PTEROPTOCHIDÆ.			
377. *Scytalopus* ...	8	Columbia & Brazil to Chili and Tierra del Fuego	
378. *Merulaxis* ...	1	Central Brazil	
379. *Rhinocrypta* ...	2	La Plata and Patagonia	
380. *Liosceles*... ...	1	Madeira Valley	
381. *Pteroptochus* ...	2	Chili and Chiloe	
382. *Hylactes*... ...	3	Chili	
383. *Acropternis* ..	1	Columbia and Ecuador	
384. *Triptorhinus*...	1	Chili	

Order, Family, and Genus.	No. of Species.	Rang within the Region.	Range beyond the Region.
PICARIÆ.			
PICIDÆ.			
385. *Picumnus* ...	14	Honduras to Brazil and Bolivia	
386. Picus	6	Mexico, Chili, La Plata, and S. Patagonia	All reg. but Austral & Ethiopian
(Sphyrapicus...	1	Mexico and Guatemala)	Nearctic genus
87. Campephilus ...	12	Mexico to Patagonia, Cuba	Nearctic
388. Dryocopus ...	4	Mexico to S. Brazil	Palæarctic
389. *Celeus*	15	Mexico and S. Brazil	
390. *Nesoceleus* ...	1	Cuba	
391. *Chrysoptilus* ...	6	Tropical S. America	
392. Centurus ...	10	Mexico to Venezuela, Antilles	Nearctic
393. *Chloronerpes* ...	35	Tropical America, Hayti	
394. *Xiphidiopicus*	1	Cuba	
395. Melanerpes ...	9	Mexico to Brazil, Porto Rico	Nearctic
396. *Leuconerpes* ...	1	Brazil, Bolivia	
397. Colaptes... ...	7	Open country of trop. America, Greater Antilles	Nearctic
398. *Hypoxanthus*...	1	Venezuela and Ecuador	
MEGALÆMIDÆ.			
399. *Capito*	10	Costa Rica to Peru and Guiana	
400. *Tetragonops* ..	2	Costa Rica and Ecuador	
RHAMPHASTIDÆ.			
401. *Rhamphastos*...	12	All tropical America	
402. *Pteroglossus* ...	16	Mexico to Guiana and Brazil	
403. *Selenidera* ...	7	Veragua to Brazil	
404. *Andigena* ...	6	Columbia to W. Ecuador, Bolivia and Brazil	
405. *Aulacorhamphus*	10	Mexico to Venezuela and Bolivia	
CUCULIDÆ.			
406. *Crotophaga* ...	3	Tropical America and Antilles	Nearctic to Pennsylvania
407. *Guira*	1	Brazil and Paraguay	
408. *Neomorphus* ...	4	Nicaragua to Brazil and Upper Amazon	
409. *Geococcyx* ...	1	Guatemala	Texas to Calfornia
410. *Dromococcyx* ...	2	Mexico to Brazil	
411. *Diplopterus* ...	1	Mexico to Ecuador and Brazil	
412. *Saurothera* ...	4	Greater Antilles	
413. *Hyetornis* ..	2	Jamaica and Hayti	
414. *Piaya*	3	Mexico to W. Ecuador & Brazil	
415. *Morococcyx* ...	1	Mexico to Costa Rica	
416. Coccygus ...	10	Tropical America and Antilles, Cocos Islands	Nearctic
BUCCONIDÆ.			
417. *Bucco*	21	Guatemala to Guiana, Paraguay and Bolivia	
418. *Malacoptila* ...	10	Guatemala to Guiana, W. Ecuador and Bolivia	
419. *Nonnula* ...	5	Columbia and Amazonia	

Order, Family, and Genus.	No. of Species.	Range within the Region.	Range beyond the Region.
420. *Monasa*	7	Costa Rica to Brazil	
421. *Chelidoptera* ...	2	Columbia to Guiana and Brazil	
GALBRILIDÆ.			
422. *Galbula*	9	Guatemala to Brazil and Bolivia	
423. *Urogalba* ...	2	Guiana to Lower Amazon	
424. *Brachygalba* ...	4	Columbia to Brazil and Bolivia	
425. *Jacamaralcyon*	1	Brazil	
426. *Jacamerops* ...	2	Columbia to Amazonia	
427. *Galbalcyrhynchus* ...	1	Upper Amazon	
TODIDÆ.			
428. *Todus*	5	Greater Antilles	
MOMOTIDÆ.			
429. *Momotus* ...	10	Mexico to W. Ecuador, Brazil and Bolivia	
430. *Urospatha* ...	1	Costa Rica to Columbia	
431. *Baryphthengus*	1	Brazil and Paraguay	
432. *Hylomanes* ...	2	Mexico and Guatemala	
433. *Prionirhynchus*	2	Guatemala to Upper Amazon	
434. *Eumomota* ...	1	Honduras to Chiriqui	
TROGONIDÆ.			
435. *Prionoteles* ...	1	Cuba	
436. *Temnotrogon* ...	1	Hayti	
437. *Trogon*	22	Mexico to W. Ecuador & Parag.	
438. *Euptilotis* ...	1	Mexico	
439. *Pharomacrus*	5	Guatemala to Upper Amazon and Bolivia	
ALCEDINIDÆ.			
440. Ceryle	8	Mexico to Brazil, Patagonia and Chili	Nearc., S.Palæarc.,Orien.
STEATORNITHIDÆ.			
441. *Steatornis* ...	1	Columb., Venezuela, & Trinidad	
CAPRIMULGIDÆ.			
442. *Nyctibius* ...	6	Brazil to Guatemala & Jamaica	
443. *Hydropsalis* ...	8	Columbia & Guiana to La Plata	
444. Antrostomus...	10	Mexico and Cuba to Bolivia and La Plata	All U. States to Canada
445. *Stenopsis* ...	4	Martinique to Columb., W. Peru and Chili	
446. *Siphonorhis* ...	1	Jamaica	
447. *Heleothreptus*	1	Central Brazil	
448. *Nyctidromus* ...	1	Central America to S. Brazil	
449. *Podager*	1	Tropical S. America	
450. *Lurocalis* ...	2	Guiana to Brazil	
451. Chordeiles ...	7	Mexico to W. Peru and Brazil Jamaica and Porto Rico	All U. States to Canada
452. *Nyctiprogne* ...	1	Amazonia	

Order, Family, and Genus.	No. of Species.	Range within the Region.	Range beyond the Region.
CYPSELIDÆ.			
453. *Cypselus* ...	3	Antilles to Guiana and Bolivia	The Eastern Hemisphere
454. *Panyptila* ...	3	Guatemala and Guiana	
455. *Chætura*... ..	9	Mexico to Ecuador and Brazil	Almost cosmopolite
456. *Hemiprocne* ...	3	Mexico to La Plata, Jamaica and Hayti	
457. *Cypseloides* ...	2	Brazil and Peru	
458. *Nephœcetes* ...	1	Jamaica	
TROCHILIDÆ.			
459. *Grypus*	1	Brazil	
460. *Androdon* ...	1	Ecuador	
461. *Eutoxeres* ...	2	Costa Rica to Ecuador	
462. *Glaucis*	2	Panama to Brazil	
463. *Phaethornis* ...	14	Tropical N. and S. America	
464. *Pygmornis* ...	8	Mexico to Guiana and Brazil	
465. *Threnetes* ...	4	Costa Rica to Amazonia and W. Ecuador	
466. *Dolerisca* ...	1	Venezuela	
467. *Eupetomena* ...	1	Guiana to Brazil	
468. *Sphenoproctus*	2	Mexico to Guatemala	
469. *Campylopterus*	9	Mexico to Amazonia	
470. *Phæochroa* ..	2	Guatemala to Columbia	
471. *Aphantochroa*	3	Ecuador and Brazil	
472. *Urochroa* ...	1	Ecuador	
473. *Sternoclyta* ...	1	Venezuela	
474. *Eugenes*	2	Mexico to Costa Rica	
475. *Cœligena* ...	1	Mexico	
476. *Lamprolœma*...	1	Mexico and Guatemala	
477. *Delattria* ...	2	Guatemala	
478. *Oreopyra* ...	4	Costa Rica to Chiriqui	
479. *Heliopœdica* ...	2	Mexico and Guatemala	
480. *Topaza*	2	Guiana	
481. *Oreotrochilus*...	6	Ecuador to Peru and Chili	
482. *Lampornis* ...	7	Mexico & W. India to Amazonia	
483. *Eulampis* ...	2	Lesser Antilles	
484. *Avocettula* ...	1	Guiana	
485. *Lafresnaya* ...	2	Venezuela and Columbia	
486. *Doryphora* ...	5	Costa Rica to Ecuador	
487. *Chalybura* ...	5	Costa Rica to Columbia	
488. *Heliodoxa* ...	5	Costa Rica to Venezue. & Boliv.	
489. *Iolœma*	2	Ecuador to Peru	
490. *Phœolœma* ...	2	Columbia and Ecuador	
491. *Eugenia*... ...	1	Ecuador	
492. *Aithurus* ...	1	Jamaica	
493. *Thalurania* ...	10	Costa Rica to Guiana, Ecuador and Brazil	
494. *Panoplites* ...	3	Columbia and Ecuador	
495. *Florisuga* ...	2	Guatemala to Brazil	
496. *Microchera* ...	2	Nicaragua to Veragua	
497. *Lophorius* ...	7	Mexico to Brazil, Peru, & Bolivia	
498. *Polemistria* ...	2	Columbia to S. Brazil	
499. *Discura*... ...	2	Brazil	
500. *Gouldia*	4	Costa Rica to Brazil & Bolivia	

Order, Family, and Genus.	No. of Species.	Range within the Region.	Range beyond the Region.
501. Trochilus ...	2	Mexico to Veragua	To Canada and Sitka
502. *Mellisuga*	1	Jamaica to Hayti	
503. *Calypte*	3	Mexico and Cuba	
504. Selasphorus ...	7	Mexico to Veragua	W. & Cen. United States
505. Atthis	1	Mexico and Guatemala	California and Colorado
506. *Stellula* ...	1	Mexico	
507. *Calothorax* ...	2	Mexico	
508. *Acestrura* ...	3	Venezuela to Ecuador & Bolivia	
509. *Chætocercus* ...	3	Venezuela and Ecuador	
510. *Myrtis*	2	Ecuador to Bolivia, W. of Andes	
511. *Thaumastura*	1	W. Peru	
512. *Rhodopis*... ...	2	W. Peru and Chili	
513. *Doricha*	5	Mexico to Veragua, Bahamas	
514. *Tilmatura* ...	1	Guatemala	
515. *Calliphlox* ...	2	Ecuador and Brazil	
516. *Loddigesia* ...	1	Peruvian Andes	
517. *Steganura* ...	6	Venezuela to Ecuador & Bolivia	
518. *Lesbia*......	6	Columbia to Peru	
519. *Cynanthus* ...	2	Venezuela to Ecuador	
520. *Sparganura* ...	4	Columbia to Bolivia & La Plata	
521. *Pterophanes* ...	1	Columbia to Peru	
522. *Aglæactis*	4	Columbia to Bolivia	
523. *Oxypogon* ...	2	Venezuela and Columbia	
524. *Oreonympha* ...	1	Peru	
525. *Rhamphomicron*	6	Columbia to Bolivia	
526. *Urosticte*	2	Ecuador	
527. *Metallura* ...	6	Columbia to Bolivia	
528. *Adelomia* ...	4	Venezuela to Peru & Bolivia	
529. *Avocettinus* ...	1	Columbia	
530. *Anthocephala*...	1	Columbia	
531. *Chrysolampis*...	1	Venezuela to Brazil	
532. *Orthorhynchus*	2	Lesser Antilles	
533. *Cephalolepis* ...	3	Brazil	
534. *Clais*	1	Venezuela and Columbia	
535. *Baucis*	1	Mexico to Veragua	
536. *Heliactin* ...	1	Brazil	
537. *Heliothrix* ...	3	Guatemala to Ecuador & Brazil	
538. *Schistes*	2	Columbia and Ecuador	
539. *Phlogophilus* ...	1	Ecuador	
540. *Augastes*... ...	2	Brazil	
541. *Petasophora* ...	5	Mexico to Peru and Brazil	
542. *Chrysobronchus*	3	Venezuela to Brazil	
543. *Patagona* ..	1	Ecuador to Bolivia and Chili	
544. *Docimastes* ...	1	Columbia and Ecuador	
545. *Helianthea* ...	7	Columbia to Bolivia	
546. *Heliotrypha* ...	2	Columbia and Ecuador	
547. *Heliangelus* ...	6	Venezuela to Peru	
548. *Diphlogœna* ...	5	Bolivia	
549. *Clytolæma* ...	2	E. Ecuador and Brazil	
550. *Bourcieria* ...	5	Venezuela to Peru	
551. *Lampropygia*...	4	Venezuela to Bolivia	
552. *Heliomastes* ...	5	Mexico to Ecuador & Venezuela	
553. *Lepidolarynx*...	1	Brazil	
554. *Calliperidia* ...	1	Central Brazil and Paraguay	
555. *Eustephanus* ..	3	Chili, S. Patagonia, and Juan Fernandez Islands,	

Order, Family, and Genus.	No. of Species.	Range within the Region.	Range beyond the Region.
556. *Eriocnemis* ...	14	Venezuela to Ecuador	
557. *Cyanomyia* ...	6	Mexico to Peru	
558. *Hemistilbon* ...	1	Mexico	
559. *Leucippus* ...	2	Peru and Bolivia	
560. *Thaumatias* ..	15	Mexico to Guiana, Upr. Amazon, and Brazil	
561. *Amazilia* ...	14	Mexico to W. Ecuador & Peru	
562. *Saucerottia* ...	7	Costa Rica to Columb. & Venezue.	
563. *Eupherusa* ...	3	Mexico to Veragua	
564. *Chrysuronia* ...	5	Guatemala to Ecuador & La Plata	
565. *Eucephala* ...	7	Venezuela to Guiana and Brazil	
566. *Panterpe*... ...	1	Costa Rica and Chiriqui	
567 *Juliamyia* ...	2	Panama to Ecuador	
568. *Circe*	3	Mexico	
569. *Phœoptila* ...	1	Mexico	
570. *Damophila* ...	1	Costa Rica to Ecuador	
571. *Hylocharis* ...	3	Amazonia and Brazil	
572. *Sapphironia* ...	2	Columbia and Veragua	
573. *Sporadinus* ...	3	Cuba, Bahamas, Hayti, Porto Rico	
574. *Chlorostilbon* ...	8	Mexico to Brazil and La Plata	
575. *Panychlora* ...	3	Venezuela and Columbia	
576. *Smaragdochrysis*	1	Brazil	
PSITTACI.			
CONURIDÆ.			
577. *Ara*...	15	Trop. North and South America, Cuba, Jamaica (extinct)	
578. *Rhyncopsitta* ...	1	Mexico	
579. *Henicognathus*	1	Chili	
580. *Conurus*	30	The whole region	S. & S.E. United States
581. *Pyrrhura* ...	16	Costa Rica to Paraguay & Bolivia	
582. *Bolborhynchus*	7	Mexico to Peru, Central Brazil, and La Plata	
583. *Brotogerys* ...	9	Trop. North and South America	
PSITTACIDÆ.			
584. *Caica*	9	Mexico to Amazonia	
585. *Chrysotis* ...	32	All the tropical sub-regions	
586. *Triclaria* ..	1	Brazil	
587. *Deroptyus* ...	1	Guiana and Rio Negro	
588. *Pionus*	9	Costa Rica to Bolivia and Brazil	
589. *Urochroma* ...	7	Venezuela to Brazil	
590. *Psittacula* ...	6	Mexico to W. Ecuador & Brazil	
COLUMBÆ.			
591. Columba ...	18	Trop. sub-regions with Chili and La Plata	All regions but Austral.
592. Zenaidura ...	2	Mexico to Veragua	Nearctic
593. Chamæpelia ...	6	Mexico to Brazil and Bolivia	S. Nearctic
594. *Columbula* ..	2	Brazil and La Plata to Chili	
595. *Scardafella* ...	2	Guatemala and Brazil	
596. *Zenaida*	10	Antilles and S. America to Chili and La Plata	

Order Family, and Genus.	No. of Species.	Range within the Region.	Range beyond the Region.
597. *Melopelia* ..	2	Mexico to Chili	South & West Nearctic
598. *Peristera...* ...	4	Mexico to Brazil	
599. *Metriopelia* ...	2	W.America from Ecuador to Chili	
600. *Gymnopelia* ...	1	West Peru and Bolivia	
601. *Leptoptila* ...	11	Tropical sub-regions	
602. *Geotrygon* ...	14	Tropical sub-regions	
603. *Starnœnas* ...	1	Cuba	
GALLINÆ.			
TETRAONIDÆ.			
604. *Odontophorus*	17	Trop. North and South America	
605. *Dendrortyx* ...	3	Mexico to Costa Rica	
606. Cyrtonyx ...	3	Mexico to Guatemala	S. Central United States
607. Ortyx	5	Mexico to Costa Rica, Cuba	Nearctic to Canada
608. *Eupsychortyx*	5	Mexico to Columbia and Guiana	
(Callipepla ...	2	Mexico)	California
PHASIANIDÆ.			
609. Meleagris ...	2	Mexico and Honduras	Nearctic
CRACIDÆ.			
610. *Crax*	8	Mexico to Venezuela & S. Brazil	
611. *Nothocrax* ...	1	Guiana and Upper Amazon	
612. *Pauxi*	1	Guiana and Venezuela	
613. *Mitua*	2	Guiana to Peru	
614. *Stegnolœma* ...	1	Columbia and Ecuador	
615. *Penelope...* ...	13	Trop. North and South America	
616. *Penelopina* ...	1	Guatemala	
617 *Pipile*	3	Venezuela to Brazil and Peru	
618. *Aburria*	1	Columbia	
619. *Chamœpetes* ...	2	Costa Rica to Peru	
620. *Ortalida...* ...	18	Trop. North and South America	New Mexico
621. *Oreophasis* ...	1	Guatemala	
TINAMIDÆ.			
622. *Tinamus* ...	7	Trop. North and South America	
623. *Nothocercus* ...	3	Costa Rica to Venezue. & Ecuador	
624. *Crypturus* ...	16	Trop. North and South America	
625. *Rhynchotus* ...	2	Brazil to Bolivia and La Plata	
626. *Nothoprocta* ...	4	Ecuador to Bolivia and Chili	
627. *Nothura...* ...	4	Brazil to Bolivia and La Plata	
628. *Taoniscus* ...	1	Brazil and Paraguay	
629. *Calodromas* ...	1	La Plata	
630. *Tinamotis* ...	1	Andes of Peru and Bolivia	
OPISTHOCOMI.			
OPISTHOCOMIDÆ.			
631. *Opisthocomus...*	1	Guiana and Lower Amazon	

Order, Family, and Genus.	No. of Species.	Range within the Region.	Range beyond the Region.
ACCIPITRES.			
VULTURIDÆ.			
(CATHARTINÆ.)			
632. *Sarcorhamphus*	2	The Andes and S. of 41° S. Lat.	
633. *Cathartes* ...	1	Mexico to 20° S. Lat.	
634. Catharista ...	1	Mexico to 40° S. Lat.	S. United States
635. Pseudogryphis	3	Mexico to Falkland Ids., Cuba, Jamaica	United States
FALCONIDÆ.			
636. Polyborus ...	2	The whole region	California and Florida
637. *Ibycter*	8	Guatemala to Terra del Fuego	
638. Circus	3	Nearly the whole region	Almost cosmopolite
639. *Micrastur* ...	7	Trop. North and South America	
640. *Geranospiza* ...	2	Trop. North and South America	
641. Antenor... ...	1	Mexico to Chili and La Plata	California and Texas
642. Astur	1	Trop. N. and S. America	Almost cosmopolite
643. Accipiter ...	15	The whole region	Almost cosmopolite
644. *Heterospizias*...	1	Trop. S. America, E. of Andes	
645. Tachytriorchis	2	Mexico to Paraguay	California
646. Buteo	9	Mexico to Patagonia	Almost cosmopolite
647. *Buteola*	1	Veragua to Amazonia	
648. Asturina ..	7	Mexico to Bolivia and La Plata	S.E. United States
649. *Busarellus* ...	1	Brazil and Guiana	
650. *Buteogallus* ...	1	Columbia and Guiana	
651. *Urubutinga* ...	12	Mexico to Brazil and Bolivia	
652. *Harpyhaliœetus*	1	Veragua to Chili & N. Patagonia	
653. *Morphnus* ...	1	Panama to Amazonia	
654. *Thrasaëtus* ...	1	Mexico to Bolivia and Paraguay	
655. Lophotriorchis	1	Bogota	Indo-Malaya
656. *Spiziastur* ...	1	Guatemala to Brazil	
657. Spizaëtus ...	4	Mexico to Paraguay	Africa, India, Malaya
658. *Herpetotheres*...	1	S. Mexico to Bolivia & Paraguay	
659. Nauclerus ...	1	Mexico to Brazil	S. United States
660. *Rostrhamus* ...	3	Antilles to Brazil and Peru	Florida
661. *Leptodon* ...	4	Central America to S. Brazil and Bolivia	
662. Elanus	1	Mexico to Chili	Califor., Old World trop.
663. *Gampsonyx* ...	1	Trinidad to Brazil	
664. *Harpagus* ...	3	Central America to Brazil & Peru	
665. Ictinia	2	Mexico to Brazil	South United States
666. *Spiziapteryx* ...	1	La Plata	
667. Falco	3	The whole region	Almost cosmopolite
668. Cerchneis ...	3	The whole region	Almost cosmopolite
PANDIONIDÆ.			
669. Pandion	1	The whole region	Cosmopolite
STRIGIDÆ.			
670. Glaucidium ...	6	The whole region	W. United Sts., Palæarc.
671. Micrathene ...	1	Mexico	Arizona, New Mexico
672. Pholeoptynx ...	1	The whole region	N.W. America & Texas
673. Bubo	1	The whole region	All regions but Austral.

Order, Family, and Genus.	No, of Species.	Range within the Region.	Range beyond the Region.
674. Scops	6	Mexico to Brazil and La Plata	Almost cosmopolite
675. *Gymnoglaux* ...	2	West India Islands	
676. *Lophostrix* ...	2	Guatemala to Lower Amazon	
677. Syrnium... ...	3	Mexico to Patagonia	All regions but Austral.
678. *Ciccaba*	10	Mexico to Peru and Paraguay	
679. *Nyctalatinus* ...	1	Columbia	
680. *Pulsatrix* ...	2	Guatemala to Brazil and Peru	
681. Asio	2	The whole region	All regions but Austral.
682. *Nyctalops* ...	1	Cuba and Mexico to Brazil	
683. *Pseudoscops* ...	1	Jamaica	
(Nyctale... ...	1	Mexico)	N. Temperate genus
684. Strix	2	The whole region	Almost cosmopolite

Peculiar or very Characteristic Genera of Wading and Swimming Birds.

GRALLÆ.

RALLIDÆ.

Aramides ...	23	The whole region	Nearctic
Heliornis ...	1	Tropical America	

SCOLOPACIDÆ.

Eureunetes ...	3	The whole region	Nearctic

CHIONIDIDÆ.

Chionis	2	Sts. of Magellan, Falkland Ids.	Kerguelen's Island

THINOCORIDÆ.

Attagis	4	Andes to Fuegia and Falkland Islands	
Thinocoris ...	2	Peru, Chili, and La Plata	

CHARADRIIDÆ.

Phœgornis ...	1	Temperate S. America	
Oreophilus ...	1	Temperate S. America	
Pluvianellus ...	1	Temperate S. America	
Aphriza	1	W. coast of S. America	W. coast of N. America

CARIAMIDÆ.

Cariama ...	2	S. Brazil and La Plata	

ARAMIDÆ.

Aramus	5	Mexico and Cuba to Brazil	

PSOPHIIDÆ.

Psophia	6	Equatorial S. America	

EURYPYGIDÆ.

Eurypyga ...	2	Tropical America	

Order, Family, and Genus.	No. of Species.	Range within the Region.	Range beyond the Region.
ARDEIDÆ.			
Tigrisoma ...	3	The whole region	
Cancroma ...	1	Tropical S. America	
PALAMEDEIDÆ.			
Palamedea ...	1	Equatorial America	
Chauna	2	Columbia, Brazil, and La Plata	
ANSERES.			
ANATIDÆ.			
Cairina	1	Tropical S. America	
Merganetta ...	3	Andes	
Micropterus ...	1	Temperate S. America	
SPHENISCIDÆ.			
Eudyptes ...	6	Temperate S. America	Antarctic shores
Aptenodytes	2	Falkland Islands	Antarctic shores
STRUTHIONES.			
STRUTHIONIDÆ.			
685. *Rhea*	3	S. Temperate America	

CHAPTER XV.

THIS region consists almost wholly of Temperate North America as defined by physical geographers. In area it is about equal to the Neotropical region. It possesses a vast mountain range traversing its entire length from north to south, comparable with, and in fact a continuation of, the Andes,—and a smaller range near the east coast, equally comparable with the mountains of Brazil and Guiana. These mountains supply its great river-system of the Mississippi, second only to that of the Amazon; and in its vast group of fresh-water lakes or inland seas, it possesses a feature unmatched by any other region, except perhaps by the Ethiopian. It possesses every variety of climate between arctic and tropical; extensive forests and vast prairies; a greatly varied surface and a rich and beautiful flora. But these great advantages are somewhat neutralized by other physical features. It extends far towards the north, and there it reaches its greatest width; while in its southern and warmest portion it suddenly narrows. The northern mass of land causes its isothermal lines to bend southwards; and its winter tempera-ture especially, is far lower than at corresponding latitudes in Europe. This diminishes the available area for supporting animal life; the amount and character of which must be, to a great extent, determined by the nature of the least favourable part of the year. Again, owing to the position of its mountain ranges and the direction of prevalent winds, a large extent of its interior, east of the Rocky Mountains, is bare and arid, and often almost desert; while the most favoured districts,—those east of

NEARCTIC REGION
Scale 1 inch—1,000 miles

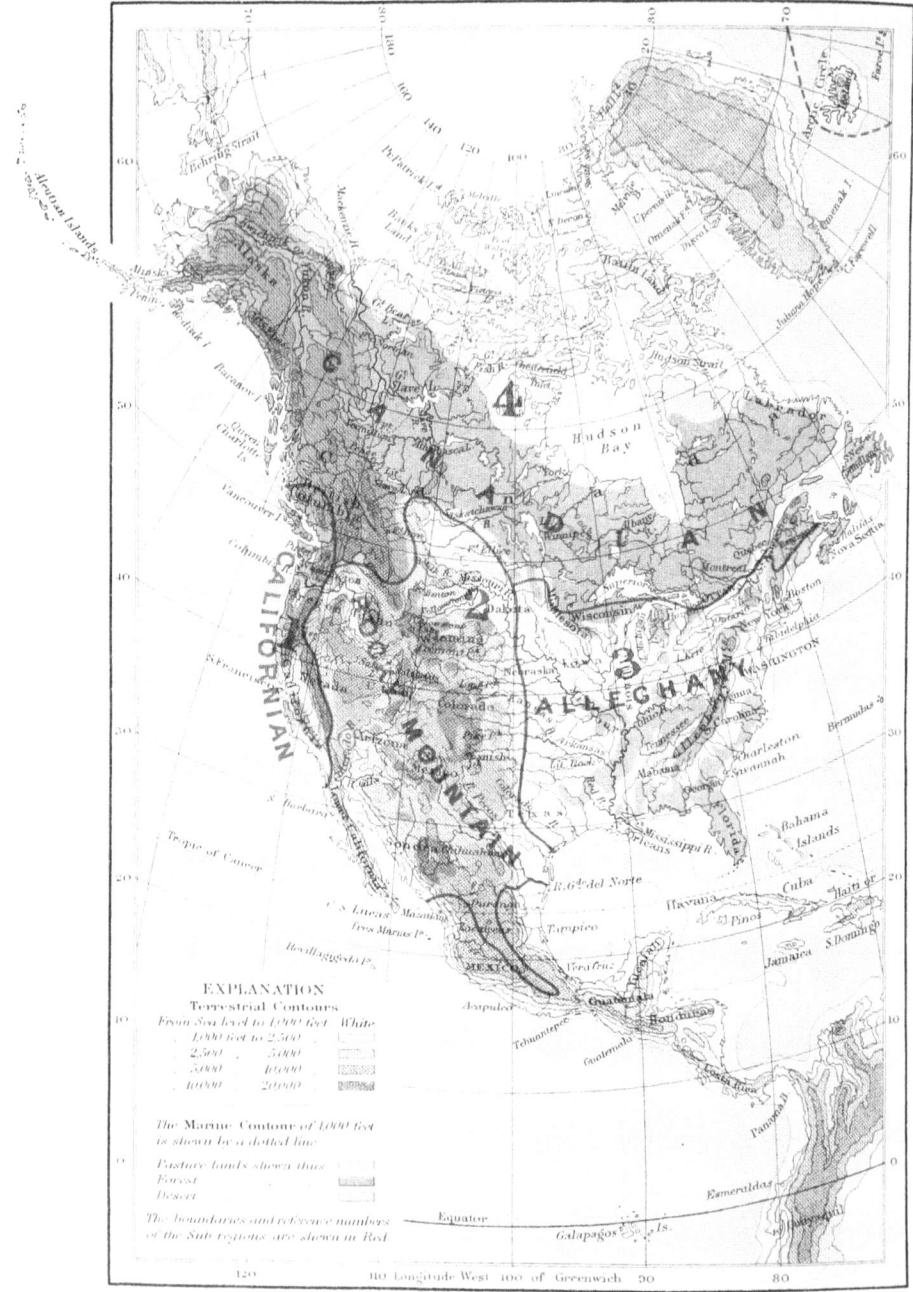

EXPLANATION

Terrestrial Contours

From Sea level to 1000 feet	White
1000 feet to 2500	
2500 , 5000	
5000 , 10000	
10000 , 20000	

The Marine Contour of 1000 feet
is shown by a dotted line

Pasture lands shown thus
Forest
Desert

The boundaries and reference numbers
of the Sub regions are shown in Red

London; Macmillan & Co.

the Mississippi and west of the Sierra Nevada, bear but a small proportion to its whole area. Again, we know that at a very recent period geologically, it was subjected to a very severe Glacial epoch, which wrapped a full half of it in a mantle of ice, and exterminated a large number of animals which previously inhabited it. Taking all this into account, we need not be surprised to find the Nearctic region somewhat less rich and varied in its forms of life than the Palæarctic or the Australian regions, with which alone it can fairly be compared. The wonder rather is that it should be so little inferior to them in this respect, and that it should possess such a variety of groups, and such a multitude of forms, in every class of animals.

Zoological characteristics of the Nearctic Region.—Temperate North America possesses representatives of 26 families of Mammalia, 48 of Birds, 18 of Reptiles, 11 of Amphibia, and 18 of Fresh-water Fish. The first three numbers are considerably less than the corresponding numbers for the Palæarctic region, while the last two are greater—in the case of fishes materially so, a circumstance readily explained by the wonderful group of fresh-water lakes and the noble southward-flowing river system of the Mississippi, to which the Palæarctic region has nothing comparable. But although somewhat deficient in the total number of its families, this region possesses its full proportion of peculiar and characteristic family and generic forms. No less than 13 families or sub-families of Vertebrata are confined to it, or just enter the adjacent Neotropical region. These are,—three of mammalia, Antilocaprinæ, Saccomyidæ and Haploodontidæ; one of birds, Chamæidæ; one of reptiles, Chirotidæ; two of amphibia, Sirenidæ and Amphiumidæ; and the remaining six of fresh-water fishes. The number of peculiar or characteristic genera is perhaps more important for our purpose; and these are very considerable, as the following enumeration will show.

Mammalia.—Of the family of moles (Talpidæ) we have 3 peculiar genera: *Condylura, Scapanus,* and *Scalops,* as well as the remarkable *Urotrichus,* found only in California and Japan. In the weasel family (Mustelidæ) we have *Latax,* a peculiar kind of otter; *Taxidea,* allied to the badgers; and one of the remarkable and characteristic skunks is separated by Dr. J. E. Gray as a genus—*Spilogale*. In the American family Procyonidæ, a peculiar genus (*Bassaris*) is found in California and Texas, extending south along the mountains of Mexico and Guatemala. *Eumetopias*, and *Halicyon*, are seals confined to the west coast of North America. The Bovidæ, or hollow-horned ruminants, contain three peculiar forms; *Antilocapra*, the remarkable prong-buck of the Rocky Mountains; *Aplocerus*, a goat-like antelope; and *Ovibos*, the musk-sheep, confined to Arctic America and Greenland. Among the Rodents are many peculiar genera: *Neotoma*, *Sigmodon*, and *Fiber*, belong to the Muridæ, or rats; *Jaculus* to the Dipodidæ, or jerboas. The very distinct family *Saccomyidæ*, or pouched rats, which have peculiar cheek pouches, or a kind of outer hairy mouth, consists of five genera all confined to this region, with one of doubtful affinities in Trinidad and Central America. In the squirrel family (Sciuridæ), *Cynomys*, the prairie-dogs, are peculiar; and *Tamias*, the ground squirrel, is very characteristic, though found also in North Asia. *Haploodon*, or sewellels, consisting of two species, forms a distinct family; and *Erethizon* is a peculiar form of tree porcupine (Cercolabidæ). True mice and rats of the genus *Mus* are not indigenous to North America, their place being supplied by a distinct genus (*Hesperomys*), confined to the American continent.

Birds.—The genera of birds absolutely peculiar to the Nearctic region are not very numerous, because, there being no boundary but one of climate between it and the Neotropical region, most of its characteristic forms enter a short distance within the limits we are obliged to concede to the latter. Owing also to the severe winter-climate of a large part of the region (which we know is a comparatively recent phenomenon), a large proportion of its birds migrate southwards, to pass the winter in the West-Indian islands or Mexico, some going as far as Guatemala, and a few even to Venezuela.

In our chapter on extinct animals, we have shown, that there is good reason for believing that the existing union of North and South America is a quite recent occurrence; and that the separation was effected by an arm of the sea across what is now Nicaragua, with perhaps another at Panama. This would leave Mexico and Guatemala joined to North America, and forming part of the Nearctic region, although no doubt containing many Neotropical forms, which they had received during earlier continental periods; and these countries might at other times have been made insular by a strait at the isthmus of Tehuantepec, and have then developed some peculiar species. The latest climatal changes have tended to restrict these Neotropical forms to those parts where the climate is really tropical; and thus Mexico has attained its present strongly marked Neotropical character, although deficient in many of the most important groups of that region.

In view of these recent changes, it seems proper not to draw any decided line between the Nearctic and Neotropical regions, but rather to apply, in the case of each genus, a test which will show whether it was probably derived at a comparatively recent date from one region or the other. The test referred to, is the existence of peculiar species of the genus, in what are undoubtedly portions of ancient North or South America. If, for example, all the species of a genus occur in North America, some, or even all, of them, migrating into the Neotropical region in winter, while there are *no peculiar Neotropical species*, then we must class that genus as strictly Nearctic; for if it were Neotropical it would certainly have developed *some* peculiar resident forms. Again, even if there should be one or two resident species peculiar to that part of Central America north of the ancient dividing strait, with an equal or greater number of species ranging over a large part of Temperate North America, the genus must still be considered Nearctic. Examples of the former case, are *Helminthophaga* and *Myiodioctes*, belonging to the Mniotiltidæ, or wood-warblers, which range over *all* Temperate North America to Canada, where *all* the species are found, but in each case one of the species is found in South America, probably as a winter migrant. Of the latter, are *Ammodramus* and *Junco* (genera of finches), which range over the whole United States, but each have one peculiar species in Guatemala. These may be claimed as exclusively Nearctic genera, on the ground that Guatemala was recently Nearctic; and is now really a transition territory, of which the lowlands have been invaded and taken exclusive possession of by a Neotropical fauna, while the highlands are still (in part at least) occupied by Nearctic forms.

In his article on "Birds," in the new edition of the "Encyclopædia Britannica" (now publishing), Professor Newton points out, that the number of *peculiar genera* of Nearctic birds is much less than in each of the various sub-divisions of the Neotropical region; and that the total number of genera is also less, while the bulk of them are common either to the Neotropical or Palæarctic regions. This is undoubtedly the case if any fixed geographical boundary is taken; and it would thus seem that the "Nearctic" should, in birds, form a sub-region only. But, if we define "Nearctic genera" as above indicated, we find a considerable amount of speciality, as the following list will show. The names not italicised are those which are represented in Mexico or Guatemala by peculiar species:—

LIST OF TYPICAL NEARCTIC GENERA OF LAND BIRDS.

1. *Oreoscoptes*	17. *Phænopepla*	33. *Empidias*
2. *Harporhynchus*	18. *Xanthocephalus*	34. *Sphyrapicus*
3. *Sialia*	19. *Scolecophagus*	35. *Hylatomus*
4. *Chamæa*	20. Pipilo	36. *Trochilus*
5. *Catherpes*	21. Junco	37. *Atthis*
6. *Salpinctus*	22. *Melospiza*	38. *Ectopistes*
7. *Psaltriparus*	23. Spizella	39. *Centrocercus*
8. *Auriparus*	24. *Passerculus*	40. *Pediocœtes*
9. *Gymnokitta*	25. *Pocæcetes*	41. *Cupidonia*
10. *Picicorvus*	26. Ammodromus	? Ortyx
11. *Mniotilta.*	27. *Cyanospiza*	42. *Oreortyx*
12. *Oporornis*	28. *Pyrrhuloxia*	43. *Lophortyx*
13. *Icteria*	29. *Calamospiza*	44. Callipepla
14. *Helmintherus*	30. Chondestes	45. Cyrtonyx
15. *Helminthophaga*	31. *Centronyx*	46. Meleagris
16. *Myiodioctes*	32. *Neocorys*	47. *Micrathene*

The above are all groups which are either wholly Nearctic or typically so, but entering more or less into the debatable ground of the Neotropical region; though none possess any peculiar species in the ancient Neotropical land south of Nicaragua. But we have, besides these, a number of genera which we are accustomed to consider as typically European, or Palæarctic, having representatives in North America; although in many cases it would be more correct to say that they are Nearctic genera, represented in Europe, since America possesses more species than Europe or North Asia. The following is a list of genera which have as much right to be considered typically Nearctic as Palæarctic:—

1. Regulus	9. Corvus	16. *Euspiza*
2. Certhia	10. *Ampelis*	17. *Plectrophanes*
3. Sitta	11. Loxia	18. Tetrao
4. Parus	12. Pinicola	19. Lagopus
5. Lophophanes	13. Linota	20. *Nyctala*
6. Lanius	14. *Passerella*	21. *Archibuteo*
7. Perisoreus	15. *Leucosticte.*	22. Haliæetus
8. Pica		

The seven genera italicized have a decided preponderance of Nearctic species, and have every right to be considered typically Nearctic; while the remainder are so well represented by peculiar species, that it is quite possible many of them may have originated here, rather than in the Palæarctic region, all alike being quite foreign to the Neotropical.

On the whole, then, we have 47 in the first and 7 in the second table, making 54 genera which we may fairly class as typically Nearctic, out of a total of 168 genera of land birds, or nearly one-third of the whole. This is an amount of peculiarity which is comparable with that of either of the less isolated regions; and, combined with the more marked and more exclusively peculiar forms in the other orders of vertebrates, fully establishes Temperate North America as a region, distinct alike from the Neotropical and the Palæarctic.

Reptiles.—Although temperate climates are always comparatively poor in reptiles, a considerable number of genera are peculiar to the Nearctic region. Of snakes, there are, *Conophis, Chilomeniscus, Pituophis*, and *Ischnognathus*, belonging to the Colubridæ; *Farancia*, and *Dimodes*, Homalopsidæ; *Lichanotus*, one of the Pythonidæ; *Cenchris, Crotalophorus, Uropsophorus*, and *Crotalus*, belonging to the Crotalidæ or rattlesnakes.

Of Lizards, *Chirotes*, forming a peculiar family; *Ophisaurus*, the curious glass-snake, belonging to the Zonuridæ; with *Phrynosoma* (commonly called horned toads), *Callisaurus*, *Uta*, *Euphryne*, *Uma*, and *Holbrookia*, genera of Iguanidæ.

Testudinidæ, or Tortoises, show a great development of the genus *Emys;* with *Aromochelys* and *Chelydra* as peculiar genera.

Amphibia.—In this class the Nearctic region is very rich, possessing representatives of nine of the families, of which two are peculiar to the region, and there are no less than fifteen peculiar genera. *Siren* forms the family Sirenidæ; *Menobranchus* belongs to the Proteidæ; *Amphiuma* is the only representative of the Amphiumidæ; there are nine peculiar genera of Salamandridæ. Among the tail-less batrachians (frogs and toads) we have *Scaphiopus*, belonging to the Alytidæ; *Pseudacris* to the Hylidæ; and *Acris* to the Polypedatidæ.

Fresh-water Fishes.—The Nearctic region possesses no less than five peculiar family types, and twenty-four peculiar genera of this class. The families are Aphredoderidæ, consisting of a single species found in the Eastern States; Percopsidæ, founded on a species peculiar to Lake Superior; Heteropygii, containing two genera peculiar to the Eastern States; Hyodontidæ and Amiidæ, each consisting of a single species. The genera are as follows: *Paralabrax*, found in California; *Huro*, peculiar to Lake Huron; *Pileoma*, *Boleosoma*, *Bryttus* and *Pomotis* in the Eastern States—all belonging to the perch family. *Hypodelus* and *Noturus*, belonging to the Siluridæ. *Thaleichthys*, one of the Salmonidæ peculiar to the Columbia river. *Moxostoma*, *Pimephales*, *Hyborhynchus*, *Rhinichthys*, in the Eastern States; *Ericymba*, *Exoglossum*, *Leucosomus*, and *Carpiodes*, more widely distributed; *Cochlognathus*, in Texas; *Mylaphorodon* and *Orthodon*, in California; *Meda*, in the river Gila; and *Acrochilus*, in the Columbia river—all belonging to the Cyprinidæ. *Scaphirhynchus*, found only in the Mississippi and its tributaries, belongs to the sturgeon family (Accipenseridæ).

Summary of Nearctic Vertebrata. — The Nearctic region possesses 24 peculiar genera of mammalia, 49 of birds, 21 of reptiles, and 29 of fresh-water fishes, making 123 in all. Of these 70 are mammals and land-birds, out of a total of 242 genera of these groups, a proportion of about two-sevenths. This is the smallest proportion of peculiar genera we have found in any of the regions; but many of the genera are of such isolated and exceptional forms that they constitute separate families, so that we have no less than 12 families of vertebrata confined to the region. The Palæarctic region has only 3 peculiar families, and even the Oriental region only 12; so that, judged by this test, the Nearctic region is remarkably well characterized. We must also remember that, owing to the migration of many of its peculiar forms during the Glacial period, it has recently lost some of its speciality; and we should therefore give some weight to the many characteristic groups it possesses, which, though not quite peculiar to it, form important features in its fauna, and help to separate it from the other regions with which it has been thought to be closely allied. It is thus well distinguished from the Palæarctic region by its Procyonidæ, or racoons, *Hesperomys*, or vesper mice, and *Didelphys*, or opossums, among Mammalia; by its Vireonidæ, or greenlets, Mniotiltidæ, or wood-warblers, Icteridæ, or hang-nests, Tyrannidæ, or tyrant shrikes, and Trochilidæ, or humming-birds, among birds, families which, extending to its extreme northern limits must be held to be as truly characteristic of it as of the Neotropical region; by its Teidæ, Iguanidæ, and *Cinosternum*, among reptiles; and by its Siluridæ, and Lepidosteidæ, among fishes. From the Neotropical region it is still more clearly separated, by its numerous insectivora; by its bears; its Old World forms of ruminants; its beaver; its numerous *Arvicolæ*, or voles; its *Sciuropterus*, or flying squirrels; *Tamias*, or ground-squirrels; and *Lagomys*, or marmots, among mammals; its numerous Paridæ, or tits, and Tetraonidæ, or grouse, among birds; its Trionychidæ among reptiles; its Proteidæ, and Salamandridæ, among Amphibia; and its Gasterosteidæ, Atherinidæ, Esocidæ, Umbridæ, Accipenseridæ, and Polydontidæ, among fishes.

These characteristic features, taken in conjunction with the absolutely peculiar groups before enumerated, demonstrate that the Nearctic region cannot with propriety be combined with any other. Though not very rich, and having many disadvantages of climate and of physical condition, it .is yet sufficiently well characterized in its zoological features to rank as one of the well-marked primary divisions of the earth's surface.

There is one other consideration bearing on this question which should not be lost sight of. In establishing our regions we have depended wholly upon their *now* possessing a sufficient number and variety of animal forms, and a fair proportion of peculiar types ; but when the validity of our conclusion on these grounds is disputed, we may supplement the evidence by an appeal to the past history of the region in question. In this case we find a remarkable support to our views. During the whole Tertiary period, North America was, zoologically, far more strongly contrasted with South America than it is now; while, during the same long series of ages, it was always clearly separated from the Eastern hemisphere or the Palæarctic region by the exclusive possession of important families and numerous genera of Mammalia, as shown by our summary of its extinct fauna in Chapter VII. Not only may we claim North America as now forming one of the great zoological regions, but as having continued to be one ever since the Eocene period.

Insects.

In describing the Palæarctic and Neotropical regions, many of the peculiarities of the insect-fauna of this region have been incidentally referred to ; and as a tolerably full account of the distribution of the several families is given in the Fourth Part of our work (Chapter XXI.), we shall treat the subject very briefly here.

Lepidoptera.—The butterflies of the Nearctic region have lately been studied with much assiduity, and we are now able to form some idea of their nature and extent. Nearly 500 species belonging to about 100 genera have been described; showing that the region, which a few years ago was thought to be very poor in species of butterflies, is really much richer than Europe, and probably about as rich as the more extensive Palæarctic region. There is, however, very little speciality in the forms. A considerable number of Neotropical types enter the southern States; but there are hardly any peculiar genera, except one of the Lycænidæ and perhaps a few among the Hesperidæ. The most conspicuous feature of the region is its fine group of Papilios, belonging to types (*P. turnus* and *P. troilus*) which are characteristically Nearctic. It is also as rich as the Palæarctic region in some genera which we are accustomed to consider as pre-eminently European; such as *Argynnis, Melitæa, Grapta, Chionabas,* and a few others. Still, we must acknowledge, that if we formed our conclusions from the butterflies alone, we could hardly separate the Nearctic from the Palæarctic region. This identity probably dates from the Miocene period; for when our existing arctic regions supported a luxuriant vegetation, butterflies would have been plentiful; and as the cold came on, these would move southwards both in America and Europe, and, owing to the long continuance of the generic types of insects, would remain little modified till now.

Coleoptera.—Only a few indications can be given of the peculiarities of the Nearctic coleoptera. In Cicindelidæ the region possesses, besides the cosmopolite *Cicindela,* four other genera, two of which—*Amblychile* and *Omus*—are peculiar to the West Coast and the Rocky Mountains. Of Carabidæ it possesses *Dicœlus, Pasimachus, Eurytrichus, Sphœroderus, Pinacodera,* and a number of smaller genera, altogether peculiar to it; *Helluomorpha, Galerita, Callida,* and *Tetragonoderus,* in common with South America; and a large number of characteristic European forms.

The Lucanidæ are all of European types. The region is poor in Cetoniidæ, but has representatives of the South American *Euphoria,* as well as of four European genera. Of Buprestidæ it has the South American *Actenodes;* a single species of the Ethiopian and Eastern *Belionota,* in California; and about a dozen other genera of European and wide distribution.

Among Longicorns it possesses fifty-nine peculiar genera, representatives of five Neotropical, and thirteen Palæarctic genera; as well as many of wider distribution. *Prionus* is the chief representative of the Prionidæ; *Leptura* and *Crossidius* of the

Cerambycidæ; *Leptostylus, Liopus, Graphidurus,* and *Tetraopes,* of the Lamiidæ, the latter genus being confined to the region.

Terrestrial and Fluviatile Mollusca.

The land-shells of temperate North America almost all belong to the Inoperculate or Pulmoniferous division; the Operculata being represented only by a few species of *Helicina* and *Truncatella,* chiefly in the Southern States. According to Mr. Binney's recent "Catalogue of the Terrestrial Air-breathing Mollusks of North America," the fauna consists of the following genera:—*Glandina* (6 sp.); *Macrocyclis* (5 sp.); *Zonites* (37 sp.); *Vitrina* (4 sp.); *Limax* (5 sp.); *Arion* (3 sp.); *Ariolimax* (3 sp.); *Prophysaon* (1 sp.); *Binneia* (1 sp.); *Hemiphillia* (1 sp.); *Patula* (16 sp.); *Helix* (80); *Holospira* (2 sp.); *Cylindrella* (2 sp.); *Macroceramus* (2 sp.); *Bulimulus* (8 sp.); *Cionella* (2 sp.); *Stenogyra* (4 sp.); *Pupa* (19 sp.); *Strophia* (1 sp.); *Vertigo* (6 sp.); *Liguus* (1 sp.); *Orthalicus* (2 sp.); *Punctum* (1 sp.); *Succinea* (26 sp.); *Tebennophorus* (1 sp.); *Pallifera* (1 sp.); *Veronicella* (2 sp.).

All the larger genera range over the whole region, but the following have a more restricted distribution; *Macrocyclis* has only one species in the East, the rest being Californian or Central; *Ariolimax, Prophysaon, Binneia,* and *Hemiphillia,* are confined to the Western sub-region. Lower California has affinities with Mexico, 18 species being peculiar to it, of which two are true *Bulimi,* a genus unknown in other parts of the region. The Central or Rocky Mountain sub-region is chiefly characterised by six peculiar species of *Patula.* The Eastern sub-region is by far the richest, nine-tenths of the whole number of species being found in it. The Alleghany Mountains form the richest portion of this sub-region, possessing nearly half the total number of species, and at least 24 species found nowhere else. The southern States have also several peculiar species, but they are not so productive as the Alleghanies. The Canadian sub-region possesses 32 species, of which nearly half are northern forms more or less common to the whole Arctic regions, and several of this character have spread southwards all over the United States. Species of *Vitrina, Zonites, Pupa*, and *Succinea*, are found in Greenland ; and Eastern Palæarctic species of *Vitrina, Patula*, and *Pupa* occur in Alaska. More than 30 species of shells living in the Eastern States, are found fossil in the Post-Pliocene deposits of the Ohio and Mississippi.

Fresh-water Shells.—North America surpasses every other part of the globe in the number and variety of its fresh-water mollusca, both univalve and bivalve. The numbers up to 1866 were as follows :—Melaniadæ, 380 species ; Paludinidæ, 58 species ; Cycladidæ, 44 species ; and Unionidæ, 552 species. The last family had, however, increased to 832 species in 1874, according to Dr. Isaac Lea, who has made them his special study ; but it is probable that many of these are such as would be considered varieties by most conchologists. Many of the species of *Unio* are very large, of varied forms, and rich internal colouring, and the group forms a prominent feature of the Nearctic fauna. By far the larger proportion of the fresh-water shells inhabit the Eastern or Alleghany sub-region ; and their great development is a powerful argument against any recent extensive submergence beneath the ocean of the lowlands of North America.

The Nearctic Sub-regions.

The sub-divisions of the Nearctic region, although pretty clearly indicated by physical features and peculiarities of climate and vegetation, are by no means so strongly marked out in their zoology as we might expect. The same genera, as a rule, extend over the whole region ; while the species of the several sub-regions are in most cases different. Even the vast range of the Rocky Mountains has not been an effectual barrier against this wide dispersal of the same forms of life ; and although some important groups are limited by it, these are exceptions to the rule. Even now, we find fertile valleys and plateaus of moderate elevation, penetrating the range on either side ; and both to the north and south there are passes which can be freely traversed by most animals during the summer. Previous to the glacial epoch there was probably a warm period, when every part of the range supported an abundant and varied fauna, which, when the cold period arrived, would descend to the lowlands, and people the country to the east, west, and south, with similar forms of life.

The first, and most important sub-division we can make, consists of the Eastern United States, extending across the Mississippi and the more fertile prairies, to about the 100°th. meridian of west longitude, where the arid and almost desert country commences. Southwards, the boundary bends towards the coast, near the line of the Brazos or Colorado rivers. To the north the limits are undefined; but as a considerable number of species and genera occur in the United States but not in Canada, it will be convenient to draw the line somewhere near the boundary of the two countries, except that the district between lakes Huron and Ontario, and probably Nova Scotia, may be included in the present sub-region. As far west as the Mississippi, this was originally a vast forest country; and it is still well wooded, and clothed with a varied and luxuriant vegetation.

The next, or Central sub-region, consists of the dry, elevated, and often arid district of the Rocky Mountains, with its great plateaus, and the barren plains of its eastern slope; extending northwards to near the commencement of the great forests north of the Saskatchewan, and southward to the Rio Grande del Norte, the Gulf of California, and to Cape St. Lucas, as shown on our maps. This sub-region is of an essentially desert character, although the higher valleys of the Rocky Mountains are often well wooded, and in these are found some northern and some western types.

The third, or Californian sub-region, is small, but very luxuriant, occupying the comparatively narrow strip of country between the Sierra Nevada and the Pacific. To the north it may include Vancouver's Island and the southern part of British Columbia, while to the south it extends to the head of the Gulf of California.

The fourth division, comprises the remainder of North America; and is a country of pine forests, and of barren wastes towards the Arctic Ocean. It has fewer peculiar species to characterise it than any other, but it possesses several characteristic arctic forms, while many of those peculiar to the south are absent; so that it is a very convenient, if it should not be considered an altogether natural, sub-region.

We will now give an outline of the most important zoological features of each of these divisions, taking them in the order in which they are arranged in the Fourth Part of this work. California comes first, as it has some tropical forms not found elsewhere, and thus forms a transition from the Neotropical region.

I. The Western or Californian Sub-region.

This small district possesses a fruitful soil and a highly favourable climate, and is, in proportion to its extent, perhaps the richest portion of the continent, both zoologically and botanically. Its winters are far milder than those of the Eastern States in corresponding latitudes ; and this, perhaps, has enabled it to support several tropical forms which give a special character to its fauna. It is here only, in the whole region, that bats of the families Phyllostomidæ and Noctilionidæ, and a serpent of the tropical family, Pythonidæ, are found, as well as several Neotropical forms of birds and reptiles.

Mammalia.—The following genera are not found in any other part of the Nearctic region. *Macrotus* (Phyllostomidæ), one species in California; *Antrozous* (Vespertilionidæ), one species on the West Coast; *Urotrichus* (Talpidæ) one species in British Columbia; sub-genus *Nesorex* (Soricidæ), one species in Oregon ; *Bassaris* (Procyonidæ), California ; *Enhydra* (Mustelidæ), Pacific Coast; *Morunga* (Phocidæ), California ; *Haploodon* (Haploodontidæ) a rat-like animal, allied to the beavers and marmots, and constituting a peculiar family found only in California and British Columbia. The following characteristic Nearctic forms also extend into this sub-region :—*Taxidea, Procyon, Didelphys, Sciuropterus, Tamias, Spermophilus, Dipodomys, Perognathus, Jaculus.*

Birds.—Few genera of birds are quite peculiar to this sub-region, since most of the Western forms extend into the central district, yet it has a few. *Glaucidium*, a genus of Owls, is confined (in the Nearctic region) to California; *Chamœa*, a singular form allied to the wrens, and forming a distinct family, is quite peculiar; *Geococcyx*, a Neotropical form of cuckoo, extends to California and Southern Texas.   The following genera are very characteristic of the sub-region, and some of them almost confined to it: *Myiadestes* (Sylviidæ); *Psaltriparus* (Paridæ); *Cyanocitta*, *Picicorvus* (Corvidæ); *Hesperiphona, Peucœa, Chondestes* (Fringillidæ); *Selasphorus, Atthis* (Trochilidæ); *Columba, Melopelia* (Columbidæ); *Oreortyx* (Tetraonidæ).

Reptiles.—The following genera are not found in any other part of the Nearctic region: *Charina* (Tortricidæ); *Lichanotus* (Pythonidæ); *Gerrhonotus* (Zonuridæ); *Phyllodactylus* (Geckotidæ); *Anolius* and *Tropidolepis* (Iguanidæ). *Sceloporus* (Iguanidæ) is only found elsewhere in Florida. All the larger North American groups of lizards and snakes are also represented here; but in tortoises it is deficient, owing to the absence of lakes and large rivers.

Amphibia.—California possesses two genera of Salamandridæ, *Aneides* and *Heredia*, which do not extend to the other sub-regions.

Fresh-water Fish.—There are two or three peculiar genera of Cyprinidæ, but the sub-region is comparatively poor in this group.

Plate XVIII. Illustrative of the Zoology of California and the Rocky Mountains.—We have chosen for the subject of this illustration, the peculiar Birds of the Western mountains. The two birds in the foreground are a species of grouse (*Pediocœtes Columbianus*), entirely confined to this sub-region; while the only other species of the genus is found in the prairies north and west of Wisconsin, so that the group is peculiar to northern and western America. The crested birds in the middle of the picture (*Oreortyx picta*), are partridges, belonging to the American subfamily Odontophorinæ. This is the only species of the genus which is confined to California and Oregon. The bird at the top is the blue crow (*Gymnokitta cyanocephala*), confined to the Rocky Mountains and Sierra Nevada from New Mexico and Arizona northwards, and more properly belonging to the Central

PLATE XVIII.

SCENE IN CALIFORNIA, WITH SOME CHARACTERISTIC BIRDS.

sub-region. It is allied to the European nutcracker; but according to the American ornithologist, Dr. Coues, has also resemblances to the jays, and certainly forms a distinct genus. The grizzly bear (*Ursus ferox*) in the background, is one of the characteristic animals of the Californian highlands.

II. The Central, or Rocky Mountain Sub-region.

This extensive district is, for the greater part of its extent, from 2,000 to 5,000 feet above the sea, and is excessively arid; and, except in the immediate vicinity of streams and on some of the higher slopes of the mountains, is almost wholly treeless. Its zoology is therefore peculiar. Many of the most characteristic genera and families of the Eastern States are absent; while a number of curious desert and alpine forms give it a character of its own, and render it very interesting to the naturalist.

Mammalia.—The remarkable prong-horned antelope (*Antilocapra*), the mountain goat (*Aplocerus*), the mountain sheep or bighorn (*Ovis montana*), and the prairie-dog (*Cynomys*), one of the Rodentia, are peculiar to this sub-region; while the family of the Saccomyidæ, or pouched rats, is represented by many forms and is very characteristic. Here is also the chief home of the bison. The glutton (*Gulo*) and marmot (*Lagomys*) enter it from the north; while it has the racoon (*Procyon*), flying squirrel (*Sciuropterus*), ground squirrel (*Tamias*), pouched marmot (*Spermophilus*) and jumping mouse (*Jaculus*) in common with the countries east or west of it.

Plate XIX. Illustrative of the Zoology of the Central Plains or Prairies.—We here introduce four of the most characteristic mammalia of the great American plains or prairies, three of them being types confined to North America. The graceful animals on the left are the prong-horned antelopes (*Antilocapra americana*), whose small horns, though hollow like those of the antelopes, are shed annually like those of the deer. To the right we have the prairie-dogs of the trappers (*Cynomys ludovicianus*) which, as will be easily seen, are rodents, and allied to the marmots of the European Alps. Their burrows are numerous on the prairies, and the manner in which they perch themselves on little mounds and gaze on intruders, is noticed by all travellers. On the left, in the foreground, is one of the extraordinary pouched rats of America (*Geomys bursarius*). These are burrowing animals, feeding on roots; and the mouth is, as it were, double, the outer portion very wide and hairy, behind which is the small inner mouth. Its use may be to keep out the earth from the mouth while the animal is gnawing roots. A mouth so constructed is found in no other animals but in these North American rats. In the distance is a herd of bisons (*Bison americanus*), the typical beast of the prairies.

Birds.—This sub-region has many peculiar forms of birds, both residents, and migrants from the south or north. Among the peculiar resident species we may probably reckon a dipper, (*Cinclus*); *Salpinctes*, one of the wrens; *Poospiza, Calamospiza,* genera of finches; *Picicorvus, Gymnokitta,* genera of the crow family; *Centrocercus* and *Pediocœtes*, genera of grouse. As winter migrants from the north it has *Leucosticte* and *Plectrophanes,* genera of finches; *Perisoreus,* a genus of the crow family; *Picoides,* the Arctic woodpecker; and *Lagopus,* ptarmigan. Its summer migrants, many of which may be resident in the warmer districts, are more numerous. Such are, *Oreoscoptes,* a genus of thrushes; *Campylorhynchus* and *Catherpes,* wrens; *Paroides,* one of the tits; *Phœnopepla,* allied to the waxwing; *Embernagra* and *Spermophila,* genera of finches; *Pyrocephalus,* one of the tyrant shrikes; *Callipepla* and *Cyrtonyx,* American partridges. Besides these, the more widely spread genera, *Harporhynchus, Lophophanes, Carpodacus, Spizella,* and *Cyanocitta,* are characteristic of the central district, and two genera of humming-birds—*Atthis* and *Selasphorus*—only occur here and in California. Prof. Baird notes 40 genera of birds which are represented by distinct allied species in the western, central, and eastern divisions of the United States, corresponding to our sub-regions.

It is a curious fact that the birds of this sub-region should extend across the Gulf of California, and that Cape St. Lucas, at the southern extremity of the peninsula, should be decidedly more "Central" than "Californian" in its ornithology. Prof.

PLATE XIX.

THE AMERICAN PRAIRIES, WITH CHARACTERISTIC MAMMALIA.

Baird says, that its fauna is almost identical with that of the
Gila River, and has hardly any relation to that of Upper
California. It possesses a considerable number (about twenty)
of peculiar species of birds, but all belong to genera characteristic of the present sub-region; and there is no resemblance to
the birds of Mazatlan, just across the gulf in the Neotropical
region.

Reptiles, Amphibia, and Fishes.—A large number of snakes
and lizards inhabit this sub-region, but they have not yet
been classified with sufficient precision to enable us to make
much use of them. Among lizards, Iguanidæ, Geckotidæ,
Scincidæ, and Zonuridæ, appear to be numerous; and many
new genera of doubtful value have been described. Among
snakes, Calamariidæ, Colubridæ, and Crotalidæ are represented.
Among Amphibia, *Siredon,* one of the Proteidæ, is peculiar.
The rivers and lakes of the Great Central Basin, and the
Colorado River, contain many peculiar forms of Cyprinidæ.

III. *The Eastern or Alleghany Sub-region.*

This sub-region contains examples of all that is most characteristic of Nearctic zoology. It is for the most part an undu-
lating or mountainous forest-clad country, with a warm or
temperate climate, but somewhat extreme in character, and
everywhere abounding in animal and vegetable life. To the west,
across the Mississippi, the country becomes more open, gradually
rises, becomes much drier, and at length merges into the arid
plains of the central sub-region. To the south, in Georgia,
Florida, and Louisiana, a sub-tropical climate prevails, and
winter is almost unknown. To the north, in Michigan and New
England, the winters are very severe, and streams and lakes are
frozen for months together. These different climates, however,
produce little effect on the forms of animal life; the species to
some extent change as we go from north to south, but the same
types everywhere prevail. This portion of the United States,
having been longest inhabited by Europeans, has been more
thoroughly explored than other parts of North America; and to
this more complete knowledge its superior zoological richness

may be to some extent due; but there can be little doubt that it is also positively, and not merely relatively, more productive in varied forms of animal life than either of the other sub-regions.

Mammalia.—There seems to be only one genus absolutely peculiar to this sub-region—the very remarkable *Condylura*, or star-nosed mole, only found from Pennsylvania to Nova Scotia, and as far as about 94° west longitude. It also has opossums (*Didelphys*) in common with California, and three out of four species of *Scalops*, a genus of moles; as well as the skunk (*Mephitis*), American badger (*Taxidea*), racoon (*Procyon*), pouched rat (*Geomys*), beaver rat (*Fiber*), jumping mouse (*Jaculus*), tree porcupine (*Erethizon*), and other characteristic Nearctic forms.

Birds.—The birds of this sub-region have been carefully studied by American naturalists, and many interesting facts ascertained as to their distribution and migrations. About 120 species of birds are peculiar to the east coast of the United States, but only about 30 of these are residents all the year round in any part of it; the bird population being essentially a migratory one, coming from the north in winter and the south in summer. The largest number of species seems to be congregated in the district of the Alleghany mountains. A considerable proportion of the passerine birds winter in Central America and the West Indian Islands, and go to the Middle States or Canada to breed; so that even the luxuriant Southern States do not possess many birds which may be called permanent residents. Thus, in East Pennsylvania there are only 52, and in the district of Columbia 54 species, found all the year round, out of about 130 which breed in these localities; very much below the number which permanently reside in Great Britain.

This sub-region is well characterised by its almost exclusive possession of *Ectopistes*, the celebrated passenger pigeon, whose enormous flocks and breeding places have been so often described; and *Cupidonia*, a remarkable genus of grouse. The only Nearctic parrot, *Conurus carolinensis*, is found in the Southern States; as well as *Crotophaga*, a South American genus usually associated with the cuckoos. *Helmintherus* and

Oporornis, genera of wood-warblers, may be considered to be peculiar to this sub-region, since in each case only one of the two species migrates as far as Central America; while two other genera of the same family, *Siurus* and *Setophaga*, as well as the finch genus, *Euspiza*, do not extend to either of the western sub-regions. *Parus*, a genus of tits, comes into the district from the north; *Otocorys*, an alpine lark, and *Coturniculus*, an American finch, from the west; and such characteristic Nearctic genera as *Antrostomus* (the whip-poor-will goatsuckers); *Helminthophaga*, *Dendrœca*, and *Myiodioctes* (wood-warblers); *Vireo* (greenlets); *Dolichonyx* (rice-bird); *Quiscalus* (troupial); *Meleagris* (turkey); and *Ortyx* (American partridge), are wide-spread and abundant. In Mr. J. A. Allen's elaborate and interesting paper on the birds of eastern North America, he enumerates 32 species which breed only in the more temperate portions of this province, and may therefore be considered to be especially characteristic of it. These belong to the following genera:—*Turdus, Galeoscoptes, Harporhynchus, Sialia, Dendrœca, Wilsonia, Pyranga, Vireo, Lanivireo, Lophophanes, Coturniculus, Ammodromus, Spizella, Euspiza, Hedymeles, Cyanospiza, Pipilo, Cardinalis, Icterus, Corvus, Centurus, Melanerpes, Antrostomus, Coccyzus, Ortyx,* and *Cupidonia.*

Reptiles.—In this class the Eastern States are rich, possessing many peculiar forms not found in other parts of the region. Among snakes it has the genera *Farancia* and *Dimodes* belonging to the fresh-water snakes (Homalopsidæ); the South American genus *Elaps;* and 3 genera of rattlesnakes, *Cenchris, Crotalophorus,* and *Crotalus.* The following genera of snakes are said to occur in the State of New York:—*Coluber, Tropidonotus, Leptophis, Calamaria, Heterodon, Trigonocephalus, Crotalus, Psammophis, Helicops, Rhinostoma, Pituophis,* and *Elaps.*

Among lizards, *Chirotes*, forming a peculiar family of Amphisbenians, inhabits Missouri and Mexico; while the remarkable glass-snake, *Ophisaurus*, belonging to the family Zonuridæ, is peculiar to the Southern States; and the South American *Sphærodactylus*, one of the gecko family, reaches Florida. Other genera which extend as far north as the State of New

York are, *Scincus, Tropidolepis, Plestiodon, Lygosoma, Ameiva,* and *Phrynosoma.*

Tortoises, especially the fresh-water kind, are very abundant; and the genera *Aromochelys, Chelydra, Terrapene,* and *Trionyx,* are nearly, if not quite, confined to this division of the region.

Amphibia.—Almost all the remarkable forms of Urodela, or tailed batrachians, peculiar to the region are found here only; such as *Siren* and *Pseudobranchus,* constituting the family Sirenidæ; *Menobranchus,* allied to the *Proteus* of Europe; *Amphiuma,* an eel-like creature with four rudimentary feet, constituting the family Amphiumidæ; *Notopthalmus, Desmognathus,* and *Menopoma,* belonging to the Salamandridæ; together with several other genera of wider range. Of Anura, or tail-less batrachians, there are no peculiar genera, but the Neotropical genus of toads, *Engystoma,* extends as far as South Carolina.

Fishes.—Owing to its possession of the Mississippi and the great lakes, almost all the peculiar forms of North American fishes are confined to this sub-region. Such are *Perca, Pileoma, Huro, Bryttus,* and *Pomotis* (Percidæ); the families Aphredoderidæ and Percopsidæ; several genera of Cyprinodontidæ and Cyprinidæ; and the family Polydontidæ.

Islands of the Alleghany Sub-region.

The Bermudas.—These islands, situated in the Atlantic, about 700 miles from the coast of Carolina, are chiefly interesting for the proof they afford of the power of a great variety of birds to cross so wide an extent of ocean. There are only 6 or 8 species of birds which are permanent residents on the islands, all common North American species; while no less than 140 species have been recorded as visiting them. Most of these are stragglers, many only noticed once; others appear frequently and in great numbers, but very few, perhaps not a dozen, come every year, and can be considered regular migrants. The permanent residents are, a greenlet (*Vireo noveboracensis*), the catbird (*Galeoscoptes carolinensis*), the blue bird (*Sialia sialis*), the cardinal (*Cardinalis virginianus*), the American crow (*Corvus americanus*), and the ground dove (*Chamœpelia passerina*). The most regular visitants are a kingfisher (*Ceryle alcyon*), the wood-wagtail (*Siurus noveboracensis*), the rice-bird (*Dolichonyx oryzivorus*), and a moorhen (*Gallinula galeata*). Besides the American species, four European birds have been taken at the Bermudas : *Saxicola œnanthe, Alauda arvensis* (perhaps introduced), *Crex pratensis,* and *Scolopax gallinago.*

A common American lizard, *Plestiodon longirostris,* is the only land reptile found on the islands.

IV. The Sub-Arctic or Canadian Sub-region.

This sub-region serves to connect together the other three, since they all merge gradually into it; while to the north it passes into the circumpolar zone which is common to the Palæ-arctic and Nearctic regions. The greater portion of it is an extensive forest-district, mostly of coniferæ; and where these cease towards the north, barren wastes extend to the polar ocean. It possesses several northern or arctic forms of Mammalia, such as the glutton, lemming, reindeer, and elk, which barely enter the more southern sub-regions ; as well as the polar bear and arctic fox ; but it also has some peculiar forms, and many of the most characteristic Nearctic types. The remarkable musk-sheep (*Ovibos*) is confined to this sub-region, ranging over a considerable extent of country north of the forests, as well as Greenland. It has been extinct in Europe and Asia since the Post-pliocene epoch. Such purely Nearctic genera as *Procyon, Latax, Erethizon, Jaculus, Fiber, Thomomys,* and *Hesperomys,* abound, many of them ranging to the shores of Hudson's Bay and the barren wastes of northern Labrador. Others, such as *Blarina, Condylura,* and *Mephitis,* are found only in Nova Scotia and various parts of Canada. About 20 species of Mammalia seem to be peculiar to this sub-region.

Plate XX. Illustrating the Zoology of Canada.—We have here a group of Mammalia characteristic of Canada and the colder parts of the United States. Conspicuous in the foreground is the skunk (*Mephitis mephitica*), belonging to a genus of the weasel family found only in America. This animal is celebrated for its power of ejecting a terribly offensive.liquid, the odour of which is almost intolerable. The skunks are nocturnal animals, and are generally marked, as in the species represented, with conspicuous bands and patches of white. This enables them to be easily seen at night, and thus serves to warn larger animals not to attack them. To the left is the curious little jumping mouse (*Jaculus hudsonius*), the American representative of the Palæarctic jerboa. Climbing up a tree on the left is the tree porcupine (*Erethizon dorsatus*), belonging to the family Cercolabidæ, which represents, on the American continent, the porcupines of the Old World. In the background is the elk or moose (*Alces americanus*), perhaps identical with the European elk, and the most striking inhabitant of the northern forests of America, as the bison is of the prairies.

Birds.—Although the Canadian sub-region possesses very few resident birds, the numbers which breed in it are perhaps greater than in the other sub-regions, because a large number of circumpolar species are found here exclusively. From a comparison of Mr. Allen's tables it appears, that more than 200 species are regular migrants to Canada in the breeding season, and nearly half of these are land-birds. Among them are to be found a considerable number of genera of the American families Tyrannidæ and Mniotiltidæ, as well as the American genera *Sialia, Progne, Vireo, Cistothorus, Junco, Pipilo, Zonotrichia, Spizella, Melospiza, Molothrus, Agelœus, Cyanura, Sphyrapicus*, and many others ; so that the ornithology of these northern regions is still mainly Nearctic in character. Besides these, it has such specially northern forms as *Surnia* (Strigidæ) ; *Picoides* (Picidæ) ; *Pinicola* (Fringillidæ) ; as well as *Leucosticte, Plectrophanes, Perisoreus*, and *Lagopus*, which extend further south, especially in the middle sub-region. No less than 212 species of birds have been collected in the new United States territory of Alaska (formerly Russian America), where a humming-bird (*Selasphorus rufus*) breeds. The great majority of these are typically American, including such forms as *Colaptes, Helminthophaga, Siurus, Dendrœca, Myiodioctes, Passerculus, Zonotrichia, Junco, Spizella, Melospizpa, Passerella, Scoleophagas, Pediocetes*, and *Bonasa ;*

PLATE XX.

A CANADIAN FOREST, WITH CHARACTERISTIC MAMMALIA.

together with many northern birds common to both continents. Yet a few Palæarctic forms, not known in other parts of the sub-region, appear here. These are *Budytes flava, Phylloscopus kennicottii,* and *Pyrrhula coccinea,* all belonging to genera not occurring elsewhere in North America. Considering the proximity of the district to North-east Asia, and the high probability that there was an actual land connection at, and south of, Behring's Straits, in late Tertiary times, it is somewhat remarkable that the admixture of Palæarctic and Nearctic groups is not greater than it is. The Palæarctic element, however, forms so small a portion of the whole fauna, that it may be satisfactorily accounted for by the establishment of immigrants since the Glacial period. The great interest felt by ornithologists in the discovery of the three genera above-named, with a wren allied to a European species, is an indication that the faunas even of the northern parts of the Nearctic and Palæarctic regions are, as regards birds, radically distinct. It may be mentioned that the birds of the Aleutian Isles are also, so far as known, almost wholly Nearctic. The number of land-birds known from Alaska is 77; and from the Aleutian Isles 16 species, all of which, except one, are North American.

Reptiles.—These are comparatively few and unimportant. There are however five snakes and three tortoises which are limited to Canada proper; while further north there are only Amphibia, represented by frogs and toads, and a salamander of the genus *Plethodon.*

Fishes.—Most of the groups of fresh-water fish of the Nearctic region are represented here, especially those of the perch, salmon, and pike families ; but there seem to be few or no peculiar genera.

Insects.—These are far less numerous than in the more temperate districts, but are still tolerably abundant. In Canada there are 53 species of butterflies, viz., Papilionidæ, 4 ; Pieridæ, 2 ; Nymphalidæ, 21 ; Satyridæ, 3 ; Lycænidæ 16, and Hesperidæ 7. Most of these are, no doubt, found chiefly in the southern parts of Canada. That Coleoptera are pretty numerous is shown, by more than 800 species having been collected on the shores of Lake Superior; 177 being Geodephaga and 39 Longicorns.

Greenland.—This great arctic island must be considered as belonging to the Nearctic region, since of its six land mammals, three are exclusively American (*Myodes torquatus, Lepus glacialis,* and *Ovibos moschatus*), while the other three (*Vulpes lagopus, Ursus maritimus,* and *Rangifer tarandus*) are circumpolar. Only fourteen land-birds are either resident in, or regular migrants to the country; and of these two are European (*Haliœetus albicilla,* and *Falco peregrinus*), while three are American (*Anthus ludovi- cianus, Zonotrichia leucophrys,* and *Lagopus rupestris*), the rest being arctic species common to both continents. The waders and aquatics (49 in number) are nearly equally divided between both continents; but the land-birds which visit Greenland as stragglers are mostly American. Yet although the Nearctic element somewhat preponderates, Greenland really belongs to that circumpolar debateable land, which is common to the two North Temperate regions.

Concluding remarks.—We have already discussed pretty fully, though somewhat incidentally, the status and relations of the Nearctic region; first in our chapter on Zoological regions, then in our review of extinct faunas, and lastly in the earlier part of this chapter. It will not therefore be necessary to go further into the question here; but we shall, in our next chapter, give a brief summary of the general conclusions we have reached as to the past history and mutual zoological relations of all the great divisions of the earth.

TABLES OF DISTRIBUTION.

In drawing up these tables, showing the distribution of various classes of animals in the Nearctic region, the following sources of information have been chiefly relied on, in addition to the general treatises, monographs, and catalogues used in the compilation of the 4th Part of this work.

Mammalia.—Professor Baird's Catalogue ; Allen's List of the Bats; Mr. Lord's List for British Columbia ; Brown, for Greenland ; Packard for Labrador.

Birds.—Baird, Cassin, and Allen's Lists for United States ; Richardson's Fauna Boreali Americana ; Jones, for Bermudas ; and papers by Brown, Coues, Lord, Packard, Dall, and Professor Newton.

TABLE I.

FAMILIES OF ANIMALS INHABITING THE NEARCTIC REGION.

EXPLANATION.

Names in *italics* show the families which are peculiar to the region.
Names inclosed thus (......) show families which barely enter the region, and are not considered properly to belong to it.
Numbers correspond to the series of numbers to the families in Part IV.

Order and Family.	Sub-regions.				Range beyond the Region.
	Cali-fornia.	Rocky Mntns.	Alle-ghanies.	Canada.	
MAMMALIA.					
CHIROPTERA.					
10. Phyllostomidæ	—				Neotropical
12. Vespertilionidæ	—	—	—	—	Cosmopolite
13. Noctilionidæ...	—				Tropical regions
INSECTIVORA.					
21. Talpidæ... ...	—	—	—	—	Palæarctic
22. Soricidæ ..	—	—	—	—	The Eastern Hemisphere, excl. Australia
CARNIVORA.					
23. Felidæ	—	—	—	—	All regions but the Australian
28. Canidæ	—	—	—	—	All regions but the Australian
29. Mustelidæ ...	—	—	—	—	All regions but the Australian
30. Procyonidæ ...	—	—	—	—	Neotropical
32. Ursidæ ..	—	—	—	—	Palæarctic, Oriental
33. Otariidæ... ...	—			—	N. and S. temperate zones
34. Trichechidæ ..				—	Arctic regions
35. Phocidæ... ...	—				N. and S. temperate zones
CETACEA.					
36 to 41.					Oceanic
UNGULATA.					
47. Suidæ			—		All other continents but Australia
50. Cervidæ... ...	—	—-	—	—	All regions but Ethiopian and Australian
52. Bovidæ	—	.--		—	Palæarctic, Ethiopian, Oriental
RODENTIA.					
55. Muridæ	—	—	—	—	Almost cosmopolite
57. Dipodidæ ...	—	—	—	—	Palæarctic, Ethiopian
59. *Saccomyidæ* ...	—	—	—	—	Mexican sub-region
60. Castoridæ ...	—	—	—	—	Palæarctic
61. Sciuridæ... ...	—	—	—	—	All regions but Australian

Order and Family.	Sub-regions.				Range beyond the Region.
	California.	Rocky Mntns.	Alleghanies.	Canada.	
62. *Haploodontidæ*	—				
66. Cercolabidæ ...	—	—	—	—	Neotropical
69. Lagomyidæ ...		—		—	Palæarctic
70. Leporidæ ...	—	—	—	—	All regions but Australian
MARSUPIALIA.					
76. Didelphyidæ...	—		—		Neotropical
BIRDS.					
PASSERES.					
1. Turdidæ... ...	—	—	—	—	Almost cosmopolite
2. Sylviidæ... ...	—	—	—	—	Almost cosmopolite
5. Cinclidæ ...		—		--	Palæarctic, Oriental, Andes
6. Troglodytidæ	—	—	—	—	All regions but Australian
7. *Chamœidæ* ...	—				
8. Certhiidæ ...	—	—	—	—	Palæarctic, Oriental, Australian
9. Sittidæ	—	—	—	—	Palæarctic, Oriental, Australian
10. Paridæ	--	—	—	—	The Eastern Hemisphere
19. Laniidæ	—	—	—	—	The Eastern Hemisphere
20. Corvidæ	—	—	—	—	Cosmopolite
26. (Cœrebidæ) ...		—			Neotropical family
27. Mniotiltidæ ...	—	—	—	—	Neotropical
28. Vireonidæ ...	—	—	—	—	Neotropical
29. Ampelidæ ...	—	—	—	—	Palæarctic, Antilles, Guatemala
30. Hirundinidæ...	—	--	—	--	Cosmopolite
31. Icteridæ... ...	—	—	—	—	Neotropical
32. Tanagridæ ...	—	—	—	—	Neotropical
33. Fringillidæ ...	—	—	—	—	All regions but Australian
37. Alaudidæ ...	—	—	—	—	All regions but Neotropical
38. Motacillidæ ...	—	—	—	—	Cosmopolite
39. Tyrannidæ ..	—	—	—	--	Neotropical
PICARIÆ.					
51. Picidæ	—	—	—	—	All regions but Australian
58. Cuculidæ ...	—	—	—		Almost cosmopolite
67. Alcedinidæ ...	--	—	—	—	Cosmopolite
73. Caprimulgidæ	—	—	—	—	Cosmopolite
74. Cypselidæ ...	—	—	—	—	Almost cosmopolite
75. Trochilidæ ...	--.	—	—	—	Neotropical
PSITTACI.					
80. Conuridæ ...		—			Neotropical
COLUMBÆ.					
84. Columbidæ ...	—	—	—	—	Cosmopolite
GALLINÆ.					
87. Tetraonidæ ...	—	—	—	—	Almost cosmopolite
88. Phasianidæ ...		—	—		Palæarctic, Oriental, Ethiopian, Honduras
91. (Cracidæ) ...		—			Neotropical

Order and Family.	Cali-fornia.	Rocky Mntns.	Alle-ghanies.	Canada.	Range beyond the Region.
ACCIPITRES.					
94. Vulturidæ ...	—	—	—		All regions but Australian
96. Falconidæ ...	—	—	—	—	Cosmopolite
97. Pandionidæ ..	—	·	—	—	Cosmopolite
98. Strigidæ ...	—	—	—	—	Cosmopolite
GRALLÆ.					
99. Rallidæ ...	—	—	—	—	Cosmopolite
100. Scolopacidæ...	—	—	—	—	Cosmopolite
105. Charadriidæ	—	—	—	—	Cosmopolite
107. Gruidæ ...	—	—	—		All regions but Neotropical
113. Ardeidæ ...	—	—	—	—	Cosmopolite
114. Plataleidæ ...	—	—	—	—	Almost cosmopolite
115. Ciconiidæ ...		—			All the regions
ANSERES.					
118. Anatidæ ...	—	—	—	—	Cosmopolite
119. Laridæ... ...	·	—	—	—	Cosmopolite
120. Procellariidæ	—	—	—	—	Cosmopolite
121. Pelecanidæ ...	—	—	—	—	Cosmopolite
123. Colymbidæ ...				—	North temperate and arctic zones
124. Podicipidæ ..	—	—	—	—	Cosmopolite
125. Alcidæ... ...	—			—	North temperate and arctic zones
REPTILIA.					
OPHIDIA.					
5. Calamariidæ ...	—	—	—		All the regions
6. Oligodontidæ...			—		Neotropical, Oriental, Japan
7. Colubridæ ...	—	—	—	—	Almost cosmopolite
8. Homalopsidæ			—		All the regions
17. Pythonidæ ...	—				All tropical regions
20. Elapidæ			—		All tropical regions, Japan
24. Crotalidæ ...	—	—	—	—	Neotropical, Palæarctic, Oriental
LACERTILIA.					
27. *Chirotidæ* ...		—	—		Mexico
32. Teidæ	—	—	—		Neotropical
34. Zonuridæ ...	—	—	—	—	All regions but Australian
35. Chalcidæ ...			—		Neotropical
45. Scincidæ ...	—	—	—		Almost cosmopolite
49. Geckotidæ ...	—	—	—		Almost cosmopolite
50. Iguanidæ ...	—	—	—		Neotropical
CROCODILIA.					
56. Alligatoridæ ...			—		Neotropical
CHELONIA.					
57. Testudinidæ ...	—	—	—	—	All continents but Australian
59. Trionychidæ ...			—		Ethiopian, Oriental, Japan
60. Cheloniidæ ...					Marine

Order and Family.	California.	Rocky Mtns.	Alle-ghanies.	Canada.	Range beyond the Region.
AMPHIBIA.					
URODELA.					
2. *Sirenidæ* ...			—		
3. Proteidæ ..			—		Palæarctic
4. *Amphiumidæ*			—		
5. Menopomidæ			—		Palæarctic
6. Salamandridæ	—	—	—	—	Andes, Palæarctic
ANOURA.					
10. Bufonidæ ...	—	—	—	—	All continents but Australia
12. Engystomidæ..			—		All regions but Nearctic
15. Alytidæ	—	—	—		All regions but Oriental
17. Hylidæ	—	—	—	—	All regions but Ethiopian
18. Polypedatidæ			—		All the regions
19. Ranidæ	—	—	—	—	Almost cosmopolite
FISHES (FRESH-WATER).					
ACANTHOPTERYGII.					
1. Gasterosteidæ	—	—	—	—	Palæarctic
3. Percidæ	—	—	—	—	Cosmopolite
4. *Aphredoderidæ*			—		
12. Sciænidæ ...	—	—	—	—	All regions but Australian
37. Atherinidæ ...	—	—	—	—	Palæarctic
PHYSOSTOMI.					
59. Siluridæ... ...	—	—	—	--	All warm regions
65. Salmonidæ ...	—	—	—	—	Palæarctic, New Zealand
66. *Percopsidæ* ...			—		
70. Esocidæ	—	—	—	—	Palæarctic
71. Umbridæ ...	—	—	—	—	Palæarctic
73. Cyprinodontidæ	—	—	—		All regions but Australian
74. *Heteropygii* ...			—		
75. Cyprinidæ ...	—	--	—	—	Not in S. America or Australia
77. *Hyodontidæ* ...	—	—	—	—	
GANOIDEI.					
93. *Amiidæ*	—	—	—	—	
95. *Lepidosteidæ* ...	--	—	—	—	
96. Accipenseridæ	—	—	—	—	Palæarctic
97. Polydontidæ ...			—		Palæarctic
INSECTS. LEPI-DOPTERA (PART)					
DIURNI (BUTTER-FLIES).					
1. Danaidæ ...	—	—	—	—	All warm regions
2. Satyridæ ...	—	—	—	—	Cosmopolite
7. (Heliconidæ)...			—		Neotropical

Order and Family.	Sub-regions.				Range beyond the Region.
	California.	Rocky Mntns.	Alleghanies.	Canada.	
8. Nymphalidæ ...	—	—	—	—	Cosmopolite
9. Libytheidæ ...		—	—	—	Not in Australia
12. Erycinidæ ...	—	—	—		Neotropical
13. Lycænidæ ...	—	—	—	—	Cosmopolite
14. Pieridæ	—	—	—	—	Cosmopolite
15. Papilionidæ ...	—	—	—	—	Cosmopolite
16. Hesperidæ ...	—	—	—	—	Cosmopolite
SPHINGIDEA.					
17. Zygænidæ ...	—	—	—	—	Cosmopolite
18. Castniidæ ...		—	—		Neotropical, Australian
22. Ægeriidæ ...	—	—	—	—	Not in Australia
23. Sphingidæ ...	—	—	—	—	Cosmopolite

TABLE II.

LIST OF GENERA OF TERRESTRIAL MAMMALIA AND BIRDS
INHABITING THE NEARCTIC REGION.

EXPLANATION.

Names in *italics* show genera peculiar to the region.
Names enclosed thus (...) indicate genera which barely enter the region, and are not
  considered properly to belong to it.|
Genera properly belonging to the region are numbered consecutively.

MAMMALIA.

Order, Family, and Genus.	No. of Species.	Range within the Region.	Range beyond the Region.
CHIROPTERA.			
PHYLLOSTOMIDÆ.			
1. Macrotus	1	California	Mexico, Antilles
VESPERTILIONIDÆ.			
2. Scotophilus ...	5	Universal, to Hudson's Bay	Neotr., Orient., Austral.
3. Vespertilio ...	6	Universal, to Hudson's Bay	Cosmopolite'
4. Nycticejus ..	1	South and East	India, Tropical Africa, temperate S. America
5. Lasiurus	3	Temp. N. Amer. to Nova Scotia	Tropical America
6. *Synotus*	2	S. E. and Central States	
7. *Autrozous* ...	1	W. Coast	
NOCTILIONIDÆ.			
8. Nyctinomus ...	1	Cal. and S. Central Sub-region	Neotropical, Oriental, S. Palæarctic
INSECTIVORA.			
TALPIDÆ.			
9. *Condylura* ...	1	Eastern N. America	
10. *Scapanus*	2	New York to San Francisco	
11. *Scalops*	3	S. of Great Lakes & Brit. Columb.	
12. Urotrichus ...	1	British Columbia	Japan
SORICIDÆ.			
13. Sorex	16	The whole region	Palæarc., Ethiop., Orien.
14. Neosorex	1	Vancouver's Island (a sub-genus)	
15. Blarina	7	Canada to Mexico (a sub-genus)	
CARNIVORA.			
FELIDÆ.			
16. Felis...	5	S. of 55° N. Latitude	All regs. but Australian
17. Lynx	3	S. of 56° N. Latitude	Palæarctic

Order, Family, and Genus.	No. of Species.	Range within the Region.	Range beyond the Region.
CANIDÆ.			
18. Lupus	6	All N. America	Palæarctic, Oriental
19. Vulpes	6	N. America to Arctic Ocean and Greenland	Palæarc , Ethiop., Orient.
MUSTELIDÆ.			
20. Martes	2	Pennsylvania to Paget's Sound	Palæarctic, Oriental
21. Mustela	11	All N. America	Peru, Palæarctic, Ethiopian, Oriental
22. Gulo...	1	Rocky Mountains and Canada	N. Palæarctic
23. Latax	2	United States and Canada	
24. Enhydris	1	Pacific coast	W. coast of S. America
25. Taxidea	2	Arkansas to 58° N. Lat.	
26. Mephitis	6	United States and Canada	Neotropical
PROCYONIDÆ.			
27. Procyon	2	Texas to Canada, California	Neotropical
28. Bassaris	1	California and Texas	Guatemala and Mexico
URSIDÆ.			
29. Ursus	3	N. America and Greenland	Palæarctic, Oriental
OTARIIDÆ.			
30. Callorhinus ...	1	Behring's Straits	Kamschatka
31. Zalophus	1	S. California to N. Pacific	Japan
Eumetopias ...	1	California to Behring's Straits	
TRICHECHIDÆ.			
32. Trichechus ...	1	Arctic Ocean to 66° N. Lat. in N. America	Palæarctic
PHOCIDÆ.			
33. Callocephalus ...	1	Greenland	Palæarctic
34. Pagomys	1	N. Atlantic and N. Pacific	Japan
35. Pagophilus	1	N. Atlantic and N. Pacific	Palæarctic
36. Halicyon	1	N. W. coast of America	
37. Phoca	1	Northern Coast	Palæarctic
38. Halichœrus ...	1	Greenland	Palæarctic
39. Morunga... ...	1	California	S. temperate shores
40. Cystophora ...		Greenland	N. Atlantic
UNGULATA.			
SUIDÆ.			
41. Dicotyles... ...	1	Texas to Red River, Arkansas	Neotropical
CERVIDÆ.			
42. Alces	1	N. E. United States & Canada	N. Palæarctic
43. Rangifer	2	Maine to Arctic Ocean & Greenl.	Arctic zone
44. Cervus	6	N. America to 57° N. Lat.	Neotr., Palæarc., Orien.
BOVIDÆ.			
45. Bison	1	Between Missouri & Rocky Mtns.	E. Europe
46. Antilocapra ...	1	Central plains from Rio Grande to British Columbia	

Order, Family, and Genus.	No. of Species.	Range within the Region.	Range beyond the Region.
47. *Aplocerus*	1	Northern Rocky Mountains	
48. Capra	1	Upper Missouri and Rocky Mountains northwards	Palæarctic
49. *Ovibos*	1	Arctic America and Greenland	
RODENTIA.			
MURIDÆ.			
50. Reithrodon ...	5	N. America to Lat. 39° N.	Neotropical
51. Hesperomys ..	16	Temperate N. America	Neotropical
52. *Neotoma*	7	Temperate N. America	
53. *Sigmodon*	2	S. and S. E. States	
54. Arvicola	27	Texas and California to Hudson's Bay	Palæarctic
55. Myodes	3	N. United States to Arctic Reg. and Greenland	N. Palæarctic
56. *Fiber*	1	All N. America	Mexico
DIPODIDÆ.			
57. *Jaculus*	1	Pennsylvania to Canada and California	
SACCOMYIDÆ.			
58. *Dipodomys*	5	New Mexico to Columbia River and Carolina	
59. *Perognathus* ...	6	New Mexico to British Columbia	
60. *Thomomys* ...	2	Upper Missouri to Hudson's Bay	
61. *Geomys*	5	New Mexico to Alabama and Nebraska	
62. *Saccomys*	1	N. America	
CASTORIDÆ.			
63. Castor	1	N. Mexico to Labrador	Palæarctic
SCIURIDÆ			
64. Sciurus	18	N. America to Labrador	All regs. but Australian
65. Sciuropterus ...	4	California & E. States northwds.	Palæarctic, Oriental
66. Tamias	4	Mexico and Virginia to Canada	Mexico, N. Asia
67. Spermophilus ...	15	N., W., & Central N. America	Palæarctic
68. *Cynomys*	2	Rio Grande to Missouri (Central)	
69. Arctomys ...	4	Virginia and Nebraska, northws.	N. Palæarctic
HAPLOODONTIDÆ.			
70. *Haploodon* ...	2	California and British Columbia	
CERCOLABIDÆ			
71. *Erethizon*	2	Pennsylvania to Canada, & Pacific coast	
LAGOMYIDÆ.			
72. Lagomys	1	Rocky Mountains, 42° to 60° N. Lat.	Palæarctic
LEPORIDÆ.			
73. Lepus	15	All N. America to Greenland	All regs. but Australian

L 2

Order, Family, and Genus.	No. of Species.	Range within the Region.	Range beyond the Region.
MARSUPIALIA.			
DIDELPHYIDÆ.			
74. Didelphys ...	2	From Hudson's River & Lower California, southward	Neotropical

BIRDS.

PASSERES.			
TURDIDÆ.			
1. Turdus	9	The whole region	Almost cosmopolite
2. Mimus	2	All U. States and to Canada	Neotropical
3. Galeoscoptes ...	1	E. of N. America	To Panama
4. *Oreoscoptes* ...	1	California and Rocky Mountains	Mexico
5. *Harporhynchus*	7	N. America, chiefly the west	Mexico
SYLVIIDÆ.			
6. Myiadestes ...	1	W. of Rocky Mountains and to Canada	Neotropical
7. *Sialia*	3	All United States and to Canada	Mexico and Guatemala
8. Regulus	3	All United States & to Labrador	Palæarc., Cent. America
9. Polioptila... ...	3	Central and Southern U. States	Neotropical
CINCLIDÆ.			
10. Cinclus	1	Rocky Mountains and British America	Andes, Palæarctic
TROGLODYTIDÆ.			
11. Troglodytes ...	3	N. America	Neotropical, Palæarctic
12. Thryophilus ...	1	N. W. America	Neotropical
13. Thryothorus ...	3	All N. America	Neotropical
14. Cistothorus ...	2	N. America	Neotropical
(Campylor-) hynchus } ...	1	Gila and Rio Grande)	Neotropical genus
15. *Salpinctes* ...	1	Rocky Mountains to Oregon	
16. *Catherpes*... ...	1	Gila and Colorado	
CHAMÆIDÆ.			
17. *Chamæa*	1	California	
CERTHIIDÆ.			
18. Certhia	2	All United States and Canada	Palæarctic, Guatemala
SITTIDÆ.			
19. Sitta...	5	All United States and Canada	Palæarctic, Mexico
PARIDÆ.			
20. Parus	8	All United States and Canada	Palæarc., Orien., Mexico
21. Lophophanes ...	4	All United States	Palæarctic, Mexico
22. *Psaltriparus* ...	3	Central & Western N. America	Mexico and Guatemala
23. *Auriparus* ...	1	Rio Grande Valley	

Order, Family, and Genus.	No. of Species.	Range within the Region.	Range beyond the Region.
LANIIDÆ.			
24. Lanius	4	All N. America	Palæarc., Ethio., Orient.
CORVIDÆ.			
25. Perisoreus ...	1	Canada and Rocky Mountains	Palæarctic
26. Cyanocitta ...	9	All United States and to Canada	Neotropical
27. *Gymnokitta* ...	1	Central and N. W. States	
28. *Picicorvus* ...	1	Central and Western States to Sitka	
29. Pica	2	Central and Western States to Arctic Ocean	Palæarctic
30. Corvus	7	All N. America	Cosmop , excl. S. Amer.
CŒREBIDÆ.			
(Certhiola ...	1	Florida ; summer migrant)	Neotropical genus
MNIOTILTIDÆ.			
31. *Mniotilta* ...	1	Eastern States	Antilles, Andes of Columbia (migrant)
32. Parula	1	Eastern States and Canada	Neotropical
33. Protonotaria ...	1	Ohio and southwards	Neotrop. to Venezuela
34. *Helminthophaga*	8	All N. America	Mexico to Columbia
35. *Helmintherus* ...	2	S. and E. States to Canada	Mexico to Veragua
36. Perissoglossa	1	Eastern United States	Antilles
37. Dendrœca ...	22	All N. America	Mex. to Ecuador & Chili
38. *Oporornis*... ...	2	Eastern States	Guatemala and Panama
39. Geothlypis ...	4	All N. America	Neotropical
40. Setophaga ...	2	E. States & Canadian sub-region	Neotropical
41. *Myiodioctes* ...	5	United States and Canada	Mex. to Columb. (migr.)
42. Siurus	3	S. and E. States to Canada	Mexico to Columbia
43. *Icteria*	2	E. and Central States to Canada	Mexico to Costa Rica
VIREONIDÆ.			
44. Vireosylvia ...	7	All N. America	Antilles and Venezuela
45. Vireo	6	All United States	Antilles and Costa Rica
AMPELIDÆ.			
46. Ampelis	2	All N. America	Palæarctic, Guatemala
47. *Phœnopepla* ...	1	Gila and Lower Colorado	Mexico
HIRUNDINIDÆ.			
48. Hirundo	3	All N. America	Almost cosmopolite
49. Petrochelidon ...	1	All N. America	Neotropical
50. Cotyle	1	All N. America	All regs. but Australian
51. Stelgidopteryx	1	Southern States	Neotropical
52. Progne	1	All N. America	Neotropical
ICTERIDÆ.			
53. Icterus	7	All United States and Canada	Neotropical
54. Dolichonyx ...	1	Eastern States and Canada	Neotropical
55. Molothrus ...	1	All United States and Canada	Neotropical
56. Agelæus	3	All United States and Canada	Neotropical

Order, Family, and Genus.	No. of Species.	Range within the Region.	Range beyond the Region.
57. *Xanthocephalus*	1	The whole region	Mexico
58. Sturnella ...	2	All United States and Canada	Neotropical
59. *Scolecophagus*	2	All United States and Canada	Mexico
60. Quiscalus ...	4	S. and E. States to Labrador	Mexico to Venezuela
TANAGRIDÆ.			
61. Pyranga... ...	4	United States and Canada	Neotropical
FRINGILLIDÆ.			
62. Chrysomitris...	7	The whole region	Neotropical, Palæarctic
63. Coccothraustes	1	W. and N. W. America	Palæarctic, Guatemala
64. Embernagra ...	1	Rocky Mountain district	Neotropical
65. *Pipilo*	9	All N. America	Mexico and Guatemala
66. *Junco*	5	All United States	Mexico and Guatemala
67. Zonotrichia ...	5	The whole region	Neotropical
68. *Melospiza* ...	7	All United States to Sitka	Mexico and Guatemala
69. *Spizella*... ...	6	N. America	Mexico and Guatemala
70. *Passerella* ...	3	The whole region	Northern Asia
71. *Passerculus* ...	6	The whole region	Mexico and Guatemala
72. *Pocecetes*... ...	1	All United States	Mexico
73. *Ammodromus*	3	All United States	Mexico and Guatemala
74. Coturniculus...	3	E. and N. of N. America	Neotropical
75. Peucæa... ...	3	S. Atlantic States and California	Mexico
76. *Cyanospiza* ...	5	All United States to Canada	Central American
77. Poospiza ...	2	California and S. Central States	Neotropical
78. Carpodacus ...	5	The whole region	Mexico, Palæarctic
79. Cardinalis ...	1	S. and S. Central States	Mexico to Venezuela
80. *Pyrrhuloxia* ..	1	Texas and Rio Grande	
81. Guiraca	1	Southern States	Neotropical
82. Hedymeles ...	2	All United States	Mexico to Columbia
(Spermophila	1	Texas)	Neotropical genus
83. Loxia	2	N. of Pennsylvania	Palæarctic
84. Pinicola... ...	1	Boreal America	Palæarctic
85. Linota	2	E. and N. of N. America	Palæarctic
86. Leucosticte ...	4	Alaska to Utah	Palæarctic
87. *Calamospiza* ...	1	Arizona and Texas to Mexico	Mexico
88. *Chondestes* . .	1	Western, Cen., & Southern States	Mexico
89. Euspiza... ...	2	S. Eastern States	Palæarc., Columb. (mig.)
90. Plectrophanes	6	Boreal America and E. side of Rocky Mountains	Palæarctic
91. *Centronyx* ...	1	Mouth of Yellowstone River	
ALAUDIDÆ.			
92. Otocorys ...	1	High central plains to E. States and Canada	Palæarc., Mexico, Andes of Columbia
MOTACILLIDÆ.			
93. Anthus	1	The whole region	Cosmopolite
94. *Neocorys*... ...	1	Nebraska	
TYRANNIDÆ.			
95. Sayornis	3	E. States to Canada, California	Mexico to Ecuador
(Pyrocephalus	1	Gila and Rio Grande)	Neotropical
96. Empidonax ...	7	The whole region	Mexico to Ecuador

Order, Family, and Genus.	No. of Species.	Range within the Region.	Range beyond the Region.
97. Contopus ...	3	N. and E. of Rocky Mountains	Mexico to Amazonia
98. Myiarchus ...	2	E. and W. coasts and Canada	Neotropical
99. *Empidias* ...	1	Eastern States	Mexico
100. Tyrannus ...	4	All United States to Canada	Neotropical
(Milvulus ...	1	Texas)	Neotropical genus
PICARIÆ.			
PICIDÆ.			
101. Picoides ...	3	Arctic zone and Rocky Mounts.	Palæarctic
102. Picus	6	All United States and Canada	All regs. but Eth. & Aus.
103. *Sphyrapicus* ...	6	Brit. Columbia and Pennsylvania southwards	Mexico and Guatemala
104. Campephilus...	2	United States and Canada	Neotropical
105. *Hylatomus* ...	1	E. and W. States and Canada	
106. Centurus ...	3	The whole region	Mexico to Venezuela
107. Melanerpes ...	3	United States and S. Canada	Neotropical
108. Colaptes ...	3	United States and Canada	Neotropical
CUCULIDÆ.			
109. Crotophaga ...	2	E. States from Pennsylvania S.	Neotropical
110. Coccyzus ...	3	S. E. and Cen. States to Canada	Neotropical
111. Geococcyx ...	1	California to New Mex. & Texas	Guatemala
ALCEDINIDÆ.			
112. Ceryle	2	The whole region	Neotropical, S. Palæarctic, Oriental
CAPRIMULGIDÆ.			
113. Chordeiles ...	3	All United States to Canada	Neotropical
114. Antrostomus...	3	All United States to Canada	Neotropical
CYPSELIDÆ.			
115. Nephœcetes ...	1	N. W. America	Jamaica
116. Chætura ...	2	All U. States & British Columbia	Almost cosmopolite
TROCHILIDÆ.			
117. *Trochilus* ..	2	The whole region	Mexico to Veragua (? mi.)
118. Selasphorus ...	2	W. coast and Centre	Mexico to Veragua
119. *Atthis*	2	California and Colorado Valley	Mexico to Guatemala
PSITTACI.			
CONURIDÆ.			
120. Conurus... ..	1	S. and S. E. States	Neotropical
COLUMBÆ.			
COLUMBIDÆ.			
121. Columba ...	3	W. and Central States to Canada	All regs. but Australian
122. *Ectopistes* ...	1	E. coast to Cen. plains, Canada and British Columbia	
123. Melopelia ...	1	W. and S. Central States	Neotropical
124. Zenaidura ...	1	All United States to Canada	Mexico to Veragua
125. Chæmepelia ..	1	California and S. E. States	Neotropical

Order, Family, and Genus.	No. of Species.	Range within the Region.	Range beyond the Region.
GALLINÆ.			
TETRAONIDÆ.			
126. Cyrotonyx ..	1	S. Central States	Mexico and Guatemala
127. Ortyx	5	All United States and to Canada	Mexico to Honduras and Costa Rica
128. Callipepla ..	1	California	Mexico
129. *Lophortyx* ...	2	Arizona and California	
130. *Oreortyx* ...	1	California and Oregon	
131. Tetrao	3	N. and N. W. America	Palæarctic
132. *Centrocercus* ...	1	Rocky Mountains	
133. *Pediocætes* ...	2	N. and N. W. America	
134. *Cupidonia* ...	1	E. & N. Cen. States and Canada	
135. Bonasa... ...	1	N. United States and Canada	Palæarctic
136. Lagopus... ...	4	Arctic zone and to 39° N. Lat. in Rocky Mountains	Palæarctic
PHASIANIDÆ.			
137. *Meleagris* ...	2	E. and Central States to Canada	Mexico, Honduras
CRACIDÆ.			
(Ortalida ...	1	New Mexico)	Neotropical genus
ACCIPITRES.			
VULTURIDÆ.			
Sub-Family (CATHARTINÆ.)			
138. Catharista ...	1	United States to 40° N. Lat.	Neotropical
139. Psuedogryphis	2	United States to 49° N. Lat.	Neotropical
FALCONIDÆ.			
140. Polyborus ...	1	S. States to Florida & California	Neotropical
141. Circus	1	All N. America	Nearly cosmopolite
142. Antenor... ...	2	California and Texas	Neotropical
143. Astur	1	All N. America	Almost cosmopolite
144. Accipiter ...	3	All temperate N. America	Almost cosmopolite
145. Tachytriorchis	1	New Mexico to California	Neotropical
146. Buteo	12	All N. America	All regs. but Australian
147. Archibuteo ...	3	All N. America	N. Palæarctic
148. Asturina ...	1	S. E. States	Neotropical
149. Aquila... ...	1	The whole region	Palæarc., Ethiop.,Indian
150. Haliæetus ...	2	All N. America	All regs. but Neotropical
151. Nauclerus ..	1	E. coast to Pennsylvania and Wisconsin	Neotropical
(Rostrhamus	1	Florida)	Neotropical
152. Elanus	1	Southern and Western States	Tropical regions
153. Ictinia	1	Southern States	Neotropical
154. Falco	7	The whole region	Almost cosmopolite
155. Hierofalco ...	2	N. of N. America	N. Palæarctic
156. Cerchneis ...	1	All N. America	Almost cosmopolite
PANDIONIDÆ.			
157. Pandion... ...	1	Temperate N. America	Cosmopolite

Order, Family, and Genus.	No. of Species.	Range within the Region.	Range beyond the Region.
STRIGIDÆ.			
158. Surnia	1	Arctic & N. Temperate America	N. Palæarctic
159. Nyctea	1	S. Carolina to Greenland	N. Palæarctic
160. Glaucidium ...	1	Oregon and California	Neotropical, Palæarctic
161. *Micrathene* ...	1	Arizona and New Mexico	Mexico
162. Pholeoptynx...	1	N. W. America, Texas	Neotropical
163. Bubo	1	All N. America	All regs. but Australian
164. Scops	2	The whole region	Almost cosmopolite
165. Syrnium ...	2	E. States, California, Canada	All regs. but Australian
166. Asio	2	The whole region	All regs. but Australian
167. Nyctale	3	All N. America	Palæarctic
168. Strix	1	Temperate N. America	Almost cosmopolite

Peculiar or very Characteristic Genera of Wading and Swimming Birds.

GRALLÆ.			
SCOLOPACIDÆ.			
Micropelma ...	1	N. America	Andes to Chili
Philohela ...	1	Eastern States to Canada	
CHARADRIIDÆ.			
Aphriza... ...	1	W. coast of America	West of S. America
ANSERES.			
ANATIDÆ.			
Aix	1	N. America	China
Bucephala ...	4	N. America	Europe
Œdemia	3	N. America	Europe
Harelda	1	Arctic	Arctic Seas
Somateria ...	5	Arctic	North Palæarctic
Camptolæmus	1	N. E. America (? extinct)	
LARIDÆ.			
Creagrus ...	1	California and N. Pacific coasts	

CHAPTER XVI.

HAVING now closed our survey of the animal life of the whole earth—a survey which has necessarily been encumbered with a multiplicity of detail—we proceed to summarize the general conclusions at which we have arrived, with regard to the past history and mutual relations of the great regions into which we have divided the land surface of the globe.

All the palæontological, no less than the geological and physical evidence, at present available, points to the great land masses of the Northern Hemisphere as being of immense antiquity, and as the area in which the higher forms of life were developed. In going back through the long series of the Tertiary formations, in Europe, Asia, and North America, we find a continuous succession of vertebrate forms, including all the highest types now existing or that have existed on the earth. These extinct animals comprise ancestors or forerunners of all the chief forms now living in the Northern Hemisphere; and as we go back farther and farther into the past, we meet with ancestral forms of those types also, which are now either confined to, or specially characteristic of, the land masses of the Southern Hemisphere. Not only do we find that elephants, and rhinoceroses, and hippopotami, were once far more abundant in Europe than they are now in the tropics, but we also find that the apes of West Africa and Malaya, the lemurs of Madagascar, the Edentata of Africa and South America, and the

Marsupials of America and Australia, were all represented in Europe (and probably also in North America) during the earlier part of the Tertiary epoch. These facts, taken in their entirety, lead us to conclude that, during the whole of the Tertiary and perhaps during much of the Secondary periods, the great land masses of the earth were, as now, situated in the Northern Hemisphere; and that here alone were developed the successive types of vertebrata from the lowest to the highest. In the Southern Hemisphere there appear to have been three considerable and very ancient land masses, varying in extent from time to time, but always keeping distinct from each other, and represented, more or less completely, by Australia, South Africa, and South America of our time. Into these flowed successive waves of life, as they each in turn became temporarily united with some part of the northern land. Australia appears to have had but one such union, perhaps during the middle or latter part of the Secondary epoch, when it received the ancestors of its Monotremata and Marsupials, which it has since developed into a great variety of forms. The South African and South American lands, on the other hand, appear each to have had several successive unions and separations, allowing first of the influx of low forms only (Edentata, Insectivora and Lemurs); subsequently of Rodents and small Carnivora, and, latest of all, of the higher types of Primates, Carnivora and Ungulata.

During the whole of the Tertiary period, at least, the Northern Hemisphere appears to have been divided, as now, into an Eastern and a Western continent; always approximating and sometimes united towards the north, and then admitting of much interchange of their respective faunas; but on the whole keeping distinct, and each developing its own special family and generic types, of equally high grade, and generally belonging to the same Orders. During the Eocene and Miocene periods, the distinction of the Palæarctic and Nearctic regions was better marked than it is now; as is shown by the floras no less than by the faunas of those epochs. Dr. Newberry, in his Report on the Cretaceous and Tertiary floras of the Yellowstone and Missouri Rivers, states, that although the Miocene flora of Central North

America corresponds generally with that of the European Miocene, yet many of the tropical, and especially the Australian types, such as *Hakea* and *Dryandra*, are absent. Owing to the recent discovery of a rich Cretaceous flora in North America, probably of the same age as that of Aix-la-Chapelle in Europe, we are able to continue the comparison; and it appears, that at this early period the difference was still more marked. The predominant feature of the European Cretaceous flora seems to have been the abundance of Proteaceæ, of which seven genera now living in Australia or the Cape of Good Hope have been recognised, besides others which are extinct. There are also several species of *Pandanus*, or screw-pine, now confined to the tropics of the Eastern Hemisphere, and along with these, oaks, pines, and other more temperate forms. The North American Cretaceous flora, although far richer than that of Europe, contains no Proteaceæ or *Pandani*, but immense numbers of forest trees of living and extinct genera. Among the former we have oaks, beeches, willows, planes, alders, dog-wood, and cypress; together with such American forms as magnolias, sassafras, and liriodendrons. There are also a few not now found in America, as *Araucaria* and *Cinnamomum*, the latter still living in Japan. This remarkable flora has been found over a wide extent of country—New Jersey, Alabama, Kansas, and near the sources of the Missouri in the latitude of Quebec—so that we can hardly impute its peculiarly temperate character to the great elevation of so large an area. The intervening Eocene flora approximates closely, in North America, to that of the Miocene period; while in Europe it seems to have been fully as tropical in character as that of the preceding Cretaceous period; fruits of *Nipa*, *Pandanus*, *Anona*, *Acacia*, and many Proteaceæ, occurring in the London clay at the mouth of the Thames.

These facts appear, at first sight, to be inconsistent, unless we suppose the climates of Europe and North America to have been widely different in these early times; but they may perhaps be harmonised, on the supposition of a more uniform and a somewhat milder climate then prevailing over the whole Northern Hemisphere; the contrast in the vegetation of these countries being due to a radical difference of type, and therefore not indicative of climate. The early European flora seems to have been a portion of that which now exists only in the tropical and sub-tropical lands of the Eastern Hemisphere; and, as much of this flora still survives in Australia, Tasmania, Japan, and the Cape of Good Hope, it does not necessarily imply more than a warm and equable temperate climate. The early North American flora, on the other hand, seems to have been essentially the same in type as that which now exists there, and which, in the Miocene period, was well represented in Europe; and it is such as now flourishes best in the warmer parts of the United States. But whatever conclusion we may arrive at on the question of climate, there can be no doubt as to the distinctness of the floras of the ancient Nearctic and Palæarctic regions; and the view derived from our study of their existing and extinct faunas—that these two regions have, in past times, been more clearly separated than they are now—receives strong support from the unexpected evidence now obtained as to the character and mutations of their vegetable forms, during so vast an epoch as is comprised in the whole duration of the Tertiary period.

The general phenomena of the distribution of living animals, combined with the evidence of extinct forms, lead us to conclude that the Palæarctic region of early Tertiary times was, for the most part, situated beyond the tropics, although it probably had a greater southward extension than at the present time. It certainly included much of North Africa, and perhaps reached far into what is now the Sahara; while a southward extension of its central mass may have included the Abyssinian highlands, where some truly Palæarctic forms are still found. This is rendered probable by the fossils of Perim Island a little further east, which show that the characteristic Miocene fauna of South Europe and North India prevailed so far within the tropics. There existed, however, at the extreme eastern and western limits of the region, two extensive equatorial land-areas, our Indo-Malayan and West African sub-regions—both of which must have been united for more or less considerable periods with the northern continent. They would then have received

from it such of the higher vertebrates as were best adapted for
the peculiar climatal and organic conditions which everywhere
prevail near the equator; and these would be preserved, under
variously modified forms, when they had ceased to exist in
the less favourable and constantly deteriorating climate of the
north.   At later epochs, both these equatorial lands became
united to some part of the great South African continent (then
including Madagascar), and we thus have explained many of
the similarities presented by the faunas of these distant, and
generally very different countries.

During the Miocene period, when a subtropical climate prevailed over much of Europe and Central Asia, there would be no
such marked contrast as now prevails between temperate and
tropical zones; and at this time much of our Oriental region,
perhaps, formed a hardly separable portion of the great Palæarctic
land.   But when, from unknown causes, the climate of Europe
became less genial, and when the elevation of the Himalayan
chain and the Mongolian plateau caused an abrupt difference of
climate on the northern and southern sides of that great mountain barrier, a tropical and a temperate region were necessarily
formed; and many of the animals which once roamed over the
greater part of the older and more extensive region, now became
restricted to its southern or northern divisions respectively.
Then came the great change we have already described (vol. i.
p. 288), opening the newly-formed plains of Central Africa to the
incursions of the higher forms of Europe; and following on this,
a still further deterioration of climate, resulting in that marked
contrast between temperate and tropical faunas, which is now one
of the most prominent features in the distribution of animal as
well as of vegetable forms.

It is not necessary to go into any further details here, as we
have already, in our discussion of the origin of the fauna of the
several regions, pointed out what changes most probably occurred
in each case.   These details are, however, to a great extent
speculative; and they must remain so till we obtain as much
knowledge of the extinct faunas and past geological history of
the southern lands, as we have of those of Europe and North

America. But the broad conclusions at which we have now arrived seem to rest on a sufficiently extensive basis of facts; and they lead us to a clearer conception of the mutual relations and comparative importance of the several regions than could be obtained at an earlier stage of our inquiries.

If our views of the origin of the several regions are correct, it is clear that no mere binary division—into north and south, or into east and west—can be altogether satisfactory, since at the dawn of the Tertiary period we still find our six regions, or what may be termed the rudiments of them, already established. The north and south division truly represents the fact, that the great northern continents are the seat and birth-place of all the higher forms of life, while the southern continents have derived the greater part, if not the whole, of their vertebrate fauna from the north; but it implies the erroneous conclusion, that the chief southern lands—Australia and South America—are more closely related to each other than to the northern continent. The fact, however, is that the fauna of each has been derived, independently, and perhaps at very different times, from the north, with which they therefore have a true genetic relation; while any intercommunion between themselves has been comparatively recent and superficial, and has in no way modified the great features of animal life in each. The east and west division, represents—according to our views—a more fundamental diversity; since we find the northern continent itself so divided in the earliest Eocene, and even in Cretaceous times; while we have the strongest proof that South America was peopled from the Nearctic, and Australia and Africa from the Palæarctic region: hence, the Eastern and Western Hemispheres are the two great branches of the tree of life of our globe. But this division, taken by itself, would obscure the facts—firstly, of the close relation and parallelism of the Nearctic and Palæarctic regions, not only now but as far back as we can clearly trace them in the past; and, secondly, of the existing radical diversity of the Australian region from the rest of the Eastern Hemisphere.

Owing to the much greater extent of the old Palæarctic region (including our Oriental), and the greater diversity of

Mammalia it appears to have produced, we can have little doubt
that here was the earliest seat of the development of the
vertebrate type; and probably of the higher forms of insects
and land-molluscs. Whether the Nearctic region ever formed
one mass with it, or only received successive immigrations from
it by northern land-connections both in an easterly and westerly
direction, we cannot decide; but the latter seems the most
probable supposition. In any case, we must concede the first
rank to the Palæarctic and Oriental regions, as representing the
most important part of what seems always to have been the
Great Continent of the earth, and the source from which all the
other regions were supplied with the higher forms of life. These
once formed a single great region, which has been since divided
into a temperate and a tropical portion, now sufficiently distinct;
while the Nearctic region has, by deterioration of climate,
suffered a considerable diminution of productive area, and
has in consequence lost a number of its more remarkable forms.
The two temperate regions have thus come to resemble each
other more than they once did, while the Oriental retains
more of the zoological aspect of the great northern regions
of Miocene times. The Ethiopian, from having been once an
insular region, where lower types of vertebrates alone prevailed,
has been so overrun with higher types from the old Palæarctic
and Oriental lands that it now rivals, or even surpasses, the
Oriental region in its representation of the ancient fauna of
the great northern continent. Both of our tropical regions of
the Eastern Hemisphere possess faunas which are, to some
extent, composite, being made up in different proportions of
the productions of the northern and southern continents,—the
former prevailing largely in the Oriental, while the latter
constitutes an important feature in the Ethiopian fauna. The
Neotropical region has probably undergone great fluctuations
in early times; but it was, undoubtedly, for long periods completely isolated, and then developed the Edentate type of
Mammals and the Formicaroid type of Passerine birds into
a variety of forms, comparable with the diversified Marsupials
of Australia, and typical Passeres of the Eastern Hemisphere.

It has, however, received successive infusions of higher types from the north, which now mingle in various degrees with its lower forms. At an early period it must have received a low form of Primates, which has been developed into the two peculiar families of American monkeys; while its llamas, tapirs, deer, and peccaries, came in at a later date, and its opossums and extinct horses probably among the latest. The Australian region alone, after having been united with the great northern continent at a very early date (probably during the Secondary period) has ever since remained more or less completely isolated; and thus exhibits the development of a primeval type of mammal, almost wholly uninfluenced by any incursions of a later and higher type. In this respect it is unique among all the great regions of the earth.

We see, then, that each of our six regions has had a history of its own, the main outlines of which we have been able to trace with tolerable certainty. Each of them is now characterised—as it seems to have been in all past time of which we have any tolerably full record—by well-marked zoological features; while all are connected and related in the complex modes we have endeavoured to unravel. To combine any two or more of these regions, on account of existing similarities which are, for the most part, of recent origin, would obscure some of the most important and interesting features of their past history and present condition. And it seems no less impracticable to combine the whole into groups of higher rank; since it has been shown that there are two opposing modes of doing this, and that each of them represents but one aspect of a problem, which can only be solved by giving equal attention to all its aspects.

For reasons which have been already stated, and which are sufficiently obvious, we have relied almost exclusively on the distribution of living and extinct mammalia, in arriving at these conclusions. But we believe they will apply equally to elucidate the phenomena presented by the distribution of all terrestrial organisms, when combined with a careful consideration of the various means of dispersal of the different groups, and the comparative longevity of their species and genera. Even insects, which are perhaps of all animals the farthest removed from mammalia in this respect, agree, in the great outlines of their distribution, with the vertebrate orders. The Regions are admittedly the same, or nearly the same for both; and the discrepancies that occur are of a nature which can be explained by two undoubted facts—the greater antiquity, and the greater facilities for dispersal, of insects.

But this principle, if sound, must be carried farther, and be applied to plants also. There are not wanting indications that this may be successfully done; and it seems not improbable, that the reason why botanists have hitherto failed to determine, with any unanimity, which are the most natural phytological regions, and to work out any connected theory of the migrations of plants, is, because they have not been furnished with the clue to the past changes of the great land masses, which could only be arrived at by such an examination of the past and present distribution of the higher animals as has been here attempted. The difficulties in the way of the study of the distribution of plants, from this point of view, will be undoubtedly very great; owing to the unusual facilities for distribution many of them possess, and the absence of any group which might take the place of the mammalia among animals, and serve as a guide and standard for the rest. We cannot expect the regions to be so well defined in the case of plants as in that of animals; and there are sure to be many anomalies and discrepancies, which will require long study to unravel. The Six Great Regions here adopted, are however, as a whole, very well characterised by their vegetable forms. The floras of tropical America, of Australia, of South Africa, and of Indo-Malaya, stand out with as much individuality as do the faunas; while the plants of the Palæarctic and Nearctic regions, exhibit resemblances and diversities, of a character not unlike those found among the animals.

This is not a mere question of applying to the vegetable kingdom a series of arbitrary divisions of the earth which have been found useful to zoologists; for it really involves a fundamental problem in the theory of evolution. The question we have to answer, is, firstly—whether the distribution of plants is, like that of animals, mainly and primarily dependent on the past revolutions of the earth's surface; or, whether other, and altogether distinct causes, have had a preponderating influence in determining the range and limits of vegetable forms; and, secondly—whether those revolutions have been, in their general outlines, correctly interpreted by means of a study of the distribution and affinities of the higher animals. The first question is one for botanists alone to answer; but, on the second point, the author ventures to hope for an affirmative reply, from such of his readers as will weigh carefully the facts and arguments he has adduced.

The remaining part of this volume, will consist, of a systematic review of the distribution of each family of animals, and an application of the principles already established to elucidate the chief phenomena they present. The present chapter must, therefore, be considered as the conclusion of the argumentative and theoretical part of the present work; but it must be read in connection with the various discussions in Parts II. and III., in which the conclusions to be drawn from the several groups of facts have been successively given;—and especially in connection with the general observations at the end of each of the six chapters on the Zoological Regions.

The hypothetical view, as to the more recent of the great Geographical changes of the Earth's surface, here set forth, is not the result of any preconceived theory, but has grown out of a careful study of the facts accumulated, and has led to a considerable modification of the author's previous views. It may be described, as an application of the general theory of Evolution, to solve the problem of the distribution of animals; but it also furnishes some independent support to that theory, both by showing what a great variety of curious facts are explained by its means, and by answering some of the objections,

M 2

which have been founded on supposed difficulties in the distribution of animals in space and time.

It also illustrates and supports the geological doctrine, of the general permanence of our great continents and oceans, by showing how many facts in the distribution of animals can only be explained and understood on such a supposition; and it exhibits, in a striking manner, the enormous influence of the Glacial epoch, in determining the existing zoological features of the various continents.

And, lastly, it furnishes a more consistent and intelligible idea than has yet been reached by any other mode of investigation, of all the more important changes of the earth's surface that have probably occurred during the entire Tertiary period; and of the influence of these changes, in bringing about the general features, as well as many of the more interesting details and puzzling anomalies, of the Geographical Distribution of Animals.

PART IV.

GEOGRAPHICAL ZOOLOGY:

A SYSTEMATIC SKETCH OF THE CHIEF FAMILIES OF LAND ANIMALS IN THEIR GEOGRAPHICAL RELATIONS.

INTRODUCTION.

In the preceding part of our work, we have discussed the geographical distribution of animals from the point of view of the geographer; taking the different regions of the earth in succession, and giving as full an account as our space would permit of their chief forms of animal life. Now, we proceed from the standpoint of the systematic zoologist; taking in succession each of the families with which we deal, and giving an account of the distribution, both of the entire family and, as far as practicable, of each of the genera of which it is composed. As in the former part, our mode of treatment led us to speculate on the past changes of the earth's surface; so here we shall endeavour to elucidate the past migrations of animals, and thus, to some extent, account for their actual distribution.

The tabular headings, showing the range of the family in each region, will enable the reader to determine at a glance the general distribution of the group, as soon as he has familiarised himself, by a study of our general and regional maps, with the limits of the regions and sub-regions, and the figures (1 to 4) by which the latter are indicated. Much pains have been taken, to give the number of the known genera and species in each family, correctly; but these numbers must, in most cases, only be looked upon as approximations; because, owing to constant accessions of fresh material on the one hand, and the discovery that many supposed species are only varieties, on the other, such statistics are in a continual state of fluctuation. In the number of genera there is the greatest uncertainty; as will be seen by the two sets of numbers sometimes given, which denote the genera according to different modern authorities.

There is also a considerable difference in the dependence to be placed on the details given in the different classes of animals. In Mammalia and Birds some degree of accuracy has, it is hoped, been attained; the classification of these groups being much advanced, and the materials for their study ample. In Reptiles this is not the case, as there is no recently published work dealing with the whole subject, or with either of the larger orders. An immense number of new species and new genera of snakes and lizards, have been described in the last twenty years; and Dr. Günther—our greatest authority on reptiles in this country—has kindly assisted me in incorporating such of these as are most trustworthy, in a general system; but until entire Orders have been described or catalogued on a uniform plan, nothing more than a general approximation to the truth can be arrived at. Still, so many of the groups are well defined, and have a clearly limited distribution, that some interesting and valuable comparisons may be made.

For Fishes, the valuable " Catalogue " of Dr. Günther was available, and it has rarely been attempted to go beyond it. A large number of new species have since been described, in all parts of the world; but it is impossible to say how many of these are really new, or what genera they actually belong to. The part devoted to this Class is, therefore, practically a summary of Dr. Günther's Catalogue; and it is believed that the discoveries since made will not materially invalidate the conclusions to be drawn from such a large number of species, which have been critically examined and classified on a uniform system by one of our most able naturalists. When a supplement to this catalogue is issued, it will be easier to make the necessary alterations in distribution, than if a mass of untrustworthy materials had been mixed up with it.

For Insects, excellent materials are furnished, in the Catalogue of Mr. Kirby for Butterflies and in that of Drs. Gemminger and Harold for Coleoptera. I have also made use of some recently published memoirs on the Insects of Japan and St. Helena, and a few other recent works; and have, I believe, elaborated a more extensive series of facts to illustrate the distribution of insects, than has been made use of by any previous writer. Several discussions on the bearing of the facts of insect distribution, will also be found under the several Regions, in the preceding part of this work.

Terrestrial Mollusca form a group, as to the treatment of which I have most misgivings; owing to my almost entire ignorance of Malacology, and the great changes recently made in the classification of shells. There is also much uncertainty as to genera and sub-genera, which is very puzzling to one who merely wishes to get at general results. Finding it impossible to incorporate the new matter with the old, or to harmonise the different classifications of modern conchologists, I thought it better to confine myself to the standard works of Martens and Pfeiffer, with such additions of new species as I could make without fear of going far wrong. In some cases I have made use of recent monographs—especially on the shells of Europe, North America, the West Indian Islands, and the Sandwich Islands; and have, I venture to hope, not fallen into much error in the general conclusions at which I have arrived.

CHAPTER XVII.

THE DISTRIBUTION OF THE FAMILIES AND GENERA OF MAMMALIA.

Order I.—PRIMATES.

FAMILY 1.—SIMIIDÆ. (4 Genera, 12 Species).

GENERAL DISTRIBUTION.					
NEOTROPICAL SUB-REGIONS.	NEARCTIC SUB-REGIONS.	PALÆARCTIC SUB-REGIONS.	ETHIOPIAN SUB-REGIONS.	ORIENTAL SUB-REGIONS.	AUSTRALIAN SUB-REGIONS.
— — — —	— — ·· —	— — — —	— 2 — —	— — 3 . 4	— — — —

THE Simiidæ, or Anthropoid Apes, comprehend those forms of the monkey-tribe which, in general organization, approach nearest to man.   They inhabit the tropics of the Old World, and are most abundant near the equator; but they are limited to certain districts, being quite unknown in eastern and southern Africa, and the whole peninsula of Hindostan.

The genus *Troglodytes* (or *Mimetes*, as it is sometimes named) comprehends the chimpanzee and gorilla.   It is confined to the West African sub-region, being found on the coast about 12° North and South of the equator, from the Gambia to Benguela, and as far inland as the great equatorial forests extend.   There are perhaps other species of chimpanzee; since Livingstone met with what he supposed to be a new species in the forest region west of Lake Tanganyika, while Dr. Schweinfurth found one in the country beyond the western watershed of the Nile.   The gorilla is confined within narrower limits on and near the equator.

We have to pass over more than 70° of longitude before we again meet with Anthropoid Apes, in the northern part of Sumatra— where a specimen of the orang-utan (*Simia satyrus*) now in the Calcutta Museum, was obtained by Dr. Abel, and described by him in the *Asiatic Researches*, vol. xv.—and in Borneo, from which latter island almost all the specimens in European museums have been derived. There are supposed to be two species of *Simia* in Borneo, a larger and a smaller ; but their distinctness is not admitted by all naturalists. Both appear to be confined to the swampy forests near the north, west, and south coasts.

The Gibbons, or long-armed apes, forming the genus *Hylobates*, (7 species) are found in all the large islands of the Indo-Malayan sub-region, except the Philippines ; and also in Sylhet and Assam south of the Brahmaputra river, eastward to Cambodja and South China to the west of Canton, and in the island of Hainan.

The Siamang (*Siamanga syndactyla*) presents some anatomical peculiarities, and has the second and third toes united to the last joint, but in general form and structure it does not differ from *Hylobates*. It is the largest of the long-armed apes, and inhabits Sumatra and the Malay peninsula.

FAMILY 2.—SEMNOPITHECIDÆ.   (2 Genera, 30 Species.)

GENERAL DISTRIBUTION.					
NEOTROPICAL SUB-REGIONS.	NEARCTIC SUB-REGIONS.	PALÆARCTIC SUB-REGIONS.	ETHIOPIAN SUB-REGIONS.	ORIENTAL SUB-REGIONS.	AUSTRALIAN SUB-REGIONS.
— — — —	— — — —	— — — 4	1 . 2 — —	1 . 2 . 3 . 4	— — — —

The Semnopithecidæ, are long-tailed monkeys without cheek-pouches, and with rather rounded faces, the muzzle not being prominent. They have nearly the same distribution as the last family, but are more widely dispersed in both Africa and Asia, one species just entering the Palæarctic region.

The Eastern genus *Presbytes* or *Semnopithecus* (29 species), is spread over almost the whole of the Oriental region wherever the forests are extensive. They extend along the Himalayas to beyond Simla, where a species has been observed at an altitude of 11,000 feet, playing among fir-trees laden with snow wreaths. On the west side of India they are not found to the north of 14° N. latitude. On the east they extend into Arakan, and to Borneo and Java, but not apparently into Siam or Cambodja. Along the eastern extension of the Himalayas they again occur in East Thibet; a remarkable species with a large upturned nose (*S. roxellana*) having been discovered by Père David at Moupin (about Lat. 32° N.) in the highest forests, where the winters are severe and last for several months, and where the vegetation, and the other forms of animal life, are wholly those of the Palæarctic region. It is very curious that this species should somewhat resemble the young state of the proboscis monkey (*S. nasalis*), which inhabits one of the most uniform, damp, and hot climates on the globe—the river-swamps of Borneo.

Colobus, the African genus (11 species), is very closely allied to the preceding, differing chiefly in the thumb being absent or rudimentary. They are confined to the tropical regions—Abyssinia on the east, and from the Gambia to Angola and the island of Fernando Po, on the west.

FAMILY 3.—CYNOPITHECIDÆ.   (7 Genera, 67 Species).

GENERAL DISTRIBUTION.					
NEOTROPICAL SUB-REGIONS.	NEARCTIC SUB-REGIONS.	PALÆARCTIC SUB-REGIONS.	ETHIOPIAN SUB-REGIONS.	ORIENTAL SUB-REGIONS.	AUSTRALIAN SUB-REGIONS.
— — — —	— — — —	— 2 — 4	1 . 2 . 3 —	1 . 2 . 3 . 4	1 — — —

This family comprehends all the monkeys with cheek pouches, and the baboons. Some of these have very long tails, some none; some are dog-faced, others tolerably round-faced; but there are so many transitions from one to the other, and such a general agreement in structure, that they are now considered to form a very natural family. Their range is more extensive than any other family of Quadrumana, since they not only occur in every part of the Ethiopian and Oriental regions, but enter the Palæarctic region in the east and west, and the Australian region as far as the islands of Timor and Batchian. The African genera are *Myiopithecus, Cercopithecus, Cercocebus, Theropithecus,* and *Cynocephalus;* the Oriental genera, *Macacus,* and *Cynopithecus.*

Myiopithecus (1 species), consisting of the talapoin monkey of West Africa, differs from the other African monkeys in the structure of the last molar tooth; in the large ears, short face, and wide internasal septum; in this respect, as well as in its grace and gentleness, resembling some of the American monkeys.

Cercopithecus (24 species), contains all the more graceful and prettily coloured monkeys of tropical Africa, and comprises the guenons, the white-nosed, and the green monkeys. They range from the Gambia to the Congo, and from Abyssinia to the Zambesi.

Cercocebus (5 species), the mangabeys, of West Africa, are very closely allied to the eastern genus *Macacus.*

Theropithecus (2 species), including the gelada of Abyssinia and an allied species, resemble in form the baboons, but have the nostrils placed as in the last genus.

Cynocephalus (10 species), the baboons, are found in all parts of Africa. They consist of animals which vary much in appearance, but which agree in having an elongated dog-like muzzle with terminal nostrils, and being of terrestrial habits. Some of the baboons are of very large size, the mandrill (*C. maimon*) being only inferior to the orang and gorilla.

Macacus (25 species), is the commonest form of eastern monkey, and is found in every part of the Oriental region, as well as in North Africa, Gibraltar, Thibet, North China, and Japan; and one of the commonest species, *M. cynomolgus,* has extended its range from Java eastward to the extremity of Timor. The tail varies greatly in length, and in the Gibraltar monkey (*M. innus*) is quite absent. A remarkable species clothed with very thick fur, has lately been discovered in the snowy mountains of eastern Thibet.

Cynopithecus (? 2 sp.).—This genus consists of a black baboon-like Ape, inhabiting Celebes, Batchian, and the Philippine Islands; but perhaps introduced by man into the latter islands and into Batchian. It is doubtful if there is more than one species. The tail of this animal is a fleshy tubercle, the nostrils as in *Macacus,* but the muzzle is very prominent; and the development of the maxillary bones into strong lateral ridges corresponds to the structure of the most typical baboons. This species extends further east than any other quadrumanous animal.

FAMILY 4.—CEBIDÆ.   (10 Genera, 78 Species.)

GENERAL DISTRIBUTION.

NEOTROPICAL SUB-REGIONS.	NEARCTIC SUB-REGIONS.	PALÆARCTIC SUB-REGIONS.	ETHIOPIAN SUB-REGIONS.	ORIENTAL SUB-REGIONS.	AUSTRALIAN SUB-REGIONS.
— 2.3 —	— — — —	— — — —	— — — —	— — — —	— — — —

The Cebidæ, which comprehend all the larger American Monkeys, differ from those of the Old World by having an additional molar tooth in each jaw, and a broad nasal septum; while they have neither cheek-pouches nor ischial callosities, and the thumb is never completely opposable. Some have prehensile tails, especially adapting them for an arboreal life. They are divided into four sub-families,—Cebinæ, Mycetinæ, Pitheciinæ, and Nyctipithecinæ. The Cebidæ are strictly confined to the forest regions of tropical America, from the southern part of Mexico to about the parallel of 30° South Latitude. The distribution of the genera is as follows:—

Sub-family, Cebinæ.—*Cebus* (18 sp.), is the largest genus of American monkeys, and ranges from Costa Rica to Paraguay. They are commonly called sapajous. *Lagothrix* (5 sp.), the woolly monkeys, are rather larger and less active than the preceding; they are confined to the forests of the Upper Amazon Valley, and along the slopes of the Andes to Venezuela and Bolivia. *Ateles* (14 sp.), the spider monkeys, have very long limbs and tail. They range over the whole area of the family, and occur on the west side of the Equatorial Andes and on the Pacific coast of Guatemala. *Eriodes* (3 sp.), are somewhat intermediate between the last two genera, and are confined to the eastern parts of Brazil south of the equator. The three last mentioned genera have very powerful prehensile tails, the end being bare beneath; whereas the species of *Cebus* have the tail completely covered with hair, although prehensile, and therefore not so perfect a grasping organ.

Sub-family, Mycetinæ, consists of but a single genus, *Mycetes* (10 sp.), the howling monkeys, characterized by having a hollow bony vessel in the throat formed by an enlargement of the hyoid bone, which enables them to produce a wonderful howling noise. They are large, heavy animals, with a powerful and perfect prehensile tail. They range from East Guatemala to Paraguay. (Plate XIV., vol. ii., p. 24.)

Sub-family, Pitheciinæ, the sakis, have a non-prehensile bushy tail. *Pithecia* (7 sp.), has the tail of moderate length; while *Brachiurus* (5 sp.) has it very short. Both appear to be restricted to the great equatorial forests of South America.

Sub-family, Nyctipithecinæ, are small and elegant monkeys, with long, hairy, non-prehensile tails. *Nyctipithecus* (5 sp.), the night-monkeys or douroucoulis, have large eyes, nocturnal habits, and are somewhat lemurine in their appearance. They range from Nicaragua to the Amazon and eastern Peru. *Saimiris* or *Chrysothrix* (3 sp.), the squirrel-monkeys, are beautiful and active little creatures, found in most of the tropical forests from Costa Rica to Brazil and Bolivia. *Callithrix* (11 sp.), are somewhat intermediate between the last two genera, and are found all over South America from Panama to the southern limits of the great forests.

FAMILY 5.—HAPALIDÆ.　(2 Genera, 32 Species.)

GENERAL DISTRIBUTION.					
NEOTROPICAL SUB-REGIONS.	NEARCTIC SUB-REGIONS.	PALÆARCTIC SUB-REGIONS.	ETHIOPIAN SUB-REGIONS.	ORIENTAL SUB-REGIONS.	AUSTRALIAN SUB-REGIONS.
— 2 — —	— — — —	— — — —	— — — —	— — —	— — — —

The Hapalidæ, or marmosets, are very small monkeys, which differ from the true Cebidæ in the absence of one premolar tooth, while they possess the additional molar tooth; so that while they have the same number of teeth (thirty-two) as the Old World monkeys, they differ from them even more than do the

Cebidæ.   The thumb is not at all opposable, and all the fingers are armed with sharp claws.   The hallux, or thumb-like great toe, is very small; the tail is long and not prehensile.   The two genera *Hapale* (9 sp.), and *Midas* (24 sp.), are of doubtful value, though some naturalists have still further sub-divided them. They are confined to the tropical forests of South America, and are most abundant in the districts near the equator.

Sub-order—LEMUROIDEA.

FAMILY 6.—LEMURIDÆ.   (11 Genera, 53 Species.)

GENERAL DISTRIBUTION.					
NEOTROPICAL SUB-REGIONS.	NEARCTIC SUB-REGIONS.	PALÆARCTIC SUB-REGIONS.	ETHIOPIAN SUB-REGIONS.	ORIENTAL SUB-REGIONS.	AUSTRALIAN SUB-REGIONS.
— — — —	— — — —	1.2.3.4	— 2.3.4	— — — —	— — — —

The Lemuridæ, comprehending all the animals usually termed Lemurs and many of their allies, are divided by Professor Mivart —who has carefully studied the group—into four sub-families and eleven genera, as follows :—

Sub-family Indrisinæ, consisting of the genus *Indris* (5 sp.), is confined to Madagascar.

Sub-family Lemurinæ, contains five genera, viz. :—*Lemur*, (15 sp.); *Hapalemur* (2 sp.); *Microcebus* (4 sp.); *Chirogaleus* (5 sp.); and *Lepilemur* (2 sp.) ;—all confined to Madagascar.

Sub-family Nycticebinæ, contains four genera, viz.:—*Nycticebus* (3 sp.)—small, short-tailed, nocturnal animals, called slow-lemurs, —range from East Bengal to South China, and to Borneo and Java; *Loris* (1 sp.)—a very small, tail-less, nocturnal lemur, which inhabits Madras, Malabar, and Ceylon; *Perodicticus* (1 sp.) —the potto—a small lemur with almost rudimentary fore-finger, found at Sierra Leone (Plate V., vol. i., p. 264); *Arctocebus* (1 sp.)—the angwantibo,—another extraordinary form in which the forefinger is quite absent and the first toe armed with a long claw,—inhabits Old Calabar.

Sub-family Galaginæ, contains only the genus *Galago* (14 sp.), which is confined to the African continent, ranging from Senegal and Fernando Po to Zanzibar and Natal.

FAMILY 7.—TARSIIDÆ.   (1 Genus, 1 Species.)

GENERAL DISTRIBUTION.

NEOTROPICAL SUB-REGIONS.	NEARCTIC SUB-REGIONS.	PALÆARCTIC SUB-REGIONS.	ETHIOPIAN SUB-REGIONS.	ORIENTAL SUB-REGIONS.	AUSTRALIAN SUB-REGIONS.
— — — —	— — — —	— — — —	— — — —	— — — 4	— — — —

The curious *Tarsius spectrum*, which constitutes this family, inhabits Sumatra, Banca, and Borneo, and is also found in some parts of Celebes, which would bring it into the Australian region; but this island is altogether so anomalous that we can only consider its productions to have somewhat more affinity with the Australian than the Oriental region, but hardly to belong to either. The Tarsier is a small, long-tailed, nocturnal animal, of curious structure and appearance; and it forms the only link of connection with the next family, which it resembles in the extraordinary development of the toes, one of which is much larger and more slender than the rest. (Plate VIII., vol. i. p. 337.)

FAMILY 8.—CHIROMYIDÆ.   (1 Genus, 1 Species.)

GENERAL DISTRIBUTION.

NEOTROPICAL SUB-REGIONS.	NEARCTIC SUB-REGIONS.	PALÆARCTIC SUB-REGIONS.	ETHIOPIAN SUB-REGIONS.	ORIENTAL SUB-REGIONS.	AUSTRALIAN SUB-REGIONS.
— — — —	— — — —	— — — —	— — — 4	— — — —	— — — —

The Aye-aye, (*Chiromys*), the sole representative of this family, is confined to the island of Madagascar. It was for a long time very imperfectly known, and was supposed to belong to the Rodentia; but it has now been ascertained to be an exceedingly specialized form of the Lemuroid type, and must be considered to be one of the most extraordinary of the mammalia now inhabiting the globe. (Plate VI., vol. i., p. 278.)

Fossil Quadrumana.

Not much progress has yet been made in tracing back the various forms of Apes and Monkeys to their earliest appearance on the globe; but there have been some interesting recent discoveries, which lead us to hope that the field is not yet exhausted. The following is a summary of what is known as to the early forms of each family :—

Simiidæ.—Two or three species of this family have been found in the Upper Miocene deposits of France and Switzerland. *Pliopithecus*, of which a species has been found at each locality, was allied to the gibbons (*Hylobates*), and perhaps to *Semnopithecus*. A more remarkable form, named *Dryopithecus*, as large as a man, and having peculiarities of structure which are thought by Gervais and Lartet to indicate a nearer approach to the human form than any existing Ape, has been found in strata of the same age in France.

Semnopithecidæ.—Species of *Semnopithecus* have been found in the Upper Miocene of Greece, and others in the Siwalik Hills of N. W. India, also of Upper Miocene age. An allied form also occurs in the Miocene of Wurtemburg. *Mesopithecus* from Greece is somewhat intermediate between *Semnopithecus* and *Macacus*.

Remains supposed to be of *Semnopithecus*, have also occurred in the Pliocene of Montpellier.

Cynopithecidæ.—*Macacus* has occurred in Pliocene deposits at Grays, Essex; and also in the South of France along with *Cercopithecus*.

Cebidæ.—In the caves of Brazil remains of the genera *Cebus*, *Mycetes*, *Callithrix*, and *Hapale*, have been found; as well as an extinct form of larger size—*Protopithecus*.

Lemuroidea.—A true lemur has recently been discovered in the Eocene of France; and it is supposed to be most nearly allied to the peculiar West African genera, *Perodicticus* and *Arctocebus*.

Cænopithecus, from the Swiss Jura, is supposed to have affinities both for the Lemuridæ and the American Cebidæ.

In the lower Eocene of North America remains have been

discovered, which are believed to belong to this sub-order: but they form two distinct families,—Lemuravidæ and Limnotheridæ. Other remains from the Miocene are believed to be intermediate between these and the Cebidæ,—a most interesting and suggestive affinity, if well founded. For the genera of these American Lemuroidea, see vol. i., p. 133.

General Remarks on the Distribution of Primates.

The most striking fact presented by this order, from our present point of view, is the strict limitation of well-marked families to definite areas. The Cebidæ and Hapalidæ would alone serve to mark out tropical America as the nucleus of one of the great zoological divisions of the earth. In the Eastern Hemisphere, the corresponding fact is the entire absence of the order from the Australian region, with the exception of one or two outlying forms, which have evidently transgressed the normal limits of their group. The separation of the Ethiopian and Oriental regions is, in this order, mainly indicated by the distribution of the genera, no one of which is common to the two regions. The two highest families, the Simiidæ and the Semnopithecidæ, are pretty equally distributed about two equatorial foci, one situated in West Africa, the other in the Malay archipelago,—in Borneo or the Peninsula of Malacca;—while the third family, Cynopithecidæ, ranges over the whole of both regions, and somewhat overpasses their limits. The Lemuroid group, on the other hand, offers us one of the most singular phenomena in geographical distribution. It consists of three families, the species of which are grouped into six sub-families and 13 genera. One of these families and two of the sub-families, comprising 7 genera, and no less than 30 out of the total of 50 species, are confined to the one island of Madagascar. Of the remainder, 3 genera, comprising 15 species, are spread over tropical Africa; while three other genera with 5 species, inhabit certain restricted portions of India and the Malay islands. These curious facts point unmistakably to the former existence of a large tract of land in what is now the Indian Ocean, connecting Madagascar on the one hand with Ceylon, and with the Malay countries on the other. About this same time (but perhaps not contemporaneously) Madagascar must have been connected with some portion of Southern Africa, and the whole of the country would possess no other Primates but Lemuroidea. After the Madagascar territory (very much larger than the existing island) had been separated, a connection appears to have been long maintained (probably by a northerly route) between the more equatorial portions of Asia and Africa; till those higher forms had become developed, which were afterwards differentiated into *Simia, Presbytes,* and *Cynopithecus,* on the one hand, and into *Troglodytes, Colobus,* and *Cynocephalus,* on the other. In accordance with the principle of competition so well expounded by Mr. Darwin, we can understand how, in the vast Asiatic and African area north of the Equator, with a great variety of physical conditions and the influence of a host of competing forms of life, higher types were developed than in the less extensive and long-isolated countries south of the Equator. In Madagascar, where these less complex conditions prevailed in a considerable land-area, the lowly organized Lemuroids have diverged into many specialized forms of their own peculiar type; while on the continents they have, to a great extent, become exterminated, or have maintained their existence in a few cases, in islands or in mountain ranges. In Africa the nocturnal and arboreal *Galagos* are adapted to a special mode of life, in which they probably have few competitors.

How and when the ancestors of the Cebidæ and Hapalidæ entered the South American continent, it is less easy to conceive. The only rays of light we yet have on the subject are, the supposed affinities of the fossil *Cœnopithecus* of the Swiss, and the Lemuravidæ of the North American Eocene, with both Cebidæ and Lemuroids, and the fact that in Miocene or Eocene times a mild climate prevailed up to the Arctic circle. The discovery of an undoubted Lemuroid in the Eocene of Europe, indicates that the great Northern Continent was probably the birthplace of this low type of mammal, and the source whence Africa and Southern Asia were peopled with them, as it was, at a later period, with the higher forms of monkeys and apes.

Order II.—CHIROPTERA.

FAMILY 9.—PTEROPIDÆ.   (9 Genera, 65 Species.)

GENERAL DISTRIBUTION.					
NEOTROPICAL SUB-REGIONS.	NEARCTIC SUB-REGIONS.	PALÆARCTIC SUB-REGIONS.	ETHIOPIAN SUB-REGIONS.	ORIENTAL SUB-REGIONS.	AUSTRALIAN SUB-REGIONS.
— — — —	— — — —	— — — 4	1.2.3.4	1.2.3.4	1.2.3 —

The Pteropidæ, or fruit-eating Bats, sometimes called flying-foxes, are pretty evenly distributed over the tropical regions of the Old World and Australia. They range over all Africa and the whole of the Oriental Region, and northward, to Amoy in China and to the South of Japan. They are also found in the more fertile parts of Australia and Tasmania, and in the Pacific Islands as far east as the Marianne and Samoa Islands; but not in the Sandwich Islands or New Zealand.

The genera of bats are exceedingly numerous, but they are in a very unsettled state, and the synonymy is exceedingly confused. The details of their distribution cannot therefore be usefully entered into here. The Pteropidæ differ so much from all other bats, that they are considered to form a distinct suborder of Chiroptera, and by some naturalists even a distinct order of Mammalia.

No fossil Pteropidæ have been discovered.

FAMILY 10.—PHYLLOSTOMIDÆ.   (31 Genera, 60 Species.)

GENERAL DISTRIBUTION.					
NEOTROPICAL SUB-REGIONS.	NEARCTIC SUB-REGIONS.	PALÆARCTIC SUB-REGIONS.	ETHIOPIAN SUB-REGIONS.	ORIENTAL SUB-REGIONS.	AUSTRALIAN SUB-REGIONS.
1.2.3 —	1 — — —	— — — —	— — — ··	·· ··· — —	— — — ··

The Phyllostomidæ, or simple leaf-nosed Bats, are confined to the Neotropical region, from Mexico and the Antilles to the

southern limits of the forest region east of the Andes, and to
about lat. 33° S. in Chili. None are found in the Nearctic
region, with the exception of one species in California (*Macrotus
Californicus*), closely allied to Mexican and West Indian forms.
The celebrated blood-sucking vampyre bats of South America
belong to this group. Two genera, *Desmodus* and *Diphylla*, form
Dr. Peters' family Desmodidæ. Mr. Dobson, in his recently
published arrangement, divides the family into five groups :—
Mormopes, Vampyri, Glossophagæ, Stenodermata, and Desmo-
dontes.

Numerous remains of extinct species of this family have been
found in the bone-caves of Brazil.

FAMILY 11.—RHINOLOPHIDÆ.   (7 Genera, 70 Species.)

GENERAL DISTRIBUTION.					
NEOTROPICAL SUB-REGIONS.	NEARCTIC SUB-REGIONS.	PALÆARCTIC SUB-REGIONS.	ETHIOPIAN SUB-REGIONS.	ORIENTAL SUB-REGIONS.	AUSTRALIAN SUB-REGIONS.
— — — —	— — — —	1 . 2 . 3 . 4	1 . 2 . 3 . 4	1 . 2 . 3 . 4	1 . 2 — —

The Rhinolophidæ, or Horse-shoe Bats (so-called from a
curiously-shaped membranous appendance to the nose), range
over all the Ethiopian and Oriental regions, the southern part
of the Palæarctic region, Australia and Tasmania. They are
most abundant and varied in the Oriental region, where twelve
genera are found; while only five inhabit the Australian and
Ethiopian regions respectively. Europe has only one genus and
four species, mostly found in the southern parts, and none going
further north than the latitude of England, where two species
occur. Two others are found in Japan, at the opposite extremity
of the Palæarctic region.

The genera *Nycteris* and *Megaderma*, which range over the
Ethiopian and Oriental regions to the Moluccas, are considered
by Dr. Peters to form a distinct family, Megadermidæ; and
Mr. Dobson in his recent arrangement (published after our first

volume was printed) adopts the same family under the name of Nycteridæ. The curious Indian genus *Rhinopoma*, which, following Dr. J. E. Gray, we have classed in this family, is considered by Mr. Dobson to belong to the Noctilionidæ.

Fossil Rhinolophidæ.—Remains of a species of *Rhinolophus* still living in England, have been found in Kent's Cavern, near Torquay.

FAMILY 12.—VESPERTILIONIDÆ.　　(18 Genera, 200 Species.)

GENERAL DISTRIBUTION.					
NEOTROPICAL SUB-REGIONS.	NEARCTIC SUB-REGIONS.	PALÆARCTIC SUB-REGIONS.	ETHIOPIAN SUB-REGIONS.	ORIENTAL SUB-REGIONS.	AUSTRALIAN SUB-REGIONS.
1.2.3.4	1.2.3.4	1.2.3.4	1.2.3.4	1.2.3.4	1.2.3.4

The small bats constituting the family Vespertilionidæ, have no nose-membrane, but an internal earlet or *tragus*, and often very large ears. They range over almost the whole globe, being apparently only limited by the necessity of procuring insect food. In America they are found as far north as Hudson's Bay and the Columbia river; and in Europe they approach, if they do not pass the Arctic circle. Such remote islands as the Azores, Bermudas, Fiji Islands, Sandwich Islands, and New Zealand, all possess species of this group of bats, some of which probably inhabit every island in warm or temperate parts of the globe.

The genus *Taphozous*, which, in our Tables of Distribution in vol. i. we have included in this family, is placed by Mr. Dobson in his family Emballonuridæ, which is equivalent to our next family, Noctilionidæ.

Fossil Vespertilionidæ.—Several living European bats of this family—*Scotophilus murinus, Plecotus auritus, Vespertilio noctula,* and *V. pipestrellus*—have been found fossil in bone-caves in various parts of Europe.

Extinct species of *Vespertilio* have occurred in the Lower Miocene at Mayence, in the Upper Miocene of the South of France, and in the Upper Eocene of the Paris basin.

FAMILY 13.—NOCTILIONIDÆ.   (14 Genera, 50 Species.)

GENERAL DISTRIBUTION.

NEOTROPICAL SUB-REGIONS.	NEARCTIC SUB-REGIONS.	PALÆARCTIC SUB-REGIONS.	ETHIOPIAN SUB-REGIONS.	ORIENTAL SUB-REGIONS.	AUSTRALIAN SUB-REGIONS.
1.2.3.4	1 — — —	— 2 — —	1.2.3.4	1.2.3.4	— — — 4

The Noctilionidæ, or short-headed Bats, are found in every region, but are very unequally distributed.   Their head-quarters is the Neotropical region, where most of the genera occur, and where they range from Mexico to Buenos Ayres and Chili, while in North America there is only one species in California.   They are unknown in Australia; but one species occurs in New Zealand, and another in Norfolk Island.   Several species of *Dysopes* (or *Molossus*) inhabit the Oriental region; one or two species being widely distributed over the continent, while two others inhabit the Indo-Malayan Islands.   A species of this same genus occurs in South Africa, and another in Madagascar and in the Island of Bourbon; while one inhabits Southern Europe and North Africa, and another is found at Amoy in China.   It will be seen therefore, that these are really South American bats, which have a few allies widely scattered over the various regions of the globe. Their affinities are, according to Mr. Tomes, with the Phyllostomidæ, a purely South American family.   The species which forms the connecting link is the *Mystacina tuberculata*, a New Zealand bat, which may, with almost equal propriety be placed in either family, and which affords an interesting illustration of the many points of resemblance between the Australian and Neotropical regions.

Dr. Peters has separated this family into three,—Mormopidæ, which is wholly Neotropical, and is especially abundant in the West Indian Islands; Molossidæ, chiefly consisting of the genus *Molossus*; and Noctilionidæ, comprising the remainder of the family, and wholly Neotropical.   Mr. Dobson, however, classes the Mormopes with the Phyllostomidæ, and reduces the

Molossi to the rank of a sub-family. In our first volume we have classed *Rhinopoma* with the Rhinolophidæ, and *Taphozous* with the Vespertilionidæ ; but according to Mr. Dobson both these genera belong to the present family.

Remarks on the Distribution of the Order Chiroptera.

Although the bats, from their great powers of flight, are not amenable to the limitations which determine the distribution of other terrestrial mammals, yet certain great facts of distribution come out in a very striking manner. The speciality of the Neotropical region is well shown, not only by its exclusive possession of one large family (Phyllostomidæ), but almost equally so by the total absence of two others (Pteropidæ and Rhinolophidæ). The Nearctic region is also unusually well marked, by the total absence of a family (Rhinolophidæ) which is tolerably well represented in the Palæarctic. The Pteropidæ well characterize the tropical regions of the Old World and Australia ; while the Vespertilionidæ are more characteristic of the Palæarctic and Nearctic regions, which together possess about 60 species of this family.

The bats are a very difficult study, and it is quite uncertain how many distinct species are really known. Schinz, in his *Synopsis Mammalium* (1844) describes 330, while the list given by Mr. Andrew Murray in his *Geographical Distribution of Mammalia* (1866), contains 400 species. A small number of new species have been since described, but others have been sunk as synonyms, so that we can perhaps hardly obtain a nearer approximation to the truth than the last number. In Europe there are 35 species, and only 17 in North America.

Fossil Chiroptera.—The fossil remains of bats that have yet been discovered, being chiefly allied to forms still existing in the same countries, throw no light on the origin or affinities of this remarkable and isolated order of Mammalia ; but as species very similar to those now living were in existence so far back as Miocene or even Eocene times, we may be sure the group is one of immense antiquity, and that there has been ample time for the amount of variation and extinction required to bring about

the limitation of types, and the peculiarities of distribution we
now find to exist.

Order III.—INSECTIVORA.

FAMILY 14.—GALEOPITHECIDÆ.   (1 Genus, 2 Species.)

GENERAL DISTRIBUTION.					
NEOTROPICAL SUB-REGIONS.	NEARCTIC SUB-REGIONS.	PALÆARCTIC SUB-REGIONS.	ETHIOPIAN SUB-REGIONS.	ORIENTAL SUB-REGIONS.	AUSTRALIAN SUB-REGIONS.
— — — —	— — — —	— — — —	— — — —	— — — 4	— — — —

The singular and isolated genus *Galeopithecus*, or flying lemur,
has been usually placed among the Lemuroidea, but it is now
considered to come best at the head of the Insectivora.   Its food
however, seems to be purely vegetable, and the very small, blind,
and naked young, closely attached to the wrinkled skin of the
mother's breast, perhaps indicates some affinity with the Marsupials.   This animal seems, in fact, to be a lateral offshoot of
some low form, which has survived during the process of development of the Insectivora, the Lemuroidea, and the Marsupials,
from an ancestral type.   Only two species are known, one
found in Malacca, Sumatra, and Borneo, but not in Java; the
other in the Philippine islands (Plate VIII. vol. i. p. 337).

FAMILY 15.—MACROSCELIDIDÆ.   (3 Genera, 10 Species.)

GENERAL DISTRIBUTION.					
NEOTROPICAL SUB-REGIONS.	NEARCTIC SUB-REGIONS.	PALÆARCTIC SUB-REGIONS.	ETHIOPIAN SUB-REGIONS.	ORIENTAL SUB-REGIONS.	AUSTRALIAN SUB-REGIONS.
— — — —	— — — —	— 2 — —	1 — 3 —	— — — —	— — — —

The Macroscelides, or elephant shrews, are extraordinary little
animals, with trunk-like snout and kangaroo-like hind-legs.
They are almost confined to South Africa, whence they extend
up the east coast as far as the Zambezi and Mozambique.   A

single outlying species of *Macroscelides* inhabits Barbary and
Algeria ; while the two genera *Petrodromus*, and *Rhyncocyon*, each
represented by a single species, have only been found at
Mozambique.

FAMILY 16.—TUPAIIDÆ.　(3 Genera, 10 species.)

GENERAL DISTRIBUTION.					
NEOTROPICAL SUB-REGIONS.	NEARCTIC SUB-REGIONS.	PALÆARCTIC SUB-REGIONS.	ETHIOPIAN SUB-REGIONS.	ORIENTAL SUB-REGIONS.	AUSTRALIAN SUB-REGIONS.
— — — —	— — — —	— — — —	— — — —	— 2 . 3 . 4	— — — —

The Tupaiidæ are squirrel-like shrews, having bushy tails,
and often climbing up trees, but also feeding on the ground and
among low bushes. The typical *Tupaia* (7 species), are called
ground squirrels by the Malays. They are most abundant in
the Malay islands and Indo-Chinese countries, but one species
is found in the Khasia Mountains, and one in the Eastern Ghauts
near Madras. The small shorter-tailed *Hylomys* (2 species) is
found from Tenasserim to Java and Borneo ; while the elegant
little *Ptilocerus* (1 species) with its long pencilled tail, is confined
to Borneo ; (Plate VIII. vol. i. p. 337). The family is therefore
especially Malayan, with outlying species in northern and continental India.

Extinct Species.—*Oxygomphus,* found in the Tertiary deposits
of Germany, is believed to belong to this family ; as is *Omomys,*
from the Pliocene of the United States.

FAMILY 17.—ERINACEIDÆ.　(2 Genera, 15 Species.)

GENERAL DISTRIBUTION.					
NEOTROPICAL SUB-REGIONS.	NEARCTIC SUB-REGIONS.	PALÆARCTIC SUB-REGIONS.	ETHIOPIAN SUB-REGIONS.	ORIENTAL SUB-REGIONS.	AUSTRALIAN SUB-REGIONS.
— — — —	— — — —	1 . 2 . 3 . 4	— — 3 —	1 . 2 — 4	— — — —

The Hedgehogs, comprised in the genus *Erinaceus* (14 species),
are widely distributed over the Palæarctic, and a part of the

Oriental regions; but they only occur in the Ethiopian region in South Africa and in the Deserts of the north, which more properly belong to the Palæarctic region. They are absent from the Malayan, and also from the Indo-Chinese sub-regions; except that they extend from the north of China to Amoy and Formosa and into the temperate highlands of the Western Himalayas. The curious *Gymnura* (1 species) is found in Borneo, Sumatra, and the Malay peninsula.

Extinct Species.—The common hedgehog has been found fossil in several Post-tertiary deposits, while extinct species occur in the lower Miocene of Auvergne and in some other parts of Europe. Many of these remains are classed in different genera from the living species;—(*Amphechinus, Tetracus, Galerix.*)

FAMILY 18.—CENTETIDÆ.   (6 Genera, 10 Species.)

GENERAL DISTRIBUTION.					
NEOTROPICAL SUB-REGIONS.	NEARCTIC SUB-REGIONS.	PALÆARCTIC SUB-REGIONS.	ETHIOPIAN SUB-REGIONS.	ORIENTAL SUB-REGIONS.	AUSTRALIAN SUB-REGIONS.
— — — 4	— — — —	— — — —	— — — 4	— — — —	— — — —

The Centetidæ are small animals, many of them having a spiny covering, whence the species of *Centetes* have been called Madagascar hedgehogs. The genera *Centetes* (2 species), *Hemicentetes* (1 species), *Ericulus* (1 species), *Echinops* (3 species), and the recently described *Oryzorictes* (1 species), are all exclusively inhabitants of Madagascar, and are almost or quite tail-less. The remaining genus, *Solenodon*, is a more slender and active animal, with a long, rat-like tail, shrew-like head, and coarse fur; and the two known species are among the very few indigenous mammals of the West Indian islands, one being found in Cuba (Plate XVII., vol. ii., p. 67), the other in Hayti. Although presenting many points of difference in detail, the essential characters of this curious animal are, according to Professors Peters and Mivart, identical with the rest of the Centetidæ. We have thus a most remarkable and well-established case of discontinuous distribution, two portions of the same family being now separated from each other by an extensive continent, as well as by a deep ocean.

Extinct Species.—Remains found in the Lower Miocene of the South of France are believed to belong to the genus *Echinops,* or one closely allied to it.

FAMILY 19.—POTAMOGALIDÆ.   (1 Genus, 1 Species,)

			GENERAL DISTRIBUTION.		
NEOTROPICAL SUB-REGIONS.	NEARCTIC SUB-REGIONS.	PALÆARCTIC SUB-REGIONS.	ETHIOPIAN SUB-REGIONS.	ORIENTAL SUB-REGIONS.	AUSTRALIAN SUB-REGIONS.
— — — —	— — — —	— — — —	— 2 — —	— — — —	— — — —

The genus *Potamogale* was founded on a curious, small, otter-like animal from West Africa, first found by M. Du Chaillu at the Gaboon, and afterwards by the Portuguese at Angola. Its affinities are with several groups of Insectivora, but it is sufficiently peculiar to require the establishment of a distinct family for its reception. (Plate V., vol. i., p. 264.)

FAMILY 20.—CHRYSOCHLORIDÆ.   (2 Genera, 3 Species.)

			GENERAL DISTRIBUTION.		
NEOTROPICAL SUB-REGIONS.	NEARCTIC SUB-REGIONS.	PALÆARCTIC SUB-REGIONS.	ETHIOPIAN SUB-REGIONS.	ORIENTAL SUB-REGIONS.	AUSTRALIAN SUB-REGIONS.
— — — —	— — — —	— — — —	— — 3 —	— — — —	— — — —

The Chrysochloridæ, or golden moles, of the Cape of Good Hope have been separated by Professor Mivart into two genera, *Chrysochloris* and *Chalcochloris.*   They are remarkable mole-like animals, having beautiful silky fur, with a metallic lustre and changeable golden tints.   They are peculiar to the Cape district, but one species extends as far north as the Mozambique territory. Their dentition is altogether peculiar, so as to completely separate them from the true moles.

FAMILY 21.—TALPIDÆ.   (8 Genera, 19 Species.

GENERAL DISTRIBUTION.					
NEOTROPICAL SUB-REGIONS.	NEARCTIC SUB-REGIONS.	PALÆARCTIC SUB-REGIONS.	ETHIOPIAN SUB-REGIONS.	ORIENTAL SUB-REGIONS.	AUSTRALIAN SUB-REGIONS.
— — — —	1.2.3.4	1.2.3.4	— — — —	— — 3 —	— — — —

The Moles comprise many extraordinary forms of small mammalia especially characteristic of the temperate regions of the
northern hemisphere, only sending out a few species of *Talpa*
along the Himalayas as far as Assam, and even to Tenasserim,
if there is no mistake about this locality ; while one species is
found in Formosa, the northern part of which is almost as much
Palæarctic as Oriental.   The genus *Talpa* (7 species), spreads
over the whole Palæarctic region from Great Britain to Japan;
Scaptochirus (1 species) is a recent discovery in North China;
Condylura (1 species), the star-nosed mole, inhabits Eastern
North America from Nova Scotia to Pennsylvania; *Scapanus*
(2 species) ranges across from New York to St. Francisco;
Scalops (3 species), the shrew-moles, range from Mexico to the
great lakes on the east side of America, but on the west only to
the north of Oregon.   An allied genus, *Myogale* (2 species), has
a curious discontinuous distribution in Europe, one species being
found in South-East Russia, the other in the Pyrenees (Plate II.,
vol. i., p. 218).   Another allied genus, *Nectogale* (1 species), has
recently been described by Professor Milne-Edwards from Thibet.
Urotrichus is a shrew-like mole which inhabits Japan,and a second
species has been discovered in the mountains of British Columbia;
an allied form, *Uropsilus*, inhabits East Thibet.   *Anurosorex*
and *Scaptonyx*, are new genera from North China.

Extinct Species.—The common mole has been found fossil in
bone-caves and diluvial deposits, and several extinct species of
mole-like animals occur in the Miocene deposits of the South of
France and of Germany.   These have been described under the
generic names *Dinylus, Geotrypus, Hyporissus, Galeospalax ;* while
Palæospalax has been found in the Pliocene forest-beds of Norfolk

and Ostend.   Species of *Myogale* also occur from the Miocene downwards.

FAMILY 22.—SORICIDÆ. (1 Genus, 11 Sub-genera, 65 Species.)

GENERAL DISTRIBUTION.

NEOTROPICAL SUB-REG ONS.	NEARCTIC SUB-REGIONS.	PALÆARCTIC SUB-REGIONS.	ETHIOPIAN SUB-REGIONS.	ORIENTAL SUB-REGIONS.	AUSTRALIAN SUB-REGIONS.
— — 3 —	1.2.3.4	1.2.3.4	1.2.3.4	1.2.3.4	— — — —

The Shrews have a wide distribution, being found throughout every region except the Australian and Neotropical; although, as a species is found in Timor and in some of the Moluccas, they just enter this part of the former region, while one found in Guatemala brings them into the latter. A number of species have recently been described from India and the Malay Islands, so that the Oriental region is now the richest in shrews, having 28 species; the Nearctic comes next with 24; while the Ethiopian has 11, and the Palæarctic 10 species. The sub-genera are *Crossopus, Amphisorex, Neosorex, Crocidura, Diplomesodon, Pinulia, Pachyura, Blarina, Feroculus, Anausorex.*

Extinct Species.—Several species of *Sorex* have been found fossil in the Miocene of the South of France, as well as the extinct genera *Mysarachne* and *Plesiosorex ;* and some existing species have occurred in Bone Caves and Diluvial deposits.

General Remarks on the Distribution of the Insectivora.

The most prominent features in the distribution of the Insectivora are,—their complete absence from South America and Australia; the presence of *Solenodon* in two of the West Indian islands while the five allied genera are found only in Madagascar ; and the absence of hedgehogs from North America. If we consider that there are only 135 known species of the order, 65 of which belong to the one genus *Sorex ;* while the remaining 26 genera contain only 70 species, which have to be classed in 8 distinct families, and present such divergent and highly specialized forms as *Galeopithecus, Erinaceus, Solenodon,* and *Condylura,* it becomes evident that we have here the detached fragments of a much more extensive group of animals, now almost extinct. Many of the forms continue to exist only in islands, removed from the severe competition of a varied mammalian population, as in Madagascar and the Antilles; while others appear to have escaped extermination either by their peculiar habits—as the various forms of Moles; by special protection—as in the Hedgehogs; or by a resemblance in form, coloration, and habits to dominant groups in their own district—as the Tupaias of Malay which resemble squirrels, and the Elephant-shrews of Africa which resemble the jerboas. The numerous cases of isolated and discontinuous distribution among the Insectivora, offer no difficulty from this point of view; since they are the necessary results of an extensive and widely-spread group of animals slowly becoming extinct, and continuing to exist only where special conditions have enabled them to maintain themselves in the struggle with more highly organized forms.

The fossil Insectivora do not throw much light on the early history of the order, since even as far back as the Miocene period they consist almost wholly of forms which can be referred to existing families. In North America they go back to the Eocene period, if certain doubtful remains have been rightly placed. The occurrence of fossil Centetidæ in Europe, supports the view we have maintained in preceding chapters, that the existing distribution of this family between Madagascar and the Antilles, proves no direct connection between those islands, but only shows us that the family once had an extensive range.

Order IV.—CARNIVORA.

FAMILY 23.—FELIDÆ.   (3 Genera, 14 Sub-genera, 66 Species.)

GENERAL DISTRIBUTION.					
NEOTROPICAL SUB-REGIONS.	NEARCTIC SUB-REGIONS.	PALÆARCTIC SUB-REGIONS.	ETHIOPIAN SUB-REGIONS.	ORIENTAL SUB-REGIONS.	AUSTRALIAN SUB-REGIONS.
1.2.3 —	1.2.3.4	1.2.3.4	1.2.3 —	1.2.3.4	— — — —

The Cats are very widely distributed over the earth—with the exception of the Australian region and the island sub-region of Madagascar and the Antilles—universally; ranging from the torrid zone to the Arctic regions and the Straits of Magellan. They are so uniform in their organization that many naturalists group them all under one genus, *Felis;* but it is now more usual to class at least the lynxes as a separate genus, while the hunting leopard, or cheetah, forms another. Dr. J. E. Gray divides these again, and makes 17 generic groups; but as this subdivision is not generally adopted, and does not bring out any special features of geographical distribution, I shall not further notice it.

The genus *Felis* (56 species) has the same general range as the whole family, except that it does not go so far north; the Amoor river in Eastern Asia, and 55° N. Lat. in America, marking its limits. *Lyncus* (10 species) is a more northern group, ranging to the polar regions in Europe and Asia, and to Lat. 66° N. in America, but not going further south than Northern Mexico and the European shores of the Mediterranean, except the caracal, which may be another genus, and which extends to Central India, Persia, North Africa and even the Cape of Good Hope. The lynxes are thus almost wholly peculiar to the Nearctic and Palæarctic regions. *Cynælurus* (1 species) the hunting leopard, ranges from Southern and Western India through Persia, Syria, Northern and Central Africa, to the Cape of Good Hope.

Extinct Felidæ.—More than twenty extinct species of true Felidæ have been described, ranging in time from the epoch of prehistoric man back to the Miocene or even the Eocene period. They occur in the south of England, in Central and South Europe, in North-West India, in Nebraska in North America, and in the caves of Brazil. Most of them are referred to the genus *Felis,* and closely resemble the existing lions, tigers, and other large cats. Another group however forms the genus *Machairodus,* a highly specialized form with serrated teeth. Five species have been described from Europe, Northern India, and both North and South America; and it is remarkable that they exhibit at least as wide a range, both in space and time, as the more numerous species referred to *Felis.* One of them undoubtedly coexisted with man in England, while another, as well as the allied *Dinictis*, has been found in the Mauvaises Terres of Nebraska, associated with *Anchitherium* and other extinct and equally remarkable forms, which are certainly Miocene if not, as some geologists think, belonging to the Eocene period. These facts clearly indicate that we have as yet made little approach to discovering the epoch when Felidæ originated, since the oldest forms yet discovered are typical and highly specialized representatives of a group which is itself the most specialized of the Carnivora. Another genus, *Pseudœlurus*, is common to the Miocene deposits of Europe and North America.

FAMILY 24.—CRYPTOPROCTIDÆ.   (1 Genus, 1 Species.)

GENERAL DISTRIBUTION.					
NEOTROPICAL SUB-REGIONS.	NEARCTIC SUB-REGIONS.	PALÆARCTIC SUB-REGIONS.	ETHIOPIAN SUB-REGIONS.	ORIENTAL SUB-REGIONS.	AUSTRALIAN SUB-REGIONS.
— — — —	— — — —	— — — —	— — — 4	— — — —	— — — —

The *Cryptoprocta ferox*, a small and graceful cat-like animal, peculiar to Madagascar, was formerly classed among the Viverridæ, but is now considered by Professor Flower to constitute a distinct family between the Cats and the Civets.

FAMILY 25.—VIVERRIDÆ.   (8–33 Genera, 100 Species.)

GENERAL DISTRIBUTION.					
NEOTROPICAL SUB-REGIONS.	NEARCTIC SUB-REGIONS.	PALÆARCTIC SUB-REGIONS.	ETHIOPIAN SUB-REGIONS.	ORIENTAL SUB-REGIONS.	AUSTRALIAN SUB-REGIONS.
— — — —	— — — —	— 2 — —	1.2.3.4	1.2.3.4	1 — — —

The Viverridæ comprise a number of small and moderate-sized carnivorous animals, popularly known as civets, genets, and ichneumons, highly characteristic of the Ethiopian and Oriental regions, several of the genera being common to both. A species of *Genetta*, and one of *Herpestes*, inhabit South Europe; while *Viverra* extends to the Moluccas, but is doubtfully indigenous. The extreme geographical limits of the family are marked by

Genetta in France and Spain, *Viverra* in Shanghae and Batchian Island, and *Herpestes* in Java and the Cape of Good Hope.

The following are the genera with their distribution as given by Dr. J. E. Gray in his latest British Museum Catalogue :

Sub-family VIVERRINÆ.—*Viverra* (3 species), North and tropical Africa, the whole Oriental region to the Moluccas ; *Viverricula* (1 species) India to Java ; *Genetta* (5 species), South Europe, Palestine, Arabia, and all Africa ; *Fossa* (1 species), Madagascar ; *Linsang* (2 species), Malacca to Java ; *Poiana* (1 species), West Africa ; *Galidia* (3 species), Madagascar ; *Hemigalea* (1 species), Malacca and Borneo ; *Arctictis* (1 species) Nepal to Sumatra and Java ; *Nandinia* (1 species), West Africa ; *Paradoxurus* (9 species), the whole Oriental region; *Paguma* (3 species), Nepal to China, Sumatra, and Borneo ; *Arctogale* (1 species), Tenasserim to Java.

Sub-family HERPESTINÆ.—*Cynogale* (1 species), Borneo ; *Galidictis* (2 species), Madagascar ; *Herpestes* (22 species), South Palæarctic, Ethiopian, and Oriental regions ; *Athylax* (3 species), Tropical and South Africa ; *Galogale* (13 species), all Africa, North India, to Cambodja ; *Galerella* (1 species), East Africa ; *Calictis* (1 species),Ceylon (?); *Ariella* (1 species), South Africa ; *Ichneumia* (4 species), Central, East, and South Africa ; *Bdeogale* (3 species), West and East Africa ; *Urva* (1 species), Himalayas to Aracan ; *Tæniogale* (1 species), Central India ; *Onychogale* (1 species), Ceylon ; *Helogale* (2 species) East and South Africa ; *Cynictis* (3 species), South Africa.

Sub-family RHINOGALIDÆ.—*Rhinogale* (1 species), East Africa ; *Mungos* (3 species), all Africa ; *Crossarchus* (1 species), Tropical Africa ; *Eupleres* (1 species), Madagascar ; *Suricata* (1 species), South Africa.

Fossil Viverridæ.—Several species of *Viverra* and *Genetta* have been found in the Upper Miocene of France, and many extinct genera have also been discovered. The most remarkable of these was *Ictitherium*, from the Upper Miocene of Greece, which has also been found in Hungary, Bessarabia, and France. Some of the species were larger than any living forms of Viverridæ, and approached the hyænas. Other extinct genera are *Thalassictis* and *Soricictis* from the Upper Miocene, the former as large as a panther; *Tylodon*, of small size, from the Upper Eocene; and *Palæonyctis* from the Lower Eocene, also small and showing a very great antiquity for this family, if really belonging to it.

FAMILY 26.—PROTELIDÆ.   (1 Genus, 1 Species.)

GENERAL DISTRIBUTION.					
NEOTROPICAL SUB-REGIONS.	NEARCTIC SUB-REGIONS.	PALÆARCTIC SUB-REGIONS.	ETHIOPIAN SUB-REGIONS.	ORIENTAL SUB-REGIONS.	AUSTRALIAN SUB-REGIONS.
— — — —	— — — —	— — — —	— — 3 —	— — — —	— — — —

The curious *Proteles* or Aard-wolf, a highly-modified form of hyæna, approaching the ichneumons, and feeding on white ants and carrion, is peculiar to South Africa.

FAMILY 27.—HYÆNIDÆ.   (1 Genus, 3 Species.)

GENERAL DISTRIBUTION.					
NEOTROPICAL SUB-REGIONS.	NEARCTIC SUB-REGIONS.	PALÆARCTIC SUB-REGIONS.	ETHIOPIAN SUB-REGIONS.	ORIENTAL SUB-REGIONS.	AUSTRALIAN SUB-REGIONS.
— — — —	— — — —	— 2 — —	1.2.3 —	1 — — —	— — — —

The Hyænas are characteristically Ethiopian, to which region two of the species are confined. The third, *Hyæna striata*, ranges over all the open country of India to the foot of the Himalayas, and through Persia, Asia Minor, and North Africa. Its fossil remains have been found in France.

Extinct Species.—The cave hyæna (*H. spelæa*) occurs abundantly in the caverns of this country and of Central Europe, and is supposed to be most nearly allied to the *H. crocuta* of South Africa. Another species is found in some parts of France. The earliest known true hyænas occur in the Pliocene formation in France, in the Red Crag (Older Pliocene) of England, and in the Upper Miocene of the Siwalik hills. In the Miocene period in Europe, quite distinct genera are found, such as *Hyænictis* and *Lycæna* from the Upper Miocene of Greece;

Ictitherium, supposed to be intermediate between Viverridæ and Hyænidæ; and *Thalassictis,* uniting the weasels and hyænas.

FAMILY 28.—CANIDÆ. (3 Genera, 17 Sub-Genera, 54 Species.)

GENERAL DISTRIBUTION.

NEOTROPICAL SUB-REGIONS.	NEARCTIC SUB-REGIONS.	PALÆARCTIC SUB-REGIONS.	ETHIOPIAN SUB-REGIONS.	ORIENTAL SUB-REGIONS.	AUSTRALIAN SUB-REGIONS.
1.2.3 –	1.2.3.4	1.2.3.4	1.2.3 –	1.2.3.4	– 2? – –

The Canidæ, comprising the animals commonly known as dogs, wolves, and foxes, have an almost universal range over the earth, being only absent from the island sub-regions of Madagascar, the Antilles, Austro-Malaya, New Zealand, and the Pacific Islands. With the exception of two remarkable forms—the hyæna dog (*Lycaon picta*), and the great-eared fox (*Megalotis Lalandei*), both from South Africa—all the species are usually placed in the genus *Canis,* the distribution of which will be the same as that of the family. Dr. J. E. Gray, in his arrangement of the family (Proc. Zool. Soc., 1868), subdivides it into fifteen genera, the names and general distribution of which are as follows:—

Icticyon (1 species), Brazil; *Cuon* (4 species), Siberia to Java; *Lupus* (5 species), North America, Europe, India to Ceylon; *Dieba* (1 species), North and West Africa; *Simenia* (1 species), Abyssinia; *Chrysocyon* (2 species), North and South America; *Canis* (4 species), India, Australia (indigenous?) *Lycalopex* (2 species), South America; *Pseudalopex* (5 species), South America and Falkland Islands; *Thous* (2 species), South America to Chili; *Vulpes* (17 species), all the great continents, except South America and Australia; *Fennecus* (4 species), all Africa; *Leucocyon* (1 species), Arctic regions; *Urocyon* (2 species), North America; *Nyctereutes* (1 species), Japan, Amoorland to Canton (Plate III., vol. i. p. 226). These are all sub-genera according to Professor Carus, except *Icticyon.* The same author makes *Lycaon* a sub-genus, while Dr. Gray makes it a sub-family!

Extinct Species.—The dog, wolf, and fox, are found fossil in caverns in many parts of Europe, and several extinct species have been found in Tertiary deposits in Europe, North India, and South America. Two species have been found so far back as the Eocene of France, but the fragments discovered are not sufficient to determine the characters with any certainty. In North America, several species of *Canis* occur in the Pliocene of Nebraska and La Plata. The genus *Galecynus*, of the Pliocene of Œninghen, and *Palæocyon*, of the Brazilian caves, are supposed to belong to the Canidæ. *Amphicyon* abounded in the Miocene period, both in Europe and North America; and some of the species were as large as a tiger. Other extinct genera are, *Cynodictis*, *Cyotherium*, and *Galethylax*, from the Eocene of France; *Pseudocyon*, *Simocyon*, and *Hemicyon*, from the Miocene; but all these show transition characters to Viverridæ or Ursidæ, and do not perhaps belong to the present family.

FAMILY 29.—MUSTELIDÆ.   (21—28 Genera, 92 Species.)

GENERAL DISTRIBUTION.					
NEOTROPICAL SUB-REGIONS.	NEARCTIC SUB-REGIONS.	PALÆARCTIC SUB-REGIONS.	ETHIOPIAN SUB-REGIONS.	ORIENTAL SUB-REGIONS.	AUSTRALIAN SUB-REGIONS.
1.2.3 —	1.2.3.4	1.2.3.4	1.2.3 —	1.2.3.4	— — — —

The Mustelidæ constitute one of those groups which range over the whole of the great continental areas. They may be divided into three sub-families—one, the Mustelinæ, containing the weasels, gluttons, and allied forms; a second, the Lutrinæ, containing the otters; and a third, often considered a distinct family, the Melininæ, containing the badgers, ratels, skunks, and their allies.

In the first group (Mustelinæ) the genera *Martes* and *Putorius* (13 species), range over all the Palæarctic region, and a considerable part of the Oriental, extending through India to Ceylon, and to Java and Borneo. Two species of *Martes* (=*Mustela* of Baird) occur in the United States. The weasels, forming the genus *Mustela* (20 species), have a still wider range, extending into tropical Africa and the Cordilleras of Peru, but not going south of the Himalayas in India. The North American species are placed in the genus *Putorius* by Professor Baird. An allied genus, *Gymnopus* (4 species), is confined to the third and fourth Oriental sub-regions. *Gulo* (1 species), the glutton, is an arctic animal keeping to the cold regions of Europe and Asia, and coming as far south as the great lakes in North America. *Galictis* (2 species), the grisons, are confined to the Neotropical region.

The Otters (Lutrinæ) range over the whole area occupied by the family. They have been subdivided into a number of groups, such as *Barangia* (1 species), found only in Sumatra; *Lontra*, containing 3 South American species; *Lutra* (7 species), ranging over the whole of the Palæarctic and Oriental regions; *Nutria* (1 species), a sea-otter confined to the west coast of America from California to Chiloe; *Lutronectes* (1 species), from Japan only; *Aonyx* (5 species), found in West and South Africa, and the third and fourth Oriental sub-regions. *Hydrogale* (1 species), confined to South Africa; *Latax* (2 species), Florida and California to Canada and British Columbia; *Pteronura* (1 species), Brazil and Surinam; and *Enhydris* (1 species), the peculiar sea-otter of California, Kamschatka and Japan. The last two are the only groups of otters, besides *Lutra*, admitted by Professor Carus as genera.

The Badgers and allies (Melininæ) have also a wide range, but with one exception are absent from South America. They comprise the following genera: *Arctonyx* (1 species), Nepal to Aracan; *Meles* (4 species), North Europe to Japan, and China as far south as Hongkong (Plate I., vol. i., p. 195); *Taxidea* (2 species), Central and Western North America to 58° N. Lat.; *Mydaus* (1 species), mountains of Java and Sumatra; *Melivora* (3 species), Tropical and South Africa and India to foot of Himalayas; *Mephitis* (12 species), America from Canada and British Columbia to the Straits of Magellan (Plate XX., vol. ii., p. 136). *Ictonyx* (2 species), Tropical Africa to the Cape; *Helictis* (4 species), Nepal to Java, Formosa and Shanghai (Plate VII., vol. i. p. 331).

Fossil Mustelidæ.—Species of otter, weasel, badger, and glutton, occur in European bone caves and other Post-tertiary deposits; and in North America *Galictis*, now found only in the Neotropical region, and, with *Mephitis*, occurring in Brazilian caves.

Species of *Mustela* have been found in the Pliocene of France and of South America; and *Lutra* in the Pliocene of North America.

In the Miocene deposits of Europe several species of *Mustela* and *Lutra* have been found; with the extinct genera *Taxodon*, *Potamotherium*, and *Palæomephitis;* as well as *Promephitis* in Greece.

In the Upper Miocene of the Siwalik Hills species of *Lutra* and *Mellivora* are found, as well as the extinct genera *Enhydrion* and *Ursitaxus*.

The family appears to have been unknown in North America during the Miocene period.

FAMILY 30.—PROCYONIDÆ.   (4 Genera, 8 Species.)

GENERAL DISTRIBUTION.					
NEOTROPICAL SUB-REGIONS.	NEARCTIC SUB-REGIONS.	PALÆARCTIC SUB-REGIONS.	ETHIOPIAN SUB-REGIONS.	ORIENTAL SUB-REGIONS.	AUSTRALIAN SUB-REGIONS.
— 2 . 3 —	1 . 2 . 3 . 4	— — — —	— — — —	— — — —	— — — —

The Procyonidæ are a small, but very curious and interesting family of bear-like quadrupeds, ranging from British Columbia and Canada on the north, to Paraguay and the limits of the tropical forests on the south.

The Racoons, forming the genus *Procyon*, are common all over North America; a well-marked variety or distinct species inhabiting the west coast, and another, most parts of South America. The genus *Nasua*, or the coatis (5 species ?), extends from Mexico and Guatemala to Paraguay. The curious arboreal prehensile-tailed kinkagou (*Cercoleptes candivolvus*) is also found in Mexico and Guatemala, and in all the great forests of Peru and North Brazil. *Bassaris* (2 species), a small weasel-like animal with a banded tail, has been usually classed with the Viverridæ or Mustelidæ, but is now found to agree closely in all important points of internal structure with this family. It is found in California, Texas, and the highlands of Mexico, and belongs therefore as much to the Nearctic as to the Neotropical region. A second species has recently been described by Professor Peters from Coban in Guatemala, in which country it has also been observed by Mr. Salvin.

Fossil Procyonidæ.—A species of *Nasua* has been found in the bone caves of Brazil, and a *Procyon* in the Pliocene or Post-pliocene deposits of Illinois and Carolina.

FAMILY 31.—ÆLURIDÆ.   (2 Genera, 2 Species.)

GENERAL DISTRIBUTION.

NEOTROPICAL SUB-REGIONS.	NEARCTIC SUB-REGIONS.	PALÆARCTIC SUB-REGIONS.	ETHIOPIAN SUB-REGIONS.	ORIENTAL SUB-REGIONS.	AUSTRALIAN SUB-REGIONS.
— — — —	— — — —	— — — 4	— — — —	— — 3 —	— — — —

The Panda (*Ælurus fulgens*), of the forest regions of the Eastern Himalayas and East Thibet, a small cat-like bear, has peculiarities of organization which render it necessary to place it in a family by itself. (Plate VII. vol. i. p. 331). An allied genus, *Æluropus*, a remarkable animal of larger size and in colour nearly all white, has recently been described by Professor Milne-Edwards, from the mountains of East Thibet; so that the family may be said to inhabit the border lands of the Oriental and Palæarctic regions. These animals have their nearest allies in the coatis and bears

FAMILY 32.—URSIDÆ.   (5 Genera, or Sub-genera, 15 Species.)

GENERAL DISTRIBUTION.

NEOTROPICAL SUB-REGIONS.	NEARCTIC SUB-REGIONS.	PALÆARCTIC SUB-REGIONS.	ETHIOPIAN SUB-REGIONS.	ORIENTAL SCB-REGIONS.	AUSTRALIAN SUB-REGIONS.
1 — — —	1.2.3.4	1.2.3.4	— — — —	1.2.3.4	— — — —

The Bears have a tolerably wide distribution, although they are entirely absent from the Australian and Ethiopian, and almost so from the Neotropical region, one species only being found in the Andes of Peru and Chili. They comprise the following groups, some of which are doubtfully ranked as genera.

Thalassarctos, the polar bear (1 species) inhabiting the Arctic regions ; *Ursus*, the true bears (12 species), which range over

all the Nearctic and Palæarctic regions as far as the Atlas Mountains, the Indo-Chinese sub-region in the mountains; and to
Hainan and Formosa; *Helarctos*, the Malay or sun-bear (1
species) confined to the Indo-Malayan sub-region; *Melursus* or
Prochilus, the honey-bear (1 species), confined to the first and
second Oriental sub-regions, over which it ranges from the
Ganges to Ceylon; and *Tremarctos*, the spectacled bear—commonly known as *Ursus ornatus*—which is isolated in the Andes
of Peru and Chili, and forms a distinct group.

Fossil Ursidæ.—Two bears (*Ursus spelæus* and *U. priscus*)
closely allied to living species, abound in the Post-tertiary deposits of Europe; and others of the same age are found in North
America, as well as an extinct genus, *Arctodus*.

Ursus arvernensis is found in the Pliocene formation of France,
and the extinct genus *Leptarchus* in that of North America.

Several species of *Amphicyon*, which appears to be an ances-
tral form of this family, are found in the Miocene deposits of
Europe and N. India; while *Ursus* also occurs in the Siwalik
Hills and Nerbudda deposits.

FAMILY 33.—OTARIIDÆ.   (4 Genera, 8 Species.)

GENERAL DISTRIBUTION.					
NEOTROPICAL SUB-REGIONS.	NEARCTIC SUB-REGIONS.	PALÆARCTIC SUB-REGIONS.	ETHIOPIAN SUB-REGIONS.	ORIENTAL SUB-REGIONS.	AUSTRALIAN SUB-REGIONS.
1 — — —	1 — — 4	— — — —	— — 3 —	— — — —	— 2 . 3 —

The Otariidæ, or Eared Seals, comprehending the sea-bears and
sea-lions, are confined to the temperate and cold shores of the
North Pacific, and to similar climates in the Southern Hemisphere,
where the larger proportion of the species are found.   They are
entirely absent from the North Atlantic shores.   Mr. J. A. Allen,
in his recent discussion of this family (Bull. Harvard Museum)
divides them into the following genera:—

Otaria (1 species), Temperate South America, from Chili to
La Plata; *Callorhinus* (1 species), Behring's Straits and Kamschatka; *Arctocephalus* (3 species), temperate regions of the

Southern Hemisphere; *Zalophus* (2 species), North Pacific, from California to Japan, and the shores of Australia and New Zealand; *Eumetopias* (1 species), Behring's Straits and California.

Fossil Otariidæ.—Remains supposed to belong to this family have been found in the Miocene of France.

FAMILY 34—TRICHECHIDÆ.   (1 Genus, 1 Species.)

GENERAL DISTRIBUTION.					
NEOTROPICAL SUB-REGIONS.	NEARCTIC SUB-REGIONS.	PALÆARCTIC SUB-REGIONS.	ETHIOPIAN SUB-REGIONS.	ORIENTAL SUB-REGIONS.	AUSTRALIAN SUB-REGIONS.
— — — —	— — — 4	1 – 3 –	— — — —	— — — —	— — — —

The Morse, or Walrus (*Trichecus rosmarus*), which alone constitutes this family, is a characteristic animal of the North Polar regions, hardly passing south of the Arctic circle except on the east and west coasts of North America, where it sometimes reaches Lat. 60°. It is most abundant on the shores of Spitzbergen, but is not found on the northern shores of Asia between Long. 80° and 160° E., or on the north shores of America from 100° to 150° west.

Its remains have been found fossil in Europe as far south as France, and in America as far as Virginia; but the small fragments discovered may render the identification uncertain.

FAMILY 35.—PHOCIDÆ.   (13 Genera, 21 Species.)

GENERAL DISTRIBUTION.					
NEOTROPICAL SUB-REGIONS.	NEARCTIC SUB-REGIONS.	PALÆARCTIC SUB-REGIONS.	ETHIOPIAN SUB-REGIONS.	ORIENTAL SUB-REGIONS.	AUSTRALIAN SUB-REGIONS.
1 – – 4?	1 – – 4	1.2.3.4	— — — —	— — — —	— 2.3 –

The earless or true Seals are pretty equally divided between the Northern and Southern Hemispheres, frequenting almost exclusively the temperate and cold regions, except two species said to occur among the West Indian islands.   The genus *Phoca* and its close allies, as well as *Halichœrus* and *Pelagius*, are northern; while *Stenorhynchus* and *Morunga*, with their allies, are mostly southern. The genera admitted by Dr. Gray in his catalogue are as follows:—

Callocephalus (3 species), Greenland, North Sea, also the Caspian Sea, and Lakes Aral and Baikal; *Pagomys* (2 species), North Sea, North Pacific, and Japan; *Pagophilus* (2 species), North Pacific and North Atlantic; *Halicyon* (1 species), North West coast of America; *Phoca* (2 species), North Atlantic and North Pacific, Japan; *Halichœrus* (1 species), Greenland, North Sea, and Baltic; *Pelagius* (2 species), Madeira, Mediterranean, Black Sea; *Stenorhynchus* (1 species), Antarctic Ocean, Falkland Islands, New Zealand; *Lobodon* (1 species), Antarctic Ocean; *Leptonyx* (1 species), Antarctic Ocean, South Australia, East Patagonia; *Ommatophoca* (1 species), Antarctic Ocean; *Morunga* (2 species), California, Falkland Islands, Temperate regions of Southern Ocean; *Cystophora* (2 species), North Atlantic, Antilles.

Fossil Seals.—Remains of living species of seals have been found in Post-tertiary deposits in many parts of Europe and in Algeria, as well as in New Zealand. *Pristiphoca occitana* is a fossil seal from the Pliocene of Montpellier, while a species of *Phoca* is said to have been found in the Miocene deposits of the United States.

General Remarks on the Distribution of the Carnivora.

Terrestrial Carnivora.—For the purposes of geographical distribution, the terrestrial and aquatic Carnivora differ too widely to be considered in one view, their areas being limited by barriers of a very different nature. The terrestrial Carnivora form a very extensive and considerably varied group of animals, having, with the doubtful exception of Australia, a world-wide distribution. Yet the range of modification of form is not very great, and the occurrence of three families consisting of but one species each, is an indication of a great amount of recent extinction. One of the most marked features presented by this group is its comparative scarcity in the Neotropical region, only four families being represented there (not counting the Ursidæ, which has only one Andean species), and both genera and species are few in number. Even the Procyonidæ, which are especially South

American, have but two genera and six species in that vast area. We might therefore, from these considerations alone, conclude that Carnivora are a development of the northern hemisphere, and have been introduced into the Neotropical region at a comparatively recent epoch. The claim of the Nearctic region to be kept distinct from the Palæarctic (with which some writers have wished to unite it) is well maintained by its possession of at least six species of *Mephitis*, or skunk, a group having no close allies in any other region,—and the genera *Procyon* and *Bassaris*,—for the latter, ranging from the high lands of Guatemala and Mexico to Texas and California, may be considered a Nearctic rather than a Neotropical form.   In the other families, the most marked feature is the total absence of Ursidæ from the Ethiopian region.   The great mass of the generic forms of Carnivora, however, are found in the Oriental and Ethiopian regions, which possess all the extensive group of Viverridæ (except a few species in the fourth Palæarctic subregion) and a large number of Felidæ and Mustelidæ.

Aquatic Carnivora.—The aquatic Carnivora present no very marked features of distribution, except their preference for cold and temperate rather than tropical seas.   Their nearest approximation to the terrestrial group, is supposed to be that of the Otariidæ to the Ursidæ; but this must be very remote, and the occurrence of both seals and bears in the Miocene period, shows, that until we find some late Secondary or early Tertiary formation rich in Mammalian remains, we are not likely to get at the transition forms indicating the steps by which the aquatic Carnivora were developed.   The most interesting special fact of distribution to be noticed, is the occurrence of seals, closely allied to those inhabiting the northern seas, in the Caspian, Lake Aral, and Lake Baikal.   In the case of the two first-named localities there is little difficulty, as they are connected with the North Sea by extensive plains of low elevation, so that a depression of less than 500 feet would open a free communication with the ocean. At a comparatively recent epoch, a great gulf of the Arctic ocean must have occupied the valley of the Irtish, and extended to the Caspian Sea; till the elevation of the Kirghiz Steppes cut off the communication with the ocean, leaving an inland sea with its seals. Lake Baikal, however, offers much greater difficulties; since it is not only a fresh-water lake, but is situated in a mountain district nearly 2,000 feet above the sea level, and entirely separated from the plains by several hundred miles of high land.   It is true that such an amount of submergence and elevation is known to have occurred in Europe so recently as during the Glacial period; but Lake Baikal is so surrounded by mountains, that it must at that time have been filled with ice, if at anything like its present elevation.   Its emergence from the sea must therefore have taken place since the cold epoch, and this would imply that an enormous extent of Northern Asia has been very recently under water.

We are accustomed to look on Seals as animals which exclusively inhabit salt water; but it is probably from other causes than its saltness that they usually keep to the open sea, and there seems no reason why fresh-water should not suit them quite as well, provided they find in it a sufficiency of food, facilities for rearing their young, and freedom from the attacks of enemies. As already remarked in vol. i. p. 218, Mr. Belt's ingenious hypothesis (founded on personal examination of the Siberian Steppes), that during the Glacial period the northern ice-cap dammed up the waters of the northward flowing Asiatic rivers, and thus formed a vast fresh-water lake which might have risen as high as Lake Baikal, seems to offer the best solution of this curious problem of distribution.

Range of Carnivora in Time.—Carnivora have been found in all the Tertiary deposits, and comprise a number of extinct genera and even families.   Several genera of Canidæ occur in the Upper Eocene of Europe; but the most remarkable fact is, that even in the Lower Eocene are found two well-marked forms, *Palæonyctis*, one of the Viverridæ, and *Arctocyon*, forming a distinct family type of very generalized characters, but unmistakably a carnivore.   This last has been found at La Fère, in the north-east of France, in a deposit which, according to M. Gaudry, is the very lowest of the Lower Eocene formation in Europe.   *Arctocyon* is therefore one of the oldest, if not the very oldest, of the higher forms of mammal yet discovered.

Order V.—CETACEA.

FAMILY 36.—BALÆNIDÆ.   (6 Genera, 14 Species.)

GENERAL DISTRIBUTION.—Temperate and Cold Seas of both Northern and Southern Hemispheres.

This family comprises the whalebone or "right" whales, the best known species being the Greenland whale (*Balæna mysticetus*). Allied species are found in all parts of the southern seas, as far north as the Cape of Good Hope; while some of the northern species are found off the coast of Spain, and even enter the Mediterranean. As most of the species indicated are imperfectly known, and their classification by no means well settled, no useful purpose will be served by enumerating the genera or sub-genera.

FAMILY 37.—BALÆNOPTERIDÆ.   (9 Genera, 22 Species.)

GENERAL DISTRIBUTION.—Cold and Temperate Seas of both Hemispheres.

This family comprises the finner whales and rorquals, and are characterised by possessing a dorsal fin and having the baleen or whalebone less developed. They are abundant in all northern seas, less so in the southern hemisphere, but they seem occasionally to enter the tropical seas. The best known genera are *Megaptera* (7 species); *Physalus* (4 species); and *Balænoptera* (2 species); all of which have species in the North Sea.

FAMILY 38.—CATODONTIDÆ.   (4 Genera, or Sub-Genera, 6 Species.)

GENERAL DISTRIBUTION.—All the Tropical Oceans, extending north and south into Temperate waters.

This family, comprising the cachalots or sperm whales, and black-fish, are separated from the true whales by having teeth in the lower jaw and no whalebone. They are pre-eminently a tropical, as distinguished from the two preceding which are arctic and antarctic families. The spermaceti whale (*Catodon macrocephalus*) abounds in the Pacific Ocean and in the deep Moluccan Sea, and also in the Indian Ocean and the Mozambique Channel. In the Atlantic it is scarce, although it occasionally comes north as far as our shores.

The genera of Catodontidæ as given by Dr. Gray are, *Catodon* (2 species ?), Warm Eastern Oceans; *Physeter* (1 species), "the black fish," North Sea; *Cogia* (2 species), South Temperate Oceans; *Euphysetes* (1 species), Coast of Australia.

FAMILY 39.—HYPEROODONTIDÆ. (9 Genera or Sub-Genera, 12 Species.)

GENERAL DISTRIBUTION.—Atlantic, Mediterranean, Indian Ocean, and Southern Ocean.

This family consists of the beaked whales, which have no permanent teeth in the upper jaw. The genera, according to Dr. Gray, are, *Hyperoodon* (2 species) "bottle-nosed whales," North Sea; *Lagenocetus* (1 species), North Sea; *Epiodon* (2 species), North and South Atlantic; *Petrorhynchus* (2 species), Mediterranean Sea and Southern Ocean; *Berardius* (1 species), New Zealand; *Xiphius* (1 species) North Atlantic; *Dolichodon* (1 species), Cape of Good Hope; *Neoziphius* (1 species) Mediterranean; *Dioplodon* (1 species), Indian Ocean.

FAMILY 40.—MONODONTIDÆ. (1 Genus, 1 Species.)

The "Narwhal" (*Monodon monoceros*) which constitutes this family, is placed by Dr. Gray along with the "white whales," in his family Belugidæ. It inhabits the North Sea.

FAMILY 41.—DELPHINIDÆ. (24 Genera or Sub-Genera, 100 Species.)

GENERAL DISTRIBUTION.—All Oceans, Seas, and Great Rivers of the globe.

This family, including the Porpoises, Dolphins, White Whales, &c., may be described as small, fish-shaped whales, having teeth in both jaws. According to Dr. Gray they form seven families and 24 genera; according to Professor Carus, four sub-families and 8 genera, but as these groups appear to be established on quite different principles, and often differ widely from each other, I shall simply enumerate Dr. Gray's genera with their distribution as given in his British Museum Catalogue.

Platanista (2 species), long-snouted porpoises, inhabiting the Ganges and Indus ; *Inia* (1 species), a somewhat similar form, inhabiting the upper waters of the Amazonian rivers : *Steno* (8 species), Indian Ocean, Cape of Good Hope, and West Pacific ; *Sotalia* (1 species), Guiana ; *Delphinus* (10 species), all the oceans ; *Clymenia* (14 species), all the oceans ; *Delphinapterus* (1 species), South Atlantic ; *Tursio* (7 species), Atlantic and Indian Oceans ; *Eutropia* (2 species), Chili, and Cape of Good Hope ; *Electra* (8 species), all the oceans ; *Leucopleurus* (1 species), North Sea ; *Lagenorhynchus* (1 species), North Sea ; *Pseudorca* (2 species), North Sea, Tasmania ; *Orcaella* (2 species), Ganges ; *Acantho-delphis* (1 species), Brazil ; *Phocæna* (2 species), North Sea ; *Neo-meris* (1 species), India ; *Grampus* (3 species), North Sea, Mediterranean, Cape of Good Hope ; *Globiocephalus* (14 species), all the oceans ; *Sphærocephalus* (1 species), North Atlantic ; *Orca* (9 species), Northern and Southern Oceans ; *Ophysia* (1 species), North Pacific ; *Beluga* (6 species), Arctic Seas, Australia ; *Pontoporia* (1 species), Monte Video.

Fossil Cetacea.

Remains of Cetacea are tolerably abundant in Tertiary deposits, both in Europe and North America. In the Lower Pliocene of England, France, and Germany, extinct species of five or six living genera of whales and dolphins have been found ; and most of these occur also in the Upper Miocene, along with many others, referred to about a dozen extinct genera.

In the Post-pliocene deposits of Vermont and South Carolina, several extinct species have been found belonging to living genera; but in the Miocene deposits of the Eastern United States cetacean remains are much more abundant, more than 30 species of extinct whales and dolphins having been described, most of them belonging to extinct genera.

The Zeuglodontidæ, an extinct family of carnivorous whales, with double-fanged serrated molar teeth, whose affinities are somewhat doubtful, are found in the older Pliocene of Europe, and in the Miocene and Eocene of the Eastern United States. *Zeuglodon* abounds in the United States, and one species reached a length of seventy feet. A species of this genus is said to have been found in Malta. *Squalodon* occurs in Europe and North America ; and in the latter country four or five other genera have been described, of which one, *Saurocetes*, has been found also at Buenos Ayres.

Order VI.—SIRENIA.

FAMILY 42.—MANATIDÆ. (3 Genera, 5 Species ?)

GENERAL DISTRIBUTION.					
NEOTROPICAL SUB-REGIONS.	NEARCTIC SUB-REGIONS.	PALÆARCTIC SUB-REGIONS.	ETHIOPIAN SUB-REGIONS.	ORIENTAL SUB-REGIONS.	AUSTRALIAN SUB-REGIONS.
— 2 — 4	— — — —	1 — 3 —	1 . 2 — —	1 . 2 — 4	1 — — —

The Sea-cows are herbivorous aquatic animals living on the coasts or in the great rivers of several parts of the globe. *Manatus* (2 species) inhabits both shores of the Atlantic, one species ranging from the Gulf of Mexico to North Brazil, and ascending the Amazon far into the interior of the continent; while the other is found on the west coast of Africa. *Halicore* (2 species ?), the Dugong, is peculiar to the Indian Ocean, extending from Mozambique to the Red Sea, thence to Western India and Ceylon, the Malay Archipelago and the north coast of Australia. *Rytina* (1 species), supposed to be now extinct, inhabited recently the North Pacific, between Kamschatka and Behring's Straits.

Fossil Sirenia.—Extinct species of *Manatus* have been found in the Post-pliocene deposits of Eastern North America from

Maryland to Florida; and an extinct genus, *Prorastomus*, in some Tertiary deposits in the Island of Jamaica.

In Post-pliocene deposits in Siberia, remains of *Rytina* have been found; while several species of the extinct genus *Halitherium*, perhaps intermediate between *Manatus* and *Halicore*, have been found in the older Pliocene and Upper Miocene of France and Germany.

Order VII.—UNGULATA.

FAMILY 43.—EQUIDÆ.   (1 Genus, 8 Species.)

GENERAL DISTRIBUTION.					
NEOTROPICAL SUB-REGIONS.	NEARCTIC SUB-REGIONS.	PALÆARCTIC SUB-REGIONS.	ETHIOPIAN SUB-REGIONS.	ORIENTAL SUB-REGIONS.	AUSTRALIAN SUB-REGIONS.
LIVING SPECIES.					
— — — —	— — — —	— 2 . 3 —	1 . 2 . 3 —	— — — —	— — — —
EXTINCT SPECIES.					
1 . 2 — —	1 . 2 . 3 —	1 . 2 . 3 . 4	— — — —	1 — 3 —	— — — —

The Horses, Asses, and Zebras form a highly specialized group now confined to the Ethiopian and Palæarctic regions, but during the middle and later tertiaries having a very extensive range. The zebras (3 species) inhabit the greater part of the Ethiopian region, while the asses (4 species) are characteristic of the deserts of the Palæarctic region from North Africa and Syria to Western India, Mongolia, and Manchuria. The domestic horse is not known in a wild state, but its remains are found in recent deposits from Britain to the Altai Mountains, so that its disappearance is probably due to human agency.

Extinct Equidæ.—Extinct forms of this family are very numerous. The genus *Equus* occurs in Post-pliocene and Pliocene deposits in Europe, North America, and South America. In North America the species are most numerous. An allied genus *Hipparion*, having rudimentary lateral toes, is represented

P 2

by several species in the Pliocene of North America, while in
Europe it occurs both in the Older Pliocene and Upper Miocene.
Various other allied forms, in which the lateral toes are more
and more developed, and most of which are now classed in a distinct family, Anchitheridæ, range back through the Miocene to
the Eocene period. A sufficient account of these has already
been given in vol. i. chap. vi. p. 135, to which the reader is
referred for the supposed origin and migrations of the horse.

FAMILY 44.—TAPIRIDÆ.   (2 Genera? 6 Species.)

GENERAL DISTRIBUTION.

NEOTROPICAL SUB-REGIONS.	NEARCTIC SUB-REGIONS.	PALÆARCTIC SUB-REGIONS.	ETHIOPIAN SUB-REGIONS.	ORIENTAL SUB-REGIONS.	AUSTRALIAN SUB-REGIONS.
— 2 . 3 —	— — — —	— — — —	— — — —	— — — 4	— — — —

The Tapirs form a small group of animals whose discontinuous
distribution plainly indicates their approaching extinction. For
a long time only two species were known, the black American,
and the white-banded Malay tapir, the former confined to the
equatorial forests of South America, the latter to the Malay
peninsula, Sumatra, and Borneo (Plate VIII. vol. i. p. 337).
Lately however another, or perhaps two distinct species (or according to Dr. J. E. Gray, four!) have been discovered in the
Andes of New Granada and Ecuador, at an elevation of from
8,000 to 12,000 feet; while one or perhaps two more, forming
the allied genus *Elasmognathus*, have been found to inhabit
Central America from Panama to Guatemala.

Extinct Tapirs.—True tapirs inhabited Western Europe, from
the latest Pliocene back to the earliest Miocene times; while
they only occur in either North or South America in the Post-
pliocene deposits and caves. The singular distribution of the
living species is thus explained, since we see that they are
an Old World group which only entered the American continent
at a comparatively recent epoch. An ancestral form of this
group—*Lophiodon*—is found in Miocene and Eocene deposits of

Europe and North America; while a still more ancient form of large size is found in the Lower Eocene of France and England, indicating an immense antiquity for this group of Mammalia. There are many other extinct forms connecting these with the Palæotheridæ, already noticed in chapter vi. (vol. i. pp. 119–125).

FAMILY 45.—RHINOCEROTIDÆ.   (1 Genus, 9 Species.)

GENERAL DISTRIBUTION.

NEOTROPICAL SUB-REGIONS.	NEARCTIC SUB-REGIONS.	PALÆARCTIC SUB-REGIONS.	ETHIOPIAN SUB-REGIONS.	ORIENTAL SUB-REGIONS.	AUSTRALIAN SUB-REGIONS.

LIVING SPECIES.

— — — —	— — — —	- — — —	1 . 2 . 3 —	— — 3 . 4	— — — —

EXTINCT SPECIES.

— — — —	1 . 2 — —	1 . 2 . 3 . 4	— — — —	1 — 3 —	— — — —

Living Rhinoceroses are especially characteristic of Africa, with Northern and Malayan India. Four or perhaps five species, all two-horned, are found in Africa, where they range over the whole country south of the desert to the Cape of Good Hope. In the Oriental region there are also four or five species, which range from the forests at the foot of the Himalayas eastwards through Assam, Chittagong, and Siam, to Sumatra, Borneo and Java. Three of these are one-horned, the others found in Sumatra, and northwards to Pegu and Chittagong, two-horned. The Asiatic differ from the African species in some dental characters, but they are in other respects so much alike that they are not generally considered to form distinct genera. In his latest catalogue however (1873), Dr. Gray has four genera, *Rhinoceros* (4 species), and *Ceratorhinus* (2 species), Asiatic; *Rhinaster* (2 species), and *Ceratotherium* (2 species), African.

Extinct Rhinocerotidæ.—Numerous species of *Rhinoceros* ranged over Europe and Asia from the Post-pliocene back to the Upper Miocene period, and in North America during the Pliocene period

only.   The hornless *Acerotherium* is Miocene only, in both
countries.   Other genera are, *Leptodon* from Greece, and *Hyracodon* from Nebraska, both of Miocene age.   More than 20
species of extinct rhinoceroses are known, and one has even been
found at an altitude of 16,000 feet in Thibet.

FAMILY 46.—HIPPOPOTAMIDÆ.   (1 Genus, 2 Species.)

GENERAL DISTRIBUTION.					
NEOTROPICAL SUB-REGIONS.	NEARCTIC SUB-REGIONS.	PALÆARCTIC SUB-REGIONS.	ETHIOPIAN SUB-REGIONS.	ORIENTAL SUB-REGIONS.	AUSTRALIAN SUB-REGIONS.
LIVING SPECIES.					
— — — —	— — — —	— — — —	1 . 2 . 3 —	— — — —	— — — —
EXTINCT SPECIES.					
— — — —	— — — —	1 . 2 — —	— — — —	1 - 3 —	— — — —

The Hippopotamus inhabits all the great rivers of Africa; a
distinct species of a smaller size being found on the west coast,
and on some of the rivers flowing into Lake Tchad.

Fossil Hippopotami.—Eight extinct species of *Hippopotamus*
are known from Europe and India, the former Post-pliocene or
Pliocene, the latter of Upper Miocene age.   They ranged as far
north as the Thames valley.   An extinct genus from the Siwalik
Hills, *Merycopotamus,* according to Dr. Falconer connects *Hippopotamus* with *Anthracotherium,* an extinct form from the Miocene
of Europe, allied to the swine.

FAMILY 47.—SUIDÆ.   (5 Genera, 22 Species.)

GENERAL DISTRIBUTION.					
NEOTROPICAL SUB-REGIONS.	NEARCTIC SUB-REGIONS.	PALÆARCTIC SUB-REGIONS.	ETHIOPIAN SUB-REGIONS.	ORIENTAL SUB-REGIONS.	AUSTRALIAN SUB-REGIONS
— 2 . 3 —	— 2 . 3 —	1 . 2 . 3 . 4	1 . 2 . 3 . 4	1 . 2 . 3 . 4	1 . — — —

The Swine may be divided into three well-marked groups,
from peculiarities in their dentition.   1. The Dicotylinæ, or

peccaries (1 genus, *Dicotyles*). These offer so many structural differences that they are often classed as a separate family. 2. The true swine (3 genera, *Sus, Potamochœrus,* and *Babirusa*) ; and, 3. The Phacochœrinæ, or wart hogs (1 genus, *Phacochœrus*). These last are also sometimes made into a separate family, but they are hardly so distinct as the Dicotylinæ.

The Peccaries (2 species), are peculiar to the Neotropical region, extending from Mexico to Paraguay. They also spread northwards into Texas, and as far as the Red River of Arkansas, thus just entering the Nearctic region ; but with this exception swine are wholly absent from this region, forming an excellent feature by which to differentiate it from the Palæarctic.

Sus (14 species), ranges over the Palæarctic and Oriental regions and into the first Australian sub-region as far as New Guinea ; but it is absent from the Ethiopian region, or barely enters it on the north-east. *Potamochœrus* (3 species ?), is wholly Ethiopian (Plate V. vol. i. p. 278). *Babirusa* (1 species), is confined to two islands, Celebes and Bouru, in the first Australian sub-region.

Phacochœrus (2 species), ranges over tropical Africa from Abyssinia to Caffraria.

Dr. J. E. Gray divides true swine (*Sus*) into 7 genera, but it seems far better to keep them as one.

Fossil Suidæ.—These are very numerous. Many extinct species of wild hog (*Sus*), are found in Europe and North India, ranging back from the Post-pliocene to the Upper Miocene formations. In the Miocene of Europe are numerous extinct genera, *Bothriodon, Anthracotherium, Palæochœrus, Hyotherium,* and some others ; while in the Upper Eocene occur *Cebochœrus, Chœropotamus,* and *Acotherium,*—these early forms having more resemblance to the peccaries.

None of these genera are found in America, where we have the living genus *Dicotyles* in the Post-pliocene and Pliocene deposits, both of North and South America ; with a number of extinct genera in the Miocene. The chief of these are, *Elotherium, Perchœrus, Leptochœrus,* and *Nanohyus,* all from Dakota, and *Thinohyus,* from Oregon. One extinct genus, *Platygonus,* closely allied to *Dicotyles,* is found in the Post-pliocene of Nebraska.

Oregon, and Arkansas. *Elotherium* is said to be allied to the peccary and hippopotamus. *Hyopotamus*, from the Miocene of Dakota, is allied to *Anthracotherium*, and forms with it (according to Dr. Leidy) a distinct family of ancestral swine.

It thus appears, that the swine were almost equally well represented in North America and Europe, during Miocene and Pliocene times, but by entirely distinct forms ; and it is a remarkable fact that these hardy omnivorous animals, should, like the horses, have entirely died out in North America, except a few peccaries which have preserved themselves in the sub-tropical parts and in the southern continent, to which they are comparatively recent emigrants. We can hardly have a more convincing proof of the vast physical changes that have occurred in the North American continent during the Pliocene and Post-pliocene epochs, than the complete extinction of these, along with so many other remarkable types of Mammalia.

According to M. Gaudry, the ancestors of all the swine, with the hippopotami and extinct *Anthracotherium, Merycopotamus,* and many allied forms,—are the *Hyracotherium* and *Pliolophus,* both found only in the London clay belonging to the Lower Eocene formation.

FAMILY 48.—CAMELIDÆ.   (2 Genera, 6 Species).

NEOTROPICAL SUB-REGIONS.	NEARCTIC SUB-REGIONS.	PALÆARCTIC SUB-REGIONS.	ETHIOPIAN SUB-REGIONS.	ORIENTAL SUB-REGIONS.	AUSTRALIAN SUB-REGIONS.
GENERAL DISTRIBUTION.					
LIVING SPECIES.					
1. — — —	— — — —	— 2 . 3 —	— — — —	— — — —	— — — —
EXTINCT SPECIES.					
1 — — —	— 2 . 3 . 4	— — 3 —	— — — —	— — 3 . —	— — — —

The Camels are an exceedingly restricted group, the majority of the species now existing only in a state of domestication. The genus *Camelus* (2 species), is a highly characteristic desert form of the Palæarctic region, from the Sahara to Mongolia as far as Lake Baikal. *Auchenia* (4 species), comprehending the Llamas and Alpacas, is equally characteristic of the mountains and deserts of the southern part of South America. Two species entirely domesticated inhabit the Peruvian and Bolivian Andes; and two others are found in a wild state, the vicuna in the Andes of Peru and Chili (Plate XVI. vol. ii. p. 40), and the guanaco over the plains of Patagonia and Tierra del Fuego.

Extinct Camelidæ.—No fossil remains of camels have been found in Europe, but one occurs in the deposits of the Siwalik Hills, usually classed as Upper Miocene, but which some naturalists think are more likely of Older Pliocene age. *Merycotherium*, teeth of which have been found in the Siberian drift, is supposed to belong to this family.

In North America, where no representative of the family now exists, the camel-tribe were once abundant. In the Post-pliocene deposits of California an *Auchenia* has been found, and in those of Kansas one of the extinct genus *Procamelus*. In the Pliocene period, this genus, which was closely allied to the living camels, abounded, six or seven species having been described from Nebraska and Texas, together with an allied form *Homocamelus*. In the Miocene period different genera appear,—*Pœbrotherium*, and *Protomeryx*,—while a *Procamelus* has been found in deposits of this age in Virginia.

In South America a species of *Auchenia* has been found in the caves of Brazil, and others in the Pliocene deposits of the pampas, together with two extinct genera, *Palæolama* and *Camelotherium*.

We thus find the ancestors of the Camelidæ in a region where they do not now exist, but which is situated so that the now widely separated living forms could easily have been derived from it. This case offers a remarkable example of the light thrown by palæontology on the distribution of living animals; and it is a warning against the too common practice of assuming the direct land connection of remote continents, in order to explain similar instances of discontinuous distribution to that of the present family.

FAMILY 49.—TRAGULIDÆ.   (2 Genera, 6 Species.)

GENERAL DISTRIBUTION.

NEOTROPICAL SUB-REGIONS.	NEARCTIC SUB-REGIONS.	PALÆARCTIC SUB-REGIONS.	ETHIOPIAN SUB-REGIONS.	ORIENTAL SUB-REGIONS.	AUSTRALIAN SUB-REGIONS.
— — — —	— — — —	— — — —	— 2 — —	1 . 2 . 3 . 4	— — — —

The Tragulidæ are a group of small, hornless, deer-like animals, with tusks in the upper jaw, and having some structural affinities with the camels.   The musk-deer was formerly classed in this family, which it resembles externally ; but a minute examination of its structure by M. Milne-Edwards, has shown it to be more nearly allied to the true deer.   The Chevrotains, or mouse-deer, *Tragulus* (5 species), range over all India to the foot of the Himalayas and Ceylon, and through Assam, Malacca, and Cambodja, to Sumatra, Borneo, and Java (Plate VIII., vol. i. p. 337).   *Hyomoschus* (1 species), is found in West Africa.

Extinct Tragulidæ.—A species of *Hyomoschus* is said to have been found in the Miocene of the South of France, as well as three extinct genera, *Dremotherium* (also found in Greece), with *Lophiomeryx* from the Upper Miocene, said to be allied to *Tragulus ;* and *Amphitragulus* from the Lower Miocene, of more remote affinities, and sometimes placed among the Deer.   There seems to be no doubt, however, that this family existed in Europe in Miocene times ; and thus another case of discontinuous distribution is satisfactorily accounted for.

FAMILY 50.—CERVIDÆ.   (8 Genera, 52 Species.)

GENERAL DISTRIBUTION.

NEOTROPICAL SUB-REGIONS.	NEARCTIC SUB-REGIONS.	PALÆARCTIC SUB-REGIONS.	ETHIOPIAN SUB-REGIONS.	ORIENTAL SUB-REGIONS.	AUSTRALIAN SUB-REGIONS.
1 . 2 . 3 —	1 . 2 . 3 . 4	1 . 2 . 3 . 4	— — — —	1 . 2 . 3 . 4	1 — — —

The Cervidæ, or deer tribe, are an extensive group of animals equally adapted for inhabiting forests or open plains, the Arctic regions or the Tropics. They range in fact over the whole of the great continents of the globe, with the one striking exception of Africa, where they are only found on the shores of the Mediterranean which form part of the Palæarctic region. The following is the distribution of the genera.

Alces (1 species), the elk or moose, ranges all over Northern Europe and Asia, as far south as East Prussia, the Caucasus, and North China ; and over Arctic America to Maine on the East, and British Columbia on the west. The American species may however be distinct, although very closely allied to that of Europe. *Tarandus* (1 species), the reindeer, has a similar range to the last, but keeps farther north in Europe, inhabiting Greenland and Spitzbergen ; and in America extends farther south, to New Brunswick and the north shore of Lake Superior. There are several varieties or species of this animal confined to special districts, but they are not yet well determined. *Cervus* (40 species), the true deer, have been sub-divided into numerous subgenera characteristic of separate districts. They range over the whole area of the family, except that they do not go beyond 57° N. in America and a little further in Europe and Asia. In South America they extend over Patagonia and even to Tierra del Fuego. They are found in the north of Africa, and over the whole of the Oriental region, and beyond it as far as the Moluccas and Timor, where however they have probably been introduced by man at an early period. *Dama* (1 species), the fallow deer, is a native of the shores of the Mediterranean, from Spain and Barbary to Syria. *Capreolus* (2 species), the roe-deer, inhabits all Temperate and South Europe to Syria, with a distinct species in N. China. *Cervulus* (4 species), the muntjacs, are found in all the forest districts of the Oriental region, from India and Ceylon to China as far north as Ningpo and Formosa, also southward to the Philippines, Borneo, and Java. *Moschus* (1 species) the musk-deer, inhabits Central Asia from the Amoor and Pekin, to the Himalayas and the Siamese mountains above 8000 ft. elevation. This is usually classed as a distinct family, but M. Milne-Edwards remarks, that it differs in no important points of organisation from the rest of the Cervidæ. *Hydropotcs*

(1 species) inhabits China from the Yang-tse Kiang northwards. This new genus has recently been discovered by Mr. Swinhoe, who says its nearest affinities are with *Moschus*. Other new forms are *Lophotragus*, and *Elaphodus*, both inhabiting North China; the former is hornless, the latter has very small horns about an inch long.

Extinct Deer.—Numerous extinct species of the genus *Cervus* are found fossil in many parts of Europe, and in all formations between the Post-pliocene and the Upper Miocene. The Elk and Reindeer are also found in caves and Post-pliocene deposits, the latter as far south as the South of France. Extinct genera only, occur in the Upper Miocene in various parts of Europe :— *Micromeryx*, *Palæomeryx*, and *Dicrocercus* have been described; with others referred doubtfully to *Moschus*, and an allied genus *Amphimoschus*.

In N. America, remains of this family are very scarce, a *Cervus* allied to the existing wapiti deer, being found in Post-pliocene deposits, and an extinct genus, *Leptomeryx*, in the Upper Miocene of Dakota and Oregon. Another extinct genus, *Merycodus*, from the Pliocene of Oregon, is said to be allied to camels and deer.

In South America, several species of *Cervus* have been found in the Brazilian caves, and in the Pliocene deposits of La Plata.

It thus appears, that there are not yet sufficient materials for determining the origin and migrations of the Cervidæ. There can be little doubt that they are an Old World group, and a comparatively recent development; and that some time during the Miocene period they passed to North America, and subsequently to the Southern continent. They do not however appear to have developed much in North America, owing perhaps to their finding the country already amply stocked with numerous forms of indigenous Ungulates.

FAMILY 51.—CAMELOPARDALIDÆ. (1 Genus, 1 Species.)

GENERAL DISTRIBUTION.					
NEOTROPICAL SUB-REGIONS.	NEARCTIC SUB-REGIONS.	PALÆARCTIC SUB-REGIONS.	ETHIOPIAN SUB-REGIONS.	ORIENTAL SUB-REGIONS.	AUSTRALIAN SUB-REGIONS.
LIVING SPECIES.					
— — — —	— — — —	— — — —	1 – 3 –	— — — —	— — — —
EXTINCT SPECIES.					
— — — —	— — — —	– 2 – –	1 – – –	– – 3 –	— — — —

The Camelopardalidæ, or giraffes, now consist of but a single species which ranges over all the open country of the Ethiopian region, and is therefore almost absent from West Africa, which is more especially a forest district. During the Middle Tertiary period, however, these animals had a wider range, over Southern Europe and Western India as far as the slopes of the Himalayas.

Extinct Species.—Species of *Camelopardalis* have been found in Greece, the Siwalik Hills, and Perim Island at the entrance to the Red Sea; and an extinct genus, *Helladotherium*, more bulky but not so tall as the giraffe, ranged from the south of France to Greece and North-west India.

FAMILY 52.—BOVIDÆ. (34 Genera, 149 Species.)

GENERAL DISTRIBUTION.					
NEOTROPICAL SUB-REGIONS.	NEARCTIC SUB-REGIONS.	PALÆARCTIC SUB-REGIONS.	ETHIOPIAN SUB-REGIONS.	ORIENTAL SUB-REGIONS.	AUSTRALIAN SUB-REGIONS.
— — — —	1 . 2 – 4	1 . 2 . 3 . 4	1 . 2 . 3 –	1 . 2 . 3 . 4	1 – – –

This large and important family, includes all the animals commonly known as oxen, buffaloes, antelopes, sheep, and goats, which have been classed by many naturalists in at least three, and sometimes four or five, distinct families. Zoologically, they are briefly and accurately defined as, " hollow-horned ruminants ;" and, although they present wide differences in external form, they grade so insensibly into each other, that no satisfactory definition of the smaller family groups can be found. As a whole they are almost confined to the great Old World continent, only a few forms extending along the highlands and prairies of the Nearctic region ; while one peculiar type is found in Celebes, an island which is almost intermediate between the Oriental and Australian regions. In each of the Old World regions there are found a characteristic set of types. Antelopes prevail in the Ethiopian region ; sheep and goats in the Palæ-arctic ; while the oxen are perhaps best developed in the Oriental region.

Sir Victor Brooke, who has paid special attention to this family, divides them into 13 sub-families, and I here adopt the arrangement of the genera and species which he has been so good as to communicate to me in MSS.

Sub-family I. BOVINÆ (6 genera, 13 species). This group is one of the best marked in the family. It comprises the Oxen and Buffaloes with their allies, and has a distribution very nearly the same as that of the entire family. The genera are as follows : *Bos* (1 sp.), now represented by our domestic cattle, the descendants of the *Bos primigenius*, which ranged over a large part of Central Europe in the time of the Romans. The Chillingham wild cattle are supposed to be the nearest approach to the original species. *Bison* (2 sp.), one still wild in Poland and the Caucasus ; the other in North America, ranging over the prairies west of the Mississippi, and on the eastern slopes of the Rocky Mountains (Plate XIX., vol. ii., p. 129). *Bibos* (3 sp.), the Indian wild cattle, ranging over a large part of the Oriental region, from Southern India to Assam, Burmah, the Malay Peninsula, Borneo, and Java. *Poephagus* (1 sp.), the yak, confined to the high plains of Western Thibet. *Bubalus* (5 sp.), the buffaloes, of which three species are African, ranging over all the continental parts of the Ethiopian region ; one Northern and Central Indian ; and the domesticated animal in South Europe and North Africa. *Anoa* (1 sp.), the small wild cow of Celebes, a very peculiar form more nearly allied to the buffaloes than to any other type of oxen.

Sub-family II. TRAGELAPHINÆ (3 genera, 11 species). The Bovine Antelopes are large and handsome animals, mostly Ethiopian, but extending into the adjacent parts of the Palæarctic and Oriental regions. The genera are: *Orcas* (2 sp.), elands, inhabiting all Tropical and South Africa. *Tragelaphus* (8 sp.), including the bosch-bok, kudu, and other large antelopes, ranges over all Tropical and South Africa (Plate IV., vol. ii., p. 261). *Portax* (1 sp.) India, but rare in Madras and north of the Ganges.

Sub-family III. ORYGINÆ (2 genera, 5 species). *Oryx* (4 sp.) is a desert genus, ranging over all the African deserts to South Arabia and Syria; *Addax* (1 sp.) inhabits North Africa, North Arabia, and Syria.

Sub-family IV. HIPPOTRAGINÆ (1 genus, 3 species). The Sable Antelopes, *Hippotragus*, form an isolated group inhabiting the open country of Tropical Africa and south to the Cape.

Sub-family V. GAZELLINÆ (6 genera, 23 species). This is a group of small or moderate-sized animals, most abundant in the deserts on the borders of the Palæarctic, Oriental, and Ethiopian regions. *Gazella* (17 sp.) is typically a Palæarctic desert group, ranging over the great desert plateaus of North Africa, from Senegal and Abyssinia to Syria, Persia, Beloochistan, and the plains of India, with one outlying species in South Africa. *Procapra* (2 sp.), Western Thibet and Mongolia to about 110° east longitude. *Antilope* (1 sp.) inhabits all the plains of India. *Æpyceros* (1 sp.) the pallah, inhabits the open country of South and South-east Africa. *Saiga* (1 sp.) a singular sheep-faced antelope, which inhabits the steppes of Eastern Europe and Western Asia from Poland to the Irtish River, south of 55° north latitude. (Plate II., vol. i., p. 218.) *Panthalops* (1 sp.) confined to the highlands of Western Thibet and perhaps Turkestan.

Sub-family VI. ANTILOCAPRINÆ (1 genus, 1 species), *Antilocapra*, the prong-horned antelope, inhabit both sides of the Rocky Mountains, extending north to the Saskatchewan and

Columbia River, west to the coast range of California, and east to the Missouri. Its remarkable deciduous horns seem to indicate a transition to the Cervidæ. (Plate XIX., vol. ii., p. 129.)

Sub-family VII. CERVICAPRINÆ (5 genera, 21 species). This group of Antelopes is wholly confined to the continental portion of the Ethiopian region. The genera are: *Cervicapra* (4 sp.), Africa, south of the equator and Abyssinia; *Kobus* (6 sp.), grassy plains and marshes of Tropical Africa; *Pelea* (1 sp.), South Africa; *Nanotragus* (9 species), Africa, south of the Sahara; *Neotragus* (1 sp.) Abyssinia and East Africa.

Sub-family VIII. CEPHALOPHINÆ (2 genera, 24 species), Africa and India; *Cephalophus* (22 sp.), continental Ethiopian region; *Tetraceros* (2 sp.) hilly part of all India, but rare north of the Ganges.

Sub-family IX. ALCEPHALINÆ (2 genera, 11 species), large African Antelopes, one species just entering the Palæarctic region. The genera are: *Alcephalus* (9 sp.) all Africa and north-east to Syria; *Catoblepas* (2 sp.), gnus, Africa, south of the Equator.

Sub-region X. BUDORCINÆ (1 genus, 2 species) *Budorcas* inhabits the high Himalayas from Nepal to East Thibet.

Sub-family XI. RUPICAPRINÆ (1 genus, 2 species) the Chamois, *Rupicapra*, inhabit the high European Alps from the Pyrenees to the Caucasus. (Plate I., vol. i., p. 195.)

Sub-family XII. NEMORHEDINÆ (2 genera, 10 species). These goat-like Antelopes inhabit portions of the Palæarctic and Oriental regions, as well as the Rocky Mountains in the Nearctic region. *Nemorhedus* (9 sp.) ranges from the Eastern Himalayas to N. China and Japan, and south to Formosa, the Malay Peninsula and Sumatra. *Aplocerus* (1 sp.), the mountain goat of the trappers, inhabits the northern parts of California and the Rocky Mountains.

Sub-family XIII. CAPRINÆ (2 genera, 23 species). The Goats and Sheep form an extensive series, highly characteristic of the Palæarctic region, but with an outlying species on the Neilgherries in Southern India, and one in the Rocky Mountains and California. The genera are *Capra* (22 sp.) and *Ovibos* (1 sp.).

The genus *Capra* consists of several sub-groups which have been named as genera, but it is unnecessary here to do more than divide them into "Goats and Ibexes" on the one hand and "Sheep" on the other—each comprising 11 species. The former range over all the South European Alps from Spain to the Caucasus; to Abyssinia, Persia, and Scinde; over the high Himalayas to E. Thibet and N. China; with an outlying species in the Neilgherries. The latter are only found in the mountains of Corsica, Sardinia, and Crete, in Europe; in Asia Minor, Persia, and in Central and North-Eastern Asia, with one somewhat isolated species in the Atlas mountains; while in America a species is found in the Rocky Mountains and the coast range of California. *Ovibos* (1 sp.), the musk-sheep, inhabits Arctic America north of lat. 60; but it occurs fossil in Post-glacial gravels on the Yena and Obi in Siberia, in Germany and France along with the Mammoth and with flint implements, and in caves of the Reindeer period; also in the brick earth in the south of England, associated with *Rhinoceros megarhinus* and *Elephas antiquus*.

Extinct Bovidæ.—In the caverns and diluviums of Europe, of the Post-Pliocene period, the remains are found of extinct species of *Bos, Bison,* and *Capra*, and in the caverns of the south of France *Rupicapra*, and an antelope near *Hippotragus*. *Bos* and *Bison* also occur in Pliocene deposits. In the Miocene of Europe, the only remains are antelopes closely allied to existing species, and these are especially numerous in Greece, where remains referred to two living and four extinct genera have been discovered. In the Miocene of India numerous extinct species of *Bos*, and two extinct genera, *Hemibos* and *Amphibos*, have been found, one of them at a great elevation in Thibet. Antelopes, allied to living Indian species, are chiefly found in the Nerbudda deposits.

In North America, the only bovine remains are those of a *Bison*, and a sheep or goat, in the Post-pliocene deposits; and of two species of musk-sheep, sometimes classed in a distinct genus *Bootherium*, from beds of the same age in Arkansas and Ohio. *Casoryx*, from the Pliocene of Nebraska, is supposed to be allied to the antelopes and to deer.

In the caves of Brazil remains of two animals said to be antelopes, have been discovered. They are classed by Gervais in the
genera *Antilope* and *Leptotherium*, but the presence of true antelopes in S. America at this period is so improbable, that there is
probably some error of identification.

The extinct family Sivatheridæ, containing the extraordinary
and gigantic four-horned *Sivatherium* and *Bramatherium*, of the
Siwalik deposits, are most nearly allied to the antelopes.

From the preceding facts we may conclude, that the great
existing development of the Bovidæ is comparatively recent.
The type may have originated early in the Miocene period, the
oxen being at first most tropical, while the antelopes inhabited
the desert zone a little further north. The sheep and goats seem
to be the most recent development of the bovine type, which
was probably long confined to the Eastern Hemisphere.

General Remarks on the Distribution of the Ungulata.

With the exception of the Australian region, from which this
order of mammalia is almost entirely wanting, the Ungulata are
almost universally distributed over the continental parts of all the
other regions. Of the ten families, 7 are Ethiopian, 6 Oriental, 5
Palæarctic, 4 Neotropical, and 3 Nearctic. The Ethiopian region
owes its superiority to the exclusive possession of the hippopotamus and giraffe, both of which inhabited the Palæarctic and
Oriental regions in Miocene times. The excessive poverty of the
Nearctic region in this order is remarkable ; the swine being
represented only by *Dicotyles* in its extreme southern portion,
while the Bovidæ are restricted to four isolated species. Deer
alone are fairly well represented. But, during the Eocene and
Miocene periods, North America was wonderfully rich in varied
forms of Ungulates, of which there were at least 8 or 9 families;
while we have reason to believe that during the same periods the
Ethiopian region was excessively poor, and that it probably received the ancestors of all its existing families from Europe or
Western Asia in later Miocene or Pliocene times. Many types that
once abounded in both Europe and North America are now preserved only in South America and Central or Tropical Asia,—as

the tapirs and camels; while others once confined to Europe and Asia have found a refuge in Africa,—as the hippopotamus and giraffe; so that in no other order do we find such striking examples of those radical changes in the distribution of the higher animals which were effected during the latter part of the Tertiary period. The present distribution of this order is, in fact, utterly unintelligible without reference to the numerous extinct forms of existing and allied families; but as this subject has been sufficiently discussed in the Second Part of this work (Chapters VI. and VII.) it is unnecessary to give further details here.

Order VIII.—PROBOSCIDEA.

FAMILY 53.—ELEPHANTIDÆ.   (1 Genus, 2 Species.)

GENERAL DISTRIBUTION.					
NEOTROPICAL SUB-REGIONS.	NEARCTIC SUB-REGIONS.	PALÆARCTIC SUB-REGIONS.	ETHIOPIAN SUB-REGIONS.	ORIENTAL SUB-REGIONS.	AUSTRALIAN SUB-REGIONS.
LIVING SPECIES.					
— — — —	— — — —	— — — —	1 . 2 . 3 —	1 . 2 . 3 . 4	— — — —
EXTINCT SPECIES.					
1 . 2 — —	1 . 2 . 3 . 4	1 . 2 . 3 . 4	1 — — —	1 — 3 —	— — — —

The elephants are now represented by two species, the African, which ranges all over that continent south of the Sahara, and the Indian, which is found over all the wooded parts of the Oriental region, from the slopes of the Himalayas to Ceylon, and eastward, to the frontiers of China and to Sumatra and Borneo. These, however, are but the feeble remnants of a host of gigantic creatures, which roamed over all the great continents except Australia during the Tertiary period, and several of which were contemporary with man.

Extinct Elephants.—At least 14 extinct species of *Elephas*, and a rather greater number of the allied genus *Mastodon* (distinguished by their less complex grinding teeth) have now been

Q 2

discovered. Elephants ranged over all the Palæarctic and Nearctic regions in Post-Pliocene times; in Europe and Central India they go back to the Pliocene; and only in India to the Upper Miocene period; the number of species increasing as we go back to the older formations.

In North America two or three species of *Mastodon* are Post-pliocene and Pliocene; and a species is found in the caves of Brazil, and in the Pliocene deposits of the pampas of La Plata, of the Bolivian Andes, and of Honduras and the Bahamas. In Europe the genus is Upper Miocene and Pliocene, but is especially abundant in the former period. In the East, it extends from Perim island to Burmah and over all India, and is mostly Miocene, but with perhaps one species Pliocene in Central India.

An account of the range of such animals as belong to extinct families of Proboscidea, will be found in Chapters VI. and VII.; from which it will be seen that, although the family Elephantidæ undoubtedly originated in the Eastern Hemisphere, it is not improbable that the first traces of the order Proboscidea are to be found in N. America.

Order IX.—HYRACOIDEA.

FAMILY 54.—HYRACIDÆ. (1 Genus. 10–12 Species.)

GENERAL DISTRIBUTION.

NEOTROPICAL SUB-REGIONS.	NEARCTIC SUB-REGIONS.	PALÆARCTIC SUB-REGIONS.	ETHIOPIAN SUB-REGIONS.	ORIENTAL SUB-REGIONS.	AUSTRALIAN SUB-REGIONS.
— — — —	— — — —	— 2. — —	1.2.3 —	— — — —	— — — —

The genus *Hyrax*, which alone constitutes this family, consists of small animals having the appearance of hares or marmots, but which more resemble the genus *Rhinoceros* in their teeth and skeleton. They range all over the Ethiopian region, except Madagascar; a peculiar species is found in Fernando Po, and they just enter the Palæarctic as far as Syria. They may therefore be considered as an exclusively Ethiopian group. In Dr. Gray's last Catalogue (1873) he divides the genus into three—*Hyrax*, *Euhyrax* and *Dendrohyrax*—the latter consisting of two species confined apparently to West and South Africa.

No extinct forms of this family have yet been discovered; the *Hyracotherium* of the London clay (Lower Eocene) which was supposed to resemble *Hyrax*, is now believed to be an ancestral type of the Suidæ or swine.

Order X.—RODENTIA.

FAMILY 55.—MURIDÆ.　(37 Genera, 330 Species.)

GENERAL DISTRIBUTION.					
NEOTROPICAL SUB-REGIONS.	NEARCTIC SUB-REGIONS.	PALÆARCTIC SUB-REGIONS.	ETHIOPIAN SUB-REGIONS.	ORIENTAL SUB-REGIONS.	AUSTRALIAN SUB-REGIONS.
1.2.3.4	1.2.3.4	1.2.3.4	1.2.3.4	1.2.3.4	— 2 — —

The Muridæ, comprising the rats and mice with their allies, are almost universally distributed over the globe (even not reckoning the domestic species which have been introduced almost everywhere by man), the exceptions being the three insular groups belonging to the Australian region, from none of which have any species yet been obtained. Before enumerating the genera it will be as well to say a few words on the peculiarities of distribution they present. The true mice, forming the genus *Mus*, is distributed over the whole of the world except N. and S. America where not a single indigenous species occurs, being replaced by the genus *Hesperomys;* five other genera, comprehending all the remaining species found in South America are peculiar to the Neotropical region. Three genera are confined to the Palæarctic region, and three others to the Nearctic. No less than twelve genera are exclusively Ethiopian, while only three are exclusively Oriental and three Australian.

Mus (100-120 sp.) the Eastern Hemisphere, but absent from the Pacific and Austro-Malayan Islands, except Celebes and Papua; *Lasiomys* (1 sp.) Guinea; *Acanthomys* (5-6 sp.) Africa, India and

N. Australia; *Cricetomys* (1 sp.) Tropical Africa; *Saccostomus* (2 sp.) Mozambique; *Cricetus* (9 sp.) Palæarctic region and Egypt; *Cricetulus* (1 sp., Milne-Edwards, 1870) Pekin; *Pseudomys* (1 sp.) Australia; *Hapalotis* (13 sp.) Australia; *Phlœomys* (1 sp.) Philippines; *Platacanthomys* (1 sp., Blyth, 1865) Malabar; *Dendromys* (2 sp.) S. Africa; *Nesomys* (1 sp. Peters, 1870) Madagascar; *Steatomys* (2 sp.) N. and S. Africa; *Pelomys* (1 sp.) Mozambique; *Reithrodon* (9 sp.) N. America, Lat. 29° to Mexico, and south to Tierra del Fuego; *Acodon* (1 sp.) Peru; *Myxomys* (1 sp.) Guatemala; *Hesperomys* (90 sp.) North and South America; *Holochilus* (4 sp.) South America; *Oxymycterus* (4 sp.) Brazil and La Plata; *Neotoma* (6 sp.) U.S., East coast to California; *Sigmodon* (2 sp.) Southern United States; *Drymomys* (1 sp.) Peru; *Neotomys* (2 sp.) S. America; *Otomys* (6 sp.) S. and E. Africa; *Meriones = Gerbillus* (20-30 sp.) Egypt, Central Asia, India, Africa; *Rhombomys* (6 sp.) S. E. Europe, N. Africa, Central Asia; *Malacothrix* (2 sp.) South Africa; *Mystromys* (1 sp.) South Africa; *Psammomys* (1 sp.) Egypt; *Spalacomys* (1 sp.) India; *Sminthus* (1-3 sp.) East Europe, Tartary, Siberia; *Hydromys* (5 sp.) Australia and Tasmania; *Hypogeomys* (1 sp., Grandidier, 1870) Madagascar; *Brachytarsomys* (1 sp., Günther, 1874) Madagascar; *Fiber* (2 sp.) N. America to Mexico; *Arvicola* (50 sp.) Europe to Asia Minor, North Asia, Himalayas, Temp. N. America; *Cuniculus* (1 sp.) N. E. Europe, Siberia, Greenland, Arctic America; *Myodes* (4 sp.) Europe, Siberia, Arctic America, and Northern United States; *Myospalax = Siphneus* (2 sp.) Altai Mountains and N. China[1]; *Lophiomys* (1 sp.) S. Arabia, and N. E. Africa; *Echiothrix* (1 sp.) Australia.

Extinct Muridœ.—Species of *Mus, Cricetus, Arvicola,* and *Myodes,* occur in the Post-Pliocene deposits of Europe; *Arvicola, Meriones,* and the extinct genus *Cricetodon,* with some others, in the Miocene.

In North America, *Fiber, Arvicola,* and *Neotoma,* occur in caves;

---

[1] Myospalax has hitherto formed part of the next family, Spalacidæ; but a recent examination of its anatomy by M. Milne-Edwards shows that it belongs to the Muridæ, and comes near Arvicola.

an extinct genus, *Eumys*, in the Upper Miocene of Dakota, and another, *Mysops*, in the Eocene of Wyoming.

In South America *Mus*, or more probably *Hesperomys*, is abundant in Brazilian caverns, and *Oxymycterus* in the Pliocene of La Plata; while *Arvicola* is said to have occurred both in the Pliocene and Eocene deposits of the same country.

FAMILY 56.—SPALACIDÆ.   (7 Genera, 17 Species.)

GENERAL DISTRIBUTION.

NEOTROPICAL SUB-REGIONS.	NEARCTIC SUB-REGIONS.	PALÆARCTIC SUB-REGIONS.	ETHIOPIAN SUB-REGIONS.	ORIENTAL SUB-REGIONS.	AUSTRALIAN SUB-REGIONS.
— — — —	— — — —	1 . 2 . 3 —	1 . 2 . 3 —	1 — 3 . 4	— — — —

The Spalacidæ, or mole-rats, have a straggling distribution over the Old World continents. They are found over nearly the whole of Africa, but only in the South-east of Europe, and West of Temperate Asia, but appearing again in North India, Malacca, and South China. *Ellobius* (1 sp.), is found in South Russia and South-west Siberia; *Spalax* (1 sp.), Southern Russia, West Asia, Hungary, Moldavia, and Greece (Plate II., vol. i. p. 218); *Rhizomys* (6 sp.), Abyssinia, North India, Malacca, South China; *Heterocephalus* (1 sp.), Abyssinia; *Bathyerges* (= *Orycterus* 1 sp.), South Africa; *Georychus* (6 sp.), South, Central, and East Africa; *Heliophobus* (1 sp.) Mozambique.

FAMILY 57.—DIPODIDÆ. · (3 Genera, 22 Species.)

GENERAL DISTRIBUTION.

NEOTROPICAL SUB-REGIONS.	NEARCTIC SUB-REGIONS.	PALÆARCTIC SUB-REGIONS.	ETHIOPIAN SUB-REGIONS.	ORIENTAL SUB-REGIONS.	AUSTRALIAN SUB-REGIONS.
— — — —	1 . 2 . 3 . 4	— 2 . 3 . 4	1 . 2 . 3 —	— — — —	— — — —

The Jerboas, or jumping mice, are especially characteristic of the regions about the eastern extremity of the Mediterranean, being found in South Russia, the Caspian district, Arabia, Egypt, and Abyssinia; but they also extend over a large part of Africa, and eastward to India; while isolated forms occur in North America, and the Cape of Good Hope. *Dipus = Gerbillus* (20 sp.), inhabits North and Central Africa, South-East Europe, and across Temperate Asia to North China, also Affghanistan, India, and Ceylon; *Pedetes* (1 sp.), South Africa to Mozambique and Angola; *Jaculus = Meriones* (1 sp.), North America, from Nova Scotia and Canada, south to Pennsylvania and west to California and British Columbia (Plate XX., vol. ii. p. 135).

Extinct Dipodidæ.—*Dipus* occurs fossil in the Miocene of the Alps; and an extinct genus, *Issiodromys*, said to be allied to *Pedetes* of the Cape of Good Hope, is from the Pliocene formations of Auvergne in France.

FAMILY 58.—MYOXIDÆ.   (1 Genus, 12 Species.)

GENERAL DISTRIBUTION.					
NEOTROPICAL SUB-REGIONS.	NEARCTIC SUB-REGIONS.	PALÆARCTIC SUB-REGIONS.	ETHIOPIAN SUB-REGIONS.	ORIENTAL SUB-REGIONS.	AUSTRALIAN SUB-REGIONS.
— — — —	— — — —	1 . 2 . 3 . 4	1 . 2 . 3 —	— — — —	— — — —

The Dormice (*Myoxus*), are small rodents found over all the temperate parts of the Palæarctic region, from Britain to Japan; and also over most parts of Africa to the Cape, but wanting in India. Some of the African species have been separated under the name of *Graphidurus*, while those of Europe and Asia form the sub-genera *Glis*, *Muscardinus*, and *Eliomys*.

Extinct Myoxidæ.—*Myoxus* ranges from the Post-pliocene of the Maltese caverns to the Miocene of Switzerland and the Upper Eocene of France; and an extinct genus *Brachymys* is found in the Miocene of Central Europe.

FAMILY 59.—SACCOMYIDÆ.   (6 Genera, 33 Species)

GENERAL DISTRIBUTION.					
NEOTROPICAL SUB-REGIONS.	NEARCTIC SUB-REGIONS.	PALÆARCTIC SUB-REGIONS.	ETHIOPIAN SUB-REGIONS.	ORIENTAL SUB-REGIONS.	AUSTRALIAN SUB-REGIONS.
— — — —	1 . 2 . 3 . 4	— — — —	— — — —	— — — —	— — — —

The Saccomyidæ, or pouched rats, are almost wholly confined to our second Nearctic sub-region, comprising the Rocky Mountains and the elevated plains of Central North America. A few species range from this district as far as Hudson's Bay on the north, to South Carolina on the east, and to California on the west, while one genus, doubtfully placed here, goes south as far as Honduras and Trinidad. The group must therefore be considered to be pre-eminently characteristic of the Nearctic region.

The genera are,—*Dipodomys* (5 sp.), North Mexico, California, the east slope of the Rocky Mountains to the Columbia River, and one species in South Carolina ; *Perognathus* (6 sp.), North Mexico, California, east slope of the Rocky Mountains to British Columbia; *Thomomys* (2 sp.), Upper Missouri, and Upper Columbia Rivers to Hudson's Bay ; *Geomys* (5 sp.), North Mexico, and east slope of Rocky Mountains to Nebraska (Plate XIX., vol. ii. p. 129); *Saccomys* (1 sp.), North America, locality unknown ; *Heteromys* (6 sp.), Mexico, Honduras, and Trinidad. *Geomys* and *Thomomys* constitute a separate family Geomyidæ, of Professor Carus ; but I follow Professor Lilljeborg, who has made a special study of the Order, in keeping them with this family.

In the Post-Pliocene deposits of Illinois and Nebraska, remains of an existing species of *Geomys* have been found.

FAMILY 60.—CASTORIDÆ. (1 Genus, 2 Species.)

GENERAL DISTRIBUTION.

NEOTROPICAL SUB-REGIONS.	NEARCTIC SUB-REGIONS.	PALÆARCTIC SUB-REGIONS.	ETHIOPIAN SUB-REGIONS.	ORIENTAL SUB-REGIONS.	AUSTRALIAN SUB-REGIONS.
— — — —	1.2.3.4	1 — 3 —	— — — —	— — — —	— — — —

The Beavers, forming the genus *Castor*, consist of two species, the American (*Castor canadensis*) ranging over the whole of North America from Labrador to North Mexico; while the European (*Castor fiber*) appears to be confined to the temperate regions of Europe and Asia, from France to the River Amoor, over which extensive region it doubtless roamed in prehistoric times, although now becoming rare in many districts.

Extinct Castoridæ.—Extinct species of *Castor* range back from the Post-pliocene to the Upper Miocene in Europe, and to the Newer Pliocene in North America. Extinct genera in Europe are, *Trogontherium*, Post-Pliocene and Pliocene; *Chalicomys*, Older Pliocene; and *Steneofiber*, Upper Miocene. In North America *Castoroides* is Post-Pliocene, and *Palæocastor*, Upper Miocene. The family thus first appears on the same geological horizon in both Europe and North America.

FAMILY 61.—SCIURIDÆ.—(8 Genera, 180–200 Species.)

GENERAL DISTRIBUTION.

NEOTROPICAL SUB-REGIONS.	NEARCTIC SUB-REGIONS.	PALÆARCTIC SUB-REGIONS.	ETHIOPIAN SUB-REGIONS.	ORIENTAL SUB-REGIONS.	AUSTRALIAN SUB-REGIONS.
— 2.3 —	1.2.3.4	1.2.3.4	1.2.3 —	1.2.3.4	— — — —

The Squirrel family, comprehending also the marmots and prairie-dogs, are very widely spread over the earth. They are especially abundant in the Nearctic, Palæarctic, and Oriental regions, and rather less frequent in the Ethiopian and Neotropical, in which last region they do not extend south of Paraguay. They are absent from the West Indian islands, Madagascar, and Australia, only occurring in Celebes which doubtfully belongs to the Australian region. The genera are as follows:—

Sciurus (100—120 sp., including the sub-genera Spermosciurus, Xerus, Macroxus, Rheithrosciurus, and Rhinosciurus), comprises the true squirrels, and occupies the area of the whole family wherever woods and forests occur. The approximate number of species in each region is as follows : Nearctic 18, Palæarctic 6, Ethiopian 18, Oriental 50, Australian (Celebes) 5, Neotropical 30. *Sciuropterus* (16—19 sp.), comprises the flat-tailed flying squirrels, which range from Lapland and Finland to North China and Japan, and southward through India and Ceylon, to Malacca and Java, with a species in Formosa ; while in North America they occur from Labrador to British Columbia, and south to Minnesota and Southern California. *Pteromys* (12 sp.), comprising the round-tailed flying squirrels, is a more southern form, being confined to the wooded regions of India from the Western Himalayas to Java and Borneo, with species in Formosa and Japan. *Tamias* (5 sp.), the ground squirrels, are chiefly North American, ranging from Mexico to Puget's Sound on the west coast, and from Virginia to Montreal on the Atlantic coast; while one species is found over all northern Asia. *Spermophilus* (26 sp.), the pouched marmots, are confined to the Nearctic and Palæarctic regions ; in the former extending from the Arctic Ocean to Mexico and the west coast, but not passing east of Lake Michigan and the lower Mississippi; in the latter from Silesia through South Russia to the Amoor and Kamschatka, most abundant in the desert plains of Tartary and Mongolia. *Arctomys* (8 sp.), the marmots, are found in the northern parts of North America as far down as Virginia and Nebraska to the Rocky Mountains and British Columbia, but not in California ; and from the Swiss Alps eastward to Lake Baikal and Kamschatka, and south as far as the Himalayas, above 8,000 feet elevation. *Cynomys* (2 sp.), the prairie-dogs, inhabit the plains east of the Rocky Mountains from the Upper Missouri to the Red River and Rio Grande (Plate XIX., vol. ii. p. 129). *Anomalurus* (5 sp.), consists of animals which resemble flying-squirrels, but differ from all other members of the family in some points of internal structure. They form a very aberrant portion of the Sciuridæ, and, according to some naturalists, a distinct family. They inhabit West Africa and the island of Fernando Po.

Extinct Sciuridœ.—These are tolerably abundant. The genus *Sciurus* appears to be a remarkably ancient form, extinct species being found in the Miocene, and even in the Upper Eocene formations of Europe. *Spermophilus* goes back to the Upper Miocene; *Arctomys* to the Newer Pliocene. Extinct genera are, *Brachymys, Lithomys* and *Plesiarctomys,* from the European Miocene, the latter said to be intermediate between marmots and squirrels.

In North America, *Sciurus, Tamias,* and *Arctomys* occur in the Post-pliocene deposits only. The extinct genera are *Ischyromys,* from the Upper Miocene of Nebraska; *Paramys,* allied to the marmots, and *Sciuravus,* near the squirrels, from the Eocene of Wyoming.

Here we have unmistakable evidence that the true squirrels (*Sciurus*) are an Old World type, which has only recently entered North America; and this is in accordance with the comparative scarcity of this group in South America, a country so well adapted to them, and their great abundance in the Oriental region, which, with the Palæarctic, was probably the country of their origin and early development. The family, however, has been traced equally far back in Europe and North America, so that we have as yet no means of determining where it originated.

FAMILY 62.—HAPLOODONTIDÆ.—(1 Genus, 2 Species.)

GENERAL DISTRIBUTION.

NEOTROPICAL SUB-REGIONS.	NEARCTIC SUB-REGIONS.	PALÆARCTIC SUB-REGIONS.	ETHIOPIAN SUB-REGIONS.	ORIENTAL SUB-REGIONS.	AUSTRALIAN SUB-REGIONS.
— — — —	1 — — —	— — — —	— — — —	— — — —	— — — —

The genus *Haploodon* or *Aplodontia,* consists of two curious rat-like animals, inhabiting the west coast of America, from the southern part of British Columbia to the mountains of California. They seem to have affinities both with the beavers and marmots, and Professor Lilljeborg constitutes a separate family to receive them.

FAMILY 63.—CHINCHILLIDÆ.   (3 Genera, 6 Species.)

GENERAL DISTRIBUTION.

NEOTROPICAL SUB-REGIONS.	NEARCTIC SUB-REGIONS.	PALÆARCTIC SUB-REGIONS.	ETHIOPIAN SUB-REGIONS.	ORIENTAL SUB-REGIONS.	AUSTRALIAN SUB-REGIONS.
1 — — —	— — — —	— — — —	— — — —	— — — —	— — — —

The Chinchillidæ, including the chinchillas and viscachas, are
confined to the alpine zones of the Andes, from the boundary of
Ecuador and Peru to the southern parts of Chili; and over the
Pampas, to the Rio Negro on the south, and the River Uruguay
on the east. *Chinchilla* (2 sp.), the true chinchillas, are found
in the Andes of Chili and Peru, south of 9° S. lat., and from
8,000 to 12,000 feet elevation (Plate XVI. vol. ii. p. 40); *Lagidium* (3 sp.), the alpine viscachas, inhabit the loftiest plateaus
and mountains from 11,000 to 16,000 feet, and extend furthest
north of any of the family; while *Lagostomus* (1 sp.), the viscacha of the Pampas, has the range above indicated. The family
is thus confined within the limits of a single sub-region.

Extinct Chinchillidæ.—*Lagostomus* has been found fossil in
the caves of Brazil, and in the Pliocene deposits of La
Plata. The only known extinct forms of this family are *Amblyrhiza* and *Loxomylus*, found in cavern-deposits in the island of
Anguilla, of Post-Pliocene age. These are very interesting, as
showing the greater range of this family so recently; though its
absence from North America and Europe indicates that it is a
peculiar development of the Neotropical region.

FAMILY 64.—OCTODONTIDÆ.   (8 Genera, 19 Species.)

GENERAL DISTRIBUTION.

NEOTROPICAL SUB-REGIONS.	NEARCTIC SUB-REGIONS.	PALÆARCTIC SUB-REGIONS.	ETHIOPIAN SUB-REGIONS.	ORIENTAL SUB-REGIONS.	AUSTRALIAN SUB-REGIONS.
1.2 — 4	— — — —	— 2 — —	1 — — —	— — — —	— — — —

The Octodontidæ include a number of curious and obscure rat-like animals, mostly confined to the mountains and open plains of South America, but having a few stragglers in other parts of the world, as will be seen by our notes on the genera. The most remarkable point in their distribution is, that two genera are peculiar to the West Indian islands, while no species of the family inhabits the northern half of South America. The distribution of the genera is as follows :—*Habrocomus* (2 sp.), Chili; *Capromys* (3 sp.), two of which inhabit Cuba, the third Jamaica (Plate XVII. vol. ii. p. 67); *Plagiodontia* (1 sp.), only known from Hayti; *Spalacopus*, including *Schizodon* (2 sp.), Chili, and east side of Southern Andes; *Octodon* (3 sp.), Peru, Bolivia, and Chili; *Ctenomys* (6 sp.), the tuco-tuco of the Pampas, the Campos of Brazil to Bolivia and Tierra del Fuego; *Ctenodactylus* (1 sp.), Tripoli, North Africa; *Pectinator* (1 sp.), East Africa, Abyssinia, 4,000 to 5,000 feet.

Capromys and *Plagiodontia*, the two West Indian genera, were classed among the Echimyidæ by Mr. Waterhouse, but Professor Lilljeborg removes them to this family.

Extinct Octodontidæ.—Species of *Ctenomys* have been found in the Pliocene of La Plata, and an extinct genus *Megamys*, said to be allied to *Capromys*, in the Eocene of the same country. In Europe, *Palæomys* and *Archæomys* from the lower Miocene of Germany and France, are also said to be allied to *Capromys*.

FAMILY 65.—ECHIMYIDÆ.   (10 Genera, 30 Species.)

GENERAL DISTRIBUTION.					
NEOTROPICAL SUB-REGIONS.	NEARCTIC SUB-REGIONS.	PALÆARCTIC SUB-REGIONS.	ETHIOPIAN SUB-REGIONS.	ORIENTAL SUB-REGIONS.	AUSTRALIAN SUB-REGIONS.
1 . 2 — —	— — — —	— — — —	1 – 3 –	— — — —	—·— — —

The Echimyidæ, or spiny rats, are a family, chiefly South American, of which the Coypu, a large beaver-like water-rat from Peru and Chili is the best known. Two of the genera are found in South Africa, but all the rest inhabit the continent of South America, East of the Andes, none being yet known north of Panama. The genera are as follows:—*Dactylomys* (2 sp.), Guiana and Brazil; *Cercomys* (1 sp.), Central Brazil; *Lasiuromys* (1 sp.), San Paulo, Brazil; *Petromys* (1 sp.), South Africa; *Myopotamus* (1 sp.), the coypu, on the East side of the Andes from Peru to 42° S. lat., on the West side from 33° to 48° S. lat.; *Carterodon* (1 sp.), Minaes Geraes, Brazil; *Aulacodes* (1. sp.), West and South Africa; *Mesomys* (1 sp.), Borba on the Amazon; *Echimys* (11 sp.), from Guiana and the Ecuadorian Andes to Paraguay; *Loncheres* (10 sp.), New Granada to Brazil.

Fossil and Extinct Echimyidæ.—The genus *Carterodon* was established on bones found in the Brazilian caves, and it was several years afterwards that specimens were obtained showing the animal to be a living species. Extinct species of *Myopotamus* and *Loncheres* have also been found in these caves, with the extinct genera *Lonchophorus* and *Phyllomys*.

No remains of this family have been discovered in North America; but in the Miocene and Upper Eocene deposits of France there are many species of an extinct genus *Theridomys*, which is said to be allied to this group or to the next (Cercolabidæ) *Aulacodon*, from the Upper Miocene of Germany, is allied to the West African *Aulacodes;* and some other remains from the lower Miocene of Auvergne, are supposed to belong to *Echimys*.

FAMILY 66.—CERCOLABIDÆ. (3 Genera, 13–15 Species.)

GENERAL DISTRIBUTION.					
NEOTROPICAL SUB-REGIONS.	NEARCTIC SUB-REGIONS.	PALÆARCTIC SUB-REGIONS.	ETHIOPIAN SUB-REGIONS.	ORIENTAL SUB-REGIONS.	AUSTRALIAN SUB-REGIONS.
— 2 . 3 —	1 . 2 . 3 . 4	— — — —	— — — —	— — — —	— — — —

The Cercolabidæ, or arboreal porcupines, are a group of rodents entirely confined to America, where they range from the northern limit of trees on the Mackenzie River, to the southern limit of forests in Paraguay. There is however an intervening district, the Southern United States, from which they are absent. *Erethizon* (3 sp.), the Canadian porcupine, is found throughout

Canada and as far south as Northern Pennsylvania, and west to
the Mississippi (Plate XX., vol. ii. p. 135); an allied species inhabiting the west coast from California to Alaska, and inland to
the head of the Missouri River; while a third is found in the
north-western part of South America; *Cercolabes* (12 sp.), ranges
from Mexico and Guatemala to Paraguay, on the eastern side of
the Andes; *Chætomys* (1 sp.), North Brazil.

Extinct Cercolabidæ.—A large species of *Cercolabes* has been
found in the Brazilian caves, but none have been discovered in
North America or Europe. We may conclude therefore that
this is probably a South American type, which has thence spread
into North America at a comparatively recent epoch. The
peculiar distribution of *Cercolabes* may be explained by suppos-
ing it to have migrated northwards along the west coast by means
of the wooded slopes of the Rocky Mountains. It could then
only reach the Eastern States by way of the forest region of the
great lakes, and then move southward. This it may be now
doing, but it has not yet reached the Southern States of Eastern
North America.

FAMILY 67.—HYSTRICIDÆ.   (3 Genera, 12 Species.)

GENERAL DISTRIBUTION.					
NEOTROPICAL SUB-REGIONS.	NEARCTIC SUB-REGIONS.	PALÆARCTIC SUB-REGIONS.	ETHIOPIAN SUB-REGIONS.	ORIENTAL SUB-REGIONS.	AUSTRALIAN SUB-REGIONS.
— — — —	— — — —	— 2 — —	|1.2.3 —	1.2.3.4	— — — —

The true Porcupines have a very compact and well-marked
distribution, over the whole of the Oriental and Ethiopian regions
(except Madagascar), and the second Palæarctic sub-region.
There is some confusion as to their sub-division into genera, but
the following are those most usually admitted :—*Hystrix* (5 sp.),
South Europe to the Cape of Good Hope, all India, Ceylon, and
South China; *Atherura* (5 sp.), "brush-tailed porcupines," inhabit West Africa, India, to Siam, Sumatra, and Borneo; *Acan-
thion* (2 sp.), Nepal and Malacca, to Sumatra, Borneo, and Java.

Extinct Hystricidæ.—Several extinct species of *Hystrix* have been found in the Pliocene and Miocene deposits of Europe, and one in the Pliocene of Nebraska in North America.

FAMILY 68.—CAVIIDÆ.　(6 Genera, 28 Species.)

GENERAL DISTRIBUTION.					
NEOTROPICAL SUB-REGIONS.	NEARCTIC SUB-REGIONS.	PALÆARCTIC SUB-REGIONS.	ETHIOPIAN SUB-REGIONS.	ORIENTAL SUB-REGIONS.	AUSTRALIAN SUB-REGIONS.
1 . 2 . 3 . 4	— — — —	— — ·· —	— — — —	— — — —	— — — —

The Cavies and Agoutis were placed in distinct families by Mr. Waterhouse, in which he is followed by Professor Carus, but they have been united by Professor Lilljeborg, and without pretending to decide which classification is the more correct I follow the latter, because there is a striking external resemblance between the two groups, and they have an identical distribution in the Neotropical region, and with one exception are all found east of the Andes. *Dasyprocta* (9 sp.), the agouti, ranges from Mexico to Paraguay, one species inhabiting the small West Indian islands of St. Vincent, Lucia, and Grenada ; *Cœlogenys* (2 sp.), the paca, is found from Guatemala to Paraguay, and a second species (somewhat doubtful) in Eastern Peru; *Hydrochœrus* (1 sp.), the capybara, inhabits the banks of rivers from Guayana to La Plata ; *Cavia* (9 sp.), the guinea-pigs, Brazil to the Straits of Magellan, and one species west of the Andes at Yça in Peru ; *Kerodon* (6 sp.), Brazil and Peru to Magellan ; *Dolichotis* (1 sp,), the Patagonian cavy from Mendoza to 48° 30′ south latitude, on sterile plains.

Extinct Caviidæ.—*Hydrochœrus, Cœlogenys, Dasyprocta,* and *Kerodon,* have occurred abundantly in the caves of Brazil, and the last-named genus in the Pliocene of La Plata. *Hydrochœrus* has been found in the Post-Pliocene deposits of South Carolina. *Cavia* and *Dasyprocta* are said to have been found in the Miocene of Switzerland and France. No well-marked extinct genera of this family have been recorded.

If the determination of the above-mentioned fossil species of *Cavia* and *Dasyprocta* are correct, it would show that this now exclusively South American family is really derived from Europe, where it has long been extinct.

FAMILY 69.—LAGOMYIDÆ.   (1 Genus, 11 Species.)

GENERAL DISTRIBUTION.					
NEOTROPICAL SUB-REGIONS.	NEARCTIC SUB-REGIONS.	PALÆARCTIC SUB-REGIONS.	ETHIOPIAN SUB-REGIONS.	ORIENTAL SUB-REGIONS.	AUSTRALIAN SUB-REGIONS.
— — — —	— 2 — 4	— — 3 —	— — — —	— — — —	— — — —

The Lagomyidæ, or pikas, are small alpine and desert animals which range from the south of the Ural Mountains to Cashmere and the Himalayas, at heights of 11,000 to 14,000 feet, and northward to the Polar regions and the north-eastern extremity of Siberia. They just enter the eastern extremity of Europe as far as the Volga, but with this exception, seem strictly limited to the third Palæarctic sub-region. In America they are confined to the Rocky Mountains from about 42° to 60' north latitude.

Extinct Lagomyidæ.—Extinct species of *Lagomys* have occurred in the southern parts of Europe, from the Post-Pliocene to the Miocene formations. *Titanomys*, an extinct genus, is found in the Miocene of France and Germany.

FAMILY 70.—LEPORIDÆ.   (1 Genus, 35–40 Species.)

GENERAL DISTRIBUTION.					
NEOTROPICAL SUB-REGIONS.	NEARCTIC SUB-REGIONS.	PALÆARCTIC SUB-REGIONS.	ETHIOPIAN SUB-REGIONS.	ORIENTAL SUB-REGIONS.	AUSTRALIAN SUB-REGIONS.
— 2 . 3 —	1 . 2 . 3 . 4	1 . 2 . 3 . 4	1 — 3 —	1 . 2 . 3 —	— — — —

The Hares and Rabbits are especially characteristic of the Nearctic and Palæarctic, but are also thinly scattered over the Ethiopian and Oriental regions. In the Neotropical region they are very scarce, only one species being found in South America, in the mountains of Brazil and various parts of the Andes, while one or two of the North American species extend into Mexico and Guatemala. In the Nearctic region, they are most abundant in the central and western parts of the continent, and they extend to the Arctic Ocean and to Greenland. They are found in every part of the Palæarctic region, from Ireland to Japan; three species range over all India to Ceylon, and others occur in Hainan, Formosa, South China, and the mountains of Pegu; the Ethiopian region has only four or five species, mostly in the southern extremity and along the East coast. An Indian species is now wild in some parts of Java, but it has probably been introduced.

Extinct Leporidæ.—Species of *Lepus* occur in the Post-Pliocene and Newer Pliocene of France; but only in the Post-Pliocene of North America, and the caves of Brazil.

General Remarks on the Distribution of the Rodentia.

With the exception of the Australian region and Madagascar, where Muridæ alone have been found, this order is one of the most universally and evenly distributed over the entire globe. Of the sixteen families which compose it, the Palæarctic region has 10; the Ethiopian, Nearctic, and Neotropical, each 9; and the Oriental only 5. These figures are very curious and suggestive. We know that the rodentia are exceedingly ancient, since some of the living genera date back to the Eocene period; and some ancestral types might thus have reached the remote South American and South African lands at the time of one of their earliest unions with the northern continents. In both these countries the rodents diverged into many special forms, and being small animals easily able to conceal themselves, have largely survived the introduction of higher Mammalia. In the Palæarctic and Nearctic regions, their small size and faculty of hibernation may have enabled them to maintain themselves during those great physical changes which resulted in the extermination or banishment of so many of the larger and more highly organised Mammalia, to which, in these regions, they now bear a somewhat inordinate proportion. The reasons why they are now less numerous and varied in the Oriental region, may be of two kinds. The comparatively small area of that region and its

R 2

uniformity of climate, would naturally lead to less development of such a group as this, than in the vastly more extensive and varied and almost equally luxuriant Palæarctic region of Eocene and Miocene times; while on the other hand the greater number of the smaller Carnivora in the tropics during the Pliocene and Post-Pliocene epochs, would be a constant check upon the increase of these defenceless animals, and no doubt exterminate a number of them.

The Rodents thus offer a striking contrast to the Ungulates; and these two great orders afford an admirable illustration of the different way in which physical and organic changes may affect large and small herbivorous Mammalia; often leading to the extinction of the former, while favouring the comparative development of the latter.

Order XI.—EDENTATA.

FAMILY 71.—BRADYPODIDÆ.   (3 Genera, 12 Species.)

GENERAL DISTRIBUTION.					
NEOTROPICAL SUB-REGIONS.	NEARCTIC SUB-REGIONS.	PALÆARCTIC SUB-REGIONS.	ETHIOPIAN SUB-REGIONS.	ORIENTAL SUB-REGIONS.	AUSTRALIAN SUB-REGIONS.
— 2 . 3 —	— — — —	— — — —	— — — —	— — — —	— — — —

The Sloths are a remarkable group of arboreal mammals, strictly confined to the great forests of the Neotropical region, from Guatemala to Brazil and Eastern Bolivia. None are found west of the Andes, nor do they appear to extend into Paraguay, or beyond the Tropic of Capricorn on the east coast. The genera as defined by Dr. Gray in 1871 are :—*Cholœpus* (2 sp.), " Sloths with two toes on fore limbs, sexes alike," Costa Rica to Brazil; *Bradypus* (2 sp.), " Sloths with three toes on fore limbs, sexes alike," Central Brazil, Amazon to Rio de Janeiro; *Arctopithecus* (8 sp.), " Sloths with three toes on fore limbs, males with a coloured patch on the back," Costa Rica to Brazil and Eastern Bolivia (Plate XIV., vol. ii. p. 24).

Extinct Bradypodidæ.—In the caves of Brazil are found three extinct genera of Sloths—*Cœlodon, Sphenodon,* and *Ochotherium.* More distantly allied, and probably forming distinct families, are *Scelidotherium* and *Megatherium,* from the caves of Brazil and the Pliocene deposits of La Plata and Patagonia.

FAMILY 72.—MANIDIDÆ.   (1 Genus, 3 Species.)

GENERAL DISTRIBUTION.					
NEOTROPICAL SUB-REGIONS.	NEARCTIC SUB-REGIONS.	PALÆARCTIC SUB-REGIONS.	ETHIOPIAN SUB-REGIONS.	ORIENTAL SUB-REGIONS.	AUSTRALIAN SUB-REGIONS.
— — — —	— — — —	— — — —	1 . 2 . 3 –	1 . 2 . 3 . 4	— — — —

The Manididæ, or scaly ant-eaters, are the only Edentate Mammalia found out of America. They are spread over the Ethiopian and Oriental regions; in the former from Sennaar to West Africa and the Cape; in the latter from the Himalayas to Ceylon, and Eastward to Borneo and Java, as well as to South China, as far as Amoy, Hainan, and Formosa. They have been sub-divided, according to differences in the scaly covering, into five groups, *Manis, Phatagin, Smutsia, Pholidotus* and *Pangolin,* the three former being confined to Africa, the last common to Africa and the East, while *Pholidotus* seems confined to Java. It is doubtful if these divisions are more than sub-genera, and as such they are treated here.

No extinct species referable to this family are yet known.

FAMILY 73.—DASYPODIDÆ.   (6 Genera, 17 Species.)

GENERAL DISTRIBUTION.					
NEOTROPICAL SUB-REGIONS.	NEARCTIC SUB-REGIONS.	PALÆARCTIC SUB-REGIONS.	ETHIOPIAN SUB-REGIONS.	ORIENTAL SUB-REGIONS.	AUSTRALIAN SUB-REGIONS.
1 . 2 . 3 –	— — — —	— — — —	— — — —	— — — —	— — — —

The Dasypodidæ, or armadillos, are a highly characteristic Neo-tropical family, ranging from the northern extremity of the region

in south Texas, to 50° south latitude on the plains of Patagonia.
The distribution of the genera is as follows:—*Tatusia* (5 sp.),
has the range of the whole family from the lower Rio Grande of
Texas to Patagonia; *Prionodontes* (1 sp.), the giant armadillo,
Surinam to Paraguay; *Dasypus* (4 sp.), Brazil to Bolivia, Chili,
and La Plata; *Xenurus* (3 sp.), Guiana to Paraguay; *Tolypeutes*
(2 sp.), the three-banded armadillos, Bolivia and La Plata;
Chlamydophorus (2 sp.), near Mendoza in La Plata, and Santa
Cruz de la Sierra in Bolivia.

Extinct Armadillos.—Many species of *Dasypus* and *Xenurus*
have been found in the caves of Brazil, together with many
extinct genera—*Hoplophorus, Euryodon, Heterodon, Pachy-
therium,* and *Chlamydotherium,* the latter as large as a rhinoceros. *Eutatus,* allied to *Tolypeutes,* is from the Pliocene deposits of La Plata.

FAMILY 74.—ORYCTEROPODIDÆ.   (1 Genus, 2 Species.)

GENERAL DISTRIBUTION.					
NEOTROPICAL SUB-REGIONS.	NEARCTIC SUB-REGIONS.	PALÆARCTIC SUB-REGIONS.	ETHIOPIAN SUB-REGIONS.	ORIENTAL SUB-REGIONS.	AUSTRALIAN SUB-REGIONS.
— — — —	— — — —	— — — —	1 – 3 –	— — — —	— — — —

The Aard-vark, or Cape ant-eater (*Orycteropus capensis*) is a
curious form of Edentate animal, with the general form of an
ant-eater, but with the bristly skin and long obtuse snout of a
pig. A second species inhabits the interior of North-East
Africa and Senegal, that of the latter country perhaps forming a
third species (Plate IV. vol. i. p. 261).

Extinct Orycteropodidæ.—The genus *Macrotherium,* remains of
which occur in the Miocene deposits of France, Germany, and
Greece, is allied to this group, though perhaps forming a separate family. The same may be said of the *Ancylotherium,* a
huge animal found only in the Miocene deposits of Greece.

FAMILY 75.—MYRMECOPHAGIDÆ.   (3 Genera, 5 Species.)

GENERAL DISTRIBUTION.					
NEOTROPICAL SUB-REGIONS.	NEARCTIC SUB-REGIONS.	PALÆARCTIC SUB-REGIONS.	ETHIOPIAN SUB-REGIONS.	ORIENTAL SUB-REGIONS.	AUSTRALIAN SUB-REGIONS.
1 . 2 . 3 —	— — — —	— — — —	— — — —	— — — —	— — — —

The true ant-eaters are strictly confined to the wooded portions of the Neotropical region, ranging from Honduras to Paraguay on the East side of the Andes. The three genera now generally admitted are : *Myrmecophaga* (1 sp.), the great ant-eater, Northern Brazil to Paraguay; *Tamandua* (2 sp.), 4-toed ant-eaters, Guatemala, Ecuador to Paraguay (Plate XIV. vol. ii. p. 24); *Cyclothurus* (2 sp.), 2-toed ant-eaters, Honduras and Costa Rica to Brazil.

Extinct Ant-eaters.—The only extinct form of this family seems to be the *Glossotherium*, found in the caves of Brazil, and the Tertiary deposits of Uruguay. It is said to be allied to *Myrmecophaga* and *Manis*.

General Remarks on the Distribution of the Edentata.

These singular animals are almost confined to South America, where they constitute an important part of the fauna. In Africa, two family types are scantily represented, and one of these extends over all the Oriental region. In Pliocene and Post-Pliocene times the Edentata were wonderfully developed in South America, many of them being huge animals, rivalling in bulk, the rhinoceros and hippopotamus. As none of these forms resemble those of Africa, while the only European fossil Edentata are of African type, it seems probable that South Africa, like South America, was a centre of development for this group of mammalia ; and it is in the highest degree probable that, should extensive fluviatile deposits of Pliocene or Miocene age be discovered in the former country, an extinct fauna, not less strange and grotesque than that of South America, will be brought to light. From the fact that so few remains of this order occur in Europe, and those of one family type, and in Miocene deposits only, it seems a fair conclusion, that this represents an incursion of an ancient Ethiopian form into Europe analogous to that which invaded North America from the south during the Post-Pliocene epoch. The extension of the Manididæ, or scaly ant-eaters, over tropical Asia may have occurred at the same, or a somewhat later epoch.

For a summary of the Numerous Edentata of North and South America which belong to extinct families, see vol. i. p. 147.

Order XII.—MARSUPIALIA.

FAMILY 76.—DIDELPHYIDÆ.   (3 Genera, 22 Species.)

GENERAL DISTRIBUTION.					
NEOTROPICAL SUB-REGIONS.	NEARCTIC SUB-REGIONS.	PALÆARCTIC SUB-REGIONS.	ETHIOPIAN SUB-REGIONS.	ORIENTAL SUB-REGIONS.	AUSTRALIAN SUB-REGIONS.
1 . 2 . 3 —	1 — 3 —	— — — —	— — — —	— — — —	— — — —

The Didelphyidæ, or true opossums, range throughout all the wooded districts of the Neotropical region from the southern boundary of Texas to the River La Plata, and on the west coast to 42° S. Lat., where a species of *Didelphys* was obtained by Professor Cunningham. One species only is found in the Nearctic region, extending from Florida to the Hudson River, and west to the Missouri. The species named *Didelphys californica* inhabits Mexico, and only extends into the southern extremity of California. The species are most numerous in the great forest region of Brazil, and they have been recently found to the west of the Andes near Guayaquil, as well as in Chili. The exact number of species is very doubtful, owing to the difficulty of determining them from dried skins. All but two belong to the genus *Didelphys*, which has the range above given for the family (Plate XIV., vol. ii. p. 24); *Chironectes* (1 sp.), the yapock or water opossum, inhabits Guiana and Brazil; *Hyracodon* (1 sp.), is a small rat-like animal discovered by Mr. Fraser in Ecuador, and which may perhaps belong to another family.

Extinct Didelphyidæ.—No less than seven species of *Didelphys* have been found in the caves of Brazil, but none in the older formations. In North America the living species only, has been found in Post-Pliocene deposits. In Europe, however, many species of small opossums, now classed as a distinct genus, *Peratherium*, have been found in various Tertiary deposits from the Upper Miocene to the Upper Eocene.

We have here a sufficient proof that the American Marsupials have nothing to do with those of Australia, but were derived from Europe, where their ancestors lived during a long series of ages.

FAMILY 77.—DASYURIDÆ.　(10 Genera, 30 Species.)

GENERAL DISTRIBUTION.					
NEOTROPICAL SUB-REGIONS.	NEARCTIC SUB-REGIONS.	PALÆARCTIC SUB-REGIONS.	ETHIOPIAN SUB-REGIONS.	ORIENTAL SUB-REGIONS.	AUSTRALIAN SUB-REGIONS.
— — — —	— — — —	— — — —	— — — —	— — — —	1 . 2 — —

The Dasyuridæ, or native cats, are a group of carnivorous or insectivorous marsupials, ranging from the size of a wolf to that of a mouse. They are found all over Australia and Tasmania, as well as in New Guinea and the adjacent Papuan islands. Several new genera and species have recently been described by Mr. G. Krefft, of the Sydney Museum, and are included in the following enumeration. *Phasgogale* (3 sp.), New Guinea, West, East, and South Australia; *Antechinomys* (1 sp.), Interior of South Australia; *Antechinus* (12 sp.), Aru Islands, all Australia, and Tasmania; *Chætocercus* (1 sp.), South Australia; *Dactylopsila* (1 sp.), Aru Islands and North Australia; *Podabrus* (5 sp.), West, East, and South Australia, and Tasmania; *Myoictis* (1 sp.), Aru Islands; *Sarcophilus* (1 sp.), Tasmania; *Dasyurus* (4 sp.), North, East, and South, Australia, and Tasmania; *Thylacinus* (1 sp.), Tasmania (Plate XI., vol. i. p. 439).

Extinct species of *Dasyurus* and *Thylacinus* have been found in the Post-Pliocene deposits of Australia.

FAMILY 78.—MYRMECOBIIDÆ.   (1 Genus, 1 Species.)

GENERAL DISTRIBUTION.

NEOTROPICAL SUB-REGIONS.	NEARCTIC SUB-REGIONS.	PALÆARCTIC SUB-REGIONS.	ETHIOPIAN SUB-REGIONS.	ORIENTAL SUB-REGIONS.	AUSTRALIAN SUB-REGIONS.
— — —	— — —	— — —	— — —	— — —	— 2 — —

The only representative of this family is the *Myrmecobius fasciatus*, or native ant-eater, a small bushy-tailed squirrel-like animal, found in the South and West of Australia.

FAMILY 79.—PERAMELIDÆ.   (3 Genera, 10 Species.)

GENERAL DISTRIBUTION.

NEOTROPICAL SUB-REGIONS.	NEARCTIC SUB-REGIONS.	PALÆARCTIC SUB-REGIONS.	ETHIOPIAN SUB-REGIONS.	ORIENTAL SUB-REGIONS.	AUSTRALIAN SUB-REGIONS.
— — —	— — —	— — —	— — .. —	— — —	1 . 2 — —

The Peramelidæ, or bandicoots, are small insectivorous Marsupials, having something of the form of the kangaroos. They range over the whole of Australia and Tasmania, as well as the Papuan Islands. The genus *Perameles* (8 sp.), has the range of the family, one species being found in New Guinea and the Aru Islands (Plate XI., vol. i. p. 440) ; *Peragalea* (1 sp.), inhabits West Australia only ; and *Chœropus* (1 sp.), a beautiful little animal with something of the appearance of a mouse-deer, is found in both South, East, and West Australia.

FAMILY 80.—MACROPODIDÆ.   (10 Genera, 56 Species.)

GENERAL DISTRIBUTION.

NEOTROPICAL SUB-REGIONS.	NEARCTIC SUB-REGIONS.	PALÆARCTIC SUB-REGIONS.	ETHIOPIAN SUB-REGIONS.	ORIENTAL SUB-REGIONS.	AUSTRALIAN SUB-REGIONS.
- — — —	— — — —	— — — —	— — — —	— — — —	1 . 2 — —

The well-known Kangaroos are the most largely developed family of Marsupials, and they appear to be the form best adapted for the present conditions of life in Australia, over every part of which they range. One genus of true terrestrial kangaroos (*Dorcopsis*), inhabits the Papuan Islands, as do also the curious tree kangaroos (*Dendrolagus*) which, without much apparent modification of form, are able to climb trees and feed upon the foliage. The genera, as established by Mr. Waterhouse, are as follows : *Macropus* (4 sp.), West, South, and East Australia, and Tasmania (Plate XII., vol. i. p. 441); *Osphranter* (5 sp.), all Australia ; *Halmaturus* (18 sp.), all Australia and Tasmania ; *Petrogale* (7 sp.), all Australia ; *Dendrolagus* (2 sp.), New Guinea (Plate X., vol. i. p. 414) ; *Dorcopsis* (2 sp.) Aru and Mysol Islands, and New Guinea ; *Onychogalea* (3 sp.), Central Australia ; *Lagorchestes* (5 sp.), North, West, and South Australia ; *Bettongia* (6 sp.), West, South, and East, Australia, and Tasmania ; *Hypsiprymnus* (4 sp.), West and East Australia, and Tasmania.

Extinct Macropodidæ.—Many species of the genera *Macropus* and *Hypsiprymnus* have been found in the cave-deposits and other Post-Tertiary strata of Australia. Among the extinct genera are *Protemnodon* and *Sthenurus*, which are more allied to the tree-kangaroos of New Guinea than to living Australian species ; the gigantic *Diprotodon*, a kangaroo nearly as large as an elephant ; and *Nototherium*, of smaller size.

FAMILY 81.—PHALANGISTIDÆ.   (8 Genera, 27 Species.)

GENERAL DISTRIBUTION.					
NEOTROPICAL SUB-REGIONS.	NEARCTIC SUB-REGIONS.	PALÆARCTIC SUB-REGIONS.	ETHIOPIAN SUB-REGIONS.	ORIENTAL SUB-REGIONS.	AUSTRALIAN SUB-REGIONS.
— — — —	— — — —	— — — —	— — — —	— — — —	1 . 2 — —

The Phalangistidæ, or phalangers, are one of the most varied and interesting groups of Marsupials, being modified in a variety of ways for an arboreal life. We have the clumsy-looking tail-less koala, or native sloth ; the prehensile-tailed opossum-like phalangers ; the beautiful flying oppossums, so closely resembling in form the flying squirrels of North America and India, but often no larger than a mouse; the beautiful dormouse-like *Dromiciæ*, one species of which is only 2¼ inches long or less than the harvest-mouse; and the little *Tarsipes*, a true honey-sucker with an extensile tongue, and of the size of a mouse. These extreme modifications and specializations within the range of a single family, are sufficient to indicate the great antiquity of the Australian fauna; and they render it almost certain that the region it occupied was once much more extensive, so as to supply the variety of conditions and the struggle between competing forms of life, which would be required to develop so many curiously modified forms, of which we now probably see only a remnant.

The Phalangistidæ not only range over all Australia and Tasmania, but over the whole of the Austro-Malayan sub-region from New Guinea to the Moluccas and Celebes. The distribution of the genera is as follows :—*Phascolarctos* (1 sp.), the koala, East Australia; *Phalangista* (5 sp.), East, South, and West Australia, and Tasmania ; *Cuscus* (8 sp.), woolly phalangers, New Guinea, North Australia, Timor, Moluccas and Celebes ; *Petaurista* (1 sp.) large flying phalanger, East Australia ; *Belideus* (5 sp.), flying opossums, South, East, and North Australia, New Guiana and Moluccas; *Acrobata* (1 sp.), pigmy flying opossum, South and East Australia; *Dromicia* (5 sp.), dormouse-phalangers, West and East Australia, and Tasmania; *Tarsipes* (1 sp.), West Australia.

Thylacoleo, a large extinct marsupial of doubtful affinities, seems to be somewhat intermediate between this family and the kangaroos. Professor Owen considered it to be carnivorous, and able to prey upon the huge *Diprotodon*, while Professor Flower and Mr. Gerard Krefft, believe that it was herbivorous.

FAMILY 82.—PHASCOLOMYIDÆ.   (1 Genus, 3 Species.)

GENERAL DISTRIBUTION.

NEOTROPICAL SUB-REGIONS.	NEARCTIC SUB-REGIONS.	PALÆARCTIC SUB-REGIONS.	ETHIOPIAN SUB-REGIONS.	ORIENTAL SUB-REGIONS.	AUSTRALIAN SUB-REGIONS.
— — — —	— — — —	— — — —	— — — —	·· — — —	— 2 — —

The Wombats are tail-less, terrestrial, burrowing animals, about the size of a badger, but feeding on roots and grass. They inhabit South Australia and Tasmania (Plate XI. vol. i. p. 439)

An extinct wombat, as large as a tapir, has been found in the Australian Pliocene deposits.

General Remarks on the Distribution of Marsupialia.

We have here the most remarkable case, of an extensive and highly varied order being confined to one very limited area on the earth's surface, the only exception being the opossums in America. It has been already shown that these are comparatively recent immigrants, which have survived in that country long after they disappeared in Europe. As, however, no other form but that of the Didelphyidæ occurs there during the Tertiary period, we must suppose that it was at a far more remote epoch that the ancestral forms of all the other Marsupials entered Australia ; and the curious little mammals of the Oolite and Trias, offer valuable indications as to the time when this really took place.

A notice of these extinct marsupials of the secondary period will be found at vol. i. p. 159.

Order XIII.—MONOTREMATA.

FAMILY 83.—ORNITHORHYNCHIDÆ.   (1 Genus, 1 Species.)

GENERAL DISTRIBUTION.					
NEOTROPICAL SUB-REGIONS.	NEARCTIC SUB-REGIONS.	PALÆARCTIC SUB-REGIONS.	ETHIOPIAN SUB-REGIONS.	ORIENTAL SUB-REGIONS.	AUSTRALIAN SUB-REGIONS.
— — — —	— — — —	— — — —	— — — —	— — — —	— 2 — —

The *Ornithorhynchus*, or duck-billed Platypus, one of the most remarkable and isolated of existing mammalia, is found in East and South Australia, and Tasmania.

FAMILY 84.—ECHIDNIDÆ.   (1 Genus, 2 Species.)

GENERAL DISTRIBUTION.

NEOTROPICAL SUB-REGIONS.	NEARCTIC SUB-REGIONS.	PALÆARCTIC SUB-REGIONS.	ETHIOPIAN SUB-REGIONS.	ORIENTAL SUB-REGIONS.	AUSTRALIAN SUB-REGIONS.
— — — —	— — — —	— — — —	— — — —	— — — —	— 2 — —

The *Echidna*, or Australian Hedgehog, although quite as remarkable in internal structure as the Ornithorhynchus, is not so
peculiar in external appearance, having very much the aspect of
a hedgehog or spiny armadillo.   The two species of this genus
are very closely allied ; one inhabits East and South Australia,
the other Tasmania.

Extinct Echidnidæ.—Remains of a very large fossil species of
Echidna have lately (1868) been discovered at Darling Downs
in Australia.

Remark on the Distribution of the Monotremata.

This order is the lowest and most anomalous of the mammalia,
and nothing resembling it has been found among the very
numerous extinct animals discovered in any other part of the
world than Australia.

CHAPTER XVIII.

THE DISTRIBUTION OF THE FAMILIES AND GENERA OF BIRDS.

Order I.—PASSERES.

FAMILY 1.—TURDIDÆ.   (21 Genera, 205 Species.)

GENERAL DISTRIBUTION.					
NEOTROPICAL SUB-REGIONS.	NEARCTIC SUB-REGIONS.	PALÆARCTIC SUB-REGIONS.	ETHIOPIAN SUB-REGIONS.	ORIENTAL SUB-REGIONS.	AUSTRALIAN SUB-REGIONS.
1.2.3.4	1.2.3.4	1.2.3.4	1.2.3.4	1.2.3 4	1.2.3 —

The extensive and familiar group of Thrushes ranges over every region and sub-region, except New Zealand. It abounds most in the North Temperate regions, and has its least development in the Australian region. Thrushes are among the most perfectly organized of birds, and it is to this cause, perhaps, as well as to their omnivorous diet, that they have been enabled to establish themselves on a number of remote islands. Peculiar species of true thrush are found in Norfolk Island, and in the small Lord Howes' Island nearer Australia; the Island of St. Thomas in the Gulf of Guinea has a peculiar species; while the Mid-Atlantic island Tristan d'Acunha,—one of the most remote and isolated spots on the globe,—has a peculiarly modified form of thrush. Several of the smaller West Indian Islands have also peculiar species or genera of thrushes.

The family is of somewhat uncertain extent, blending insensibly with the warblers (Sylviidæ) as well as with the Indian bulbuls

(Pycnonotidæ), while one genus, usually placed in it (*Myiophonus*) seems to agree better with *Enicurus* among the Cinclidæ. The genera here admitted into the thrush family are the following, the numbers prefixed to some of the genera indicating their position in Gray's *Hand List of the Genera and Species of Birds* :—

([1143]) *Brachypteryx* (8 sp.), Nepaul to Java and Ceylon (this may belong to the Timaliidæ); *Turdus* (100 sp.) has the range of the whole family, abounding in the Palæarctic, Oriental and Neotropical regions, while it is less plentiful in the Nearctic and Ethiopian, and very scarce in the Australian; ([934]) *Oreocincla* (11 sp.), Palæarctic and Oriental regions, Australia and Tasmania; ([942]) *Rhodinocichla* (1 sp.), Venezuela; ([946]) *Melanoptila* (1 sp.), Honduras; ([947 948]) *Catharus* (10 sp.) Mexico to Equador; ([949 950]) *Margarops* (4 sp.), Hayti and Porto Rico to St. Lucia · ([951]) *Nesccichla* (1 sp.), Tristan d'Acunha; ([952]) *Geocichla* (8 sp.), India to Formosa and Celebes, Timor and North Australia; ([954 955]) *Monticola* (8 sp.), Central Europe to South Africa and to China, Philippine Islands, Gilolo and Java; ([956]) *Orocætes* (3 sp.), Himalayas and N. China; *Zoothera* (3 sp.) Himalayas, Aracan, Java, and Lombok ; *Mimus* (20 sp.) Canada to Patagonia, West Indies and Galapagos ; ([962]) *Oreoscoptes* (1 sp.), Rocky Mountains and Mexico; ([963]) *Melanotis* (2 sp.), South Mexico and Guatemala; ([964]) *Galeoscoptes* (1 sp.), Canada and Eastern United States to Cuba and Panama ; ([965 966]) *Mimocichla* (5 sp.), Greater Antilles; ([967 968]) *Harporhynchus* (7 sp.), North America, from the great lakes to Mexico ; *Cinclocerthia* (3 sp.), Lesser Antilles ; ([970]) *Rhamphocinclus* (1 sp.), Lesser Antilles ; *Chætops* (3 sp.), South Africa ; *Cossypha* = *Bessonornis* (15 sp.) Ethiopian region and Palestine.

FAMILY 2.—SYLVIIDÆ.   (74 Genera, 640 Species.)

GENERAL DISTRIBUTION.					
NEOTROPICAL SUB-REGIONS.	NEARCTIC SUB-REGIONS.	PALÆARCTIC SUB-REGIONS.	ETHIOPIAN SUB-REGIONS.	ORIENTAL SUB-REGIONS.	AUSTRALIAN SUB-REGIONS.
– 2 . 3 . 4	1 . 2 . 3 . 4	1 . 2 . 3 . 4	1 . 2 . 3 . 4	1 . 2 . 3 . 4	1 . 2 . 3 . 4

This immense family, comprising all the birds usually known as " warblers," is, as here constituted, of almost universal distribution. Yet it is so numerous and preponderant over the whole Eastern Hemisphere, that it may be well termed an Old-World group; only two undoubted genera with very few species belonging to the Nearctic region, while two or three others whose position is somewhat doubtful, are found in California and the Neotropical region.

Canon Tristram, who has paid great attention to this difficult group, has kindly communicated to me a MSS. arrangement of the genera and species, which, with a very few additions and alterations, I implicitly follow. He divides the Sylviidæ into seven sub-families, as follows :

1. Drymœcinæ (15 genera 194 sp.), confined to the Old World and Australia, and especially abundant in the three Tropical regions. 2. Calamoherpinæ (11 genera, 75 sp.), has the same general distribution as the last, but is scarce in the Australian and abundant in the Palæarctic region ; 3. Phylloscopinæ (11 genera, 139 sp.), has the same distribution as the entire family, but is most abundant in the Oriental and Palæarctic regions. 4. Sylviinæ (6 genera, 33 sp.), most abundant in the Palæarctic region, very scarce in the Australian and Oriental regions, absent from America. 5. Ruticillinæ (10 genera, 50 sp.); entirely absent from America and Australia ; abounds in the Oriental and Palæarctic regions. 6. Saxicolinæ (12 genera, 126 sp.), absent from America (except the extreme north-west), abundant in the Oriental region and moderately so in the Palæarctic, Ethiopian, and Australian. 7. Accentorinæ (6 genera, 21 sp.), absent from the Ethiopian region and South America, most abundant in Australia, one small genus (*Sialia*), in North America.

The distribution of the several genera arranged under these sub-families, is as follows:

1. DRYMŒCINÆ.—[736] *Orthotomus* (13 sp.), all the Oriental region; [737] *Prinia* (11 sp.), all the Oriental region; [738 740 742 746] *Drymœca* (83 sp.), Ethiopian and Oriental regions, most abundant in the former; [743 to 745 and 749 to 752] *Cisticola* (32 sp.) Ethiopian and Oriental regions, with South Europe, China

and Australia; (^{741}) *Suya* (5 sp.), Nepal to South China and
Formosa; (^{773}) *Sphenœacus* (7 sp.), Australia, New Zealand, and
Chatham Island, with one species (?) in South Africa; ($^{770\ 772}$)
Megalurus (4 sp.), Central India to Java and Timor; ($^{774\ 775}$)
Poodytes (2 sp.), Australia; (^{766}) *Amytis* (3 sp.), Australia; (^{768})
Sphenura (4 sp.), Australia; (^{764}) *Malurus* (16 sp.), Australia and
Tasmania; ($^{762\ 763}$) *Chthonicola* (3 sp.), Australia; (^{761}) *Calaman-*
thus (2 sp.), Australia and Tasmania; (^{759}) *Camaroptera* (5 sp.),
Africa and Fernando Po; (^{753}) *Apalis* (1 sp.), South Africa.

2. CALAMOHERPINÆ.—($^{777\ to\ 781\ and\ sp.\ 2968}$) *Acrocephalus* (35 sp.),
Palæarctic, Ethiopian, continental part of Oriental region, Mo-
luccas, Caroline Islands, and Australia; ($^{782\ 818}$) *Dumeticola* (4 sp.),
Nepal to East Thibet, Central Asia, high regions; ($^{783\ 790}$) *Pota-*
modus (3 sp.), Central and South Europe, and East Thibet;
($^{789\ and\ sp.\ 2969}$) *Lusciniola* (1 sp.), South Europe; ($^{791\ 792}$) *Locus-*
tella (8 sp.), Palæarctic region to Central India and China; (^{739})
Horites (5 sp.), Nepal to North-west China and Formosa; (^{784})
$-$ ^{786}) *Bradypletus* = *Cettia* (10 sp.), South Europe, Palestine, and
South Africa; ($^{747\ 748}$) *Catriscus* (3 sp.), Tropical and South
Africa; *Bernieria* (2 sp.), and (^{756}) *Ellisia* (3 sp.), Madagascar;
($^{832\ *}$) *Mystacornis* (1 sp.), Madagascar; (^{787}) *Calamodus* (2 sp.),
Europe and Palestine; (^{734}) *Tatare* (2 sp.) Samoa to Marquesas
Islands.

3. PHYLLOSCOPINÆ.[1]—*Phylloscopus* (18 sp.), all Palæarctic and
Oriental regions to Batchian; ($^{757\ 758\ 820}$) *Eremomela* (16 sp.), Tropical and South Africa; (^{754}) *Eroessa* (1 sp.), Madagascar;[1] *Hy-*
polais (12 sp.), Palæarctic region, all India, Timor, North and
South Africa; ($^{815\ 816\ 819}$) *Abrornis* (26 sp.), Oriental region; (^{814})
Reguloides (4 sp.), Palæarctic and continental Oriental regions;
(^{822}) *Sericornis* (7 sp.), Australia and Tasmania ($^{823\ 824\ 1451}$) *Acan-*
thiza (14 sp.), Australia and New Caledonia; (^{821}) *Regulus* (7 sp.),
all Palæarctic and Nearctic regions and south to Guatemala;
(^{890}) *Polioptila* (13 sp.), Paraguay to New Mexico; (^{825}) *Gerygone*
(22 sp.), Australia, Papuan and Timor groups, New Zealand
and Norfolk Island.

---

[1] The species of the genera *Phylloscopus* and *Hypolais* are so mixed up in
the *Hand List*, that Mr. Tristram has furnished me with the following

4. SYLVIINÆ.—([793]) *Aedon* (9 sp.), Spain and Palestine, to East and South Africa; ([858]) *Drymodes* (2 sp.), Australia; ([800]) *Pyrophthalma* (2 sp.), South Europe and Palestine; ([801]) *Melizophilus* (3 sp.), South-west Europe and North-east Africa; ([802 804]) *Sylvia* = *Alsecus* (8 sp.), Palæarctic region to India and Ceylon, and North-east Africa; ([806 809]) *Curruca* (7 sp.), Central and South Europe, Madeira, Palestine, Central India, North-east Africa, and South Africa.

5. RUTICILLINÆ.—([827]) *Luscinia* (2 sp.), West Asia, Europe, North Africa; ([839]) *Cyanecula* (3 sp.), Europe, North-east Africa, India, Ceylon, and China; ([840]) *Calliope* (2 sp.), North Asia, Himalayas, Central India, and China; ([838]) *Erithacus* (3 sp.), Europe, North-east Africa, Japan, and North China; ([828 830 837]) *Ruticilla* (20 sp.), Palæarctic and Oriental regions to Senegal and Abyssinia, and east to Timor; abounds in Himalayas; ([829]) *Chœmarrhornis* (1 sp.), Himalayas; ([831 832 834]) *Larvivora* (10 sp.), Oriental region and Japan; ([833]) *Notodela* (3 sp.), Himalayas, Pegu, Formosa, Java; ([835]) *Tarsiger* (2 sp.), Nepal; ([841]) *Grandala* (1 sp.), High Himalayas of Nepal.

6. SAXICOLINÆ.—([975]) *Copsychus* (7 sp.), all Oriental region and Madagascar; ([976]) *Kittacincla* (5 sp.), Oriental region to

enumeration of the species which in his view properly belong to them, by the numbers in that work :—

Phylloscopus.	*Hypolais.*
3032	3026
3033	3028
3048 = 3038	3029
3039	3054 = 3031 = 3036
3063 = 3047 = 3054 = 3061	3042
3048	3043
3049	304
3050	3062 = 3047
3051	3046 = 2932
3052	3035
3053	2976
3056 = 3081	
3057	
3059	
3060	

s 2

Ceylon, Andaman Islands, Formosa, and Borneo ; ($^{794} - {}^{799}$) *Tham-nobia* (10 sp.), Ethiopian region and India to foot of Himalayas; (^{977}) *Gervasia* (2 sp.), Madagascar and Seychelle Islands ; ($^{845\ 847}$) *Dromolæa* (18 sp.), Africa to South Europe, Palestine, North-west India, and North China ; ($^{842\ 843\ 846}$) *Saxicola* (36 sp.), Africa, North-west India, whole Palæarctic region, migrating to Alaska and Greenland ; ($^{848\ 849}$) *Oreicola* (5 sp.), Timor, Lombok, and Burmah ; (^{844}) *Cercomela* (6 sp.), North-east Africa to North-west India ; (^{850}) *Pratincola* (15 sp.), Europe, Ethiopian, and Oriental regions to Celebes and Timor ; (^{917}) *Ephthianura* (3 sp.), Australia ; ($^{851} - {}^{856}$) *Petrœca* (17 sp.), Australian region, Papua to New Zealand, Chatham and Auckland Islands, and Samoa ; (^{857}) *Miro* (2 sp.), New Zealand (doubtfully placed here).

7. ACCENTORINÆ.—(^{771}) *Cinclorhamphus* (2 sp.), Australia ; (^{860}) *Origma* (1 sp.), East Australia ; (^{859}) *Sialia* (8 sp.), United States to Guatemala ; (^{861}) *Accentor* (12 sp.), Palæarctic region to Himalayas and North-west China ; (^{703}) *Orthonyx* (4 sp.), East Australia and New Zealand (doubtfully placed here).

The following two genera, which have been usually classed as Ampelidæ, are arranged by Messrs. Sclater and Salvin in the Sylviidæ :—

(^{1362}) *Myiadestes* (8 sp.), Peru and Bolivia, along the Andes to Mexico and California, also the Antilles ; (^{1364}) *Cichlopsis* (1 sp.), Brazil.

FAMILY 3.—TIMALIIDÆ.   (35 Genera, 240 Species.)

GENERAL DISTRIBUTION.					
NEOTROPICAL SUB-REGIONS.	NEARCTIC SUB-REGIONS.	PALÆARCTIC SUB-REGIONS.	ETHIOPIAN SUB-REGIONS.	ORIENTAL SUB-REGIONS.	AUSTRALIAN SUB-REGIONS.
— — — —	— — — —	— 2 — 4	1 . 2 . 3 . 4	1 . 2 . 3 . 4	1 . 2 — 4

The Timaliidæ, or babbling thrushes, are a group of small strong-legged active birds, mostly of dull colours which are especially characteristic of the Oriental region, in every part of which they abound, while they are much less plentiful in

Australia and Africa.   The Indo-Chinese sub-region is the head quarters of the family, whence it diminishes rapidly in all directions in variety of both generic and specific forms.   Viscount Walden has kindly assisted me in the determination of the limits of this family, as to which there is still much difference of opinion.   The distribution of the genera here admitted is as follows; and as the genera are widely scattered in the *Hand List*, reference numbers are prefixed in every case.

([1023] — [1026] [1008]) *Pomatorhinus* (27 sp.), the whole Oriental region (excluding Philippines), Australia and New Guinea; ([1027]) *Pterorhinus* (3 sp.), North China, East Thibet; ([1029] [1030]) *Malacocircus* (9 sp.), Continental India and Ceylon, Arabia, Nubia; ([1031]) *Chatarrhaa* (5 sp.), Abyssinia, Palestine, India, Nepal, Burmah, and Philippines; ([1032]) *Layardia* (3 sp.), India and Ceylon; ([1033]) *Acanthoptila* (1 sp.), Nepal; ([1034]) *Cinclosoma* (4 sp.), Australia and Tasmania: ([1035] [1036]) *Crateropus* (18 sp.), all Africa, Persia; ([1037]) *Hypergerus* (1 sp.), West Africa: ([1038]) *Cichladusa* (3 sp.), Tropical Africa; ([1039]) *Garrulax* (23 sp.), the Oriental region (excluding Philippines); ([1040]) *Janthocincla* (10 sp.), Nepal, to East Thibet, Sumatra, Formosa; ([1041] [1042]) *Gampsorhynchus* (2 sp.), Himalayas; ([1049]) *Grammatoptila* (1 sp.) North India; ([1043] – [1045]) *Trochalopteron* (24 sp.), all India to China and Formosa; ([1046]) *Actinodura* (4 sp.), Nepal to Burmah, 3,000 - 10,000 feet; ([1047]) *Pellorneum* (4 sp.), Nepal to Ceylon, Tenasserim; ([1158] [1159]) *Timalia* (12 sp.), Malaya;[1] ([1160]) *Dumetia* (2 sp.), Central India and Ceylon; ([1162]) *Stachyris* (6 sp.), Nepal to Assam, Sumatra, Formosa; ([1164]) *Pyctorhis* (3 sp.), India to Ceylon and Burmah; ([1165]) *Mixornis* (8 sp.), Himalayas and Malaya; ([1167]) *Malacopteron* (3 sp.), Malaya; ([1168] [1169]) *Alcippe* (15 sp.), Ceylon and South India, Himalayas to Aracan, Malaya, Formosa, New Guinea; ([1170]) *Macronus* (2 sp.), Malaya; ([1171]) *Cacopitta* (5 sp.), Malaya; ([1172]) *Trichastoma* (11 sp.), Nepal, Burmah, Malaya, Celebes; ([1173]) *Napothera* (6 sp.), Malaya; ([1174]) *Drymocataphus* (8 sp.), Burmah, Malaya, Ceylon,

---

[1] The term "Malaya" is used here to include the Malay Peninsula, Sumatra, Borneo, and Java, a district to which many species and genera are confined. "Malay Archipelago" will be used to include both Indo-Malaya and Austro-Malaya.

Timor; ([1175]) *Turdinus* (5 sp.), Khasia Hills, Malacca, Tenasserim; ([1176]) *Trichixos* (1 sp.), Borneo, Malacca; ([1004]) *Sibia* (6 sp.), Nepal to Assam, Tenasserim, Formosa; ([1177 1178]) *Alethe* (4 sp.), West Africa; ([1178 a]) *Oxylabes* (1 sp.), Madagascar; ([1050]) *Psophodes* (2 sp.), South, East, and West Australia; ([1048]) *Turnagra* (3 sp.), New Zealand.

FAMILY 4.—PANURIDÆ.   (4 Genera, 13 Species).

GENERAL DISTRIBUTION.					
NEOTROPICAL SUB-REGIONS.	NEARCTIC SUB-REGIONS.	PALÆARCTIC SUB-REGIONS.	ETHIOPIAN SUB-REGIONS.	ORIENTAL SUB-REGIONS.	AUSTRALIAN SUB-REGIONS.
— — — —	— — ·· —	1 . 2 — 4	— — — —	— — 3 —	— — — —

This new family is adopted, at the suggestion of Professor Newton, to include some peculiar groups of Himalayan birds whose position has usually been among the Timaliidæ or the Paridæ, but which are now found to be allied to our Bearded Reedling. The supposed affinity of this bird for the Tits has been long known to be erroneous, and the family Panuridæ was formed for its reception (Yarrell's *British Birds*, 4th edit. p. 512). The genera having hitherto been widely scattered in systematic works, are referred to by the numbers of Mr. G. R. Gray's *Hand List*.

([1901]) *Paradoxornis* (3 sp.), Himalayas and East Thibet; ([1904]) *Conostoma* (1 sp.), Himalayas and East Thibet; ([876]) *Suthora* (8 sp.), Himalayas to North-west China, Formosa; ([877]) *Chlenasicus* (1 sp.), Darjeeling; ([887]) *Panurus* (1 sp.), Central and Southern Europe; ([1902]) *Heteromorpha* (1 sp.), Nepal, 10,000 feet altitude; *Cholornis* (1 sp.), Moupin in East Thibet.

FAMILY 5.—CINCLIDÆ.   (4 Genera, 27 Species.)

GENERAL DISTRIBUTION.					
NEOTROPICAL SUB-REGIONS.	NEARCTIC SUB-REGIONS.	PALÆARCTIC SUB-REGIONS.	ETHIOPIAN SUB-REGIONS.	ORIENTAL SUB-REGIONS.	AUSTRALIAN SUB-REGIONS.
— 2 . 3 —	— 2 — 4	1 . 2 . 3 . 4	— — — ?4	1 . 2 . 3 . 4	1 — —

The Cinclidæ consist of a number of more or less thrush-like ground-birds, of which the most remarkable are the Dippers, forming the genus *Cinclus*. These are curiously distributed, from the Palæarctic region as a centre, to the alpine districts of North and South America; while the three genera which are here included as somewhat allied to *Cinclus*, all inhabit the Oriental region. The genera which I class in this family are the following :—

(⁹⁷⁸) *Cinclus* (9 sp.), Palæarctic region to West China and Formosa, Rocky Mountains, and Mexico in North America, and southward to the Andes of Peru; (⁹¹⁶) *Enicurus* (9 sp.), Himalayas to Java and West China; (⁹⁷⁹) *Eupetes* (4 sp.), Indo-Malay sub-region and New Guinea; (⁹⁷¹) *Myiophonus* (5 sp.), Himalayas to Ceylon, Java, South China, and Formosa.

(⁹⁸¹) *Mesites* (1 sp.), Madagascar, is an anomalous bird placed with *Eupetes* by Mr. G. R. Gray, but of very uncertain affinities.

FAMILY 6—TROGLODYTIDÆ. (17 Genera, 94 Species.)

GENERAL DISTRIBUTION.					
NEOTROPICAL SUB-REGIONS.	NEARCTIC SUB-REGIONS.	PALÆARCTIC SUB-REGIONS.	ETHIOPIAN SUB-REGIONS.	ORIENTAL SUB-REGIONS.	AUSTRALIAN SUB-REGIONS.
1 . 2 . 3 . 4	1 . 2 . 3 . 4	1 . 2 . 3 . 4	1 . 2 . 3 —	— — 3 . 4	1 — — —

The Troglodytidæ, or Wrens, are small birds, rather abundant and varied in the Neotropical region, with a few species scattered through the Nearctic, Palæarctic, and parts of the Oriental regions, and one doubtful genus in Africa. The constitution of the family is by no means well determined. The South American genera are taken from Messrs. Sclater and Salvin's *Nomenclator Avium Neotropicalium.*

Tesia (2 sp.), Eastern Himalayas; *Pnoepyga* (6 sp.), Himalayas to East Thibet, Java; (⁷¹⁶ ᵃⁿᵈ ⁷²³) *Troglodytes* (15 sp.), Neotropical, Nearctic, and Palæarctic regions to the Higher Himalayas; (⁶⁹⁷) *Rimator* (1 sp.), Darjeeling; *Thryothorus* (13 sp.), South Brazil to Mexico, Martinique, and Nearctic region; *Thryophilus* (13 sp.), Brazil to Mexico, and North-west America; *Cistothorus*

(5 sp.), Patagonia to Greenland; *Uropsila* (1 sp.), Mexico; *Donacobius* (2 sp.), Tropical America; *Campylorhynchus* (18 sp.), Brazil, and Bolivia to Mexico and the Gila valley; *Cyphorhinus* (5 sp.), Equatorial South America to Costa Rica; *Microcerculus* (5 sp), Brazil and Peru to Mexico; *Henicorhina* (2 sp.), Peru and Guiana to Costa Rica; *Salpinctes* (1 sp.), High Plains of Rocky Mountains; *Catherpes* (1 sp.), Mexico and Rio Grande; *Cinnicerthia* (2 sp.), Ecuador and Columbia. (^{700}) *Sylvietta* (2 sp.), Tropical and South Africa,—is placed in this family by Mr. Tristram.

FAMILY 7.—CHAMÆIDÆ.   (1 Genus, 1 Species).

GENERAL DISTRIBUTION.

NEOTROPICAL SUB-REGIONS.	NEARCTIC SUB-REGIONS.	PALÆARCTIC SUB-REGIONS.	ETHIOPIAN SUB-REGIONS.	ORIENTAL SUB-REGIONS.	AUSTRALIAN SUB-REGIONS.
— — — —	1 — — —	— — — —	— — — —	— — — —	— — — —

The bird which forms the genus *Chamæa* inhabits California; and though allied to the wrens it has certain peculiarities of structure which, in the opinion of many ornithologists, require that it should be placed in a distinct family.

FAMILY 8.—CERTHIIDÆ.   (6 Genera, 18 Species.)

GENERAL DISTRIBUTION.

NEOTROPICAL SUB-REGIONS.	NEARCTIC SUB-REGIONS.	PALÆARCTIC SUB-REGIONS.	ETHIOPIAN SUB-REGIONS.	ORIENTAL SUB-REGIONS.	AUSTRALIAN SUB-REGIONS.
— — — —	— — 3 —	1.2.3.4	— — — —	1 — 3.4	1.2 — —

The Certhiidæ, or Creepers, form a small family whose species are thinly scattered over North America from Mexico, the Palæarctic region, parts of the Oriental region, and Australia, where they are somewhat more abundant. The distribution of the genera is as follows:

Certhia (6 sp.), Nearctic and Palæarctic regions, Nepal, and Sikhim; *Salpornis* (1 sp.), Central India; *Tichodroma* (1 sp.), South

Europe to Abyssinia, Nepal, and North China; *Rhabdornis* (1 sp.), Philippine Islands; *Climacteris* (8 sp.), Australia and New Guinea.

FAMILY 9.—SITTIDÆ.   (6 Genera, 31 Species.)

GENERAL DISTRIBUTION.					
NEOTROPICAL SUB-REGIONS.	NEARCTIC SUB-REGIONS.	PALÆARCTIC SUB-REGIONS.	ETHIOPIAN SUB-REGIONS.	ORIENTAL SUB-REGIONS.	AUSTRALIAN SUB-REGIONS.
— — — —	1.2.3.4	1.2.3.4	— — — —	1.2.3.4	1.2 — 4

The Sittidæ, or Nuthatches, are another small family of tree-creeping birds, whose distribution is very similar to that of the Certhiidæ, but with a more uniform range over the Oriental region, and extending to New Zealand and Madagascar. The genera are as follows:—

Sitta (17 sp.), Palæarctic and Nearctic regions to South India and Mexico; *Dendrophila* (2 sp.), Ceylon and India to Burmah and Malaya; *Hypherpes* (1 sp.), Madagascar; *Sittella* (6 sp.), Australia and New Guinea. *Acanthisitta* (1 sp.) and *Xenicus* (4 sp.), New Zealand, are placed with some doubt in this family.

FAMILY 10.—PARIDÆ.   (14 Genera, 92 Species.)

GENERAL DISTRIBUTION.					
NEOTROPICAL SUB-REGIONS.	NEARCTIC SUB-REGIONS.	PALÆARCTIC SUB-REGIONS.	ETHIOPIAN SUB-REGIONS.	ORIENTAL SUB-REGIONS.	AUSTRALIAN SUB-REGIONS.
— — 3 —	1.2.3.4	1.2.3.4	1.2.3 —	1.2.3.4	— 2 — 4

The Paridæ, or Tits, are very abundant in the Nearctic and Palæarctic regions; many fine species are found in the Himalayas, but they are sparingly scattered through the Ethiopian, Oriental, and Australian regions. The genera usually admitted into this family are the following, but the position of some of them, especially of the Australian forms, is doubtful.

([864 — 867 870]) *Parus* (46 sp.), North America, from Mexico, Palæarctic, and Oriental regions, Tropical and South Africa;

($^{868\ 869}$) *Lophophanes* (10 sp.), Europe, the Higher Himalayas to Sikhim, North America to Mexico; *Acredula = Orites* (6 sp.), Palæarctic region; *Melanochlora* (2 sp.), Nepal to Sumatra; *Psaltria* (1 sp.), Java; *Psaltriparus* (3 sp.), Guatemala to California, and Rocky Mountains; *Auriparus* (1 sp.), Rio Grande; ($^{881\ 882}$) *Parisoma* (5 sp.), Tropical and South Africa; ($^{883\ 884}$) *Ægithalus* (6 sp.), South-east Europe to South Africa; ($^{885\ 889}$) *Ægithaliscus* (6 sp.), Afghanistan and Himalayas to Amoy; *Cephalopyrus* (1 sp.), North-west Himalayas; *Sylviparus* (1 sp.), Himalayas and Central India; *Certhiparus* (2 sp.), New Zealand; ($^{879\ 880}$) *Sphenostoma* (2 sp.), East and South Australia.

FAMILY 11.—LIOTRICHIDÆ.   (11 Genera, 35 Species.)

GENERAL DISTRIBUTION.					
NEOTROPICAL SUB-REGIONS.	NEARCTIC SUB-REGIONS.	PALÆARCTIC SUB-REGIONS.	ETHIOPIAN SUB-REGIONS.	ORIENTAL SUB-REGIONS.	AUSTRALIAN SUB-REGIONS.
— — — —	— — — —	— — — —	— — — —	— — 3 . 4	— — — —

The Liotrichidæ, or Hill-Tits, are small, active, delicately-coloured birds, almost confined to the Himalayas and their extension eastward to China. They are now generally admitted to form a distinct family. The genera are distributed as follows:

(^{1146}) *Liothrix* (3 sp.), Himalayas to China; *Siva* (3 sp.), Himalayas; *Minla* (4 sp.), Himalayas and East Thibet; *Proparus* (7 sp.), Nepal to East Thibet and Aracan; (^{1153}) *Pteruthius* (6 sp.), Himalayas to Java and West China; (^{1155}) *Cutia* (2 sp.), Nepal; (^{1019}) *Yuhina* (3 sp.), High Himalayas and Moupin; (^{1020}) *Ixulus* (3 sp.), Himalayas to Tenasserim; (^{1021}) *Myzornis* (1 sp.), Darjeeling.

FAMILY 12.—PHYLLORNITHIDÆ.   (3 Genera, 14 Species.)

GENERAL DISTRIBUTION.					
NEOTROPICAL SUB-REGIONS.	NEARCTIC SUB-REGIONS.	PALÆARCTIC SUB-REGIONS.	ETHIOPIAN SUB-REGIONS.	ORIENTAL SUB-REGIONS.	AUSTRALIAN SUB-REGIONS.
— — — —	— — — —	— — — —	— — — —	1 . 2 . 3 . 4	— — — —

The Phyllornithidæ, or " Green Bulbuls," are a small group of fruit-eating birds, strictly confined to the Oriéntal region, and ranging over the whole of it, with the one exception of the Philippine Islands. The genera are :—

([1022]) *Phyllornis* (12 sp.), India to Java, Ceylon, and Hainan ; ([1166]) *Iora* (4 sp.), the whole Oriental region ; ([1163]) *Erpornis* (2 sp.), Himalayas, Hainan, and Formosa.

FAMILY 13.—PYCNONOTIDÆ. (9 Genera, 139 Species.)

GENERAL DISTRIBUTION.					
NEOTROPICAL SUB-REGIONS.	NEARCTIC SUB-REGIONS.	PALÆARCTIC SUB-REGIONS.	ETHIOPIAN SUB-REGIONS.	ORIENTAL SUB-REGIONS.	AUSTRALIAN SUB-REGIONS.
— — — —	— — — —	— 2 — 4	1.2.3.4	1.2.3.4	1 — — —

The Pycnonotidæ, Bulbuls, or fruit-thrushes, are highly characteristic of the Oriental region, in every part of which they abound ; less plentiful in the Ethiopian region, and extending to Palestine and Japan in the Palæarctic, and to the Moluccas in the Australian region, but absent from the intervening island of Celebes. The genera are :—

Microscelis (6 sp.), Burmah, the Indo-Malay Islands, and Japan ; *Pycnonotus* (52 sp., in many sub-genera), Palestine to South Africa, the whole Oriental region, China and Japan ; *Alcurus* (1 sp.), Himalayas ; *Hemixus* (2 sp.), Nepal, Bootan, Hainan ; *Phyllastrephus* (4 sp.), West and South Africa ; *Hypsipetes* (20 sp.), the whole Oriental region, Madagascar and the Mascarene Islands ; *Tylas* (1 sp.), Madagascar ; *Criniger* (30 sp.), the whole Oriental region (excluding Philippines), West and South Africa, Moluccas ; *Ixonotus* (7 sp.), West Africa ; ([1015 1017]) *Setornis* (3 sp.), Malacca, Sumatra, and Borneo ; *Iole* (4 sp.), Aracan and Malaya ; *Andropadus* (9 sp.), Tropical Africa ; ([1157]) *Lioptilus* (1 sp.), South Africa.

FAMILY 14.—ORIOLIDÆ.   (5 Genera, 40 Species.)

GENERAL DISTRIBUTION					
NEOTROPICAL SUB-REGIONS.	NEARCTIC SUB-REGIONS.	PALÆARCTIC SUB-REGIONS.	ETHIOPIAN SUB-REGIONS.	ORIENTAL SUB-REGIONS.	AUSTRALIAN SUB-REGIONS.
— — — —	— — — —	1.2 – 4	1.2.3.4	1.2.3.4	1.2 — —

The Orioles, or Golden Thrushes, are a small group charac-teristic of the Oriental and Ethiopian regions, migrating into the western Palæarctic region, and with some of the less typical forms in Australia.   The genera are :—

Oriolus (24 sp.), Central Europe, throughout Africa, and the whole Oriental region, northward to Pekin, and eastward to Flores ; ([1073]) *Analcipus* (3 sp.), Himalayas, Formosa, Java and Borneo; *Mimeta* (9 sp.), the Moluccas and Australia ; *Sphecotheres* (3 sp.), Timor and Australia.   *Artamia* (1 sp.), Madagascar,— perhaps belongs to the next family or to Laniidæ.

FAMILY 15.—CAMPEPHAGIDÆ.   (3 Genera, 100 Species.)

GENERAL DISTRIBUTION.					
NEOTROPICAL SUB-REGIONS.	NEARCTIC SUB-REGIONS.	PALÆARCTIC SUB-REGIONS.	ETHIOPIAN SUB-REGIONS.	ORIENTAL SUB-REGIONS.	AUSTRALIAN SUB-REGIONS.
— — — —	— — — —	— — — —	1.2.3.4	1.2.3.4	1.2.3 –

The Campephagidæ, or Cuckoo Shrikes, (Campephaginæ of the *Hand List*, with the addition of *Cochoa*) are most abundant in the Australian region (especially in the Austro-Malay sub-region) less so in the Oriental, and still ess in the Ethiopian region.   The genera, for the most part as adopted by Dr. Hart-laub, are as follows :—

Pericrocotus (22 sp.), the whole Oriental region, extending north to Pekin, and east to Lombok ; ([1242 – 1244]) *Lanicterus* (4 sp.), West and South Africa ; ([1245 1246]) *Graucalus* (25 sp.), the whole Oriental region, and eastward to Austro-Malaya, the New

Hebrides, and Tasmania; *Artamides* (1 sp.), Celebes; *Pteropo docys* (1 sp.), Australia; ([1248] [1250] [1257] [1253]) *Campephaga* (16 sp.), Austro-Malaya, and New Caledonia, Philippines, the Ethiopian region; *Volvocirora* (8 sp.) the Oriental region (excluding Philippines); *Lalage* (18 sp.), the whole Malay Archipelago to New Caledonia and Australia; *Symmorphus* (1 sp.), Australia; *Oxynotus* (2 sp.), Mauritius and Bourbon; ([1204]) *Cochoa* (3 sp.), Himalayas, Java. The position of this last genus is doubtful. Jerdon puts it in the Liotrichidæ; Sundeval in the Sturnidæ; Bonaparte in the Dicruridæ; Professor Newton suggests the Pycnonotidæ; but it seems on the whole best placed here.

FAMILY 16.—DICRURIDÆ.　　(6 Genera, 58 Species.)

GENERAL DISTRIBUTION.					
NEOTROPICAL SUB-REGIONS.	NEARCTIC SUB-REGIONS.	PALÆARCTIC SUB-REGIONS.	ETHIOPIAN SUB-REGIONS.	ORIENTAL SUB-REGIONS.	AUSTRALIAN SUB-REGIONS.
— — — —	— — — —	— — — —	1 . 2 . 3 . 4	1 . 2 . 3 . 4	1 . 2 — —

The Dicruridæ, or Drongo Shrikes (Dicruridæ of the *Hand List,* omitting the genus *Melœnornis*), have nearly the same distribution as the last family, with which they are sometimes united. They are, however, most abundant and varied in the Oriental region, much less so both in the Australian and Ethiopian regions. The distribution of the genera is as follows :—

Dicrurus (46 sp., in several sub-genera), has the range of the whole family, extending east to New Ireland, and one species in Australia; *Chœtorhynchus* (1 sp.), New Guinea; *Bhringa* (2 sp.), Himalayas to Borneo (Plate IX. vol. i. p. 339); *Chibia* (2 sp.) Himalayas eastward to North China; *Chaptia* (3 sp.), all India to Malacca and Formosa; *Irena* (4 sp.), Central India, Assam, and Burmah to Borneo and the Philippine Islands. This last genus is placed by Jerdon among the Pycnonotidæ, but seems to come most naturally here or in the last family.

FAMILY 17.—MUSCICAPIDÆ.   (44 Genera, 283 Species.)

GENERAL DISTRIBUTION.

NEOTROPICAL SUB-REGIONS.	NEARCTIC SUB-REGIONS.	PALÆARCTIC SUB-REGIONS.	ETHIOPIAN SUB-REGIONS.	ORIENTAL SUB-REGIONS.	AUSTRALIAN SUB-REGIONS.
— — — —	— — — —	1.2.3.4	1.2.3.4	1.2.3.4	1.2.3.4

The Muscicapidæ, or Flycatchers (Muscicapinæ and Myiagrinæ of the *Hand List*, omitting *Cochoa* and including *Pogonocichla*) form an extensive family of usually small-sized and often bright-coloured birds, very abundant in the warmer regions of the Old World and Australia, but becoming scarce as we approach the temperate and colder regions. They are wholly absent from North and South America. The genera, many of which are not well defined, are distributed as follows :—

Peltops (1 sp.), Papuan Islands ; *Monarcha* (28 sp.), Moluccas to the Carolines and Marquesas Islands, Australia and Tasmania ; *Leucophantes* (1 sp.), New Guinea ; *Butalis* (4 sp.), Ethiopian and Palæarctic regions, Moluccas and Formosa ; *Muscicapa* (12 sp.), Europe and Africa ; *Muscicapula* (6 sp.), India to Western China ; *Alseonax* (1 sp.), South Africa ; *Erythrosterna* (7 sp.), Europe to China and Java ; *Newtonia* (1 sp.), Madagascar ; *Xanthopygia* (2 sp.), Japan, China, Malacca ; *Hemipus* (1 sp.), India and Ceylon ; *Pycnophrys* (1 sp.), Java ; *Hyliota* (2 sp.), West Africa ; *Erythrocercus* (2 sp.), West Africa and Zambesi ; *Micrœca* (6 sp.), Australia, Timor, and Papuan Islands ; *Artomyias* (2 sp.), West Africa ; *Pseudobias* (1 sp.), Madagascar ; *Hemichelidon* (3 sp.), the Oriental region and North China ; *Smithornis* (2 sp.), West and South Africa ; *Megabias* (1 sp.), West Africa ; *Cassinia* (2 sp.), West Africa ; *Bias*, (1 sp.), Tropical Africa ; *Niltava* (3 sp.), Himalayas to West China ; *Cyornis* (16 sp.), the whole Oriental region ; *Cyanoptila* (1 sp.), Japan, China, Hainan ; *Eumyias* (7 sp.), India to South China, Ceylon, and Sumatra ; (1213 and 1216) *Siphia* (8 sp.), North India, Formosa, Timor ; *Anthipes* (1 sp.), Nepal ; *Scisura* (5 sp.), Australia and Austro-

Malaya (excluding Celebes); (*Myiagra* (16 sp.), Australia and Moluccas to Caroline and Samoa Islands: *Hypothymis* (2 sp.), Oriental region and Celebes ; *Elminia* (2 sp.), Tropical Africa ; *Muscitodus* (2 sp.), Fiji Islands ; *Machærirhynchus* (4 sp.), Papuan Islands and North Australia ; *Platystira* (12 sp.), Tropical and South Africa ; *Rhipidura* (45 sp.), the Oriental and Australian regions to the Samoa Islands and Tasmania ; *Chelidorynx* (1 sp.), North India ; *Myialestes* (2 sp.), India to Ceylon, China, Java and Celebes ; *Tchitrea* (26 sp.), the entire Ethiopian and Oriental regions, and to North China and Japan ; *Philentoma* (4 sp.) Malacca, Sumatra, Borneo, and Philippine Islands ; *Todopsis* (6 sp.), Papuan Islands ; ([836]) *Pogonocichla* (1 sp.), South Africa ; ([1061] – [1063])*Bradyornis* (7 sp.), Tropical and South Africa ; ([1460]) *Chasiempis* (2 sp.), Sandwich Islands.

FAMILY 18.—PACHYCEPHALIDÆ. (5 Genera, 62 Species.)

GENERAL DISTRIBUTION.					
NEOTROPICAL SUB-REGIONS.	NEARCTIC SUB-REGIONS.	PALÆARCTIC SUB-REGIONS.	ETHIOPIAN SUB-REGIONS.	ORIENTAL SUB-REGIONS.	AUSTRALIAN SUB-REGIONS.
— — — —	— — — —	— — — —	— — — 4	— — 3.4	1.2.3 —

The Pachycephalidæ, or Thick-headed Shrikes (Pachycephalinæ of the *Hand List* omitting *Colluricincla, Cracticus,* and *Pardalotus*) are almost confined to the Australian region, a single species extending to Java and Aracan, and another (?) to Madagascar. The family has generally been united with the Laniidæ, but most modern ornithologists consider it to be distinct. The distribution of the genera is as follows :—

Oreœca (1 sp.), Australia ; *Falcunculus* (2 sp.), Australia ; *Pachycephala* (44 sp.), Sula Islands (east of Celebes) to the Fiji Islands, and Australia ; *Hylocharis* (4 sp.), Timor, Celebes, Indo-Malaya, and Aracan ; *Calicalicus* (1 sp.), Madagascar ; *Eopsaltria* (14 sp.), Australia, New Caledonia, and the New Hebrides ; *Artamia* (4 sp.), Madagascar,—may belong to this family, or to Laniidæ, Oriolidæ, or Artamidæ, according to different authors.

FAMILY 19.—LANIIDÆ.　(19 Genera, 145 Species.)

GENERAL DISTRIBUTION.

NEOTROPICAL SUB-REGIONS.	NEARCTIC SUB-REGIONS.	PALÆARCTIC SUB-REGIONS.	ETHIOPIAN SUB-REGIONS.	ORIENTAL SUB-REGIONS.	AUSTRALIAN SUB-REGIONS.
— — — —	1.2.3.4	1.2.3.4	1.2.3.4	1.2.3.4	1.2.3—

The Laniidæ, or Shrikes (Laniinæ and Malaconotinæ of the *Hand List*, and including *Colluricincla*), are most abundant and varied in Africa, less plentiful in the Oriental, Australian, and Palæarctic regions, with a few species in the Nearctic region as far as Mexico. The constitution of the family is, however, somewhat uncertain. The genera here admitted are :—

Colluricincla (4 sp.), Australia and Tasmania; *Rectes* (18 sp.), Papuan Islands, North Australia, to Pelew and Fiji Islands; (1462 — 1464 1466 1470 1471 — 1473) *Lanius* (50 sp.), the whole Nearctic, Palæarctic, Ethiopian, and Oriental regions, one species reaching Timor, none in Madagascar; *Laniellus* (1 sp.), Java; *Hypocolius* (1 sp.), Abyssinia and Upper Nile; *Corvinella* (1 sp.), South and West Africa; *Urolestes* (1 sp.), South and East Africa; *Tephrodornis* (4 sp.), Oriental region to Hainan and Java; *Hypodes* (1 sp.), West Africa; *Fraseria* (2 sp.), West Africa; *Cuphopterus* (1 sp.), Princes' Island; *Nilaus* (1 sp.), South and West Africa; *Prionops* (9 sp.), Tropical Africa; *Eurocephalus* (2 sp.), North, East, and South Africa, and Abyssinia; *Chaunonotus* (1 sp.), West Africa; *Vanga* (4 sp.), Madagascar (Plate VI. vol. i. p. 278); *Laniarius* (36 sp.), the whole Ethiopian region; *Telephonus* (10 sp.), all Africa and South Europe; *Meristes* (2 sp.), Tropical and South Africa; *Nicator* (1 sp.), East Africa.

FAMILY 20.—CORVIDÆ.　(24 Genera, 190 Species.)

GENERAL DISTRIBUTION.

NEOTROPICAL SUB-REGIONS.	NEARCTIC SUB-REGIONS.	PALÆARCTIC SUB-REGIONS.	ETHIOPIAN SUB-REGIONS.	ORIENTAL SUB-REGIONS.	AUSTRALIAN SUB-REGIONS.
1.2.3.4	1.2.3.4	1.2.3.4	1.2.3.4	1.2.3.4	1.2.3—

The Corvidæ, or Crows, Jays, &c., form an extensive and somewhat heterogeneous group, some members of which inhabit almost every part of the globe, although none of the genera are cosmopolitan. The true crows are found everywhere but in South America; the magpies, choughs, and nutcrackers are characteristic of the Palæarctic region; the jays are Palæarctic, Oriental, and American; while the piping crows are peculiarly Australian. The more detailed distribution of the genera is as follows:—

Sub-family I. Gymnorhininæ (Piping Crows).—*Strepera* (4 sp.), and *Gymnorhina* (3 sp.), are Australian only; *Cracticus* (9 sp.), ranges from New Guinea to Tasmania (this is usually put with the Shrikes, but it has more affinity with the preceding genera); *Pityriasis* (1 sp.), Borneo (an extraordinary bird of very doubtful affinities); *Grallina* (1 sp.), Australia, is put here by Sundevall,—among Motacillidæ, by Gould.

Sub-family II. Garrulinæ (Jays).—*Platylophus* = *Lophocitta* (4 sp.), Malaya; *Garrulus* (12 sp.), Palæarctic region, China and Himalayas; *Perisoreus* (2 sp.), North of Palæarctic and Nearctic regions; *Cyanurus* (22 sp.), American, from Bolivia to Canada, most abundant in Central America, but absent from the Antilles; *Cyanocorax* (15 sp.), La Plata to Mexico; *Calocitta* (2 sp.), Guatemala and Mexico; *Psilorhinus* (3 sp.), Costa Rica to Texas; *Urocissa* (6 sp.), Western Himalayas to China and Formosa; *Cissa* (3 sp.), South-eastern Himalayas to Tenasserim, Ceylon, Sumatra, and Java.

Sub-family III. Dendrocittinæ (Tree Crows).—*Temnurus* (3 sp.), Cochin China, Malacca to Borneo (not Java); *Dendrocitta* (9 sp.), the Oriental region to Sumatra, Hainan, and Formosa; *Crypsirhina* (3 sp.), Pegu, Siam, and Java; *Ptilostomus* (2 sp.), West, East, and South Africa.

Sub-family IV. Corvinæ (Crows and Magpies).—*Nucifraga* (4 sp.), Palæarctic region to the Himalayas and North China; *Picicorvus* (1 sp.), the Rocky Mountains and California; *Gymnokitta* (1 sp.), Rocky Mountains and Arizona (Plate XVIII., Vol. II., p. 128); *Pica* (9 sp.), Palæarctic region, Arctic America, and California; *Cyanopica* (3 sp.), Spain, North-east Asia, Japan;

Streptocitta (2 sp.), Celebes ; *Charitornis* (1 sp.), Sula Islands;
Corvus (55 sp.), universally distributed except South America
and New Zealand, but found in Guatemala and the Antilles
to Porto Rico ; reaches the extreme north of Europe and Asia ;
Gymnocorvus (2 sp.), Papuan Islands ; *Picathartes* (1 sp.), West
Africa ; *Corvultur* (2 sp.), Tropical and South Africa.

Sub-family V. Fregilinæ (Choughs).—*Fregilus* (3 sp.), mountains and cliffs of Palæarctic region from West Europe to the
Himalayas and North China, Abyssinia (Plate I., Vol. I., p.
195) ; *Corcorax* (1 sp.), Australia.

FAMILY 21.—PARADISEIDÆ.   (19 Genera, 34 Species.)

GENERAL DISTRIBUTION.					
NEOTROPICAL SUB-REGIONS.	NEARCTIC SUB-REGIONS.	PALÆARCTIC SUB-REGIONS.	ETHIOPIAN SUB-REGIONS.	ORIENTAL SUB-REGIONS.	AUSTRALIAN SUB-REGIONS.
— — — —	— — — —	— — — —	— — — —	— — — —	1 . 2 — —

The Paradiseidæ, or " Birds of Paradise," form one of the most
remarkable families of birds, unsurpassed alike for the singularity
and the beauty of their plumage.   Till recently the family was restricted to about eight species of the more typical Paradise birds,
but in his splendid monograph of the group, Mr. Elliot has
combined together a number of allied forms which had been
doubtfully placed in several adjacent families.   The various
species of true Paradise birds, having ornamental plumes developed from different parts of the body, are almost wholly confined
to New Guinea and the adjacent Papuan Islands, one species
only being found in the Moluccas and one in North Australia ;
while the less typical Bower-birds, having no such developments
of plumage, are most characteristic of the north and east of
Australia, with a few species in New Guinea.   The distribution
of the genera according to Mr. Elliot's monograph is as follows :—

Sub-family I. Paradiseinæ.—*Paradisea* (4 sp.), Papuan Is-
lands ; *Manucodia* (3 sp.), Papuan Islands and North Australia;
Astrapia (1 sp.), New Guinea ; *Parotia* (1 sp.), New Guinea;
Lophorhina (1 sp.), New Guinea ; *Diphyllodes* (3 sp.), Papuan

Islands; *Xanthomelus* (1 sp.), New Guinea; *Cicinnurus* (1 sp.), Papuan Islands; *Paradigalla* (1 sp.), New Guinea; *Semioptera* (1 sp.), Gilolo and Batchian.

Sub-family II. Epimachinæ.—*Epimachus* (1 sp.), New Guinea; *Drepanornis* (1 sp.), New Guinea; *Seleucides* (1 sp.), New Guinea (Plate X., Vol. I., p. 414); *Ptilorhis* (4 sp.), New Guinea and North Australia.

Sub-family III. Tectonarchinæ (Bower-birds).—*Sericulus* (1 sp.), Eastern Australia; *Ptilonorhynchus* (1 sp.), Eastern Australia; *Chlamydodera* (4 sp.), North and East Australia; *Æluroedus* (3 sp.), Papuan Islands and East Australia; *Amblyornis* (1 sp.), New Guinea.

FAMILY 22.—MELIPHAGIDÆ.   (23 Genera, 190 Species.)

GENERAL DISTRIBUTION.					
NEOTROPICAL SUB-REGIONS.	NEARCTIC SUB-REGIONS.	PALÆARCTIC SUB-REGIONS.	ETHIOPIAN SUB-REGIONS.	ORIENTAL SUB-REGIONS.	AUSTRALIAN SUB-REGIONS.
— — — —	— — — —	— — — —	— — — —	— — — —	1 . 2 . 3 . 4

(As in the *Hand List*, but omitting Zosterops, and slightly altering the arrangement.)

The extensive group of the Meliphagidæ, or Honey-suckers, is wholly Australian, for the genus *Zosterops*, which extends into the Oriental and Ethiopian regions, does not naturally belong to it. Several of the genera are confined to Australia, others to New Zealand, while a few range over the whole Australian region. The genera are distributed as follows:—

Myzomela (18 sp.), has the widest range, extending from Celebes to the Samoa Islands, and to Timor and Eastern Australia; *Entomophila* (4 sp.), Australia and New Guinea; *Gliciphila* (10 sp.), Australia, Timor, New Guinea, and New Caledonia; *Acanthorhynchus* (2 sp.), Australia and Tasmania; *Meliphaga* (1 sp.), Australia; *Ptilotis* (40 sp.), Gilolo and Lombok to Australia and Tasmania, and to the Samoa and Tonga Islands; *Meliornis* (5 sp.). Australia and Tasmania; *Prosthemadera* (1 sp.), *Pogonornis* (1 sp.), New Zealand; *Anthornis* (4 sp.), New Zealand and Chatham Islands; *Anthochoera* (4 sp.), Australia and Tasmania; *Xan-*

thotis (4 sp.), Papuan Islands and Australia; *Leptornis* (2 sp.), Samoa Islands and New Caledonia; *Philemon* = *Tropidorhyncus* (18 sp.), Moluccas and Lombok to New Guinea, Australia, Tasmania and New Caledonia; *Entomiza* (2 sp.), Australia; *Manorhina* (5 sp.), Australia and Tasmania; *Euthyrhynchus* (3 sp.), New Guinea; *Melirrhophetes* (2 sp.), New Guinea; *Melidectes* (1 sp.), New Guinea; *Melipotes* (1 sp.), New Guinea; *Melithreptus* (8 sp.), New Guinea, Australia, and Tasmania; ([397]) *Moho* (3 sp.), Sandwich Islands; *Chætoptila* (1 sp.), Sandwich Islands.

FAMILY 23.—NECTARINIIDÆ.   (16 Genera, 122 Species.)

GENERAL DISTRIBUTION.

NEOTROPICAL SUB-REGIONS.	NEARCTIC SUB-REGIONS.	PALÆARCTIC SUB-REGIONS.	ETHIOPIAN SUB-REGIONS.	ORIENTAL SUB-REGIONS.	AUSTRALIAN SUB-REGIONS.
— — — —	— — — —	— 2 — —	1.2.3.4	1.2.3.4	1.2 — —

The Nectariniidæ, or Sun-birds, form a rather extensive group of insectivorous honey-suckers, often adorned with brilliant metallic plumage, and bearing a superficial resemblance to the American humming-birds, although not in any way related to them. They abound in the Ethiopian, Oriental, and Australian regions, as far east as New Ireland, and south to Queensland, while one species inhabits the hot Jordan Valley in the Palæarctic region. For the Eastern genera I follow Lord Walden's classification (Ibis, 1870); the African species not having been so carefully studied are mostly placed in one genus. The genera adopted are as follows:—

Promerops (1 sp.), South Africa; *Nectarinia* (60 sp.), the whole Ethiopian region; *Cinnyricinclus* (5 sp.), West Africa; *Drepanornis* (1 sp.), Madagascar; *Arachnecthra* (13 sp.), Palestine, all India to Hainan, the Papuan Islands, and North-east Australia; *Æthopyga* (15 sp,), Himalayas and Central India to West China, Hainan, Java, and Northern Celebes; *Nectarophila* (5 sp.), Central India and Ceylon, Assam and Aracan to Java, Celebes and the Philippines; *Chalcostetha* (6 sp.), Malay Peninsula to New Guinea; *Anthreptes* (1 sp.), Siam, Malay Peninsula to

Sula Islands, and Flores; *Cosmeteira* (1 sp.), Papuan Islands; *Arachnothera* (15 sp.), the Oriental region (excluding Philippines) Celebes, Lombok, and Papuan Islands.

FAMILY 24.—DICÆIDÆ. (5 Genera, 107 Species.)

GENERAL DISTRIBUTION.					
NEOTROPICAL SUB-REGIONS.	NEARCTIC SUB-REGIONS.	PALÆARCTIC SUB-REGIONS.	ETHIOPIAN SUB-REGIONS.	ORIENTAL SUB-REGIONS.	AUSTRALIAN SUB-REGIONS.
— — — —	— — — —	— — — 4	1.2.3.4	1.2.3.4	1.2.3.4

The Dicæidæ, or Flower-peckers, consist of very small, gaily-coloured birds, rather abundant over the whole Oriental and much of the Australian regions, and one genus extending over the Ethiopian region. The genera here adopted are the following :—

([622]) *Zosterops* (68 sp.), the whole Ethiopian, Oriental, and Australian regions, as far east as the Fiji Islands, and north to Pekin and Japan; ([400 — 403]) *Dicæum* (25 sp.), the whole Oriental region, except China, with the Australian region as far as the Solomon Islands; ([404]) *Pachyglossa* (2 sp. [1437 1442]), Nepal and Northern Celebes; ([405]) *Piprisoma* (2 sp), Himalayas to Ceylon and Timor; ([1450]) *Pardalotus* (10 sp.), Australia and Tasmania; ([407 — 409]) *Prionochilus* (5 sp.), Indo-Malay sub-region and Papuan Islands.

FAMILY 25.—DREPANIDIDÆ. (4 Genera, 8 Species.)

GENERAL DISTRIBUTION.					
NEOTROPICAL SUB-REGIONS.	NEARCTIC SUB-REGIONS.	PALÆARCTIC SUB-REGIONS.	ETHIOPIAN SUB-REGIONS.	ORIENTAL SUB-REGIONS.	AUSTRALIAN SUB-REGIONS.
— — — —	— — — —	— — — —	— — — —	— — — —	— — 3 —

The Drepanididæ are confined to the Sandwich Islands, and I follow Mr. Sclater's suggestion in bringing together the following genera to form this family :—

Drepanis (3 sp.); *Hemignathus* (3 sp.); *Loxops* (1 sp.); *Psittirostra* (1 sp.). If these are correctly associated, the great differences in the bill indicate that they are the remains of a larger and more varied family, once inhabiting more extensive land surfaces in the Pacific.

FAMILY 26.—CŒREBIDÆ.   (11 Genera, 55 Species.)

GENERAL DISTRIBUTION.					
NEOTROPICAL SUB-REGIONS.	NEARCTIC SUB-REGIONS.	PALÆARCTIC SUB-REGIONS.	ETHIOPIAN SUB-REGIONS.	ORIENTAL SUB-REGIONS.	AUSTRALIAN SUB-REGIONS.
— 2 . 3 . 4	— — 3 —	— — — —	— — — —	— — — —	— — — —

(According to the arrangement of Messrs. Sclater and Salvin.)

The Cœrebidæ, or Sugar-birds, are delicate little birds allied to the preceding families, but with extensile honey-sucking tongues. They are almost wholly confined to the tropical parts of America, only one species of *Certhiola* ranging so far north as Florida. The following is the distribution of the genera :—

Diglossa (14 sp.), Peru and Bolivia to Guiana and Mexico; *Diglossopis* (1 sp.), Ecuador to Venezuela; *Oreomanes* (1 sp.), Ecuador; *Conirostrum* (6 sp.), Bolivia to Ecuador and Columbia; *Hemidacnis* (1 sp.), Upper Amazon and Columbia; *Dacnis* (13 sp.), Brazil to Ecuador and Costa Rica ; *Certhidea* (2 sp.), Galapagos Islands ; *Chlorophanes* (2 sp.), Brazil to Central America and Cuba ; *Cœreba* (4 sp.), Brazil to Mexico; *Certhiola* (10 sp.), Amazon to Mexico, West Indies, and Florida ; *Glossoptila* (1 sp.), Jamaica.

FAMILY 27.—MNIOTIITIDÆ.   (18 Genera, 115 Species.)

GENERAL DISTRIBUTION.					
NEOTROPICAL SUB-REGIONS.	NEARCTIC SUB-REGIONS.	PALÆARCTIC SUB-REGIONS.	ETHIOPIAN SUB-REGIONS.	ORIENTAL SUB-REGIONS.	AUSTRALIAN SUB-REGIONS.
— 2 . 3 . 4	1 . 2 . 3 . 4	— — — —	— — — —	— — — —	— — — —

(Messrs. Sclater and Salvin are followed for the Neotropical, Baird and Allen for the Nearctic region.)

The Mniotiltidæ, or Wood-warblers, are an interesting group of small and elegant birds, allied to the preceding family and to the greenlets, and perhaps also to the warblers and tits of Europe.

They range over all North America from Panama to the Arctic regions, but do not extend far beyond the tropic in Southern America. They are almost as abundant in the Nearctic as in the Neotropical region ; and considering the favourable conditions of existence in Tropical America, this fact, in connection with their absence from the South Temperate zone would lead us to suppose that they originated in North Temperate America, and subsequently spread southward into the tropics. This supposition is strengthened by the fact that their metropolis, in the breeding season, is to the north of the United States. The genera adopted by Messrs. Sclater and Salvin are as follows:—

([918]) *Siurus* (4 sp.), Venezuela and West Indies to Eastern States and Canada; *Mniotilta* (1 sp.), Venezuela, Mexico, and Antilles to the Eastern States ; *Parula* (5 sp.), Brazil to Mexico, and the Eastern States, and Canada; *Protonotaria* (1 sp.), Antilles to Ohio ; *Helminthophaga* (8 sp.), Columbia to Arctic America *Helmintherus* (2 sp.), Central America to Eastern States; *Perissoglossa* (1 sp.), Antilles and Eastern States ; *Dendrœca* (33 sp.), Amazon to Antilles, and Arctic America, and south to Chili ; *Oporornis* (2 sp.), Guatemala to Eastern States; *Geothlypis* (11 sp.), all North America and Brazil ; *Myiodioctes* (5 sp.), all North America and Columbia ; *Basileuterus* (22 sp.), Bolivia and Brazil to Mexico ; *Setophaga* (15 sp.), Brazil to Canada ; *Ergaticus* (2 sp.), Guatemala and Mexico ; *Cardellina* (1 sp.), Guatemala and Mexico ; ([1440]) *Granatellus* (3 sp.), Amazon to Mexico ; ([1441]) *Teretristis* (2 sp.), Cuba; ([1439]) *Icteria* (2 sp.), Costa Rica and United States to Canada.

FAMILY 28.—VIREONIDÆ. (7 Genera, 63 Species.)

GENERAL DISTRIBUTION.

NEOTROPICAL SUB-REGIONS.	NEARCTIC SUB-REGIONS.	PALÆARCTIC SUB-REGIONS.	ETHIOPIAN SUB-REGIONS.	ORIENTAL SUB-REGIONS.	AUSTRALIAN SUB-REGIONS.
— 2 . 3 . 4	1 . 2 . 3 . 4	— — — —	— — — —	— — — —	— — — —

(Messrs. Sclater and Salvin are followed for the Neotropical genera ; Professor Baird and Mr. Allen for those of the Nearctic region.)

The Vireonidæ, or Greenlets, are a family of small fly-catching birds wholly restricted to the American continent, where they range from Paraguay to Canada. They are allied to the Mniotiltidæ and perhaps also to the Australian Pachycephalidæ. Only two of the genera, with about a dozen species, inhabit the Nearctic region. The distribution of the genera is as follows :—
Vireosylvia (13 sp.), Venezuela to Mexico, the Antilles, the Eastern States and Canada ; *Vireo* (14 sp.), Central America and the Antilles to Canada; *Neochloe* (1 sp.), Mexico ; *Hylophilus* (20 sp.), Brazil to Mexico ; *Laletes* (1 sp.), Jamaica ; *Vireolanius* (5 sp.), Amazonia to Mexico; *Cychlorhis* (9 sp.), Paraguay to Mexico.

FAMILY 29.—AMPELIDÆ.   (4 Genera, 9 Species.)

GENERAL DISTRIBUTION.					
NEOTROPICAL SUB-REGIONS.	NEARCTIC SUB-REGIONS.	PALÆARCTIC SUB-REGIONS.	ETHIOPIAN SUB-REGIONS.	ORIENTAL SUB-REGIONS.	AUSTRALIAN SUB-REGIONS.
— — 3 4	1.2.3.4	1.2.3.4	— — — —	— — — —	— — — —

The Ampelidæ, represented in Europe by the waxwing, are a small family, characteristic of the Nearctic and Palæarctic regions, but extending southward to Costa Rica and the West Indian islands. The genera are distributed as follows :—
([1539]) *Ampelis* (3 sp.), the Palæarctic and Nearctic regions, and southward to Guatemala; ([1360]) *Ptilogonys* (2 sp.), Central America ; ([1442]) *Dulus* (2 sp.), West Indian Islands; ([1361]) *Phœnopepla* (1 sp.), Mexico and the Gila Valley.

FAMILY 30.—HIRUNDINIDÆ.   (9 Genera, 91 Species.)

GENERAL DISTRIBUTION.					
NEOTROPICAL SUB-REGIONS.	NEARCTIC SUB-REGIONS.	PALÆARCTIC SUB-REGIONS.	ETHIOPIAN SUB-REGIONS.	ORIENTAL SUB-REGIONS.	AUSTRALIAN SUB-REGIONS.
1.2.3.4	1.2,3.4	1.2.3.4	1.2.3.4	1.2.3.4	1.2.3.4

The Hirundinidæ, or Swallows, are true cosmopolites. Although they do not range quite so far north (except as stragglers) as a few of the extreme polar birds, yet they pass beyond the Arctic Circle both in America and Europe, *Cotyle riparia* having been observed in the Parry Islands, while *Hirundo rustica* has been seen both in Spitzbergen and Nova Zembla. *Cotyle riparia* and *Chelidon urbica* also breed in great numbers in northern Lapland, latitude 67° to 70° north. Many of the species also, have an enormous range, the common swallow (*Hirundo rustica*) inhabiting Europe, Asia and Africa, from Lapland to the Cape of Good Hope and to the Moluccas. The genera of swallows are not well determined, a number having been established of which the value is uncertain. I admit the following, referring by numbers to the *Hand List* :—

([215] — [221] [226] — [228]) *Hirundo* (40 sp.), the range of the entire family ; ([222] [223]) *Psalidoprogne* (10 sp.), Tropical and South Africa ; ([224]) *Phedina* (1 sp.), Madagascar and Mascarene Islands ; ([225]) *Petrochelidon* (5 sp.), North and South America and Cape of Good Hope; ([229] — [232] ? [234]) *Atticora* (8 sp.), the Neotropical region and ? Australia ; ([235] [237]) *Cotyle* (11 sp.), Europe, India, Africa, North America, Antilles and Ecuador ; ([236]) *Stelgidopteryx* (5 sp.), La Plata to United States ; ([238 and 239]) *Chelidon* (6 sp.), Palæarctic region, Nepal, Borneo ; ([240] — [242]) *Progne* (5 sp.), all North and South America.

FAMILY 31.—ICTERIDÆ.   (24 Genera, 110 Species.)

GENERAL DISTRIBUTION.					
NEOTROPICAL SUB-REGIONS.	NEARCTIC SUB-REGIONS.	PALÆARCTIC SUB-REGIONS.	ETHIOPIAN SUB-REGIONS.	ORIENTAL SUB-REGIONS.	AUSTRALIAN SUB-REGIONS.
1.2.3.4	1.2.3.4	— — — —	— — — —	— — — —	— — — —

The Icteridæ, or American hang-nests, range over the whole continent, from Patagonia and the Falkland Islands to the Arctic Circle. Only about 20 species inhabit the Nearctic region, while, as usual with exclusively American families, the larger proportion of the genera and species are found in the tropical parts of South America. The genera adopted by Messrs. Sclater and Salvin are the following :—

Clypeicterus (1 sp.), Upper Amazon; *Ocyalus* (2 sp.), Upper Amazon to Mexico; *Ostinops* (8 sp.), Brazil and Bolivia to Mexico; *Cassiculus* (1 sp.), Mexico; *Cassicus* (10 sp.), South Brazil and Bolivia to Costa Rica; *Icterus* (34 sp.), La Plata to the Antilles and United States; *Dolichonyx* (1 sp.), Paraguay to Canada; *Molothrus* (8 sp.), La Plata to Northern United States; *Agelœus* (7 sp.), La Plata and Chili to Northern United States; *Xanthocephalus* (1 sp), Mexico to California and Canada; *Xanthosomus* (4 sp.), La Plata to Venezuela; *Amblyrhamphus* (1 sp.), La Plata and Bolivia; *Gymnomystax* (1 sp.), Amazonia and Guiana; *Pseudoleistes* (2 sp.), La Plata and Brazil; *Leistes* (3 sp.), La Plata to Venezuela; *Sturnella* (5 sp.), Patagonia and Falkland Islands to Middle United States; *Curœus* (1 sp.), Chili; *Nesopsar* (1 sp.), Jamaica; *Scolecophgaus* (2 sp.), Mexico to Arctic Circle; *Lampropsar* (4 sp.), Amazonia and Ecuador to Mexico; *Quiscalus* (10 sp.), Venezuela and Columbia to South and Central United States; *Hypopyrrhus* (1 sp.), Columbia; *Aphobus* (1 sp.), Brazil and Bolivia; *Cassidix* (2 sp.), Brazil to Mexico and Cuba.

FAMILY 32.—TANAGRIDÆ. (43 Genera, 304 Species.)

GENERAL DISTRIBUTION.					
NEOTROPICAL SUB-REGIONS.	NEARCTIC SUB-REGIONS.	PALÆARCTIC SUB-REGIONS.	ETHIOPIAN SUB-REGIONS.	ORIENTAL SUB-REGIONS.	AUSTRALIAN SUB-REGIONS.
1.2.3.4	— 2.3 —	— — — —	— — — —	— — — —	— — — —

The Tanagers are an extensive family of varied and beautiful fruit-eating birds, almost peculiar to the Neotropical region, only four species of a single genus (*Pyranga*) extending into the Eastern United States and Rocky Mountains. Southward they range to La Plata. They are especially abundant in the forest regions of South America east of the Andes, where no less than 40 out of the 43 genera occur; 23 of the genera are peculiar to this sub-region, while only 1 (*Phlogothraupis*) is peculiar to Central America and Mexico, and 2 (*Spindalis* and *Phœnicophilus*) to the West Indian islands. The genera adopted by Messrs. Sclater and Salvin with their distribution will be found at Vol. II., p. 99, in our account of Neotropical Zoology.

FAMILY 33.—FRINGILLIDÆ.   (74 Genera, 509 Species.)

GENERAL DISTRIBUTION.					
NEOTROPICAL SUB-REGIONS.	NEARCTIC SUB-REGIONS.	PALÆARCTIC SUB-REGIONS.	ETHIOPIAN SUB-REGIONS.	ORIENTAL SUB-REGIONS.	AUSTRALIAN SUB-REGIONS.
1 . 2 . 3 . 4	1 . 2 . 3 . 4	1 . 2 . 3 . 4	1 . 2 . 3 . 4	1 . 2 . 3 . 4	— — — —

The great family of the Fringillidæ, or finches, is in a very unsettled state as regards their division into genera, the most divergent views being held by ornithologists as to the constitution and affinities of many of the groups. All the Australian finchlike birds appear to belong to the Ploceidæ, so that the finches, as here constituted, are found in every region and sub-region, except the Australian region from which they are entirely absent —a peculiar distribution hardly to be found in any other family of birds.

Many European ornithologists separate the Emberizidæ, or buntings, as a distinct family, but as the American genera have not been so divided I am obliged to keep them together; but the genera usually classed as "buntings" are placed last, as a sub-family. In the following arrangement of the genera, I have done what I could to harmonize the views of the best modern writers. For convenience of reference the succession of the genera is that of the *Hand List*, and the numbers of the sub-genera are given whenever practicable :—

([1793 1795]) *Fringilla* (6 sp.), the whole Palæarctic region, including the Atlantic Islands ; ([1794]) *Acanthis* (3 sp.), Europe to Siberia, Persia, and North-West Himalayas ; ([1796]) *Procarduelis* (1 sp.), High Himalayas and East Thibet ; ([1797 – 1803]) *Chrysomitris* (18 sp.), Neotropical and Nearctic regions, Europe, and Siberia ; ([1804]) *Metoponia* (1 sp.), East Europe to North West Himalayas ; ([1805 and 1809]) *Chlorospiza* (9 sp.), Palæarctic region and Africa to the

Cape of Good Hope; ([1806] — [1809]) *Dryospiza* (14 sp.), South Europe, Palestine, Canaries, and all Africa; ([1810]) *Sycalis* (18 sp.), the whole Neotropical region; ([1811] — [1813] [1816] — [1819]) *Pyrgita* (34 sp.), Palæarctic and Oriental regions, and all Africa; ([1814]) *Montifringilla* (4 sp.), Palæarctic region; ([1815]) *Fringillauda* (2 sp.), North-West Himalayas to East Thibet; ([1820] — [1822]) *Coccothraustes* (6 sp.), Palæarctic region and Nepal, Nearctic region to Mexico; ([1823]) *Eophona* (2 sp.), China and Japan; ([1824]) *Mycerobas* (2 sp.), Central Asia to Persia, High Himalayas, and East Thibet; ([1825]) *Chaunoproctus* (1 sp.), Bonin Islands, south-east of Japan, (probably Palæarctic); ([1826]) *Geospiza* (7 sp.) ,Galapagos Islands; ([1827]) *Camarhynchus* (5 sp.), Galapagos Islands; ([1828]) *Cactornis* (4 sp.), Galapagos Islands; ([1830] — [1832]) *Phrygilus* (10 sp.), Columbia to Fuegia and the Falkland Islands; ([1833]) *Xenospingus* (1 sp.), Peru; ([1834]) *Diuca* (3 sp.), Peru to Chili and Patagonia; ([1835 and 1837]) *Emberizoides* (3 sp.), Venezuela to Paraguay; ([1836]) *Donacospiza* (1 sp.), South Brazil and La Plata; ([1839]) *Chamœospiza* (1 sp.), Mexico; ([1838 and 1840]) *Embernagra* (9 sp.), Arizona to La Plata; ([1841]) *Hœmophila* (6 sp.), Mexico to Costa Rica; ([1842]) *Atlapetes* (1 sp.), Mexico; ([1843]) *Pyrgisoma* (5 sp.), Mexico to Costa Rica; ([1844 and 1845]) *Pipilo* (12 sp.), all North America to Guatemala; ([1846]) *Junco* (6 sp.), all the United States to Guatemala; ([1847]) *Zonotrichia* (9 sp.), the whole Nearctic and Neotropical regions; ([1848 1849]) *Melospiza* (7 sp.), Sitka and United States to Guatemala; ([1850]) *Spizella* (7 sp.), Canada to Guatemala; ([1851]) *Passerella* (4 sp.), the Nearctic region and Northern Asia; ([1852]) *Passerculus* (6 sp.), Nearctic region and to Guatemala; ([1853]) *Poœcetes* (1 sp.), all United States and Mexico; ([1854]) *Ammodromus* (4 sp.), all United States to Guatemala; ([1855]) *Coturniculus* (6 sp.), north and east of North America to Jamaica and Bolivia; ([1856]) *Peucœa* (6 sp.), South Atlantic States and California to Mexico; ([1857]) *Tiaris* (1 sp.), Brazil; ([1858]) *Volatinia* (1 sp.), Mexico to Brazil and Bolivia; ([1859]) *Cyanospiza* (5 sp.), Canada to Guatemala; ([1860 1861]) *Paroaria* (6 sp.), Tropical South America, east of the Andes; ([1862]) *Coryphospingus* (4 sp.), Tropical South America; ([1863]) *Haplospiza* (2 sp.), Mexico and Brazil; ([1864 1891]) *Phonipara* (8 sp.), Mexico to Columbia, the greater Antilles; ([1865]) *Poospiza*

(13 sp.), California and South Central States to Bolivia and La Plata; ([424]) *Spodiornis* (1 sp.), Andes of Quito; ([1866 1867]) *Pyrrhula* (9 sp.), the whole Palæarctic region to the Azores and High Himalayas; ([1868]) *Crithagra* (17 sp.), Tropical and South Africa, Mauritius, Syria; ([1869]) *Ligurnus* (2 sp.), West Africa; ([1870 1871]) *Carpodacus* (18 sp.), Nearctic and Palæarctic regions to Mexico and Central India; ([1872 – 1874]) *Erythrospiza* (6 sp.), Southern parts of Palæarctic region; ([1875]) *Uragus* (2 sp.), Siberia and Japan; ([1876]) *Cardinalis* (2 sp.), South and Central States to Venezuela: ([1877]) *Pyrrhuloxia* (1 sp.), Texas and Rio Grande; ([1878 1879]) *Guiraca* (6 sp.), Southern United States to La Plata; ([1880]) *Amaurospiza* (2 sp.), Costa Rica and Brazil; ([1881]) *Hedymeles* (2 sp.), all United States to Columbia; ([1882]) *Pheucticus* (5 sp.), Mexico to Peru and Bolivia; ([1883]) *Oryzoborus* (6 sp.), Mexico to Ecuador and South Brazil; ([1884]) *Melopyrrha* (1 sp.), Cuba; ([1885]) *Loxigilla* (4 sp.), Antilles; ([1886 1887]) *Spermophila* (44 sp.), Texas to Bolivia and Uruguay; ([1888]) *Catamenia* (4 sp.), Columbia to Bolivia; ([1889]) *Neorhynchus* (3 sp.), West Peru; ([1892]) *Catamblyrhyncus* (1 sp.), Columbia; ([1893]) *Loxia* (7 sp.), Europe to North-west India and Japan, Arctic America to Pennsylvania, Mexico; ([1894]) *Pinicola* (3 sp.), Arctic America, North-east Europe to the Amoor, Camaroons Mountains West Africa; ([1895]) *Propyrrhula* (1 sp.), Darjeeling in the winter, ? Thibet; ([1896]) *Pyrrhospiza* (1 sp.), Snowy Himalayas; ([1897]) *Hæmatospiza* (1 sp.), South-east Himalayas, 5,000 - 10,000 feet; ([1898 1899]) *Linota* (12 sp.), Europe to Central Asia, north and east of North America; ([1900]) *Leucosticte* (7 sp.), Siberia and Thibet to Kamschatka, and from Alaska to Utah.

Sub-family Emberizinæ.—([1995]) *Calamospiza* (1 sp.), Arizona and Texas to Mexico; ([1906]) *Chondestes* (2 sp.), Western, Central, and Southern States to Mexico and Nicaragua; ([1907 – 1910]) *Euspiza* (9 sp.), Palæarctic region, India, Burmah, and South China, South-east United States to Columbia; ([1911 – 1920]) *Emberiza* (28 sp.), the whole Palæarctic region (continental), to Central India in winter; ([1921]) *Gubernatrix* (1 sp.), Paraguay and La Plata, (according to Messrs. Sclater and Salvin this comes next to *Pipilo*); ([1922]) *Fringillaria* (8 sp.), Africa and South Europe;

(^{1923} — ^{1925}) *Plectrophanes* (6 sp.), Arctic Zone to Northern Europe and North China, Arctic America, and east side of Rocky Mountains; (^{1926}) *Centronyx* (1 sp.), Mouth of Yellowstone River.

FAMILY 34.—PLOCEIDÆ.　(29 Genera, 252 species.)

		GENERAL DISTRIBUTION.			
NEOTROPICAL SUB-REGIONS.	NEARCTIC SUB-REGIONS.	PALÆARCTIC SUB-REGIONS.	ETHIOPIAN SUB-REGIONS.	ORIENTAL SUB-REGIONS.	AUSTRALIAN SUB-REGIONS.
— — — —	— — — —	— — — —	1 . 2 . 3 . 4	1 . 2 . 3 . 4	1 . 2 . 3 —

The Ploceidæ, or Weaver-finches, are especially characteristic of the Ethiopian region, where most of the genera and nearly four-fifths of the species are found; the remainder being pretty equally divided between the Oriental and Australian regions. Like the true finches these have never been properly studied, and it is exceedingly difficult to ascertain what genera are natural and how far those of Australia and Africa are distinct. The following enumeration must therefore be taken as altogether tentative and provisional. When the genera adopted differ from those of the *Hand List* they will be referred to by numbers.

Textor (5 sp.), Tropical and South Africa; (^{1650} — $^{1654\ 1657}$) *Hyphantornis* (32 sp.), Tropical and South Africa; ($^{1655\ 1656}$) *Symplectes* (8 sp), Tropical and South Africa; *Malimbus* (9 sp.), West Africa; ($^{1659\ 1661}$) *Ploceus* (6 sp.), West and East Africa, the Oriental region (excluding Philippines); (^{1660}) *Nelicurvius* (1 sp.), Madagascar; *Foudia* (12 sp.), Madagascar and Mascarene Islands, Tropical Africa; ($^{1663\ 1664}$) *Sporopipes* (2 sp.), Tropical and South Africa; ($^{1665\ -\ 1667}$) *Pyromelana* (14 sp.), Tropical and South Africa, Abyssinia to 10,500 feet; *Philetœrus* (1 sp.), South Africa; *Nigrita* (7 sp.), West Africa to Upper Nile; *Plocepasser* (4 sp.), East and South Africa; (^{1672} — ^{1674}) *Vidua* (7 sp.), Tropical and South Africa (Plate V., Vol. I., p. 264); (^{1675} — ^{1677}) *Coliuspasser* (9 sp.), Tropical and South Africa; *Chera* (1 sp.), South Africa; *Spermospiza* (2 sp.), West Africa; *Pyrenestes* (6 sp.), Tropical and South Africa; (^{1682} — $^{1687\ 1689\ 1692\ 1693\ 1698}$) *Estrilda* (26 sp.), Tropical and South Africa, India, Burmah, and Java to Australia; ($^{1688\ 1690}$

[1691] [1695] [1696]) *Pytelia* (24 sp.), Tropical and South Africa; ([1694])
Hypargos (2 sp.), Mozambique and Madagascar; ([1697]) *Emblema*
(1 sp.), North-west Australia ([1690] [1712] − [1717]) *Amadina* (15 sp.),
Tropical and South Africa, Moluccas to Australia and the Samoa
Islands; ([1700] [1701] [1710]) *Spermestes* (8 sp.), Tropical Africa and Mada-
gascar; ([1702]) *Amauresthes* (1 sp.), East and West Africa; ([1703]
[1707] − [1709] [1711]) *Munia* (30 sp.), Oriental region to Timor and
New Guinea; ([1704]) *Donacola* (3 sp.), Australia; ([1705] [1706]) *Poephila*
(6 sp.), Australia; ([1718] − [1721]) *Erythrura* (7 sp.), Sumatra to
Java, Moluccas, Timor, New Guinea, and Fiji Islands; ([1722])
Hypochera (3 sp.), Tropical and South Africa.

FAMILY 35.—STURNIDÆ.   (29 Genera, 124 Species.)

GENERAL DISTRIBUTION.					
NEOTROPICAL SUB-REGIONS.	NEARCTIC SUB-REGIONS.	PALÆARCTIC SUB-REGIONS.	ETHIOPIAN SUB-REGIONS.	ORIENTAL SUB-REGIONS.	AUSTRALIAN SUB-REGIONS.
— — — —	— — — —	1 . 2 . 3 . 4	1 . 2 . 3 . 4	1 . 2 . 3 . 4	1 − 3 . 4

The Sturnidæ, or Starlings, are a highly characteristic Old-
World group, extending to every part of the great Eastern continent and its islands, and over the Pacific Ocean to the Samoa
Islands and New Zealand, yet wholly absent from the mainland
of Australia. The family appears to be tolerably well-defined,
and the following genera are generally considered to belong to it :
([1558] [1559] [1562]) *Eulabes* (13 sp.), the Oriental region to South-west
China, Hainan, and Java,—and Flores, New Guinea and the Solomon Islands in the Australian region ; *Ampeliceps* (1 sp.), Tenasserim, Burmah, and Cochin China; *Gymnops* (1 sp.), Philippine
Islands ; *Basilornis* (2 sp.), Celebes and Ceram ; *Pastor* (1 sp.),
South-east Europe to India, Ceylon, and Burmah ; *Acridotheres*
(7 sp.), the whole Oriental region and Celebes ; ([1568] [1569]) *Sturnia*
(12 sp.), the whole Oriental region, North China, Japan, and
Siberia, Celebes ; *Dilophus* (1 sp.) South Africa ; *Sturnus* (6 sp.),
Palæarctic region, to India and South China in winter; *Sturno-
pastor* (4 sp.), India to Burmah and East Java ; *Creadion* (2 sp.)
New Zealand ; *Heterolocha* (1 sp.), New Zealand ; ([1520]) *Callæas*

(2 sp.), New Zealand ; *Buphaga* (2 sp.), Tropical and South Africa ; *Euryceros* (1 sp.), Madagascar (see Plate VI., Vol. I., p. 278.) This genus and the last should perhaps form distinct families. ([1577]) *Juida* (5 sp.), Central, West, and South Africa ; ([578]) *Lamprocolius* (20 sp.), Tropical and South Africa ; *Cinnyricinclus* (2 sp.), Tropical and South Africa ; *Onychognathus* (2 sp.), West Africa ; ([1581]) *Spreo* (4 sp.), Tropical and South Africa ; ([1582 – 1585]) *Amydrus* (7 sp.), South and East Africa, Palestine ; *Aplonis* (9 sp.), New Caledonia to the Tonga Islands ; ([1587 – 1589]) *Calornis* (18 sp.), the whole Malay Archipelago and eastward to the Ladrone and Samoa Islands ; ([1590]) *Enodes* (1 sp.), Celebes ; *Scissirostrum* (1 sp.), Celebes ; ([1592]) *Saroglossa* (1 sp.), Himalayas ; ([1593]) *Hartlaubius* (1 sp.), Madagascar ; *Fregilupus* (1 sp.), Bourbon, but it has recently become extinct ; ([363]) *Falculia* (1 sp)., Madagascar.

FAMILY 36.—ARTAMIDÆ.   (1 Genus, 17 Species.)

GENERAL DISTRIBUTION.					
NEOTROPICAL SUB-REGIONS.	NEARCTIC SUB-REGIONS.	PALÆARCTIC SUB-REGIONS.	ETHIOPIAN SUB-REGIONS.	ORIENTAL SUB-REGIONS.	AUSTRALIAN SUB-REGIONS.
— — — —	— — — —	— — — —	— — — —?	1.2.3.4	1.2.3 —

The Artamidæ, or Swallow-shrikes, are a curious group of birds, ranging over the greater part of the Oriental and Australian regions as far east as the Fiji Islands and south to Tasmania. Only a single species inhabits India, and they are more plentiful in Australia than in any other locality. The only well-marked genus is *Artamus.*

There are a few Madagascar birds belonging to the genus *Artamia,* which some ornithologists place in this family, others with the Laniidæ, but which are here classed with the Oriolidæ.

FAMILY 37.—ALAUDIDÆ.   (15 Genera, 110 Species.)

GENERAL DISTRIBUTION.					
NEOTROPICAL SUB-REGIONS.	NEARCTIC SUB-REGIONS.	PALÆARCTIC SUB-REGIONS.	ETHIOPIAN SUB-REGIONS.	ORIENTAL SUB-REGIONS.	AUSTRALIAN SUB-REGIONS.
— 2 . 3 —	— 2 . 3 . 4	1 . 2 . 3 . 4	1 . 2 . 3 . 4	1 . 2 . 3 . 4	1 . 2 — —

The Alaudidæ, or Larks, may be considered as exclusively belonging to the great Eastern continent, since the Nearctic, Neotropical, and Australian regions have each only a single species. They abound most in the open plains and deserts of Africa and Asia, and are especially numerous in South Africa. The genera, including those recently established by Mr. Sharpe, are as follows:—

Otocorys (8 sp.) ; the Palæarctic region, North America and south to the Andes of Columbia, North India; ([1928] [1929]) *Alauda* (17 sp.), Palæarctic region, all Africa, the Peninsula of India, and Ceylon ; ([1931]) *Galerita* (10 sp.), Central Europe to Senegal and Abyssinia, Persia, India and North China ; ([1932]) *Calendula* (2 sp.), Abyssinia and South Africa ; ([1933] [1934]) *Calandrella* (6 sp.), Europe, North Africa, India, Burmah, North China, and Mongolia; ([1935] — [1937]) *Melanocorypha* (7 sp.), South Europe to Tartary, Abyssinia, and North-west India ; *Pallasia* ([sp. 7781]), East Asia ; ([1938]) *Certhilauda* (4 sp.), South Europe, South Africa ; *Heterocorys* ([sp. 7792]) South Africa ; ([1939]) *Alæmon* (3 sp.), South-east Europe to Western India, and South Africa ; ([1940]) *Mirafra* (25 sp.), the Oriental and Ethiopian regions to Australia ; ([1941]) *Ammomanes* (10 sp.), South Europe to Palestine and Central India, and to Cape Verd Islands and South Africa ; ([1942] [1943]) *Megalophonus* (6 sp.), Tropical and South Africa; *Tephrocorys* (1 sp.), South Africa ; *Pyrrhulauda* (9 sp.), all Africa, Canary Islands, India and Ceylon.

FAMILY 38.— MOTACILLIDÆ.   (9 Genera, 80 Species.)

GENERAL DISTRIBUTION.

NEOTROPICAL SUB-REGIONS.	NEARCTIC SUB-REGIONS.	PALÆARCTIC SUB-REGIONS.	ETHIOPIAN SUB-REGIONS.	ORIENTAL SUB-REGIONS.	AUSTRALIAN SUB-REGIONS.
1 . 2 . 3 . 4	1 . 2 . 3 . 4	1 . 2 . 3 . 4	1 . 2 . 3 . 4	1 . 2 . 3 . 4	1 . 2 . — 4

The Motacillidæ, or Wagtails and Pipits, are universally distributed, but are most abundant in the Palæarctic, Ethiopian,
and Oriental regions, to which the true wagtails are almost confined.   The following genera are usually adopted, but some of
them are not very well defined:—

Motacilla (15 sp.), ranges over the greater part of Europe,
Asia, and Africa, and to Alaska in North-west America ; *Budytes*
(10 sp.), Europe, Africa, Asia to Philippines, Moluccas, Timor,
and North Australia; *Calobates* (3 sp.), South Palæarctic and
Oriental regions to Java; *Nemoricola* (1 sp.), Oriental region;
Anthus (30 sp.), all the great continents ; *Neocorys* (1 sp.), Cen-
tral North America; *Coryddlla* (14 sp.), South Europe to India,
China, the Malay Islands, Australia, New Zealand and the Auck-
land Islands: *Macronyx* (5 sp.), Tropical and South Africa;
Heterura (1 sp.), Himalayas.

FAMILY 39.—TYRANNIDÆ.   (71 Genera, 329 Species.)

GENERAL DISTRIBUTION.

NEOTROPICAL SUB-REGIONS.	NEARCTIC SUB-REGIONS.	PALÆARCTIC SUB-REGIONS.	ETHIOPIAN SUB-REGIONS.	ORIENTAL SUB-REGIONS.	AUSTRALIAN SUB-REGIONS.
1 . 2 . 3 , 4	1 . 2 . 3 . 4	— — — —	— — — —	— — — —	— — — —

The Tyrannidæ, or Tyrant Shrikes, form one of the most extensive and truly characteristic American families of birds ; as
they extend over the whole continent from Patagonia to the
Arctic regions, and are found also in all the chief American
islands—the Antilles, the Galapagos, the Falkland Islands, and

Juan Fernandez. As the genera are all enumerated in the table, at p. 101 of this volume, I shall here confine myself to the distribution of the sub-families, only referring to such genera as are of special geographical interest.

Sub-family I. CONOPHAGINÆ (2 genera, 13 species). Confined to tropical South America, from Brazil and Bolivia to Guiana and Columbia.

Sub-family II. TÆNIOPTERINÆ (19 genera, 76 species). This group ranges from Patagonia and the Falkland Islands to the northern United States; yet it is almost wholly South American, only 2 genera and 4 species passing north of Panama, and none inhabiting the West Indian islands. *Sayornis* has 3 species in North America, while *Tænioptera, Cnipolegus, Muscisaxicola,* and *Centrites,* range south to Patagonia.

Sub-family III. PLATYRHYNCHINÆ (16 genera, 60 species). This sub-family is wholly Neotropical and mostly South American, only 7 of the genera passing Panama and but 3 reaching Mexico, while there are none in the West Indian islands. Only 3 genera extend south to the temperate sub-region, and one of these, *Anæretes,* has a species in Juan Fernandez.

Sub-family IV ELAINEINÆ (17 genera, 91 species). This sub-family is more exclusively tropical, only two genera extending south as far as Chili and La Plata, while none enter the Nearctic region. No less than 10 of the genera pass north of Panama, and one of these, *Elainea,* which ranges from Chili to Costa Rica has several species in the West Indian islands. About one fourth of the species of this sub-family are found north of Panama.

Sub-family V. TYRANNINÆ (17 genera, 89 species). This sub-family is that which is best represented in the Nearctic region, where 6 genera and 24 species occur. *Milvulus* reaches Texas; *Tyrannus* and *Myiarchus* range over all the United States; *Sayornis,* the Eastern States and California; *Contopus* extends to Canada; *Empidonax* ranges all over North America; and *Pyrocephalus* reaches the Gila Valley as well as the Galapagos Islands. No less than 5 genera of this sub-family occur in the West Indian islands.

FAMILY 39a.—OXYRHAMPHIDÆ.   (1 Genus, 2 Species.)

GENERAL DISTRIBUTION.

NEOTROPICAL SUB-REGIONS.	NEARCTIC SUB-REGIONS.	PALÆARCTIC SUB-REGIONS.	ETHIOPIAN SUB-REGIONS.	ORIENTAL SUB-REGIONS.	AUSTRALIAN SUB-REGIONS.
— 2.3 —	— — — —	— — — —	— — — —	— — — —	— — — —

The genus *Oxyrhamphus* (2 sp.) which ranges from Brazil to Costa Rica, has usually been placed in the Dendrocolaptidæ; but Messrs Sclater and Salvin consider it to be the type of a distinct family group, most allied to the Tyrannidæ.

FAMILY 40.—PIPRIDÆ.   (15 Genera, 60 Species.)

GENERAL DISTRIBUTION.

NEOTROPICAL SUB-REGIONS.	NEARCTIC SUB-REGIONS.	PALÆARCTIC SUB-REGIONS.	ETHIOPIAN SUB-REGIONS.	ORIENTAL SUB-REGIONS.	AUSTRALIAN SUB-REGIONS.
2 . 3 —	— — — —	— — — —	— — — —	— — — —	— — — —

The Pipridæ, or Manakins, have generally been associated with the next family, and they have a very similar distribution. The great majority of the genera and species are found in the equatorial regions of South America, only 9 species belonging to 5 genera ranging north of Panama, while 2 or 3 species extend to the southern limit of the tropical forests in Paraguay and Brazil. The genera which go north of Panama are *Piprites, Pipra, Chiroxiphia, Chiromachœris,* and *Hetoropelma. Pipra* is the largest genus, containing 19 species, and having representatives throughout the whole range of the family. As in all the more extensive families peculiar to the Neotropical region, the distribution of the genera will be found in the tables appended to the chapter on the Neotropical region in the Third Part of this work. (Vol. II. p. 103).

FAMILY 41.—COTINGIDÆ.　(28 Genera, 93 Species.)

GENERAL DISTRIBUTION.

NEOTROPICAL SUB-REGIONS.	NEARCTIC SUB-REGIONS.	PALÆARCTIC SUB-REGIONS.	ETHIOPIAN SUB-REGIONS.	ORIENTAL SUB-REGIONS.	AUSTRALIAN SUB-REGIONS.
— 2 . 3 . 4	— — — —	— — — —	— — — —	— — — —	— — — —

The Cotingidæ, or Chatterers, comprise some of the most beautiful and some of the most remarkable of American birds, for such we must consider the azure and purple Cotingas, the wine-coloured white-winged Pompadour, the snowy carunculated Bell-birds, the orange-coloured Cocks-of-the-Rock, and the marvellously-plumed Umbrella-birds, (Plate XV. Vol. II. p. 28). The Cotingidæ are also one of the most pre-eminently Neotropical of all the Neotropical families, the great mass of the genera and species being concentrated in and around the vast equatorial forest region of the Amazon. Only 13 species extend north of Panama, one to the Antilles, and not more than 20 are found to the south of the Amazon Valley. Messrs. Sclater and Salvin divide the family into six sub-families, the distribution of which will be briefly indicated.

Sub-family I. TITYRINÆ (3 genera, 22 species). Ranges from Brazil to Mexico, one species of *Hadrostomus* inhabiting Jamaica.

Sub-family II. LIPAUGINÆ (4 genera, 14 species) also ranges from Brazil to Mexico ; one genus (*Ptilochloris*) is confined to Brazil.

Sub-family III. ATTALINÆ (2 genera, 10 species). Ranges from Paraguay to Costa Rica ; one genus (*Casiornis*) is confined to South Brazil and Paraguay.

Sub-family IV. RUPICOLINÆ (2 genera, 5 species). This sub-family is restricted to the Amazonian region and Guiana, with one species extending along the Andean valleys to Bolivia. The genera are *Rupicola* (3 species) and *Phœnicocercus* (2 species).

Sub-family V. COTINGINÆ (10 genera, 28 species). Ranges from Southern Brazil and Bolivia to Nicaragua ; only two species (belonging to the genera *Carpodectes* and *Cotinga*) are found north of Panama, and there are none in the West Indian islands. The great majority of these, the true Chatterers, are from the regions about the Equator.

Sub-family VI. GYMNODERINÆ (7 genera, 14 species). Ranges from Brazil to Costa Rica; two species, of the genera *Chasmorhynchus* and *Cephalopterus*, are found north of Panama, while there are none in the West Indian islands. Only 2 species are found south of the Amazon valley.

FAMILY 42.—PHYTOTOMIDÆ. (1 Genus, 3 Species.)

GENERAL DISTRIBUTION.					
NEOTROPICAL SUB-REGIONS.	NEARCTIC SUB-REGIONS.	PALÆARCTIC SUB-REGIONS.	ETHIOPIAN SUB-REGIONS.	ORIENTAL SUB-REGIONS.	AUSTRALIAN SUB-REGIONS.
1 — — —	— — — —	— — — —	— — — —	— — — —	— — — —

The Phytotomidæ, or Plant-cutters, are singular thick-billed birds, strictly confined to the temperate regions of South America. The single genus, *Phytotoma*, is found in Chili, La Plata, and Bolivia. Their affinities are uncertain, but they are believed to be allied to the series of families with which they are here associated. (Plate XVI. Vol. II. p. 128).

FAMILY 43.—EURYLÆMIDÆ. (6 Genera, 9 Species.

GENERAL DISTRIBUTION.					
NEOTROPICAL SUB-REGIONS.	NEARCTIC SUB-REGIONS.	PALÆARCTIC SUB-REGIONS.	ETHIOPIAN SUB-REGIONS.	ORIENTAL SUB-REGIONS.	AUSTRALIAN SUB-REGIONS.
— — — —	— — — —	— — — —	— — — —	— — 3 . 4	— — — —

The Eurylæmidæ, or Broad-bills, form a very small family of birds, often adorned with striking colours, and which have their nearest allies in the South American Cotingidæ. They have a very limited distribution, from the lower slopes of the Himalayas through Burmah and Siam, to Sumatra, Borneo, and Java. They are evidently the remains of a once extensive group, and from the small number of specific forms remaining, seem to be on the road to extinction. Thus we may understand their isolated geographical position. The following are the names and distribution of the genera :—

Eurylæmus (2 species), Malay Peninsula, Sumatra, Java, and Borneo; *Corydon* (1 species), Malacca, Sumatra and Borneo (Plate IX. Vol. I. p. 339); *Psarisomus* (1 species), Himalayas to Burmah, up to 6,000 feet ; *Serilophus* (2 species), Nepal to Tenasserim; *Cymbirhynchus* (2 species), Siam to Sumatra and Borneo ; *Calyptomena* (1 species), Penang to Sumatra and Borneo.

FAMILY 44.—DENDROCOLAPTIDÆ. (43 Genera, 217 Species.)

GENERAL DISTRIBUTION.					
NEOTROPICAL SUB-REGIONS.	NEARCTIC SUB-REGIONS.	PALÆARCTIC SUB-REGIONS.	ETHIOPIAN SUB-REGIONS.	ORIENTAL SUB-REGIONS.	AUSTRALIAN SUB-REGIONS.
1 . 2 . 3 —	— — — — ...	— — — —	— — — —	— — — —	— — — —

The Dendrocolaptidæ, or American Creepers, are curious brown-coloured birds with more or less rigid tail feathers, strictly confined to the continental Neotropical region, and very numerous in its south-temperate extremity. They are divided by Messrs. Sclater and Salvin into five sub-families, to which I shall confine my remarks on their distribution. The details of the numerous genera, being only interesting to specialists, will be given in the table of genera of the Neotropical region. No less than 13 of the genera are confined to South-Temperate America and the High Andes; 14 are restricted to Tropical South America, while not one is peculiar to Tropical North America, and only 15 of the 43 genera extend into that sub-region, showing that this is one of the pre-eminently South American groups.

Sub-family I. FURNARIINÆ (8 genera, 30 species). Ranges over all South America, 4 genera and 18 species being restricted to the temperate sub-region; one species is found in the Falkland Islands.

Sub-family II. SCLERURINÆ (1 genus, 6 species). Brazil to Guiana, Columbia, and north to Mexico.

Sub-family III. SYNALLAXINÆ (12 genera, 78 species). Ranges from Patagonia to Mexico ; 7 genera and 28 species are confined to the temperate sub-region; species occur in the islands of Mas-a-fuera, Trinidad, and Tobago.

Sub-family IV. PHILYDORINÆ (6 genera, 35 species). Confined to Tropical America from Brazil to Mexico; 4 genera and 8 species occur in Tropical North America.

Sub-family V. DENDROCOLAPTINÆ (14 genera, 59 species). Ranges from Chili and La Plata to Mexico; only 3 species occur in the South Temperate sub-region, while 9 of the genera extend into Tropical North America. Two of the continental species occur in the island of Tobago, which, together with Trinidad, forms part of the South American rather than of the true Antillean sub-region.

FAMILY 45.—FORMICARIIDÆ. (32 Genera, 211 Species.)

GENERAL DISTRIBUTION.					
NEOTROPICAL SUB-REGIONS.	NEARCTIC SUB-REGIONS.	PALÆARCTIC SUB-REGIONS.	ETHIOPIAN SUB-REGIONS.	ORIENTAL SUB-REGIONS.	AUSTRALIAN SUB-REGIONS.
— 2 . 3 —	— — — —	— — — —	— — — —	— — — —	— — — —

The Formicariidæ, comprising the Bush-Shrikes and Antthrushes, form one of the most exclusively Neotropical families; and the numerous species are rigidly confined to the warm and wooded districts, only a single species extending to La Plata, and none to the Antilles or to the Nearctic region. Less than 30 species are found north of Panama. Messrs. Sclater and Salvin divide the group into three sub-families, whose distribution may be conveniently treated, as in the Dendrocolaptidæ, without enumerating the genera.

Sub-family I. THAMNOPHILINÆ.—(10 genera, 70 species.) One species of *Thamnophilus* inhabits La Plata; only 3 genera and 12 species are found north of Panama, the species of this sub-family being especially abundant in the Equatorial forest districts.

Sub-family II. FORMICIVORINÆ.—(14 genera, 95 species.) Only 8 species occur north of Panama, and less than one-third of the species belong to the districts south of the Equator.

Sub-family III. FORMICARIINÆ.—(8 genera, 46 species.) About 12 species occur north of Panama, and only 5 south of the Equatorial district.

It appears, therefore, that this extensive family is especially characteristic of that part of South America from the Amazon valley northwards.

FAMILY 46.—PTEROPTOCHIDÆ.   (8 Genera, 19 Species.)

GENERAL DISTRIBUTION.					
NEOTROPICAL SUB-REGIONS.	NEARCTIC SUB-REGIONS.	PALÆARCTIC SUB-REGIONS.	ETHIOPIAN SUB-REGIONS.	ORIENTAL SUB-REGIONS.	AUSTRALIAN SUB-REGIONS.
1 . 2 — —	— — — —	— — — —	— — — —	— — — —	— — — —

The Pteroptochidæ are a group of curious Wren-like birds, almost confined to the temperate regions of South America, extending along the Andes beyond the Equator, and with a few species in South-east Brazil, and one in the valley of the Madeira. The genera are as follows :—

Scytalopus (8 sp.), Chili and West Patagonia to the Andes of Columbia; *Merulaxis* (1 sp.), South-east Brazil; *Rhinocrypta* (2 sp.), Northern Patagonia and La Plata ; *Lioscelis* (1 sp.), Madeira valley ; *Pteroptochus* (2 sp.), Chili ; *Hylactes* (3 sp.), Western Patagonia and Chili; *Acropternis* (1 sp.), Andes of Ecuador and Columbia; *Triptorhinus* (1 sp.), Chili.

FAMILY 47.—PITTIDÆ.   (4 Genera, 40 Species.)

GENERAL DISTRIBUTION.					
NEOTROPICAL SUB-REGIONS.	NEARCTIC SUB-REGIONS.	PALÆARCTIC SUB-REGIONS.	ETHIOPIAN SUB-REGIONS.	ORIENTAL SUB-REGIONS.	AUSTRALIAN SUB-REGIONS.
— — — —	— — — —	— — — 4	— 2 — —	1 . 2 . 3 . 4	1 . 2 — —

The Pittas comprise a number of beautifully-coloured Thrush-like birds, which, although confined to the Old World, are more nearly allied to the South American Pteroptochidæ than to any other family. They are most abundant in the Malay Archipelago, between the Oriental and Australian divisions of which they are pretty equally divided. They seem, however, to attain their maximum of beauty and variety in the large islands of Borneo and Sumatra; from whence they diminish in numbers in every direction till we find single species only in North China, West Africa, and Australia. The genera here adopted are the following :—

(^{1087} ^{1088} ^{1090} ^{1092} ^{1093}) *Pitta* (33 sp.), has the range of the family ; (^{1089}) *Hydrornis* (3 sp.), Himalayas and Malaya; *Eucichla* (3 sp.), Malaya; *Melampitta* (1 sp.), recently discovered in New Guinea.

FAMILY 48.—PAICTIDÆ.   (1 Genus, 2 Species.)

GENERAL DISTRIBUTION.					
NEOTROPICAL SUB-REGIONS.	NEARCTIC SUB-REGIONS.	PALÆARCTIC SUB-REGIONS.	ETHIOPIAN SUB-REGIONS.	ORIENTAL SUB-REGIONS.	AUSTRALIAN SUB-REGIONS.
— — — —	— — — —	— — — —	— — — 4	— — — —	— — — —

This family was established by Professor Sundevall, for an anomalous bird of Madagascar, which he believes to have some affinity for the American Formicariidæ, but which perhaps comes best near the Pittas. The only genus is *Philepitta*, containing two species.

FAMILY 49.—MENURIDÆ.   (1 Genus, 2 Species.)

GENERAL DISTRIBUTION.					
NEOTROPICAL SUB-REGIONS.	NEARCTIC SUB-REGIONS.	PALÆARCTIC SUB-REGIONS.	ETHIOPIAN SUB-REGIONS.	ORIENTAL SUB-REGIONS.	AUSTRALIAN SUB-REGIONS.
— — — —	— — — —	— — — —	— — — —	— — — —	— 2 — —

The Menuridæ, or Lyre Birds, remarkable for the extreme elegance of the lyre-shaped tail in the species first discovered, are birds of a very anomalous structure, and have no near affinity to any other family. Two species of *Menura* are known, confined to South and East Australia (Plate XII. Vol. I. p. 441).

FAMILY 50.—ATRICHIIDÆ.　(1 Genus, 2 Species.)

GENERAL DISTRIBUTION.					
NEOTROPICAL SUB-REGIONS.	NEARCTIC SUB-REGIONS.	PALÆARCTIC SUB-REGIONS.	ETHIOPIAN SUB-REGIONS.	ORIENTAL SUB-REGIONS.	AUSTRALIAN SUB-REGIONS.
— — — —	— — — —	— — — —	— — — —	— — — —	— 2 — —

The genus *Atrichia*, or Scrub-birds of Australia, have been formed into a separate family by Professor Newton, on account of peculiarities in the skeleton which separate them from all other Passeres. Only two species are known, inhabiting East and West Australia respectively. They are very noisy, brown-coloured birds, and have been usually classed with the warblers, near *Amytis* and other Australian species.

General remarks on the distribution of the Passeres.

The order Passeres, is the most extensive among birds, comprehending about 5,700 species grouped in 870 genera, and 51 families. The distribution of the genera, and of the families considered individually, has been already sufficiently given, and we now have to consider the peculiarities of distribution of the families collectively, and in their relations to each other, as representing well-marked types of bird-structure. The first thing to be noted is, how very few of these families are truly cosmopolitan; for although there are seven which are found in each of the great regions, yet few of these are widely distributed throughout all the regions, and we can only find three that inhabit every sub-region, and are distributed with tolerable uniformity; these are the Hirundinidæ, or swallows, the Motacillidæ or wagtails and pipits, and the Corvidæ or crows,—but the latter is a family of so heterogeneous a nature, that it possibly contains the materials of several natural families, and if so divided, the parts would probably all cease to be cosmopolitan. The Sylviidæ, the

Turdidæ, and the Paridæ, are the only other families that approach universality of distribution, and all these are wanting in one or more sub-regions. If, now, we divide the globe into the New and the Old World, the former including the whole American continent, the latter all the rest of the earth, we find that the Old World possesses exclusively 23 families, the New World exclusively 14, of which 5 are common to North and South America. But if we take the division proposed by Professor Huxley—a northern world, comprising our first four regions (from Nearctic to Oriental), and a southern world comprising our last two regions (the Australian and Neotropical)—we find that the northern division possesses only 5 families exclusively, and the southern division 13 exclusively, of which not one is common to Australia and South America. This plainly indicates that, as far as the Passeres are concerned, the latter bipartite division is not so natural as the former. Again, if we compare temperate with tropical families (not too rigidly, but as regards their general character), we find in the northern hemisphere only two families that have the character of being typically temperate—the Cinclidæ, and in a less degree the Ampelidæ—both of small extent. In the southern hemisphere we have also two, the Phytotomidæ, and in a less degree, the Pteroptochidæ; making two wholly and two mainly temperate families. Of exclusively tropical families on the other hand, we have about 12, and several others that are mainly tropical.

The several regions do not differ greatly in the number of families found in each. The Nearctic has 19, the Palæarctic 21, the Ethiopian 23, the Oriental 28, the Australian 29, and the Neotropical 23. But many of these families are only represented by a few species, or in limited districts ; and if we count only those families which are tolerably well represented, and help to form the ornithological character of the region, the richness of the several tropical regions will appear to be (as it really is) comparatively much greater. The families that are confined to single regions are not very numerous, except in the case of the Neotropical region, which has 5. The Australian has only

3, the Oriental 1, the Ethiopian 1, and the other regions have no peculiar families.

The distribution of the Passeres may be advantageously considered as divided into the five series of Turdoid, Tanagroid, Sturnoid, Formicarioid, and Anomalous Passeres. The Turdoid Passeres, consisting of the first 23 families, are especially characteristic of the Old World, none being found exclusively in America, and only two or three being at all abundant there. The Tanagroid Passeres (Families 24-33) are very characteristic of the New World, five being confined to it, and three others being quite as abundant there as in the Old World; while there is not a single exclusively Old World family in the series, except the Drepanididæ confined to the Sandwich Islands. The Sturnoid Passeres (Families 34-38) are all exclusively Old World, except that two larks inhabit parts of North America, and a few pipits South America. The Formicarioid Passeres (Families 39-48) are strikingly characteristic of the New World, to which seven of the families exclusively belong; the two Old World groups being small, and with a very restricted distribution. The Anomalous Passeres (Families 49-50) are confined to Australia.

The most remarkable feature in the geographical distribution of the Passeres is the richness of the American continent, and the large development of characteristic types that occurs there. The fact that America possesses 14 altogether peculiar families, while no less than 23 Old-World families are entirely absent from it, plainly indicates, that, if this division does not represent the most ancient and radical separation of the land surface of the globe, it must still be one of very great antiquity, and have modified in a very marked way the distribution of all living things. Not less remarkable is the richness in specific forms of the 13 peculiar American families. These contain no less than 1,570 species, leaving only about 500 American species in the 13 other Passerine families represented in the New World. If we make a deduction for those Nearctic species which occur only north of Panama, we may estimate the truly Neotropical species of Passerine birds at 1,900, which is almost exactly *one-third* of the total number of Passeres ; a wonderful illustration of the Ornithological riches of South America.

Order II.—PICARIÆ.

FAMILY 51.—PICIDÆ.   (36 Genera, 320 Species.)

GENERAL DISTRIBUTION.					
NEOTROPICAL SUB-REGIONS.	NEARCTIC SUB-REGIONS.	PALÆARCTIC SUB-REGIONS.	ETHIOPIAN SUB-REGIONS.	ORIENTAL SUB-REGIONS.	AUSTRALIAN SUB-REGIONS.
1.2.3.4	1.2.3.4	1.2.3.3	1.2.3 —	1.2.3.4	1 — — —

The Woodpeckers are very widely distributed, being only absent from the Australian region beyond Celebes and Flores. They are most abundant in the Neotropical and Oriental regions, both of which possess a number of peculiar genera ; while the other regions possess few or no peculiar forms, even the Ethiopian region having only three genera not found elsewhere. The soft-tailed Picumninæ inhabit the tropical regions only, *Picumnus* being Neotropical, *Vivia* and *Sasia* Oriental, and *Verreauxia* Ethiopian. *Picoides*, or *Apternus*, is an Arctic form peculiar to the Nearctic and Palæarctic regions. *Celeus, Chrysoptilus, Chloronerpes,* and some smaller genera, are Neotropical exclusively, and there are two peculiar forms in Cuba. *Yungipicus, Chrysocolaptes, Hemicercus, Mulleripicus, Brachypternus, Tiga,* and *Micropternus,* are the most important of the peculiar Oriental genera. *Dendropicus* and *Geocolaptes* are Ethiopian ; but there are no woodpeckers in Madagascar. The Palæarctic woodpeckers belong to the genera *Picus*—which is widely distributed, *Gecinus* —which is an Oriental form, and *Dryocopus*—which is South American. Except *Picoides,* the Nearctic woodpeckers are mostly of Neotropical genera ; but *Sphyrapicus* and *Hylatomus* are peculiar. The geological record is, as yet, almost silent as to this family ; but remains doubtfully referred to it have been found in the Miocene of Europe and the Eocene of the United States. Yet the group is evidently one of very high antiquity, as is shown by its extreme isolation, its great specialization of structure, its abundant generic forms, and its wide distribution. It originated, probably, in Central Asia, and passed through the Nearctic region to South America, in whose rich and varied forests it found the conditions for rapid development, and for the specialization of the many generic forms now found there.

A large number of genera have been established by various authors, but their limitations and affinities are not very well made out. Those which seem best established are the following :—

([2107] — [2112]) *Picumnus* (22 sp.), Tropical South America to Honduras ; ([2113]) *Vivia* (1 sp.), Himalayas to East Thibet ; ([2114]) *Sasia* (2 sp.), Nepal to Java ; ([2115]) *Verreauxia* (1 sp.), West Africa ; *Picoides* (5 sp.), northern parts of Nearctic and Palæarctic regions, and Mountains of East Thibet ; *Picus* (42 sp.), the whole Palæarctic, Oriental, Nearctic, and Neotropical regions ; ([2123]) *Hyopicus* (2 sp.), Himalayas and North China ; ([2124]) *Yungipicus* (16 sp.), Oriental region, and to Flores, Celebes, North China, and Japan ; ([2127] — [2129]) *Sphyrapicus* (7 sp.), Nearctic region, Mexico, and Bolivia ; ([2130] — [2133] [2139]) *Campephilus* (14 sp.), Neotropical and Nearctic regions ; *Hylatomus* (1 sp.), Nearctic region ; ([2137] [2140]) *Dryocopus* (5 sp.), Mexico to South Brazil, Central and Northern Europe ; ([2134]) *Reinwardtipicus* (1 sp.), Penang to Borneo ; ([2135] [2136]) *Venilia* (2 sp.), Nepal to Borneo ; *Chrysocolaptes* (8 sp.), India and Indo-Malaya ; *Dendropicus* (16 sp.), Tropical and South Africa ; *Hemicercus* (5 sp.), Malabar and Pegu to Malaya ; *Gecinus* (18 sp.), Palæarctic and Oriental regions to Java ; ([2151] — [2156]) *Dendromus* (15 sp.), West and South Africa, Zanzibar, and Abyssinia ; ([2157] — [2159]) *Mulleripicus* (6 sp.), Malabar, Pegu, Indo-Malaya, and Celebes ; *Celeus* (17 sp.), Paraguay to Mexico ; *Nesoceleus* ([sp. 8833]) Cuba ; ([2162]) *Chrysoptilus* (9 sp.), Chili and South Brazil to Mexico ; *Brachypternus* (5 sp.), India, Ceylon, and China ; ([2165] [2166]) *Tiga* (5 sp.), all India to Malaya ; ([2167]) *Gecinulus* (2 sp.), South-east Himalayas to Burmah ; *Centurus* (13 sp.), Nearctic Region to Antilles and Venezuela ; *Chloronerpes* (35 sp.), Tropical America, Hayti ; ([2171]) *Xiphidiopicus* (1 sp.), Cuba ; *Melanerpes* (11 sp.), Brazil to

Canada, Porto Rico ; *Leuconerpes* (1 sp.), Bolivia to North
Brazil; *Colaptes* (9 sp), La Plata and Bolivia to Arctic America,
Greater Antilles; *Hypoxanthus* (1 sp.), Venezuela and Ecuador;
([2187]) *Geocolaptes* (1 sp.), South Africa; *Miglyptes* (3 sp.),
Malaya ; *Micropternus* (8 sp.), India and Ceylon to South China,
Sumatra and Borneo.

FAMILY 52.—YUNGIDÆ.　(1 Genus, 5 Species.)

GENERAL DISTRIBUTION.

NEOTROPICAL SUB-REGIONS.	NEARCTIC SUB-REGIONS.	PALÆARCTIC SUB-REGIONS.	ETHIOPIAN SUB-REGIONS.	ORIENTAL SUB-REGIONS.	AUSTRALIAN SUB-REGIONS.
— — — —	— — — —	1.2.3.4	1 — 3 —	1 — — —	— — — —

The Wrynecks (*Yunx*), which constitute this family, are
small tree-creeping birds characteristic of the Palæarctic region,
but extending into North and East Africa, over the greater part
of the peninsula of India (but not to Ceylon), and just reaching
the lower ranges of the Himalayas. There is also one species
isolated in South Africa.

FAMILY 53.—INDICATORIDÆ.　(1 Genus, 12 Species.)

GENERAL DISTRIBUTION.

NEOTROPICAL SUB-REGIONS.	NEARCTIC SUB-REGIONS.	PALÆARCTIC SUB-REGIONS.	ETHIOPIAN SUB-REGIONS.	ORIENTAL SUB-REGIONS.	AUSTRALIAN SUB-REGIONS.
— — — —	— — — —	— — — —	1.2.3—	— — 3.4	— — — —

The Honey-guides (*Indicator*) constitute a small family of
doubtful affinities ; perhaps most nearly allied to the woodpeckers and barbets. They catch bees and sometimes kill small
birds ; and some of the species are parasitical like the cuckoo.
Their distribution is very interesting, as they are found in every
part of the Ethiopian region, except Madagascar, and in the
Oriental region only in Sikhim and Borneo, being absent from
the peninsula of India which is nearest, both geographically and
zoologically, to Africa.

FAMILY 54.—MEGALÆMIDÆ.   (13 Genera, 81 Species.)

GENERAL DISTRIBUTION.

NEOTROPICAL SUB-REGIONS.	NEARCTIC SUB-REGIONS.	PALÆARCTIC SUB-REGIONS.	ETHIOPIAN SUB-REGIONS.	ORIENTAL SUB-REGIONS.	AUSTRALIAN SUB-REGIONS.
— 2 . 3 —	— — — —	— — — —	1 . 2 . 3 —	1 . 2 . 3 . 4	— — — —

The Megalæmidæ, or Barbets, consist of rather small, fruit-eating birds, of heavy ungraceful shape, but adorned with the most gaudy colours, especially about the head and neck.   They form a very isolated family ; their nearest allies being, perhaps, the still more isolated Toucans of South America.   Barbets are found in all the tropics except Australia, but are especially characteristic of the great Equatorial forest-zone ; all the most remarkable forms being confined to Equatorial America, West Africa, and the Indo-Malay Islands.   They are most abundant in the Ethiopian and Oriental regions, and in the latter are universally distributed.

In the beautiful monograph of this family by the Messrs. Marshall, the barbets are divided into three sub-families, as follows :—

Pogonorhynchinæ (3 genera, 15 sp.), which are Ethiopian except the 2 species of *Tetragonops*, which are Neotropical ; Megalæminæ (6 genera, 45 sp.), which are Oriental and Ethiopian ; and Capitoninæ (4 genera, 18 sp.), common to the three regions.

The genera are each confined to a single region.   Africa possesses the largest number of peculiar forms, while the Oriental region is richest in species.

This is probably a very ancient group, and its existing distribution may be due to its former range over the Miocene South Palæarctic land, which we know possessed Trogons, Parrots, Apes, and Tapirs, groups which are now equally abundant in Equatorial countries.

The following is a tabular view of the genera with their
distribution :—

Genera	Ethiopian Region.	Oriental Region.	Neotropical Region.
POGONORHYNCHINÆ.			
Tricholæma ... 1 sp.	W. Africa		
Pogonorhynchus 12 ,,	All Trop. & S. Af.		
Tetragorops ... 2 ,,	... ...	... ..	Peru & Costa Rica
MEGALÆMINÆ.			
Megalæma ... 29 ,,	... ...	The whole region	
Xantholæma ... 4 ,,	... ...	The whole region	
Xylobucco ... 2 ,,	W. Africa		
Barbatula ... 9 ,,	Trop. & S. Africa		
Psilopogon ... 1 ,,	... ...	Sumatra	
Gymnobucco ... 2 ,,	W. Africa		
CAPITONINÆ.			
Trachyphonus... 5 ,,	Trop. & S. Africa		
Capito ... ... 10 ,,	... ...	... ...	Equatorial Amer. to Costa Rica
Calorhamphus... 2 ,,	... ..	Malay Pen., Sumatra, Borneo	
Stactolæma ... 1 ,,	W. Africa		

FAMILY 55.—RHAMPHASTIDÆ.   (5 Genera, 51 Species.)

GENERAL DISTRIBUTION.					
NEOTROPICAL SUB-REGIONS.	NEARCTIC SUB-REGIONS.	PALÆARCTIC SUB-REGIONS.	ETHIOPIAN SUB-REGIONS.	ORIENTAL SUB-REGIONS.	AUSTRALIAN SUB-REGIONS.
— 2 . 3 —	— — — —	— — — —	— — — —	— — — —	— — — —

The Toucans form one of the most remarkable and characteristic families of the Neotropical region, to which they are
strictly confined.   They differ from all other birds by their long
feathered tongues, their huge yet elegant bills, and the peculiar
texture and coloration of their plumage.   Being fruit-eaters, and
strictly adapted for an arboreal life, they are not found beyond
the forest regions; but they nevertheless range from Mexico to
Paraguay, and from the Atlantic to the Pacific.   One genus,

Andigena, is confined to the forest slopes of the South American Andes. The genera are :—

Rhamphastos (12 sp.), Mexico to South Brazil; *Pteroglossus* (16 sp.), Nicaragua to South Brazil (Plate XV. Vol. II. p. 28); *Selenidera* (7 sp.), Veragua to Brazil, east of the Andes; *Andigena* (6 sp.), the Andes, from Columbia to Bolivia, and West Brazil; *Aulacorhamphus* (10 sp.), Mexico to Peru and Bolivia.

FAMILY 56.—MUSOPHAGIDÆ. (2 Genera, 18 Species.)

GENERAL DISTRIBUTION.					
NEOTROPICAL SUB-REGIONS.	NEARCTIC SUB-REGIONS.	PALÆARCTIC SUB-REGIONS.	ETHIOPIAN SUB-REGIONS.	ORIENTAL SUB-REGIONS.	AUSTRALIAN SUB-REGIONS.
— — — —	— — — —	— — — —	1 . 2 . 3 —	— — — —	— — — —

The Musophagidæ, or Plantain-eaters and Turacos, are handsome birds, somewhat intermediate between Toucans and Cuckoos. They are confined to the Ethiopian region and are most abundant in West Africa. The Plantain eaters (*Musophaga*, 2 sp.), are confined to West Africa; the Turacos (*Turacus*, 16 sp., including the sub-genera *Corythaix* and *Schizorhis*) range over all Africa from Abyssinia to the Cape (Plate V. Vol. I. p. 264).

FAMILY 57.—COLIIDÆ. (1 Genus, 7 Species.)

GENERAL DISTRIBUTION.					
NEOTROPICAL SUB-REGIONS.	NEARCTIC SUB-REGIONS.	PALÆARCTIC SUB-REGIONS.	ETHIOPIAN SUB-REGIONS.	ORIENTAL SUB-REGIONS.	AUSTRALIAN SUB-REGIONS.
— — — —	— — — —	— — — —	1 . 2 . 3 —	— — ·· —	— — — —

The Colies, consisting of the single genus *Colius*, are an anomalous group of small finch-like birds, occuping a position between the Picariæ and Passeres, but of very doubtful affinities. Their range is nearly identical with that of the Musophagidæ, but they are most abundant in South and East Africa.

FAMILY 58.—CUCULIDÆ.   (35 Genera, 180 Species.)

GENERAL DISTRIBUTION.

NEOTROPICAL SUB-REGIONS.	NEARCTIC SUB-REGIONS.	PALÆARCTIC SUB-REGIONS.	ETHIOPIAN SUB-REGIONS.	ORIENTAL SUB-REGIONS.	AUSTRALIAN SUB-REGIONS.
1 . 2 . 3 . 4	1 . 2 . 3 —	1 . 2 . 3 . 4	1 . 2 . 3 . 4	1 . 2 . 3 . 4	1 . 2 . 3 . 4

The Cuculidæ, of which our well-known Cuckoo is one of the most widely distributed types, are essentially a tropical group of weak insectivorous birds, abounding in varied forms in all the warmer parts of the globe, but very scarce or only appearing as migrants in the temperate and colder zones. Many of the smaller Eastern species are adorned with the most intense golden or violet metallic lustre, while some of the larger forms have gaily-coloured bills or bare patches of bright red on the cheeks. Many of the cuckoos of the Eastern Hemisphere are parasitic, laying their eggs in other birds' nests ; and they are also remarkable for the manner in which they resemble other birds, as hawks, pheasants, or drongo-shrikes. The distribution of the Cuckoo family is rather remarkable. They abound most in the Oriental region, which produces no less than 18 genera, of which 11 are peculiar ; the Australian has 8, most of which are also Oriental, but 3 are peculiar, one of these being confined to Celebes and closely allied to an Oriental group ; the Ethiopian region has only 7 genera, all of which are Oriental but three, 2 of these being peculiar to Madagascar, and the other common to Madagascar and Africa. America has 11 genera, all quite distinct from those of the Eastern Hemisphere, and only three enter the Nearctic region, one species extending to Canada.

Remembering our conclusions as to the early history of the several regions, these facts enable us to indicate, with considerable probability, the origin and mode of dispersal of the cuckoos. They were almost certainly developed in the Oriental and Palæarctic regions, but reached the Neotropical at a very early date, where they have since been completely isolated. Africa must have long remained without cuckoos, the earliest immigration being to Madagascar at the time of the approximation of that sub-region to Ceylon and Malaya. A later infusion of Oriental forms took place probably by way of Arabia and Persia, when those countries were more fertile and perhaps more extensive. Australia has also received its cuckoos at a somewhat late date, a few having reached the Austro-Malay Islands somewhat earlier.

The classification of the family is somewhat unsettled. For the American genera I follow Messrs. Sclater and Salvin; and, for those of the Old World, Mr. Sharpe's suggestive paper in the *Proceedings of the Zoological Society*, 1873, p. 600. The following is the distribution of the various genera :—

([2195]) *Phœnicophäes* (1 sp.), Ceylon ; ([2196]) *Rhamphococcyx* (1 sp.), Celebes ; ([2196]) *Rhinococcyx* (1 sp.), Java ; ([2196 pt. aud 2203]) *Rhopodytes* (6 sp.), Himalayas to Ceylon, Hainan, and Malaya; ([2203 pt.]) *Poliococcyx* (1 sp.), Malacca, Sumatra, and Borneo ; ([2197]) *Dasylophus* (1 sp.), Philippine Islands ; ([2198]) *Lepidogrammus* (1 sp.), Philippine Islands ; ([2200]) *Zanclostomus* (1 sp.), Malaya; ([2201]) *Ceuthmochares* (2 sp.), Tropical and South Africa and Madagascar ; ([2202]) *Taccocua* (4 sp.), Himalayas to Ceylon and Malacca ; ([2204]) *Rhinortha* (1 sp.), Malacca, Sumatra, Borneo ; ([2199]) *Carpococcyx* (1 sp.), Borneo and Sumatra ; ([2220]) *Neomorphus* (4 sp.), Brazil to Mexico ; ([2205 2206]) *Coua* (10 sp.), Madagascar; ([2207]) *Cochlothraustes* (1 sp.), Madagascar ; ([2221]) *Centropus* (35 sp.), Tropical and South Africa, the whole Oriental region, Austro-Malaya and Australia ; ([2213]) *Crotophaga* (3 sp.), Brazil to Antilles and Pennsylvania ; ([2212]) *Guira* (1 sp.), Brazil and Paraguay ; ([2209]) *Geococcyx* (2 sp.), Guatemala to Texas and California; ([2211]) *Dromococcyx* (2 sp.), Brazil to Mexico; ([2210]) *Diplopterus* (1 sp.), Mexico to Ecuador and Brazil ; ([2208]) *Saurothera* (4 sp.), Greater Antilles ; ([2219]) *Hyetornis* (2 sp.), Jamaica and Hayti ; ([2215]) *Piaya* (3 sp.), Mexico to West Ecuador and Brazil; ([2218]) *Morococcyx* (1 sp.), Costa Rica to Mexico ; ([2214]) *Coccygus* (10 sp.), La Plata to Antilles, Mexico and Pennsylvania, Cocos Island ; ([2227]) *Cuculus* (22 sp.), Palæarctic, Ethiopian, and Oriental regions, to Moluccas and Australia ; ([2229]) *Caliecthrus* (1 sp.), Papuan Islands ; ([2230–2232]) *Cacomantis* (15 sp.), Oriental and Australian regions to Fiji Islands and Tasmania; ([2233–2237]) *Chrysococcyx* (16 sp.), Tropical and South Africa, the Oriental and Australian regions to New Zealand and Fiji Islands; ([2238]) *Surniculus* (2 sp.), India, Ceylon, and Malaya; ([2239]) *Hierococcyx* (7 sp.), the Oriental region to Amoorland and Celebes; ([2240 2241]) *Coccystes* (6 sp.), Tropical and South Africa, the Oriental region, excluding Philippines; ([2242]) *Eudynamis* (8 sp.), the Oriental and Australian regions, excluding Sandwich Islands; ([2243]) *Scythrops* (1 sp.), East Australia to Moluccas and North Celebes.

FAMILY 59.—LEPTOSOMIDÆ.   (1 Genus, 1 Species.)

GENERAL DISTRIBUTION.					
NEOTROPICAL SUB-REGIONS.	NEARCTIC SUB-REGIONS.	PALÆARCTIC SUB-REGIONS.	ETHIOPIAN SUB-REGIONS.	ORIENTAL SUB-REGIONS.	AUSTRALIAN SUB-REGIONS.
— — — —	— — — —	— — — —	— — — 4	— — — —	— — — —.

The *Leptosomus discolor*, which constitutes this family, is a bird of very abnormal characters, having some affinities both with Cuckoos and Rollers. It is confined to Madagascar (Plate VI. Vol. I. p. 278).

FAMILY 60.—BUCCONIDÆ.   (5 Genera, 43 Species.)

GENERAL DISTRIBUTION.					
NEOTROPICAL SUB-REGIONS.	NEARCTIC SUB-REGIONS.	PALÆARCTIC SUB-REGIONS.	ETHIOPIAN SUB-REGIONS.	ORIENTAL SUB-REGIONS.	AUSTRALIAN SUB-REGIONS.
- 2 . 3 -	— — — —	— — — —	— — — —	— — — —	— — — —

The Bucconidæ, or Puff-birds, are generally of small size and dull colours, with rather thick bodies and dense plumage. They form one of the characteristic Neotropical families, being most abundant in the great Equatorial forest plains, but extending as far north as Guatemala, though absent from the West Indian Islands.

The genera are:—*Bucco* (21 sp.), Guatemala to Paraguay, and West of the Andes in Ecuador; *Malacoptila* (10 sp.), Guatemala to Bolivia and Brazil; *Nonnula* (3 sp.), Amazon and Columbia; *Monasa* (7 sp.), Costa Rica to Brazil; *Chelidoptera* (2 sp.), Columbia and Guiana to Brazil.

FAMILY 61.—GALBULIDÆ.   (6 Genera, 19 Species.)

GENERAL DISTRIBUTION.

NEOTROPICAL SUB-REGIONS.	NEARCTIC SUB-REGIONS.	PALÆARCTIC SUB-REGIONS.	ETHIOPIAN SUB-REGIONS.	ORIENTAL SUB-REGIONS.	AUSTRALIAN SUB-REGIONS.
— 2 . 3 —	— — — —	— — — —	— — — —	— — — —	— — — —

The Galbulidæ, or Jacamars, are small slender birds, of generally metallic plumage; somewhat resembling in form the Bee-eaters of the Old World but less active. They have the same general distribution as the last family, but they do not occur west of the Equatorial Andes.   The genera are:—

Galbula (9 sp.), Guatemala to Brazil and Bolivia; *Urogalba* (2 sp.), Guiana and the lower Amazon; *Brachygalba* (4 sp.), Venezuela to Brazil and Bolivia; *Jacamaralcyon* (1 sp.), Brazil; *Jacamerops* (2 sp.), Panama to the Amazon; *Galbalcyrhynchus* (1 sp.), Upper Amazon.

FAMILY 62.—CORACIIDÆ.   (3 Genera, 19 Species.)

GENERAL DISTRIBUTION.

NEOTROPICAL SUB-REGIONS.	NEARCTIC SUB-REGIONS.	PALÆARCTIC SUB-REGIONS.	ETHIOPIAN SUB-REGIONS.	ORIENTAL SUB-REGIONS.	AUSTRALIAN SUB-REGIONS.
— — — —	— — — —	1 . 2 . 3 . 4	1 . 2 . 3 . 4	1 . 2 . 3 . 4	1 . 2 — —

The Rollers are a family of insectivorous birds allied to the Bee-eaters, and are very characteristic of the Ethiopian and Oriental regions; but one species (*Coracias garrula*) spreads over the Palæarctic region as far north as Sweden and the Altai mountains, while the genus *Eurystomus* reaches the Amoor valley, Australia, and the Solomon Islands.   The distribution of the genera is as follows:—

Coracias (8 sp.), the whole Ethiopian region, the Oriental region except Indo-Malaya, the Palæarctic to the above-named limits, and the island of Celebes on the confines of the Australian region; *Eurystomus* (8 sp.), West and East Africa and Madagascar, the whole Oriental region except the Peninsula of India, and the Australian as far as Australia and the Solomon Islands; *Brachypteracias* (possibly allied to *Leptosomus* ?) (4 sp.), Madagascar only, but these abnormal birds form a distinct sub-family, and according to Mr. Sharpe, three genera, *Brachypteracias*, *Atelornis*, and *Geobiastes*.

A most remarkable feature in the distribution of this family is the occurrence of a true roller (*Coracias temminckii*) in the island of Celebes, entirely cut off from the rest of the genus, which does not occur again till we reach Siam and Burmah.

The curious *Pseudochelidon* from West Africa may perhaps belong to this family or to the Cypselidæ. (Ibis. 1861, p. 321.)

FAMILY 63.—MEROPIDÆ.   (5 Genera, 34 Species.)

GENERAL DISTRIBUTION.					
NEOTROPICAL SUB-REGIONS.	NEARCTIC SUB-REGIONS.	PALÆARCTIC SUB-REGIONS.	ETHIOPIAN SUB-REGIONS.	ORIENTAL SUB-REGIONS.	AUSTRALIAN SUB-REGIONS.
— — .. —	— — — —	1 . 2 — —	1 . 2 . 3 . 4	1 . 2 . 3 . 4	1 . 2 — —

The Meropidæ, or Bee-eaters, have nearly the same distribution as the Rollers, but they do not penetrate quite so far either into the Eastern Palæarctic or the Australian regions. The distribution of the genera is as follows :—

Merops (21 sp.), has the range of the family extending on the north to South Scandinavia, and east to Australia and New Guinea; *Nyctiornis* (3 sp.), the Oriental region, except Ceylon and Java ; *Meropogon* (1 sp.), Celebes ; *Meropiscus* (3 sp.), West Africa; *Melittophagus* (6 sp.), Ethiopian region, except Madagascar.

FAMILY 64.—TODIDÆ.   (1 Genus, 5 Species.)

GENERAL DISTRIBUTION.

NEOTROPICAL SUB-REGIONS.	NEARCTIC SUB-REGIONS.	PALÆARCTIC SUB-REGIONS.	ETHIOPIAN SUB-REGIONS.	ORIENTAL SUB-REGIONS.	AUSTRALIAN SUB-REGIONS.
— — — 4	— — ·· —	— ·· — —	— — — —	— — — —	— — — —

The Todies are delicate, bright-coloured, insectivorous birds, of small size, and allied to the Motmots, although externally more resembling flycatchers. They are wholly confined to the greater Antilles, the islands of Cuba, Hayti, Jamaica, and Porto Rico having each a peculiar species of *Todus*, while another species, said to be from Jamaica, has been recently described (Plate XVI. Vol. II. p. 67).

FAMILY 65.—MOMOTIDÆ.   (6 Genera, 17 Species.)

GENERAL DISTRIBUTION.

NEOTROPICAL SUB-REGIONS.	NEARCTIC SUB-REGIONS.	PALÆARCTIC SUB-REGIONS.	ETHIOPIAN SUB-REGIONS.	ORIENTAL SUB-REGIONS.	AUSTRALIAN SUB-REGIONS.
— 2 . 3 —	— — — —	— — ·· —	— — — —	— — — —	— — — —

The Motmots range from Mexico to Paraguay and to the west coast of Ecuador, but seem to have their head-quarters in Central America, five of the genera and eleven species occurring from Panama northwards, two of the genera not occurring in South America. The genera are as follows:—

Momotus (10 sp.), Mexico to Brazil and Bolivia, one species extending to Tobago, and one to Western Ecuador ; *Urospatha* (1 sp.), Costa Rica to the Amazon ; *Baryphthengus* (1 sp.), Brazil and Paraguay ; *Hylomancs* (2 sp.), Guatemala ; *Prionirhynchus* (2 sp.), Guatemala to Upper Amazon ; *Eumomota* (1 sp.), Honduras to Chiriqui.

FAMILY 66.—TROGONIDÆ.   (7 Genera, 44 Species.)

GENERAL DISTRIBUTION.					
NEOTROPICAL SUB-REGIONS.	NEARCTIC SUB-REGIONS.	PALÆARCTIC SUB-REGIONS.	ETHIOPIAN SUB-REGIONS.	ORIENTAL SUB-REGIONS.	AUSTRALIAN SUB-REGIONS.
— 2 . 3 . 4	— — — —	— — — —	1 . 2 . 3 —	1 . 2 . 3 . 4	— — — —

The Trogons form a well-marked family of insectivorous
forest-haunting birds, whose dense yet puffy plumage exhibits
the most exquisite tints of pink, crimson, orange, brown, or
metallic green, often relieved by delicate bands of pure white.
In one Guatemalan species the tail coverts are enormously
lengthened into waving plumes of rich metallic green, as graceful and marvellous as those of the Paradise-birds.   Trogons are
tolerably abundant in the Neotropical and Oriental regions, and
are represented in Africa by a single species of a peculiar
genus.   The genera now generally admitted are the following :—

   Trogon (24 sp.), Paraguay to Mexico and west of the Andes
in Ecuador; *Temnotrogon* (1 sp.), Hayti; *Prionoteles* (1 sp.),
Cuba (Plate XVII. Vol. II. p. 67) ; *Apaloderma* (2 sp.), Tropical
and South Africa; *Harpactes* (10 sp.), the Oriental region, excluding China; *Pharomacrus* (5 sp.), Amazonia to Guatemala ;
Euptilotis (1 sp.), Mexico.

   Remains of *Trogon* have been found in the Miocene deposits
of France ; and we are thus able to understand the existing
distribution of the family.   At that exceptionally mild period in
the northern hemisphere, these birds may have ranged over all
Europe and North America; but, as the climate became more
severe they gradually became restricted to the tropical regions,
where alone a sufficiency of fruit and insect-food is found all the
year round.

FAMILY 67.—ALCEDINIDÆ. (19 Genera, 125 Species.)

GENERAL DISTRIBUTION.					
NEOTROPICAL SUB-REGIONS.	NEARCTIC SUB-REGIONS.	PALÆARCTIC SUB-REGIONS.	ETHIOPIAN SUB-REGIONS.	ORIENTAL SUB-REGIONS.	AUSTRALIAN SUB-REGIONS.
1.2.3.4	1.2.3.4	1.2.3.4	1.2.3.4	1.2.3.4	1.2.3.4

The Kingfishers are distributed universally, but very unequally, over the globe, and in this respect present some of the
most curious anomalies to be found among birds. They have
their metropolis in the eastern half of the Malay Archipelago
(our first Australian sub-region), from Celebes to New Guinea, in
which district no less than 13 out of the 19 genera occur, 8 of them
being peculiar; and it is probable that in no other equally varied
group of universal distribution, is so large a proportion of the
generic forms confined to so limited a district. From this centre
kingfishers decrease rapidly in every direction. In Australia
itself there are only 4 genera with 13 species; the whole Oriental
region has only 6 genera, 1 being peculiar; the Ethiopian also
6 genera, but 3 peculiar; and each of these have less than half
the number of species possessed by the Australian region. The
Palæarctic region possesses only 3 genera, all derived from the
Oriental region; but the most extraordinary deficiency is shown
by the usually rich Neotropical region, which possesses but a
single genus, common to the larger part of the Eastern Hemisphere, and the same genus is alone found in the Nearctic region,
the only difference being that the former possesses eight, while
the latter has but a single species. These facts almost inevitably
lead to the conclusion that America long existed without kingfishers; and that in comparatively recent times (perhaps during
the Miocene or Pliocene period), a species of the Old World
genus, *Ceryle*, found its way into North America, and spreading
rapidly southward along the great river-valleys has become
differentiated in South America into the few closely allied forms
that alone inhabit that vast country—the richest in the world in

fresh-water fish, and apparently the best fitted to sustain a varied and numerous body of kingfishers.

The names of the genera, with their distribution and the number of species in each, as given by Mr. Sharpe in his excellent monograph of the family, is as follows :—

Alcedo (9 sp.), Palæarctic, Ethiopian, and Oriental regions (but absent from Madagascar), and extending into the Austro-Malayan sub-region ; *Corythornis* (3 sp.), the whole Ethiopian region ; *Alcyone* (7 sp.), Australia and the Austro-Malayan sub-region, with one species in the Philippine Islands ; *Ceryle* (13 sp.), absent only from Australia, the northern half of the Palæarctic region, and Madagascar ; *Pelargopsis* (9 sp.), the whole Oriental region, and extending to Celebes and Timor in the Austro-Malayan sub-region ; *Ceyx* (11 sp.), the Oriental region and Austro-Malayan sub-region, but absent from Celebes, and only one species in continental India and Ceylon ; *Ceycopsis* (1 sp.), Celebes ; *Myioceyx* (2 sp.), West Africa ; *Ipsidina* (4 sp.), Ethiopian region ; *Syma* (2 sp.), Papua and North Australia ; *Halcyon* (36 sp.), Australian, Oriental, and Ethiopian regions, and the southern part of the Palæarctic ; *Dacelo* (6 sp.), Australia and New Guinea ; *Todirhamphus* (3 sp.), Eastern Pacific Islands only ; *Monachalcyon* (1 sp.), Celebes ; *Caridonax* (1 sp.), Lombok and Flores ; *Carcineutes* (2 sp.), Siam to Borneo and Java ; *Tanysiptera* (14 sp.), Moluccas New Guinea, and North Australia (Plate X. Vol. I. p. 414) ; *Cittura* (2 sp.), Celebes group ; *Melidora* (1 sp.), New Guinea.

FAMILY 68.—BUCEROTIDIÆ.   (12 Genera, 50 Species.)

GENERAL DISTRIBUTION.					
NEOTROPICAL SUB-REGIONS.	NEARCTIC SUB-REGIONS.	PALÆARCTIC SUB-REGIONS.	ETHIOPIAN SUB-REGIONS.	ORIENTAL SUB-REGIONS.	AUSTRALIAN SUB-REGIONS.
— — — —	— — — —	— — — —	1 . 2 . 3 —	1 . 2 . 3 . 4	1 — — —

The Hornbills form an isolated group of generally large-sized birds, whose huge bills form their most prominent feature. They are popularly associated with the American Toucans, but have no close relationship to them, and are now generally considered to show most resemblance, though still a very distant one, to the kingfishers.   They are abundant in the Ethiopian and Oriental regions, and extend eastward to the Solomon Islands.   Their classification is very unsettled, for though they have been divided into more than twenty genera they have not yet been carefully studied.   The following grouping of the genera—referring to the numbers in the *Hand List*—must therefore be considered as only provisional :—

(1957 1958 1963) *Buceros* (6 sp.), all Indo-Malaya, Arakan, Nepal and the Neilgherries (Plate IX. Vol. I. p. 339) ; (1959 — 1961) *Hydrocissa* (7 sp.), India and Ceylon to Malaya and Celebes ; (1962) *Berenicornis* (2 sp.), Sumatra and West Africa ; (1964) *Calao* (3 sp.), Tennaserim, Malaya, Moluccas to the Solomon Islands ; (1965) *Aceros* (1 sp.), South-east Himalayas ; (1966 1967) *Cranorrhinus* (3 sp.), Malacca, Sumatra, Borneo, Philippines, Celebes ; (1968) *Penelopides* (1 sp.), Celebes ; (1969 — 1971) *Tockus* (15 sp.), Tropical and South Africa ; (1972) *Rhinoplax* (1 sp.), Sumatra and Borneo ; (1973 — 1975) *Bycanistes* (6 sp.), West Africa with East and South Africa ; (1976 1977) *Meniceros* (3 sp.), India and Ceylon to Tenasserim ; (1978) *Bucorvus* (2 sp.), Tropical and South Africa.

FAMILY 69.—UPUPIDÆ.   (1 Genus, 6 Species.)

GENERAL DISTRIBUTION.					
NEOTROPICAL SUB-REGIONS.	NEARCTIC SUB-REGIONS.	PALÆARCTIC SUB-REGIONS.	ETHIOPIAN SUB-REGIONS.	ORIENTAL SUB-REGIONS.	AUSTRALIAN SUB-REGIONS.
— — — —	— — — —	— 2 — 4	1.2.3.4	1.2.3 —	— — — —

The Hoopoes form a small and isolated group of semi-terrestrial insectivorous birds, whose nearest affinities are with the Hornbills.   They are most characteristic of the Ethiopian region, but extend into the South of Europe and into all the continental divisions of the Oriental region, as well as to Ceylon, and northwards to Pekin and Mongolia.

FAMILY 70.—IRRISORIDÆ.   (1 Genus, 12 Species.)

GENERAL DISTRIBUTION.					
NEOTROPICAL SUB-REGIONS.	NEARCTIC SUB-REGIONS.	PALÆARCTIC SUB-REGIONS.	ETHIOPIAN SUB-REGIONS.	ORIENTAL SUB-REGIONS.	AUSTRALIAN SUB-REGIONS.
— — — —	— — — —	— — — —	1 . 2 . 3 –	— — — —	— — — —

The Irrisors are birds of generally metallic plumage, which have often been placed with the Epunachidæ and near the Sunbirds, or Birds of Paradise, but which are undoubtedly allied to the Hoopoes.   They are strictly confined to the continent of Africa, ranging from Abyssinia to the west coast, and southward to the Cape Colony.   They have been divided into several subgenera which it is not necessary here to notice (Plate IV. Vol. I. p. 261).

FAMILY 71.—PODARGIDÆ.   (3 Genera, 20 Species.)

GENERAL DISTRIBUTION.					
NEOTROPICAL SUB-REGIONS.	NEARCTIC SUB-REGIONS.	PALÆARCTIC SUB-REGIONS.	ETHIOPIAN SUB-REGIONS.	ORIENTAL SUB-REGIONS.	AUSTRALIAN SUB-REGIONS.
— — — —	— — — —	— — — —	— — — —	1 . 2 . 3 . 4	1 . 2 – –

The Podargidæ, or Frog-mouths, are a family of rather large-sized nocturnal insectivorous birds, closely allied to the Goatsuckers, but distinguished by their generally thicker bills, and especially by hunting for their food on trees or on the ground, instead of seizing it on the wing. They abound most in the Australian region, but one genus extends over a large part of the Oriental region.   The following are the genera with their distribution :—

Podargus (10 sp.), Australia, Tasmania, and the Papuan Islands (Plate XII. Vol. I. p. 441); *Batrachostomus* (6 sp.), the Oriental region (excluding Philippine Islands and China) and the northern Moluccas ; *Ægotheles* (4 sp.), Australia, Tasmania, and Papuan Islands.

FAMILY 72.—STEATORNITHIDÆ.　(1 Genus, 1 Species.)

GENERAL DISTRIBUTION.

NEOTROPICAL SUB-REGIONS.	NEARCTIC SUB-REGIONS.	PALÆARCTIC SUB-REGIONS.	ETHIOPIAN SUB-REGIONS.	ORIENTAL SUB-REGIONS.	AUSTRALIAN SUB-REGIONS.
— 2 — —	— — — —	— — — —	— — — —	— — — —	— — — —

This family contains a single bird—the Guacharo—forming the genus *Steatornis*, first discovered by Humboldt in a cavern in Venezuela, and since found in deep ravines near Bogota, and also in Trinidad. Although apparently allied to the Goat-suckers it is a vegetable-feeder, and is altogether a very anomalous bird whose position in the system is still undetermined.

FAMILY 73.—CAPRIMULGIDÆ.　(17 Genera, 91 Species.)

GENERAL DISTRIBUTION.

NEOTROPICAL SUB-REGIONS.	NEARCTIC SUB-REGIONS.	PALÆARCTIC SUB-REGIONS.	ETHIOPIAN SUB-REGIONS.	ORIENTAL SUB-REGIONS.	AUSTRALIAN SUB-REGIONS.
1.2.3.4	1.2.3.4	1.2.3.4	1.2.3.4	1.2.3.4	1.2 — —

The Goat-suckers, or Night-jars, are crepuscular insectivorous birds, which take their prey on the wing, and are remarkable for their soft and beautifully mottled plumage, swift and silent flight, and strange cries often imitating the human voice. They are universally distributed, except that they do not reach New Zealand or the remoter Pacific Islands. The South American genus, *Nyctibius*, differs in structure and habits from the other goat-suckers and should perhaps form a distinct family. More than half the genera inhabit the Neotropical region. The genera are as follows :—

Nyctibius (6 sp.), Brazil to Guatemala, Jamaica ; *Caprimulgus* (35 sp.), Palæarctic, Oriental, and Ethiopian regions, with the Austro-Malay Islands and North Australia ; *Hydropsalis* (8 sp.), Tropical South America to La Plata ; *Antrostomus* (10 sp.), La Plata and Bolivia to Canada, Cuba; *Stenopsis* (4 sp.), Martinique to Columbia, West Peru and Chili; *Siphonorhis* (1 sp.), Jamaica; *Heleothreptus* (1 sp.), Demerara; *Nyctidromus* (2 sp.), South Brazil to Central America; *Scortornis* (3 sp.), West and East Africa; *Macrodipteryx* (2 sp.), West and Central Africa; *Cosmetornis* (1 sp.), all Tropical Africa; *Podager* (1 sp.), Tropical South America to La Plata; *Lurocalis* (2 sp.), Brazil and Guiana; *Chordeiles* (8 sp.), Brazil and West Peru to Canada, Porto Rico, Jamaica; *Nyctiprogne* (1 sp.), Brazil and Amazonia; *Eurostopodus* (2 sp.), Australia and Papuan Islands; *Lyncornis* (4 sp.), Burmah, Philippines, Borneo, Celebes.

FAMILY 74.—CYPSELIDÆ.   (7 Genera, 53 Species.)

GENERAL DISTRIBUTION.					
NEOTROPICAL SUB-REGIONS.	NEARCTIC SUB-REGIONS.	PALÆARCTIC SUB-REGIONS.	ETHIOPIAN SUB-REGIONS.	ORIENTAL SUB-REGIONS.	AUSTRALIAN SUB-REGIONS.
1.2.3.4	1.2.3.4	1.2.3.4	1.2.3.4	1.2.3.4	1.2.3 –

The Swifts can almost claim to be a cosmopolitan group, but for their absence from New Zealand. They are most abundant both in genera and species in the Neotropical and Oriental regions. The following is the distribution of the genera:—

Cypselus (1 sp.), absent only from the whole of North America and the Pacific; *Panyptila* (3 sp.), Guatemala and Guiana, and extending into North-west America; *Collocalia* (10 sp.), Madagascar, the whole Oriental region and eastward through New Guinea to the Marquesas Islands; *Dendrochelidon* (5 sp.), Oriental region and eastward to New Guinea; *Chætura* (15 sp.), Continental America (excluding South Temperate), West Africa and Madagascar, the Oriental region, North China and the Amoor, Celebes, Australia; *Hemiprocne* (3 sp.), Mexico to La Plata, Jamaica and Hayti; *Cypseloides* (2 sp.), Brazil and Peru; *Nephœcetes* (2 sp.), Cuba, Jamaica, North-west America.

FAMILY 75.—TROCHILIDÆ.   (118 Genera, 390 Species.)

GENERAL DISTRIBUTION.					
NEOTROPICAL SUB-REGIONS.	NEARCTIC SUB-REGIONS.	PALÆARCTIC SUB-REGIONS.	ETHIOPIAN SUB-REGIONS.	ORIENTAL SUB-REGIONS.	AUSTRALIAN SUB-REGIONS.
1.2.3.4	1.2.3.4	— — — —	— — — —	— — — —	— — — —

The wonderfully varied and beautiful Humming-Birds are confined to the American continent, where they range from Sitka to Cape Horn, while the island of Juan Fernandez has two peculiar species.   Only 6 species, belonging to 3 genera, are found in the Nearctic region, and most of these have extended their range from the south.   They are excessively abundant in the forest-clad Andes from Mexico to Chili, some species extending up to the limits of perpetual snow; but they diminish in number and variety in the plains, however luxuriant the vegetation.   In place of giving here the names and distribution of the numerous genera into which they are now divided (which will be found in the tables of the genera of the Neotropical region), it may be more useful to present a summary of their distribution in the sub-divisions of the American continent, as follows:—

	Sub-region I. (Patagonia & S. Andes.)	Sub-region II. (Tropical S. Amer.)	Sub. region III. (Tropical N. Amer.)	Sub-region IV. (Antilles.)	Nearctic region. (Temp. N. Amer.)
Genera in each Sub-region	10	90	41	8	3
Peculiar Genera	3	58	14	5	0
Species in each Sub-region	15	275	100	15	6

The island of Juan Fernandez has two species, and Masafuera, an island beyond it, one; the three forming a peculiar genus. The island of Tres Marias, about 60 miles from the west coast of Mexico, possesses a peculiar species of humming-bird, and the Bahamas two species; but none inhabit either the Falkland Islands or the Galapagos.

Like most groups which are very rich in species and in generic forms, the humming-birds are generally very local, small generic groups being confined to limited districts; while single mountains, valleys, or small islands, often possess species found nowhere else. It is now well ascertained that the Trochilidæ are really insectivorous birds, although they also feed largely, but probably never exclusively, on the nectar of flowers. Their nearest allies are undoubtedly the Swifts; but the wide gap that now separates them from these, as well as the wonderful variety of form and of development of plumage, that is found among them, alike point to their origin, at a very remote period, in the forests of the once insular Andes. There is perhaps no more striking contrast of the like nature, to be found, than that between the American kingfishers—confined to a few closely allied forms of one Old World genus—and the American humming-birds with more than a hundred diversified generic forms unlike everything else upon the globe; and we can hardly imagine any other cause for this difference, than a (comparatively) very recent introduction in the one case, and a very high antiquity in the other.

General Remarks on the Distribution of the Picariæ.

The very heterogeneous mass of birds forming the Order Picariæ, contains 25 families, 307 genera and 1,604 species. This gives about 64 species to each family, while in the Passeres the proportion is nearly double, or 111 species per family. There are, in fact, only two very large families in the Order, which happen to be the first and last in the series—Picidæ and Trochilidæ. Two others—Cuculidæ and Alcedinidæ—are rather large; while the rest are all small, seven of them consisting only of a single genus and from one to a dozen species. Only one of the families—Alcedinidæ—is absolutely cosmopolitan, but three others are nearly so, Caprimulgidæ and Cypselidæ being only absent from New Zealand, and Cuculidæ from the Canadian sub-region of North America. Eleven families inhabit the Old World only, while seven are confined to the New World, only one of these—Trochilidæ—being common to the Neotropical and Nearctic regions.

The Picariæ are highly characteristic of tropical faunas for while no less than 15 out of the 25 families are exclusively tropical, none are confined to, or have their chief development in, the temperate regions. They are best represented in the Ethiopian region, which possesses 17 families, 4 of which are peculiar to it; while the Oriental region has only 14 families, none of which are peculiar. The Neotropical region has also 14 families, but 6 of them are peculiar. The Australian region has 8, the Palæarctic 9 and the Nearctic 6 families, but none of these are peculiar. We may see a reason for the great specialization of this tropical assemblage of birds in the Ethiopian and Neotropical regions, in the fact of the large extent of land on both sides of the Equator which these two regions alone possess, and their extreme isolation either by sea or deserts from other regions,—an isolation which we know was in both cases much greater in early Tertiary times. It is, perhaps, for a similar reason that we here find hardly any trace of the connection between Australia and South America which other groups exhibit; for that connection has most probably been effected by a former communication between the temperate southern extremities of those two continents. The most interesting and suggestive fact, is that presented by the distribution of the Megalæmidæ and Trogonidæ over the tropics of America, Africa, and Asia. In the absence of palæontological evidence as to the former history of the Megalæmidæ, we are unable to say positively, whether it owes its present distribution to a former closer union between these continents in intertropical latitudes, or to a much greater northern range of the group at the period when a luxuriant sub-tropical vegetation extended far toward the Arctic regions; but the discovery of *Trogon* in the Miocene deposits of the South of France renders it almost certain that the latter is the true explanation in the case of both these families.

The Neotropical region, owing to its enormous family of humming-birds, is by far the richest in Picariæ, possessing nearly half the total number of species, and a still larger proportion of genera. Three families, the Bucerotidæ, Meropidæ and Coraciidæ are equally characteristic of the Oriental and

Ethiopian regions, a few outlying species only entering the Australian or the Palæarctic regions. One family (Todidæ) is confined to the West Indian Islands ; and another (Leptosomidæ) consisting of but a single species, to Madagascar; parallel cases to the Drepanididæ among the Passeres, peculiar to the Sandwich Islands, and the Apterygidæ among the Struthiones, peculiar to New Zealand.

Order III.—PSITTACI.

The Parrots have been the subject of much difference of opinion among ornithologists, and no satisfactory arrangement of the order into families and genera has yet been reached. Professor Garrod has lately examined certain points in the anatomy of a large number of genera, and proposes to revolutionize the ordinary classifications. Until, however, a general examination of their whole anatomy, internal and external, has been made by some competent authority, it will be unsafe to adopt the new system, as we have as yet no guide to the comparative value of the characters made use of. I therefore keep as much as possible to the old groups, founded on external characters, only using the indications furnished by Professor Garrod's paper, to determine the position of doubtful genera.

FAMILY 76.—CACATUIDÆ.   (5 Genera, 35 Species.)

GENERAL DISTRIBUTION.					
NEOTROPICAL SUB-REGIONS.	NEARCTIC SUB-REGIONS.	PALÆARCTIC SUB-REGIONS.	ETHIOPIAN SUB-REGIONS.	ORIENTAL SUB-REGIONS.	AUSTRALIAN SUB-REGIONS
— — — —	— — — —	— — — —	— — — —	— — — 4	1 . 2 — —

The Cacatuidæ, Plyctolophidæ, or Camptolophidæ, as they have been variously termed, comprise all those crested parrots usually termed Cockatoos, together with one or two doubtful forms. They are very abundant in the Australian region, more especially in the Austro-Malayan portion of it, one species inhabiting the Philippine Islands; but they do not pass further east than the Solomon Islands and are not found in New Zealand. The distribution of the genera is as follow:—

Cacatua (18 sp.) ranges from the Philippine Islands, Celebes and Lombok, to the Solomon Islands and to Tasmania; *Calopsitta* (1 sp.) Australia; *Calyptorhynchus* (8 sp.) is confined to Australia and Tasmania; *Microglossus* (2 sp.) (perhaps a distinct family) to the Papuan district and North Australia; *Licmetis* (3 sp.) Australia, Solomon Islands, and (?) New Guinea; *Nasiterna* (3 sp.), a minute form, the smallest of the whole order, and perhaps not belonging to this family, is only known from the Papuan and Solomon Islands.

FAMILY 77.—PLATYCERCIDÆ.   (11 Genera, 57 Species.)

GENERAL DISTRIBUTION.					
NEOTROPICAL SUB-REGIONS.	NEARCTIC SUB-REGIONS.	PALÆARCTIC SUB-REGIONS.	ETHIOPIAN SUB-REGIONS.	ORIENTAL SUB-REGIONS.	AUSTRALIAN SUB-REGIONS.
— — — —	— — — —	— — — —	— — — —	— — -- —	1 . 2 . 3 . 4

The Platycercidæ comprise a series of large-tailed Parrots, of weak structure and gorgeous colours, with a few ground-feeding genera of more sober protective tints; the whole family being confined to the Australian region. The genera are:—

([1996 1999 2000]) *Platycercus* (14 sp.), Australia, Tasmania, and Norfolk Island; *Psephotus* (6 sp.), Australia; *Polytelis* (3 sp.), Australia; *Nymphicus* (1 sp.), Australia and New Caledonia; ([2002 2003]) *Aprosmictus* (6 sp.), Australia, Papua, Timor, and Moluccas; *Pyrrhulopsis* (3 sp.), Tonga and Fiji Islands; *Cyanoramphus* (14 sp.), New Zealand, Norfolk Island, New Caledonia, and Society Islands; *Melopsittacus* (1 sp.), Australia; *Euphema* (7 sp.), Australia; *Pezoporus* (1 sp.), Australia and Tasmania; *Geopsittacus* (1 sp.), West Australia. The four last genera are ground-feeders, and are believed by Professor Garrod to be allied to the Owl-Parrot of New Zealand (*Stringops*).

FAMILY 78.—PALÆORNITHIDÆ.   (8 Genera, 65 Species.)

		GENERAL DISTRIBUTION.			
NEOTROPICAL SUB-REGIONS.	NEARCTIC SUB-REGIONS.	PALÆARCTIC SUB-REGIONS.	ETHIOPIAN SUB-REGIONS.	ORIENTAL SUB-REGIONS.	AUSTRALIAN SUB-REGIONS.
— — — —	— — — —	— — — —	1.2 — 4	1.2.3.4	1.2 — —

I class here a group of birds brought together, for the most part, by geographical distribution as well as by agreement in internal structure, but which is nevertheless of a very uncertain and provisional character.

Palæornis (18 sp.), the Oriental region, Mauritius, Rodriguez, and Seychelle Islands, and a species in Tropical Africa, apparently identical with the Indian *P. torquatus*, and therefore— considering the very ancient intercourse between the two countries, and the improbability of the *species* remaining unchanged if originating by natural causes—most likely the progeny of domestic birds introduced from India. *Prioniturus* (3 sp.), Celebes and the Philippine Islands ; ([2061]) *Geoffroyus* (5 sp.), Bouru to Timor and the Solomon Islands ; *Tanygnathus* (5 sp.), Philippines, Celebes, and Moluccas to New Guinea ; *Eclectus* (8 sp.), Moluccas and Papuan Islands ; *Psittinus* (1 sp.), Tenasserim to Sumatra and Borneo ; *Cyclopsitta* (8 sp.), Papuan Islands, Philippines and North-east Australia ; *Loriculus* (17 sp.), ranges over the whole Oriental region to Flores, the Moluccas, and the Papuan island of Mysol; but most of the species are concentrated in the district including the Philippines, Celebes, Gilolo, and Flores, there being 1 in India, 1 in South China, 1 in Ceylon, 1 in Java, 1 in Malacca, Sumatra, and Borneo, 3 in Celebes, 5 in the Philippines, and the rest in the Moluccas, Mysol, and Flores. This genus forms a transition to the next family.

FAMILY 79.—TRICHOGLOSSIDÆ. (6 Genera, 57 Species.)

GENERAL DISTRIBUTION.					
NEOTROPICAL SUB-REGIONS.	NEARCTIC SUB-REGIONS.	PALÆARCTIC SUB-REGIONS.	ETHIOPIAN SUB-REGIONS.	ORIENTAL SUB-REGIONS.	AUSTRALIAN SUB-REGIONS.
— — — —	— — — —	— — — —	— — — —	— — — —	1 . 2 . 3 —

The Trichoglossidæ, or Brush-tongued Paroquets, including the Lories, are exclusively confined to the Australian region, where they extend from Celebes to the Marquesas Islands, and south to Tasmania. The genus *Nanodes* (= *Lathamus*) has been shown by Professor Garrod to differ from *Trichoglossus* in the position of the carotid arteries. I therefore make it a distinct genus but do not consider that it should be placed in another family. The genera here admitted are as follows :—

Trichoglossus (29 sp.), ranges over the whole Austro-Malay and Australian sub-regions, and to the Society Islands; ([2047]) *Nanodes* (1 sp.), Australia and Tasmania; *Charmosyna* (1 sp.), New Guinea (Plate X. Vol. I. p. 414); *Eos* (9 sp.), Bouru and Sanguir Island north of Celebes, to the Solomon Islands, and in Puynipet Island to the north-east of New Ireland; ([2039 2040]) *Lorius* (13 sp.), Bouru and the Solomon Islands ; ([2041 2043]) *Coriphilus* (4 sp.), Samoa, Tonga, Society and Marquesas Islands.

FAMILY 80.—CONURIDÆ. (7 Genera, 79 Species.)

GENERAL DISTRIBUTION.					
NEOTROPICAL SUB-REGIONS.	NEARCTIC SUB-REGIONS.	PALÆARCTIC SUB-REGIONS.	ETHIOPIAN SUB-REGIONS.	ORIENTAL SUB-REGIONS.	AUSTRALIAN SUB-REGIONS.
1 . 2 . 3 . 4	— — 3 —	— — — —	— — — —	— — — —	— — — —

The Conuridæ, which consist of the Macaws and their allies, are wholly confined to America, ranging from the Straits of Magellan to South Carolina and Nebraska, with Cuba and Jamaica. Professor Garrod places *Pyrrhura* (which has generally been classed as a part of the genus *Conurus*) in a separate family, on account of the absence of the ambiens muscle of the knee, but as we are quite ignorant of the classificational value of this character, it is better for the present to keep both as distinct genera of the same family.   The genera are :—

Ara (15 sp.), Paraguay to Mexico and Cuba; *Rhyncopsitta* (1 sp.), Mexico ; *Henicognathus* (1 sp.), Chili ; *Conurus* (30 sp.), the range of the family; *Pyrrhura* (16 sp.), Paraguay and Bolivia to Costa Rica ; *Bolborhynchus* (7 sp.), La Plata, Bolivia and West Peru, with one species in Mexico and Guatemala; *Brotogerys* (9 sp.), Brazil to Mexico.

FAMILY 81.—PSITTACIDÆ.—(12 Genera, 87 Species.)

GENERAL DISTRIBUTION.					
NEOTROPICAL SUB-REGIONS.	NEARCTIC SUB-REGIONS.	PALÆARCTIC SUB-REGIONS.	ETHIOPIAN SUB-REGIONS.	ORIENTAL SUB-REGIONS.	AUSTRALIAN SUB-REGIONS.
– 2 . 3 . 4	– – – –	– – ·· – –	1 . 2 . 3 . 4	– – – –	– – – –

The Psittacidæ comprise a somewhat heterogeneous assemblage of Parrots and Paroquets of the Neotropical and Ethiopian regions, which are combined here more for convenience than because they are believed to form a natural group.   The genera *Chrysotis* and *Pionus* have no oil-gland, while *Psittacula* and *Agapornis* have lost the furcula, but neither of these characters are probably of more than generic value.   The genera are :—

Psittacus (2 sp.), West Africa; *Coracopsis* (5 sp.), Madagascar, Comoro, and Seychelle Islands ; *Pæocephalus* (9 sp.), all Tropical and South Africa ; ([2063 – 2066]) *Caica* (9 sp.), Mexico to Amazonia ; *Chrysotis* (32 sp.), Paraguay to Mexico and the West Indian Islands; *Triclaria* (1 sp.), Brazil ; *Deroptyus* (1 sp.), Amazonia ; *Pionus* (9 sp.), Paraguay to Mexico; *Urochroma* (7 sp.), Tropical South America ; *Psittacula* (6 sp.), Brazil to Mexico; *Poliopsitta* (2 sp.), Madagascar and West Africa; *Agapornis* (4 sp.), Tropical and South Africa.

FAMILY 82.—NESTORIDÆ. (? 2 Genera, 6 Species.)

GENERAL DISTRIBUTION.

NEOTROPICAL SUB-REGIONS.	NEARCTIC SUB-REGIONS.	PALÆARCTIC SUB-REGIONS.	ETHIOPIAN SUB-REGIONS.	ORIENTAL SUB-REGIONS.	AUSTRALIAN SUB-REGIONS.
— — — —	— — — —	— — — —	— — — —	— — — —	1 — — 4

The present family is formed to receive the genus *Nestor* (5 sp.), confined to New Zealand and Norfolk Island. Its affinities are doubtful, but it appears to have relations with the American Conuridæ and the Australian Trichoglossidæ. With it is placed the rare and remarkable *Dasyptilus* (1 sp.), of New Guinea, of which however very little is known.

FAMILY 83.—STRINGOPIDÆ. (1 Genus, 2 Species.)

GENERAL DISTRIBUTION.

NEOTROPICAL SUB-REGIONS.	NEARCTIC SUB-REGIONS.	PALÆARCTIC SUB-REGIONS.	ETHIOPIAN SUB-REGIONS.	ORIENTAL SUB-REGIONS.	AUSTRALIAN SUB-REGIONS.
— — — —	— — — —	— — — —	— — ——	— — — ‥	— — — 4

This family contains only the curious owl-like nocturnal Parrot of New Zealand, *Stringops habroptilus* (Plate XIII. Vol .I. p. 455). An allied species is said to inhabit the Chatham Islands, if not now extinct.

General Remarks on the Distribution of the Psittaci.

Although the Parrots are now generally divided into several distinct families, yet they form so well marked and natural a group, and are so widely separated from all other birds, that we may best discuss their peculiarities of geographical distribution by treating them as a whole. By the preceding enumeration we find that there are about 386 species of known parrots, which are divided into 52 genera. They are preeminently a tropical group, for although a few species extend a considerable distance into the temperate zone, these are

marked exceptions to the rule which limits the parrot tribe to
the tropical and sub-tropical regions, roughly defined as extending about 30° on each side of the equator. In America a species
of *Conurus* reaches the straits of Magellan on the south, while
another inhabits the United States, and once extended to the
great lakes, although now confined to the south-eastern districts.
In Africa parrots do not reach the northern tropic, owing to the
desert nature of the country; and in the south they barely reach
the Orange River. In India they extend to about 35° N. in the
western Himalayas; and in the Australian region, not only to
New Zealand but to Macquarie Islands in 54° S., the farthest
point from the equator reached by the group. But although
found in all the tropical regions they are most unequally distributed. Africa is poorest, possessing only 6 genera and 25
species; the Oriental region is also very poor, having but 6
genera and 29 species; the Neotropical region is much richer,
having 14 genera and 141 species; while the smallest in area
and the least tropical in climate—the Australian region, possesses 31 genera and 176 species, and it also possesses exclusively
5 of the families, Trichoglossidæ, Platycercidæ, Cacatuidæ,
Nestoridæ, and Stringopidæ. The portion of the earth's surface
that contains the largest number of parrots in proportion to its
area is, undoubtedly, the Austro-Malayan sub-region, including
the islands from Celebes to the Solomon Islands. The area of
these islands is probably not one-fifteenth of that of the four
tropical regions, yet they contain from one-fifth to one-fourth of
all the known parrots. In this area too are found many of the
most remarkable forms,—all the crimson lories, the great black
Cockatoos, the pigmy *Nasiterna*, the raquet-tailed *Prioniturus*,
and the bareheaded *Dasyptilus*.

The almost universal distribution of Parrots wherever the
climate is sufficiently mild or uniform to furnish them with a
perennial supply of food, no less than their varied details of
organization, combined with a great uniformity of general type,
—tell us, in unmistakable language, of a very remote antiquity.
The only early record of extinct parrots is, however, in the
Miocene of France, where remains apparently allied to the West

African *Psittacus,* have been found.　But the origin of so wide-spread, isolated, and varied a group, must be far earlier than this, and not improbably dates back beyond the dawn of the Tertiary period.　Some primeval forms may have entered the Australian region with the Marsupials, or not long after them; while perhaps at a somewhat later epoch they were introduced into South America.　In these two regions they have greatly flourished, while in the two other tropical regions only a few types have been found, capable of maintaining themselves, among the higher forms of mammalia, and in competition with a more varied series of birds.　This seems much more probable than the supposition that so highly organized a group should have originated in the Australian region, and subsequently become so widely spread over the globe.

Order IV.—COLUMBÆ.

FAMILY 84.—COLUMBIDÆ.　(44 Genera, 355 Species.)

GENERAL DISTRIBUTION.					
NEOTROPICAL SUB-REGIONS.	NEARCTIC SUB-REGIONS.	PALÆARCTIC SUB-REGIONS.	ETHIOPIAN SUB-REGIONS.	ORIENTAL SUB-REGIONS.	AUSTRALIAN SUB-REGIONS.
1.2.3.4	1.2.3.4	1.2.3.4	1.2.3.4	1.2.3.4	1.2.3.4

The Columbidæ, or Pigeons and Doves, are almost universally distributed, but very unequally in the different regions.　Being best adapted to live in warm or temperate climates, they diminish rapidly northwards, reaching about 62° N. Latitude in North America, but considerably farther in Europe.　Both the Nearctic and Palæarctic regions are very poor in genera and species of pigeons, those of the former region being mostly allied to Neotropical, and those of the latter to Oriental and Ethiopian types.　The Ethiopian region is, however, itself very poor, and several of its peculiar forms are confined to the Madagascar sub-region.　The Neotropical region is very rich in peculiar genera, though but moderately so in number of species.　The Oriental region closely approaches it in both respects; but the Australian region is by far the richest, possessing nearly double the genera and species of any other region, and abounding in remarkable forms quite unlike those of any other part of the globe. The following table gives the number of genera and species in each region, and enables us readily to determine the comparative richness and isolation of each, as regards this extensive family :—

Regions.			No. of Genera.	Peculiar Genera.	No. of Species.
Neotropical	...	...	13	9	75
Nearctic	...	...	5	1	7
Palæarctic	...	...	3	0	9
Ethiopian	...	...	6	1	37
Oriental	...	...	12	1	66
Australian	...	...	24	14	148

With the exception of *Columba* and *Turtur*, which have a wide range, *Treron*, common to the Oriental and Ethiopian regions, and *Carpophaga*, to the Oriental and Australian, most of the genera of pigeons are either restricted to or very characteristic of a single region.

The distribution of the genera here admitted is as follows :—

Treron (37 sp.), the whole Oriental region, and eastward to Celebes, Amboyna and Flores, also the whole Ethiopian region to Madagascar; *Ptilopus* (52 sp.), the Australian region (excluding New Zealand) and the Indo-Malay sub-region; *Alectrœnas* (4 sp.), Madagascar and the Mascarene Islands: *Carpophaga* (50 sp.), the whole Australian and Oriental regions, but much the most abundant in the former; ([2274]) *Ianthœnas* (11 sp.), Japan, Andaman, Nicobar, and Philippine Islands, Timor and Gilolo to Samoa Islands; ([2278]) *Leucomelœna* (1 sp.), Australia; *Lopholaimus* (1 sp.), Australia; ([2279 and 2283]) *Alsœcomus* (2 sp.), Himalayas to Ceylon and Tenasserim; *Columba* (46 sp.), generally distributed over all the regions except the Australian, one species however in the Fiji Islands; *Ectopistes* (1 sp.), east of North America with British Columbia; *Zenaidura* (2 sp.), Veragua to Canada and British Columbia ; *Œna* (1 sp.), Tropical and South Africa; *Geopelia* (6 sp.), Philippine Islands and Java to Australia; *Macropygia* (14 sp.), Nepal, Hainan, Nicobar, Java, and Philippines to Australia and New Ireland; *Turacœna* (3 sp.), Celebes, Timor, and Solomon Islands; *Reinwardtœnas* (1 sp.), Celebes to New Guinea; *Turtur* (24 sp.), Palæarctic, Ethiopian and Oriental regions with Austro-Malaya; *Chœmepelia* (7 sp.), Brazil and Bolivia to Jamaica, California, and South-east United States; *Columbula* (2 sp.), Brazil and La Plata to Chili; *Scardafella* (2 sp.), Brazil and Guatemala; *Zenaida* (10 sp.), Chili and La Plata to Columbia and the Antilles, Fernando Noronha; *Melopelia* (2 sp.), Chili to Mexico and California; *Peristera* (4 sp.), Brazil to Mexico; *Metriopelia* (2 sp.), West America from Ecuador to Chili; *Gymnopelia* (1 sp.), West Peru and Bolivia; *Leptoptila* (11 sp.), Paraguay to Mexico and the Antilles; ([2317 2318 and 2320]) *Geotrygon* (14 sp.), Paraguay to Mexico and the Antilles; *Aplopelia* (5 sp.), Tropical and South Africa, St. Thomas and Princes Island; *Chalocopelia* (4 sp.), Tropical and South Africa; *Starnœnas* (1 sp.), Cuba; *Ocyphaps* (1 sp.), Australia (Plate XII. Vol. I. p. 441); *Petrophassa* (1 sp.), North-west Australia; *Chalocophaps* (8 sp.), the Oriental region to New Guinea and Australia; *Trugon* (1 sp.), New Guinea; *Henicophaps* (1 sp.), Waigiou and New Guinea; *Phaps* (3 sp.), Australia and Tasmania; *Leucosarcia* (1 sp.), East Australia; *Phapitreron* (2 sp.), Philippine Islands; *Geophaps* (2 sp.), North and East Australia; *Lophophaps* (3 sp.), Australia; *Calœnas* (1 sp.), scattered on the smaller islands from the Nicobars and Philippines to New Guinea; *Otidiphaps* (1 sp.), New Guinea; *Phlogœnas* (7 sp.), Philippine Islands and Celebes to the Marquesas Islands; *Goura* (2 sp.), New Guinea and the islands on the north-east (Plate X. Vol. I. p 414).

FAMILY 84*a*.—DIDUNCULIDÆ.   (1 Genus, 1 Species.)

GENERAL DISTRIBUTION.					
NEOTROPICAL SUB-REGIONS.	NEARCTIC SUB-REGIONS.	PALÆARCTIC SUB-REGIONS.	ETHIOPIAN SUB-REGIONS.	ORIENTAL SUB-REGIONS.	AUSTRALIAN SUB-REGIONS.
— — —	— — —	— — —	— — —	— — —	— — 3 —

The *Didunculus stigirostris*, a hook-billed ground-pigeon, found only in the Samoa Islands, is so peculiar in its structure that it is considered to form a distinct family.

FAMILY 85.—DIDIDÆ.—(2 Genera, 3 Species.)

GENERAL DISTRIBUTION.					
NEOTROPICAL SUB-REGIONS.	NEARCTIC SUB-REGIONS.	PALÆARCTIC SUB-REGIONS.	ETHIOPIAN SUB-REGIONS.	ORIENTAL SUB-REGIONS.	AUSTRALIAN SUB-REGIONS.
— — — —	— — — —	— — — —	— — — 4	— — — —	— — — —

The birds which constitute this family are now all extinct; but as numerous drawings are in existence, taken from living birds some of which were exhibited in Europe, and a stuffed specimen, fragments of which still remain, was in the Ashmolean Museum at Oxford down to 1755, they must be classed among recent, as opposed to geologically extinct species. The Dodo (*Didus ineptus*) a large, unwieldy, flightless bird, inhabited Mauritius down to the latter part of the 17th century; and an allied form, the Solitaire (*Pezophaps solitaria*), was found only in the island of Rodriguez, where it survived about a century later. Old voyagers mention a Dodo also in Bourbon, and a rude figure of it exists; but no remains of this bird have been found. Almost complete skeletons of the Dodo and Solitaire have, however, been recovered from the swamps of Mauritius and the caves of Rodriguez, proving that they were both extremely modified forms of pigeon. These large birds were formerly very abundant, and being excellent eating and readily captured, the early voyagers to these islands used them largely for food. As they could be caught by man, and very easily by dogs, they were soon greatly diminished in numbers; and the introduction of swine, which ran wild in the forests and fed on the eggs and young birds, completed their extermination.

The existence in the Mascarene Islands of a group of such remarkable terrestrial birds, with aborted wings, is parallel to that of the *Apteryx* and *Dinornis* in New Zealand, the Cassowaries of Austro-Malaya, and the short-winged Rails of New

Zealand, Tristan d'Acunha, and other oceanic islands; and the phenomenon is clearly dependent on the long-continued absence of enemies, which allowed of great increase of bulk and the total loss of the power of flight, without injury. In some few cases (the Ostrich for example) birds incapable of flight co-exist with large carnivorous mammalia; but these birds are large and powerful, as well as very swift, and are thus able to escape from some enemies and defend themselves against others. The entire absence of the smaller and more defenceless ground-birds from the adjacent island of Madagascar, is quite in accordance with this view, because that island has several small but destructive carnivorous animals.

General Remarks on the Distribution of the Columbæ.

The striking preponderance of Pigeons, both as to genera and species, in the Australian region, would seem to indicate that at some former period it possessed a more extensive land area in which this form of bird-life took its rise. But there are other considerations which throw doubt upon this view. The western half of the Malay Archipelago, belonging to the Oriental region, is also rich in pigeons, since it has 43 species belonging to 11 genera, rather more than are found in all the rest of the Oriental region. Again, we find that the Mascarene Islands and the Antilles both possess more pigeons than we should expect, in proportion to those of the regions to which they belong, and to their total amount of bird-life. This looks as if islands were more favourable to pigeon-development than continents; and if we group together the Pacific and the Malayan Islands, the Mascarene group and the Antilles, we find that they contain together about 170 species of pigeons belonging to 24 out of the 47 genera here adopted; while all the great continents united only produce about the same number of species belonging (if we omit those peculiar to Australia) to only 20 genera. The great development of the group in the Australian region may, therefore, be due to its consisting mainly of islands, and not to the order having originated there, and thus having had a longer period in which to develop. I have elsewhere suggested (*Ibis* 1865, p. 366) a physical cause for this peculiarity of distribution. Pigeons build rude, open nests, and their young remain helpless for a considerable period. They are thus exposed to the attacks of such arboreal quadrupeds or other animals as feed on eggs or young birds. Monkeys are very destructive in this respect; and it is a noteworthy fact that over the whole Australian region, the Mascarene Islands and the Antilles, monkeys are unknown. In the Indo-Malay sub-region, where monkeys are generally plentiful, the greatest variety of pigeons occurs in the Philippines, where there is but a single species in one island; and in Java, where monkeys are far less numerous than in Sumatra or Borneo. If we add to this consideration the fact, that mammalia and rapacious birds are, as a rule, far less abundant in islands than on continents; and that the extreme development of pigeon-life is reached in the Papuan group of islands, in which mammalia (except a few marsupials, bats, and pigs) are wholly absent, we see further reason to adopt this view. It is also to be noted that in America, comparatively few pigeons are found in the rich forests (comparable to those of the Australian insular region in which they abound), but are mostly confined to the open campos, the high Andes, and the western coast districts, from which the monkey-tribe are wholly absent.

This view is further supported by the great development of colour that is found in the pigeons of these insular regions, culminating in the golden-yellow fruit-dove of the Fiji Islands, the metallic green Nicobar-pigeon of Malaya, and the black and crimson *Alectrœnas* of Mauritius. Here also, alone, we meet with crested pigeons, rendering the possessors more conspicuous; such as the *Lopholaimus* of Australia and the crowned *Goura* of New Guinea; and here too are more peculiar forms of terrestrial pigeons than elsewhere, though none have completely lost the power of flight but the now extinct Dididæ.

The curious liking of pigeons for an insular habitat is well shown in the genera *Ianthœnas* and *Calœnas*. The former, containing 11 species, ranges over a hundred degrees of longitude, and forty-five of latitude, extending into three regions, yet nowhere inhabits a continent or even a large island. It is found in the Andaman and Nicobar Islands ; in the Philippines, Gilolo, and the smaller Papuan Islands, and in Japan ; yet not in any of the large Malay Islands or in Australia. The other genus, *Calœnas*, consists of but a single species, yet this ranges from the Nicobar Islands to New Guinea. It is not, however, as far as known, found on any of the large islands, but seems to prefer the smaller islands which surround them. We here have the general preference of pigeons for islands, further developed in these two genera into a preference for small islands ; and it is probable that the same cause—the greater freedom from danger— has produced both phenomena.

Of the geological antiquity of the Columbæ we have no evidence ; but their wide distribution, their varied forms, and their great isolation, all point to an origin, at least as far back as that we have assigned as probable in the case of the Parrots.

Order V.—*GALLINÆ*.

FAMILY 86.—PTEROCLIDÆ.   (2 Genera, 16 Species.)

GENERAL DISTRIBUTION.					
NEOTROPICAL SUB-REGIONS.	NEARCTIC SUB-REGIONS.	PALÆARCTIC SUB-REGIONS.	ETHIOPIAN SUB-REGIONS.	ORIENTAL SUB-REGIONS.	AUSTRALIAN SUB-REGIONS.
— — — —	— — — —	— 2 . 3 . 4	1 — 3 . 4	1 — — —	— — — —

The Pteroclidæ, or Sand-grouse, are elegantly formed birds with pointed tails, and plumage of beautifully varied protective tints, characteristic of the Ethiopian region and Central Asia, though extending into Southern Europe and Hindostan. Being preeminently desert-birds, they avoid the forest-districts of all these countries, but abound in the most arid situations and on the most open and barren plains. The distribution of the genera is as follows :—

Pterocles (14 sp.), has the same range as the family ; *Syrrhaptes* (2 sp.), normally inhabits Tartary, Thibet, and Mongolia to the country around Pekin, and occasionally visits Eastern Europe. But a few years back (1863) great numbers suddenly appeared in

Europe and extended westward to the shores of the Atlantic, while some even reached Ireland and the Færoes. (Plate III. Vol. I. p. 226.)

FAMILY 87.—TETRAONIDÆ.   (29 Genera, 170 Species.)

GENERAL DISTRIBUTION.					
NEOTROPICAL SUB-REGIONS.	NEARCTIC SUB-REGIONS.	PALÆARCTIC SUB-REGIONS.	ETHIOPIAN SUB-REGIONS.	ORIENTAL SUB-REGIONS.	AUSTRALIAN SUB-REGIONS.
— 2.3.4	1.2.3.4	1.2.3.4	1.2.3.4	1.2.3.4.	1.2 — 4

The Tetraonidæ, including the Grouse, Partridges, Quails, and allied forms, abound in all parts of the Eastern continents; they are less plentiful in North America and comparatively scarce in South America, more than half the Neotropical species being found north of Panama; and in the Australian region there are only a few of small size. The Ethiopian region probably contains most species; next comes the Oriental—India proper from the Himalayas to Ceylon having twenty; while the Australian region, with 15 species, is the poorest. These facts render it probable that the Tetraonidæ are essentially denizens of the great northern continents, and that their entrance into South America, Australia, and even South Africa, is, comparatively speaking, recent. They have developed into forms equally suited to the tropical plains and the arctic regions, some of them being among the few denizens of the extreme north, as well as of the highest alpine snows. The genera are somewhat unsettled, and there is even some uncertainty as to the limits between this family and the next; but the following are those now generally admitted:—

Ptilopachus (1 sp.), West Africa; *Francolinus* (34 sp.), all Africa, South Europe, India to Ceylon, and South China; *Ortygornis* (3 sp.), Himalayas to Ceylon, Sumatra, and Borneo: *Peliperdix* (1 sp.), West Africa; *Perdix* (3 sp.), the whole Continental Palæarctic region; *Margaroperdix* (1 sp.), Madagascar; *Oreoperdix* (1 sp.), Formosa; *Arborophila* (8 sp.), the Oriental Continent and the Philippines; *Peloperdix* (4 sp.), Tenasserim and Malaya; *Coturnix* (21 sp.), Temperate Palæarctic, Ethiopian and

Oriental regions, and the Australian to New Zealand; *Rollulus* (2 sp.), Siam to Sumatra, Borneo, and Philippines; *Caloperdix* (1 sp.), Malacca and Sumatra; *Odontophorus* (17 sp.), Brazil and Peru to Mexico; *Dendrortyx* (3 sp.), Guatemala and Mexico; *Cyrtonyx* (3 sp.), Guatemala to New Mexico; *Ortyx* (8 sp.), Honduras and Cuba to Canada; *Eupsychortyx* (6 sp.), Brazil and Ecuador to Mexico; *Callipepla* (3 sp.), Mexico to California; *Lophortyx* (2 sp.), Arizona and California; *Oreortyx* (1 sp.), California and Oregon (Plate XVIII., Vol. II. p. 128); *Lerwa* (1 sp.), Snowy Himalayas and East Thibet; *Caccabis* (10 sp.), Palæarctic region to Abyssinia, Arabia and the Punjaub; *Tetraogallus* (4 sp.), Caucasus and Himalayas to Altai Mountains; *Tetrao* (7 sp.), northern parts of Palæarctic and Nearctic regions; *Centrocercus* (1 sp.), Rocky Mountains; *Pediocœtes* (2 sp.), North and North-west America (Plate XVIII. Vol. II. p. 128); *Cupidonia* (1 sp.), East and North-Central United States and Canada; *Bonasa* (3 sp.), north of Nearctic and Palæarctic regions; *Lagopus* (6 sp.), Arctic Zone and northern parts of Nearctic and Palæarctic regions.

FAMILY 88.—PHASIANIDÆ.   (18 Genera, 75 Species.)

GENERAL DISTRIBUTION.					
NEOTROPICAL SUB-REGIONS.	NEARCTIC SUB-REGIONS.	PALÆARCTIC SUB-REGIONS.	ETHIOPIAN SUB-REGIONS.	ORIENTAL SUB-REGIONS.	AUSTRALIAN SUB-REGIONS.
— — 3 —	— 2 . 3 —	— 2 . 3 . 4	1 . 2 . 3 . 4	1 . 2 . 3 . 4	1 — — —

The Phasianidæ, including the Pea-fowl, Pheasants, and Jungle-fowl, the Turkeys, and the Guinea-fowl, are very widely distributed, but are far more abundant than elsewhere in the Eastern parts of Asia, both tropical and temperate. Leaving out the African guinea-fowls and the American turkeys, we have 13 genera and 63 species belonging to the Oriental and Palæarctic regions. These are grouped by Mr. Elliot (whose arrangement we mainly follow) in 5 sub-families, of which 3—Pavonniæ, Euplocaminæ, and Gallinæ—are chiefly Oriental, while the Lophophorniæ and Phasianinæ are mostly Palæarctic or from the highlands on the borders of the two regions. The genera adopted by Mr. Elliot in his *Monograph* are the following :—

PAVONINÆ, 4 genera.—*Pavo* (2 sp.), Himalayas to Ceylon, Siam, to South-west China and Java; *Argusianus* (4 sp.), Siam, Malay Peninsula, and Borneo (Plate IX. Vol. I. p. 339); *Polyplectron* (5 sp.), Upper Assam to South-west China and Sumatra; *Crossoptilon* (4 sp.), Thibet and North China. (Plate III. Vol. I. p. 226.)

LOPHOPHORINÆ, 4 genera.—*Lophophorus* (3 sp.), High woody region of Himalayas from Cashmere to West China; *Tetraophasis* (1 sp.), East Thibet; *Ceriornis* (5 sp.), Highest woody Himalayas from Cashmere to Bhotan and Western China (Plate VII. Vol. I. p. 331); *Pucrasia* (3 sp.), Lower and High woody Himalayas from the Hindoo Koosh to North-west China.

PHASIANINÆ, 2 genera.—*Phasianus* (12 sp.), Western Asia to Japan and Formosa, south to near Canton and Yunan, and the Western Himalayas, north to the Altai Mountains; *Thaumalea* (3 sp.), North-western China and Mongolia. (Plate III. Vol. I. p. 226.)

EUPLOCAMINÆ, 2 genera.—*Euplocamus* (12 sp.), Cashmere, along Southern Himalayas to Siam, South China and Formosa, and to Sumatra and Borneo; *Ithaginis* (2 sp.), High Himalayas from Nepal to North-west China.

GALLINÆ, 1 genus.—*Gallus* (4 sp.), Cashmere to Hainan, Ceylon, Borneo, Java, and eastwards to Celebes and Timor. (Central India, Ceylon, and East Java, have each a distinct species of Jungle-fowl.)

MELEAGRINÆ, 1 genus.—*Meleagris* (3 sp.), Eastern and Central United States and south to Mexico, Guatemala and Yucatan.

AGELASTINÆ, 2 genera. — *Phasidus* (1 sp.), West Africa; *Agelastes* (1 sp.), West Africa.

NUMIDINÆ, 2 genera.—*Acryllium* (1 sp.), West Africa; *Numida* (9 sp.), Ethiopian region, east to Madagascar, south to Natal and Great Fish River.

FAMILY 89.—TURNICIDÆ. (2 Genera, 24 Species.)

GENERAL DISTRIBUTION.					
NEOTROPICAL SUB-REGIONS.	NEARCTIC SUB-REGIONS.	PALÆARCTIC SUB-REGIONS.	ETHIOPIAN SUB-REGIONS.	ORIENTAL SUB-REGIONS.	AUSTRALIAN SUB-REGIONS.
— — — —	— — — —	— 2 — 4	1.2.3.4	1.2.3.4	1.2 — —

The Turnicidæ are small Quail-like birds, supposed to have
remote affinities with the American Tinamous, and with sufficient distinctive peculiarities to constitute a separate family.
They range over the Old World, from Spain all through Africa
and Madagascar, and over the whole Oriental region to Formosa,
and then north again to Pekin, as well as south-eastward to Australia and Tasmania. The genus *Turnix* (23 sp.), has the range
of the family; *Ortyxelos* (1 sp.), inhabits Senegal; but the
latter genus may not belong to this family.

FAMILY 90.—MEGAPODIIDÆ. (4 Genera, 20 Species.)

GENERAL DISTRIBUTION.					
NEOTROPICAL SUB-REGIONS.	NEARCTIC SUB-REGIONS.	PALÆARCTIC SUB-REGIONS.	ETHIOPIAN SUB-REGIONS.	ORIENTAL SUB-RIGIONS.	AUSTRALIAN SUB-REGIONS.
— — — —	— — — —	— — — —	— — — —	1.2.3 —	— — — 4

The Megapodiidæ, or Mound-makers and Brush-turkeys, are
generally dull-coloured birds of remarkable habits and economy,
which have no near allies, but are supposed to have a remote
affinity with the South American Curassows. They are highly
characteristic of the Australian region, extending into almost
every part of it except New Zealand and the remotest Pacific
islands, and only sending two species beyond its limits,—a
Megapodius in the Philippine Islands and North-west Borneo,
and another in the Nicobar Islands, separated by about 1,800
miles from its nearest ally in Lombok. The Philippine species
offers little difficulty, for these birds are found on the smallest

islands and sand-banks, and can evidently pass over a few miles
of sea with ease; but the Nicobar bird is a very different case,
because none of the numerous intervening islands offer a single
example of the family.   Instead of being a well-marked and
clearly differentiated form, as we should expect to find it if its
remote and isolated habitat were due to natural causes, it so
nearly resembles some of the closely-allied species of the Moluccas and New Guinea, that, had it been found with them, it would
hardly have been thought specifically extinct.   I therefore
believe that it is probably an introduction by the Malays, and
that, owing to the absence of enemies and general suitability of
conditions, it has thriven in the islands and has become slightly
differentiated in colour from the parent stock.   The following is
the distribution of the genera at present known :—

 Talegallus (2 sp.), New Guinea and East Australia ; *Megace-*
phalon (1 sp.), East Celebes ; *Lipoa* (1 sp.), South Australia ;
Megapodius (16 sp.), Philippine Islands and Celebes, to Timor,
North Australia, New Caledonia, the Marian and Samoa Islands,
and probably every intervening island,—also a species (doubtfully
indigenous) in the Nicobar Islands.

FAMILY 91.—CRACIDÆ.   (12 Genera, 53 Species.)

GENERAL DISTRIBUTION.					
NEOTROPICAL SUB-REGIONS.	NEARCTIC SUB-REGIONS.	PALÆARCTIC SUB-REGIONS.	ETHIOPIAN SUB-REGIONS.	ORIENTAL SUB-REGIONS.	AUSTRALIAN SUB-REGIONS.
— 2 . 3 —	— 2 — —	— — — —	— — — —	— — — —	— — — —

 (Messrs. Sclater and Salvin's arrangement is here followed).
The Cracidæ, or Curassows and Guans, comprise the largest
and handsomest game-birds of the Neotropical region, where
they take the place of the grouse and pheasants of the Old
World.   They are almost all forest-dwellers, and are a strictly
Neotropical family, only one species just entering the Nearctic
region as far as New Mexico.   They extend southward to Paraguay and the extreme south of Brazil, but none are found in the

Antilles, nor west of the Andes south of the bay of Guayaquil. The sub-families and genera are as follows :—

CRACINÆ, 4 genera.—*Crax* (8 sp.), Mexico to Paraguay (Plate XV., Vol. II. p. 28); *Nothocrax* (1 sp.), Guiana, Upper Rio Negro, and Upper Amazon; *Pauxi* (1 sp.), Guiana to Venezuela; *Mitua* (2 sp.), Guiana and Upper Amazon.

PENELOPINÆ, 7 genera.—*Stegnolœma* (1 sp.), Columbia and Ecuador; *Penelope* (14 sp.), Mexico to Paraguay and to western slope of Ecuadorian Andes; *Penelopina* (1 sp.), Guatemala ; *Pipile* (3 sp.), Venezuela to Eastern Brazil ; *Aburria* (1 sp), Columbia ; *Chamœpetes* (2 sp.), Costa Rica to Peru ; *Ortalida* (18 sp.), New Mexico to Paraguay, also Tobago.

OREOPHASINÆ, 1 genus.—*Oreophasis* (1 sp.), Guatemala.

It thus appears that the Cracinæ are confined to South America east of the Andes, except one species in Central America; whereas nine Penelopinæ and *Oreophasis* are found north of Panama. The species of the larger genera are strictly representative, each having its own distinct geographical area, so that two species of the same genus are rarely or never found in the same locality.

FAMILY 92.—TINAMIDÆ.   (9 Genera, 39 Species.)

GENERAL DISTRIBUTION.					
NEOTROPICAL SUB-REGIONS.	NEARCTIC SUB-REGIONS.	PALÆARCTIC SUB-REGIONS.	ETHIOPIAN SUB-REGIONS.	ORIENTAL SUB-REGIONS.	AUSTRALIAN SUB-REGIONS.
1.2.3 –	— — — —	— — — —	— — — —	— — — —	— — — —

The Tinamous are a very remarkable family of birds, with the general appearance of partridges or hemipodes, but with the tail either very small or entirely wanting. They differ greatly in their organization from any of the Old World Gallinæ, and approach, in some respects, the Struthiones or Ostrich tribe. They are very terrestrial in their habits, inhabiting the forests, open plains, and mountains of the Neotropical region, from Patagonia and Chili to Mexico ; but, like the Cracidæ, they are absent from the Antilles. Their colouring is very sober and protective, as is the case with so many ground-birds, and they are seldom adorned with crests or other ornamental plumes, so prevalent in the order to which they belong. The sub-families and genera, according to the arrangement of Messrs. Sclater and Salvin, are as follows :—

TINAMINÆ, 7 genera.—*Tinamus* (7 sp.), Mexico to Paraguay; *Nothocercus* (3 sp.), Costa Rica to Venezuela and Ecuador; *Crypturus* (16 sp.), Mexico to Paraguay and Bolivia ; *Rhynchotus* (2 sp.), Bolivia and South Brazil to La Plata; *Nothoprocta* (4 sp.), Ecuador to Bolivia and Chili ; *Nothura* (4 sp.), Brazil and Bolivia to Patagonia ; *Taoniscus* (1 sp.), Brazil to Paraguay.

TINAMOTINÆ, 2 genera.—*Calodromas* (1 sp.), La Plata and Patagonia; *Tinamotis* (1 sp.), Andes of Peru and Bolivia.

General Remarks on the Distribution of Gallinæ.

There are about 400 known species of Gallinaceous birds grouped into 76 genera, of which no less than 65 are each restricted to a single region. The Tetraonidæ are the only cosmopolitan family, and even these do not extend into Temperate South America, and are very poorly represented in Australia. The Cracidæ and Tinamidæ are strictly Neotropical, the Megapodiidæ almost as strictly Australian. There remains the extensive family of the Phasianidæ, which offers some interesting facts. We have first the well-marked sub-families of the Numidinæ and-Meleagrinæ, confined to the Ethiopian and Nearctic regions respectively, and we find the remaining five sub-families, comprising about 60 species, many of them the most magnificent of known birds, spread over the Oriental and the south-eastern portion of the Palæarctic regions. This restriction is remarkable, since there is no apparent cause in climate or vegetation why pheasants should not be found wild throughout southern Europe, as they were during late Tertiary and Post-Tertiary times. We have also to notice the remarkable absence of the Pheasant tribe from Hindostan and Ceylon, where the peacock and jungle-fowl are their sole representatives. These two forms also alone extend to Java, whereas in the adjacent islands of Borneo and Sumatra we have *Argusianus, Polyplectron,* and *Euplocamus.* The common jungle-fowl (the origin of our domestic poultry) is the only species which enters the Australian region as far as Celebes and Timor, and another species (*Gallus æneus*) as far as Flores, and it is not improbable that these may have been introduced by man and become wild.

We have very little knowledge of the extinct forms of Gallinæ, but what we have assures us of their high antiquity, since we find such distinct groups as the jungle-fowl, partridges, and *Pterocles*, represented in Europe in the Miocene period ; while the Turkey, then as now, appears to have been a special American type.

Order VI.—OPISTHOCOMI.

FAMILY 93.—OPISTHOCOMIDÆ.　(1 Genus, 1 Species.)

GENERAL DISTRIBUTION.					
NEOTROPICAL SUB-REGIONS.	NEARCTIC SUB-REGIONS.	PALÆARCTIC SUB-REGIONS.	ETHIOPIAN SUB-REGIONS.	ORIENTAL SUB-REGIONS.	AUSTRALIAN SUB-REGIONS.
— 2 — —	— — — —	— — — —	— — — —	— — — —	— — — —

The Hoazin (*Opisthocomus cristatus*) is the sole representative of this family and of the order Opisthocomi. It inhabits the eastern side of Equatorial America in Guiana and the Lower Amazon ; and at Pará is called " Cigana " or gipsy. It is a large, brown, long-legged, weakly-formed and loosely-crested bird, having such anomalies of structure that it is impossible to class it along with any other family. It is one of those survivors, which tell us of extinct groups, of whose past existence we should otherwise, perhaps, remain for ever ignorant.

Order VII.—ACCIPITRES.

FAMILY 94.—VULTURIDÆ.　(10 Genera, 25 Species.)

GENERAL DISTRIBUTION.					
NEOTROPICAL SUB-REGIONS.	NEARCTIC SUB-REGIONS.	PALÆARCTIC SUB-REGIONS.	ETHIOPIAN SUB-REGIONS.	ORIENTAL SUB-REGIONS.	AUSTRALIAN SUB-REGIONS.
1.2.3.4	1.2.3.4	1.2.3.4	1.2.3 —	1.2.3 —	— — — —

Vultures range over all the great continents south of the Arctic Circle, being only absent from the Australian region, the Malay Islands, Ceylon, and Madagascar. The Old and New World forms are very distinct, belonging to two well-marked divisions, often ranked as families. The distribution of the genera is as follows :—

Sub-family I. VULTURINÆ (6 genera, 16 species), confined to the Old World.—*Vultur* (1 sp.), Spain and North Africa through Nepal to China north of Ningpo ; *Gyps* (5 sp.), Europe south of 59°, Africa, except the western sub-region, India, Siam, and Northern China ; *Pseudogyps* (2 sp.), North-east Africa and Senegal, India and Burmah ; *Otogyps* (2 sp.), South Europe, North-east and South Africa, India, and Siam ; *Lophogyps* (1 sp.), North-east and South Africa and Senegal ; *Neophron* (4 sp.), South Europe, India and the greater part of Africa.

Sub-family II. SARCORHAMPHINÆ (4 genera, 9 species), confined to the New World.—*Sarcorhamphus* (2 sp.), " The Condor," Andes of South America, and southern extremity below 41° south latitude ; *Cathartes* (1 sp.), America from 20° south latitude to Trinidad and Mexico ; *Catharistes* (1 sp.), America from 40° north to 40° south latitude, but not on Pacific coast of United States ; *Pseudogryphis* (5 sp.), South America and Falkland Islands, and to 49° north latitude in North America, also Cuba and Jamaica.

FAMILY 95.—SERPENTARIIDÆ.   (1 Genus, 1 Species.)

GENERAL DISTRIBUTION.					
NEOTROPICAL SUB-REGIONS.	NEARCTIC SUB-REGIONS.	PALÆARCTIC SUB-REGIONS.	ETHIOPIAN SUB-REGIONS.	ORIENTAL SUB-REGIONS.	AUSTRALIAN SUB-REGIONS.
— — — —	— — — —	— — — —	1 . 2 . 3 —	— — — —	— — — —

The singular Secretary Bird (*Serpentarius*) is found over a large part of Africa. Its position is uncertain, as it has affinities both with the Accipitres, through *Polyboroides* (?) and with *Cariama*, which we place near the Bustards. (Plate IV. Vol. I. p. 261.)

FAMILY 96.—FALCONIDÆ. (69 Genera, 325 Species.)

GENERAL DISTRIBUTION.

NEOTROPICAL SUB-REGIONS.	NEARCTIC SUB-REGIONS.	PALÆARCTIC SUB-REGIONS.	ETHIOPIAN SUB-REGIONS.	ORIENTAL SUB-REGIONS.	AUSTRALIAN SUB-REGIONS.
1.2.3.4	1.2.3.4	1.2.3.4	1.2.3.4	1.2.3.4	1.2.3.4

The Falconidæ, including the various groups of Hawks, Kites, Buzzards, Eagles, and Falcons, are absolutely cosmopolitan, ranging far into the arctic zone and visiting the most remote oceanic islands. They are abundant in all the great continents and larger islands, preferring open to woody regions. They are divided into several sub-families, the range of some of which are restricted. For this family as well as the preceding I follow the arrangement of Mr. Sharpe's *British Museum Catalogue*, and shall give the approximate distribution of each sub-family, as well as of the several genera.

Sub-family I. POLYBORINÆ (2 genera, 10 species), the Neotropical region with California and Florida, Tropical and South Africa.—*Polyborus* (2 sp.), South America, and to California and Florida; *Ibycter* (8 sp.), Tierra del Fuego to Honduras and Guatemala.

Cariama and *Serpentarius*, which Mr. Sharpe puts here, are so anomalous that I think it better to class them in separate families—Serpentariidæ among the Accipitres, and Cariamidæ near the Bustards.

Sub-family II. ACCIPITRINÆ (10 genera, 87 species).—Cosmopolitan.—*Polyboroides* (2 sp.), Africa and Madagascar ; *Circus* (15 sp.), Old and New Worlds, widely scattered, but absent from Eastern Equatorial America, and the Malay Archipelago except Celebes ; *Micrastur* (7 sp.), and *Geranospiza* (2 sp.), Tropical parts of Neotropical region ; *Urotriorchis* (1 sp.), West Africa ; *Erythrocnema* (1 sp.), Chili and La Plata to California and Texas ; *Melierax* (5 sp.), Africa except West African sub-region ; *Astur* (30 sp.), cosmopolitan, except the Temperate South American sub-region ;

Nisoides (1 sp.), Madagascar; *Eutriorchis* (1 sp.), Madagascar; *Accipiter* (23 sp.), cosmopolitan, except Eastern Oceania.

Sub-family III. BUTEONINÆ (13 genera, 51 sp.), cosmopolitan, except the Malay and Pacific Islands.—*Urospizias* (1 sp.), East and Central Australia; *Heterospizias* (1 sp.), Tropical South America east of the Andes; *Tachytriorchis* (2 sp,), Paraguay to California; *Buteo* (18 sp.), cosmopolitan, except the Australian region and the Indo-Malayan sub-region; *Archibuteo* (4 sp.), North America to Mexico and the cooler parts of the Palæarctic region; *Buteola* (1 sp.), Veragua to the Amazon Valley; *Asturina* (7 sp.), Paraguay and Bolivia to South-east United States; *Busarellus* (1 sp.), Brazil to Guiana; *Buteogallus* (1 sp.), Guiana and Columbia; *Urubutinga* (12 sp.), South Brazil and Bolivia to Mexico; *Harpyhaliœetus* (1 sp.), Chili and North Patagonia to Veragua; *Morphnus* (1 sp.), Amazonia to Panama; *Thrasaëtus* (1 sp.), Paraguay and Bolivia to Mexico.

Sub-family IV. AQUILINÆ (31 genera, 94 species), cosmopolitan.—*Gypaëtus* (2 sp.), south of Palæarctic region from Spain to North China, Abyssinia, and South Africa; *Uroaëtus* (1 sp.), Australia and Tasmania; *Aquila* (9 sp.), Nearctic, Palæarctic, and Ethiopian regions and India; *Nisaëtus* (4 sp.), Africa and South Europe, India, Ceylon, and Australia; *Lophotriorchis* (2 sp.), Indo-Malay sub-region, and Bogotá in South America; *Neopus* (1 sp.), India and Ceylon to Burmah, Java, Celebes and Ternate; *Spiziastur* (1 sp.), Guatemala to Brazil; *Spizaëtus* (10 sp.), Central and South America, Africa, India, and Ceylon, to Celebes and New Guinea, Formosa, and Japan; *Lophoaëtus* (1 sp.), all Africa; *Asturinula* (1 sp.), Africa, except extreme south; *Herpetotheres* (1 sp.), Bolivia and Paraguay to Southern Mexico; *Dryotriorchis* (1 sp.), West Africa; *Circaëtus* (5 sp.) Africa to Central Europe, the Indian Peninsula, Timor; *Spilornis* (6 sp.), Oriental region and Celebes; *Butastur* (4 sp.), Oriental region to New Guinea and North-east Africa; *Helotarsus* (2 sp.), Africa south of the Sahara; *Haliœetus* (7 sp.), cosmopolitan, except the Neotropical region; *Gypohierax* (1 sp.), West Africa and Zanzibar; *Haliastur* (2 sp.), Indian Peninsula to Ceylon, New Cale-

donia, and Australia; *Nauclerus* (= *Elanoides*) (1 sp.), Brazil to
Southern United States; *Elanoides* (= *Nauclerus*) (1 sp.), Western and North-eastern Africa; *Milvus* (6 sp.), the Old World
and Australia; *Lophoictinia* (1 sp.), Australia; *Rostrhamus* (3
sp.), Antilles and Florida to Brazil and Peru; *Leptodon* (4
sp.), Central America to South Brazil and Bolivia; *Gypoictinia*
(1 sp.), South and West Australia; *Elanus* (5 sp.), Africa, India,
and Malay Archipelago to Australia, South America to California;
Gampsonyx (1 sp.), Trinidad to Brazil; *Henicopernis* (1 sp.),
Papuan Islands; *Machœrhamphus* (2 sp.), South-west Africa,
Madagascar, and Malacca; *Pernis* (3 sp.), Palæarctic, Oriental,
and Ethiopian regions.

Sub-family V. FALCONINÆ (11 genera, 80 species), cosmopolitan.
—*Baza* (10 sp.), India and Ceylon to the Moluccas and North
Australia, West Coast of Africa, Natal, and Madagascar; *Har-
pagus* (3 sp.), Central America to Brazil and Peru; *Ictinia* (2 sp.),
Brazil to Southern United States; *Hierax* (=*Microhierax*, Sharpe),
(4 sp.), Eastern Himalayas to Borneo and Philippines; *Polio-
hierax* (2 sp.), East Africa and Burmah; *Spiziapteryx* (1 sp.), La
Plata; *Harpa* (1 sp.), New Zealand and the Auckland Islands;
Falco (27 sp.), cosmopolitan, except the Pacific Islands; *Hierofalco*
(6 sp.), Nearctic and Palæarctic regions; *Hieracidea* (2 sp.),
Australia; *Cerchneis* (22 sp.), cosmopolitan, except Oceania.

FAMILY 97.—PANDIONIDÆ.   (2 Genera, 3 Species.)

GENERAL DISTRIBUTION.					
NEOTROPICAL SUB-REGIONS.	NEARCTIC SUB-REGIONS.	PALÆARCTIC SUB-REGIONS.	ETHIOPIAN SUB-REGIONS.	ORIENTAL SUB-REGIONS.	AUSTRALIAN SUB-REGIONS.
— 2.3.4	1.2.3.4	1.2.3.4	1.2.3.4	1.2.3.4	1.2.3.4

The Pandionidæ, or Fishing Hawks, are universally distributed,
with the exception of the Southern Temperate parts of South
America. The genera are:—

Pandion (1 sp.), the range of the entire family; *Polioaëtus* (2
sp.), India through Malay Archipelago to Celebes and Sandwich
Islands.

FAMILY 98.—STRIGIDÆ.   (23 Genera, 180 Species.)

GENERAL DISTRIBUTION.

NEOTROPICAL SUB-REGIONS.	NEARCTIC SUB-REGIONS.	PALÆARCTIC SUB-REGIONS.	ETHIOPIAN SUB-REGIONS.	ORIENTAL SUB-REGIONS.	AUSTRALIAN SUB-REGIONS.
1.2.3.4	1.2.3.4	1.2.3.4	1.2.3.4	1.2.3.4	1.2.3.4

The Strigidæ, or Owls, form an extensive and well-known family of nocturnal birds, which, although invariably placed next the Hawks, are now believed to be not very closely allied to the other Accipitres. They range over the whole globe, extending to the extreme polar regions and to the remotest oceanic islands. Their classification is very unsettled, and we therefore place the genera, for convenience, in the order in which they follow each other in the *Hand List of Birds.* Those adopted by most ornithologists are the following :—

Surnia (1 sp.), the Arctic regions of both hemispheres; *Nyctea* (1 sp.), South Carolina to Greenland and Northern Europe; *Athene* (40 sp.), the Eastern hemisphere to New Zealand and the Solomon Islands; *Ninox* (7 sp.), the Oriental region, North China and Japan; *Glaucidium* (7 sp.), Neotropical region, California, and Oregon, Europe to North China; *Micrathene* (1 sp.), Mexico and Arizona; *Pholeoptynx* (2 sp.), Neotropical region, Texas, and North-west America; *Bubo* (16 sp.), universally distributed, excluding the Australian region; *Ketupa* (3 sp.), the Oriental region, Palestine; *Scotopelia* (2 sp.), West and South Africa; *Scops* (30 sp.), universally distributed, excluding Australia and Pacific Islands; *Gymnoglaux* (2 sp.), Antilles; *Lophostrix* (2 sp.), Lower Amazon to Guatemala; *Syrnium* (22 sp.), all regions but the Australian; *Ciccaba* (10 sp.), Paraguay to Mexico; *Nyctalatinus* (1 sp.), Columbia; *Pulsatrix* (2 sp.), Brazil and Peru to Guatemala; *Asio* (6 sp.), all regions but the Australian, Sandwich Islands; *Nyctalops* (1 sp.), Cuba and Mexico to Brazil and Monte Video; *Pseudoscops* (1 sp.), Jamaica; *Nyctala* (4 sp.), the North Temperate zone; *Strix* (18 sp), universally distributed; *Phodilus* (1 sp.), Himalayas and Malaya.

In Mr. Sharpe's Catalogue (published while this work was passing through the press) the genera of Owls are reduced to 19, arranged in two families—Strigidæ, containing our last two genera, and Bubonidæ, comprising the remainder. The species are increased to 190; but some genera are reduced, as *Strix*, which is said to contain only 5 species.

General Remarks on the Distribution of the Accipitres.

The Birds of Prey are so widely distributed over the world's surface that their general distribution calls for few remarks. Of the four families all but one are cosmopolites, Vultures alone being absent from the Australian region, as well as from Indo-Malaya and Madagascar. If we take the sub-families, we find that each region has several which are confined to it. The only parts of the world where there is a marked deficiency of Accipitres is in the islands of the Pacific ; and it may be noted, as a rule, that these birds are more abundant in continents than in islands. There is not so much difference between the number of Birds of Prey in tropical and temperate regions, as is found in most other groups of land-birds. North America and Europe have about 60 species each, while India has about 80, and South America about 120. The total number of Accipitres is 550 comprised in 104 genera, and 4 (or perhaps more properly 5) families. In this estimate I have not included the Serpentariidæ, containing the Secretary Bird of Africa, as there is some doubt whether it really belongs to the Order.

Order VIII.—GRALLÆ.

FAMILY 99.—RALLIDÆ.   (18 Genera, 153 Species.)

GENERAL DISTRIBUTION.					
NEOTROPICAL SUB-REGIONS.	NEARCTIC SUB-REGIONS.	PALÆARCTIC SUB-REGIONS.	ETHIOPIAN SUB-REGIONS.	ORIENTAL SUB-REGIONS.	AUSTRALIAN SUB-REGIONS.
1.2.3.4	1.2.3.4	1.2.3.4	1.2.3.4	1.2.3.4	1.2.3.4

The Rails are among the most widely distributed families of birds, many of the genera being cosmopolitan, and several of the species ranging over half the globe. They are found in many remote islands ; and in some of these—as the *Gallinula* of Tristan d'Acunha, and the *Notornis* of Lord Howe's Island and New Zealand,—they have lost the power of flight. The classification of the Rallidæ is not satisfactory, and the following enumeration of the genera must only be taken as affording a provisional sketch of the distribution of the group :—

Rallus (18 sp.), *Porzana* (24 sp.), *Gallinula* (17 sp.), and *Fulica* (10 sp.), have a world-wide range ; *Ortygometra* (1 sp.), ranges over the whole North Temperate zone ; *Porphyrio* (14 sp.), is more especially Oriental and Australian, but occurs also in South America, in Africa, and in South Europe ; *Eulabeornis* (15 sp.), is Ethiopian, Malayan, and Australian ; *Himantornis* (1 sp.), is West African only ; *Aramides* (24 sp.), is North and South American ; *Rallina* (16 sp.), is Oriental, but ranges eastward to Papua ; *Habroptila* (1 sp.), is confined to the Moluccas ; *Pareudiastes* (1 sp.), the Samoa Islands ; *Tribonyx* (4 sp.), is Australian, and has recently been found also in New Zealand ; *Ocydromus* (4 sp.) ; *Notornis* (2 sp.), (Plate XIII. Vol. I. p. 455) ; and *Cabalus* (1 sp.), are peculiar to the New Zealand group.

The sub-family, Heliornithinæ (sometimes classed as a distinct family) consists of 2 genera, *Heliornis* (1 sp.), confined to the Neotropical region ; and *Podica* (4 sp.), the Ethiopian region excluding Madagascar, and with a species (perhaps forming another genus) in Borneo.

Extinct Rallidæ.—Remains of some species of this family have been found in the Mascarene Islands, and historical evidence shows that they have perhaps been extinct little more than a century. They belong to the genus *Fulica*, and to two extinct genera, *Aphanapteryx* and *Erythromachus*. The *Aphanapteryx* was a large bird of a reddish colour, with loose plumage, and perhaps allied to *Ocydromus*. *Erythromachus* was much smaller, of a grey-and-white colour, and is said to have lived chiefly on the eggs of the land-tortoises. (See *Ibis*, 1869, p. 256 ; and *Proc. Zool. Soc.*, 1875, p. 40.)

FAMILY 100.—SCOLOPACIDÆ.   (21 Genera, 121 Species.)

GENERAL DISTRIBUTION.					
NEOTROPICAL SUB-REGIONS.	NEARCTIC SUB-REGIONS.	PALÆARCTIC SUB-REGIONS.	ETHIOPIAN SUB-REGIONS.	ORIENTAL SUB-REGIONS.	AUSTRALIAN SUB-REGIONS.
1.2.4.4	1.2.3.4	1.2.3.4	1.2.3.4	1.2.3.4	1.2.3.4

The Scolopacidæ, comprehending the Snipes, Sandpipers, Curlews, and allied genera, are perhaps as truly cosmopolitan as any family of birds, ranging to the extreme north and visiting the remotest islands.   The genera of universal distribution are the following :—

Numenius (16 sp.) ; *Limosa* (6 sp.) ; *Totanus* (12 sp.) ; *Tringoides*, (6 sp.) ; *Himantopus* (6 sp.) ; *Tringa* (20 sp.) ; and *Gallinago* (24 sp.).   Those which have a more or less restricted distribution are :—

Ibidorhyncha (1 sp.), Central Asia and the Himalayas (Plate VII. Vol. I. p. 331) ; *Helodromas* (1 sp.), Palæarctic region and North India ; *Terekia* (1 sp.), East Palæarctic, wandering to India and Australia ; *Recurvirostra* (6 sp.), Nearctic region to the High Andes, South Palæarctic, East and South Africa, Hindostan and Australia ; *Micropelama* (1 sp.), North America to Chili ; *Machetes* (1 sp.), Palæarctic region and Hindostan (Plate I. Vol. I. p. 195) ; *Ereunetes* (3 sp.), Nearctic and Neotropical ; *Eurinorhynchus* (1 sp.), North-east Asia and Bengal ; *Calidris* (1 sp.), all regions but Australian ; *Macrorhamphus* (3 sp.), Palæarctic and Nearctic, visits Brazil and India ; *Scolopax* (4 sp.), the whole Palæarctic region, to India, Java, and Australia ; *Philohela* (1 sp.), East Nearctic ; *Rhynchœa* (4 sp.), Ethiopian and Oriental, Australia, and Temperate South America ; *Phalaropus* (3 sp.), North Temperate zone, and West Coast of America to Chili.

FAMILY 101 —CHIONIDIDÆ. (1 Genus, 2 Species.)

GENERAL DISTRIBUTION.

NEOTROPICAL SUB-REGIONS.	NEARCTIC SUB-REGIONS.	PALÆARCTIC SUB-REGIONS.	ETHIOPIAN SUB-REGIONS.	ORIENTAL SUB-REGIONS.	AUSTRALIAN SUB-REGIONS.
1 — — —	— — — —	- — — —	— — — —	— — — —	— — — —

The Sheath-bills, *Chionis* (2 sp.), are curious white birds, whose thick bill has a horny sheath at the base. Their nearest ally is *Hæmatopus*, a genus of Charadriidæ. These birds are confined to the Antarctic Islands, especially the Falkland Islands, the Crozets and Kerguelen's Land.

FAMILY 102.—THINOCORIDÆ. (2 Genera, 6 Species.)

GENERAL DISTRIBUTION.

NEOTROPICAL SUB-REGIONS.	NEARCTIC SUB-REGIONS.	PALÆARCTIC SUB-REGIONS.	ETHIOPIAN SUB-REGIONS.	ORIENTAL SUB-REGIONS.	AUSTRALIAN SUB-REGIONS.
1 — —	— — — —	— — — —	— — — —	— — — —	— — — —

The Thinocoridæ, or Quail-snipes, are small birds, confined to Temperate South America. They have much the appearance of Quails but are more nearly allied to Plovers. The two genera are :—

Attagis (4 sp.), Falkland Islands, Straits of Magellan, Chili, Bolivia, and the High Andes of Peru and Ecuador; *Thinocorus* (2 sp.), La Plata, Chili, and Peru. (Plate XVI. Vol. II. p. 40.)

FAMILY 103.—PARRIDÆ. (2 Genera, 11 Species.)

GENERAL DISTRIBUTION.

NEOTROPICAL SUB-REGIONS.	NEARCTIC SUB-REGIONS.	PALÆARCTIC SUB-REGIONS.	ETHIOPIAN SUB-REGIONS.	ORIENTAL SUB-REGIONS.	AUSTRALIAN SUB-REGIONS.
— 2.3 —	— — — —	— — — —	1.2.3.4	1.2.3.4	1.2 — —

The Parridæ, or Jacanas, are remarkable long-toed birds, often
of elegant plumage, frequenting swamps and marshes, and walk-
ing on the floating leaves of aquatic plants.   They are found in
all the tropics.   *Parra* (10 sp.), has the distribution of the family ;
Hydrophasianus (1 sp.), is confined to the Oriental region.

FAMILY 104.—GLAREOLIDÆ.   (3 Genera, 20 Species.)

GENERAL DISTRIBUTION.					
NEOTROPICAL SUB-REGIONS.	NEARCTIC SUB-REGIONS.	PALÆARCTIC SUB-REGIONS.	ETHIOPIAN SUB-REGIONS.	ORIENTAL SUB-REGIONS.	AUSTRALIAN SUB-REGIONS.
— — — —	— — — —	1.2.3.4	1.2.3.4	1.2.3.4	1.2 — —

This family, comprising the Pratincoles and Coursers, is
universally distributed over the Old World and to Australia.

Glareola (9 sp.), has the distribution of the family; *Pluxia-
nus* (1 sp.), is confined to North Africa; *Cursorius* (10 sp.),
ranges over Africa, South Europe and India.

The position of the genus *Glareola* is uncertain, for though
generally classed here, Prof. Lilljeborg considers it to be an
aberrant form of the Caprimulgidæ! It differs, in its insectivorous habits and in many points of external structure, from all
its allies, and should probably form a distinct family.

FAMILY 105.—CHARADIIDÆ.   (19 Genera, 101 Species.)

GENERAL DISTRIBUTION.					
NEOTROPICAL SUB-REGIONS.	NEARCTIC SUB-REGIONS.	PALÆARCTIC SUB-REGIONS.	ETHIOPIAN SUB-REGIONS.	ORIENTAL SUB-REGIONS.	AUSTRALIAN SUB-REGIONS.
1.2.3.4	1.2.3.4	1.2.3.4	1.2.3.4	1.2.3.3	1.2.3.4

The extensive family of the Plovers and their numerous allies,
ranges over the whole globe.   The genera now usually admitted
into this family are the following :—

Œdicnemus (9 sp.), is only absent from North America ;
Æsacus (2 sp.), India to Ceylon, Malay Islands and Australia ;

A A 2

Vanellus (3 sp.), Palæarctic and Neotropical regions; *Chœtusia*
(15 sp.), the whole Eastern Hemisphere; *Erythrogonys* (1 sp.),
Australia; *Hoplopterus* (10 sp.), widely scattered, but absent
from North America; *Squatarola* (1 sp), all the regions; *Cha-
radrius* (14 sp.), cosmopolitan; *Eudromias* (5 sp.), Eastern Hemisphere and South Temperate America; *Ægialitis* (22 sp.), cosmopolitan; *Oreophilus* (1 sp.), South Temperate America;
Thinornis (2 sp.), New Zealand; *Anarhynchus* (1 sp.), New
Zealand (Plate XIII. Vol I. p. 455); *Hæmatopus* (9 sp.), cosmopolitan; *Strepsilas* (2 sp.) almost cosmopolitan; *Aphriza* (1
sp.), West Coast of America; *Pluvianellus* (1 sp.), Straits of
Magellan; *Dromas* (1 sp.), India, Madagascar, and North-east
Africa; *Pedionomus* (1 sp.), Australia. This last genus has
usually been placed with the Turnicidæ.

FAMILY 106.—OTIDIDÆ.   (2 Genera, 26 Species.)

GENERAL DISTRIBUTION.					
NEOTROPICAL SUB-REGIONS.	NEARCTIC SUB-REGIONS.	PALÆARCTIC SUB-REGIONS.	ETHIOPIAN SUB-REGIONS.	ORIENTAL SUB-REGIONS.	AUSTRALIAN SUB-REGIONS.
— — — —	— — — —	1.2.3.4	1.2.3 —	1.2.3 —	— 2 — —

The Otididæ, or Bustards, occur in all parts of the Old World
and Australia where there are open tracts, being only absent from
Madagascar and the Malay Archipelago.

Otis (2 sp.), ranges over most of the Palæarctic region; while
Eupodotis (24 sp.), has the range of the family, but is most abundant in the Ethiopian region, which contains three-fourths of the
whole number of species.

FAMILY 107.—GRUIDÆ.   (3 Genera, 16 Species.)

GENERAL DISTRIBUTION.					
NEOTROPICAL SUB-REGIONS.	NEARCTIC SUB-REGIONS.	PALÆARCTIC SUB-REGIONS.	ETHIOPIAN SUB-REGIONS.	ORIENTAL SUB-REGIONS.	AUSTRALIAN SUB-REGIONS.
— — — —	1.2.3 —	1.2.3.4	1.2.3 —	1.2.3—	— 2 — —

The Gruidæ, or Cranes, are found in all the regions except the Neotropical.

Grus (12 sp.) inhabits the southern and western United States, the whole Palæarctic region, South-east Africa, India, and Australia; *Anthropoides* (2 sp.), Europe, North and South Africa and India; *Balearica* (2 sp.), the Ethiopian region (except Madagascar).

FAMILY 108.—CARIAMIDÆ.   (1 Genus, 2 Species.)

GENERAL DISTRIBUTION.

NEOTROPICAL SUB-REGIONS.	NEARCTIC SUB-REGIONS.	PALÆARCTIC SUB-REGIONS.	ETHIOPIAN SUB-REGIONS.	ORIENTAL SUB-REGIONS.	AUSTRALIAN SUB-REGIONS.
1 . 2 – –	– – – –	– – – –	– – – –	– – – –	– – – –

The genus *Cariama* (2 sp.), consists of remarkable crested birds inhabiting the mountains and open plains of Brazil and La Plata. In the British Museum Catalogue of the Birds of Prey, they are classed as aberrant Falconidæ, but their anomalous characters seem to require them to be placed in a distinct family, which seems better placed among the Waders.

FAMILY 109.—ARAMIDÆ.   (1 Genus, 2 Species.)

GENERAL DISTRIBUTION.

NEOTROPICAL SUB-REGIONS.	NEARCTIC SUB-REGIONS.	PALÆARCTIC SUB-REGIONS.	ETHIOPIAN SUB-REGIONS.	ORIENTAL SUB-REGIONS.	AUSTRALIAN SUB-REGIONS.
– 2 . 3 . 4	– – – –	– – – –	– – – –	– – – –	– – – –

The Guaraünas are birds which have somewhat the appearance of Herons, but which are usually classed with the Rails. They are now, however, considered to form a distinct family. The only genus, *Aramus* (2 sp.), inhabits the Neotropical region, from Mexico and Cuba to Central Brazil.

FAMILY 110.—PSOPHIIDÆ.   (1 Genus, 6 Species.)

GENERAL DISTRIBUTION.

NEOTROPICAL SUB-REGIONS.	NEARCTIC SUB-REGIONS.	PALÆARCTIC SUB-REGIONS.	ETHIOPIAN SUB-REGIONS.	ORIENTAL SUB-REGIONS.	AUSTRALIAN SUB-REGIONS.
— 2 — —	... — — —	— — — —	— — — —	— — — —	— — — —

The remarkable and beautiful birds called Trumpeters, are confined to the various parts of the Amazon valley ; and it is an interesting fact, that the range of each species appears to be bounded by some of the great rivers. Thus, *Psophia crepitans* inhabits the interior of Guiana as far as the south bank of the Rio Negro ; on the opposite or north bank of the Rio Negro *Psophia ochroptera* is found ; beyond the next great rivers, Japura and Ica, *Psophia napensis* occurs ; on the south bank of the Amazon, west of the Madeira, we have the beautiful *Psophia leucoptera* ; east of the Madeira this is replaced by *Psophia viridis*, while near Pará, beyond the Tapajoz, Xingu and Tocantins, there is another species, *Psophia obscura*. Other species may exist in the intervening river districts ; but we have here, apparently, a case of a number of well-marked species of birds capable of flight, yet with their range in certain directions accurately defined by great rivers. (Plate XV. Vol. II. p. 28.)

FAMILY 111.—EURYPYGIDÆ.   (1 Genus, 2 Species.)

GENERAL DISTRIBUTION.

NEOTROPICAL SUB-REGIONS.	NEARCTIC SUB-REGIONS.	PALÆARCTIC SUB-REGIONS.	ETHIOPIAN SUB-REGIONS.	ORIENTAL SUB-REGIONS.	AUSTRALIAN SUB-REGIONS.
— 2 . 3 —	— — — —	— — — —	— — — —	— — — ⸗	— — — —

The Eurypygidæ, or Sun-Bitterns, are small heron-like birds with beautifully-coloured wings, which frequent the muddy and wooded river-banks of tropical America. The only genus, *Eurypyga* (2 sp.), ranges from Central America to Brazil.

FAMILY 112.—RHINOCHETIDÆ.   (1 Genus, 1 Species.)

GENERAL DISTRIBUTION.

NEOTROPICAL SUB-REGIONS.	NEARCTIC SUB-REGIONS.	PALÆARCTIC SUB-REGIONS.	ETHIOPIAN SUB-REGIONS.	ORIENTAL SUB-REGIONS.	AUSTRALIAN SUB-REGIONS.
— — 3 —	— — — —	— — — —	— — — —	— .. — —	— — — —

The genus *Rhinochetus* (1 sp.), consists of a singular bird
called the Kagu, which inhabits New Caledonia, an island
which may be placed with almost equal propriety in our 1st,
2nd, or 3rd Australian sub-regions. It is a bird of a bluish
ash-colour, with a loose plumage, partaking something of the
appearance of Rail, Plover, and Heron, but with peculiarities of
structure which require it to be placed in a distinct family.
Its anatomy shows that its nearest allies are the South American
genera, *Eurypyga* and *Psophia*.

FAMILY 113.—ARDEIDÆ.   (5 Genera, 80 Species.)

GENERAL DISTRIBUTION.

NEOTROPICAL SUB-REGIONS.	NEARCTIC SUB-REGIONS.	PALÆARCTIC SUB-REGIONS.	ETHIOPIAN SUB-REGIONS.	ORIENTAL SUB-REGIONS.	AUSTRALIAN SUB-REGIONS.
1.2.3.4	1.2.3.4	1.2.3.4	1.2.3.4	1.2.3.4	1.2.3.4

The well-known Herons and Bitterns are found in every
part of the globe, and everywhere closely resemble each other.
Omitting the minuter sub-divisions, the genera are as follows:—
Ardea (60 sp.), cosmopolitan; *Botaurus* (6 sp.), almost cos-
mopolitan; *Tigrisoma* (4 sp.), Tropical America and West Africa;
Nycticorax (9 sp.), cosmopolitan; *Cancroma* (1 sp.), Tropical
America.

FAMILY 114.—PLATALEIDÆ.   (6 Genera, 30 Species.)

GENERAL DISTRIBUTION.					
NEOTROPICAL SUB-REGIONS.	NEARCTIC SUB-REGIONS.	PALÆARCTIC SUB-REGIONS.	ETHIOPIAN SUB-REGIONS.	ORIENTAL SUB-REGIONS.	AUSTRALIAN SUB-REGIONS.
1.2.3.4	1.2.3.4	1.2.3.4	1.2.3.4	1.2.3.4	1.2 - -

The Plataleidæ, including the Spoonbills and Ibises, have been classed either with the Herons or the Storks, but have most affinity with the latter.   Though not very numerous they are found over the greater part of the globe, except the colder zones and the Pacific Islands.   The following is the distribution of the genera :—

Platalea (6 sp.), all the warmer parts of the globe except the Moluccas and Pacific Islands; *Ibis* (2 sp.), Temperate North America and Tropical South America ; *Falcinellus* (2 sp.), almost cosmopolitan ; *Geronticus* (19 sp.), all Tropical countries and Temperate South America; *Scopus* (1 sp.), Tropical and South Africa; *Balæniceps* (1 sp.), the Upper Nile.   This last genus the " Shoe-bird," or boat-billed heron, perhaps forms a distinct family.

FAMILY 115.—CICONIIDÆ.   (5 Genera, 20 Species.)

GENERAL DISTRIBUTION.					
NEOTROPICAL SUB-REGIONS.	NEARCTIC SUB-REGIONS.	PALÆARCTIC SUB-REGIONS.	ETHIOPIAN SUB-REGIONS.	ORIENTAL SUB-REGIONS.	AUSTRALIAN SUB-REGIONS.
1.2.3 -	- - 3 --	1.2.3.4	1.2.3.4	1.2.3.4	1.2 - -

The Ciconiidæ, or Storks, are mostly an Old World family, only three species inhabiting the Neotropical, and one, the Nearctic region.   They are also absent from the islands of the Pacific, the Antilles, and, with one exception, from Madagascar. The genera are as follows :—

Ciconia (6 sp.), ranges through the Palæarctic, Ethiopian and

Oriental regions as far as Celebes, and in South America ; *Mycteria* (4 sp.), inhabits Africa, India, Australia and the Neotropical region; *Leptopiltus* (3 sp.), the Ethiopian and Oriental regions to Java ; *Tantalus* (5 sp.), the Ethiopian, Oriental and Neotropical regions, and the South east of North America ; *Anastomus* (2 sp.), the Ethiopian region, and India to Ceylon.

FAMILY 116.—PALAMEDEIDÆ.   (2 Genera, 3 Species.)

GENERAL DISTRIBUTION.					
NEOTROPICAL SUB-REGIONS.	NEARCTIC SUB-REGIONS.	PALÆARCTIC SUB-REGIONS.	ETHIOPIAN SUB-REGIONS.	ORIENTAL SUB-REGIONS.	AUSTRALIAN SUB-REGIONS.
1.2 – –	– – – –	– – – –	– – – –	– – – –	– – – –

The Palamedeidæ, or Screamers, are curious semi-aquatic birds of doubtful affinities, perhaps intermediate between Gallinæ and Anseres. They are peculiar to South America. The genera are:—

Palamedea (1 sp.), which inhabits the Amazon valley ; *Chauna* (2 sp.), La Plata, Brazil and Columbia.

FAMILY 117.—PHŒNICOPTERIDÆ.   (1 Genus, 8 Species.)

GENERAL DISTRIBUTION.					
NEOTROPICAL SUB-REGIONS.	NEARCTIC SUB-REGIONS.	PALÆARCTIC SUB-REGIONS.	ETHIOPIAN SUB-REGIONS.	ORIENTAL SUB-REGIONS.	AUSTRALIAN SUB-REGIONS.
1 – 3.4	– – – –	– 2 – –	1.2.3.4	1.2 – –	– – – –

The Flamingoes (*Phœnicopterus*) seem peculiar to the Ethiopian and Neotropical regions, ranging from the former into India and South Europe. America has four species, inhabiting Chili and La Plata, the Galapagos, Mexico and the West Indian islands ; the others range over all Africa, South Europe, India and Ceylon. These singular birds are placed by some authors near the Spoonbills and Ibises, by others with the Geese. Professor Huxley considers them to be " completely

intermediate between the Anserine birds on the one side and
the Storks and Herons on the other." The pterolysis according
to Nitzsch is "completely stork-like."

*General Remarks on the Distribution of the Grallæ, or Wading
and Running Birds.*

The Waders, as a rule, are birds of very wide distribution,
the four largest families Rallidæ, Scolopacidæ, Charadriidæ and
Ardeidæ, being quite cosmopolitan, as are many of the genera.
But there are also a number of small families of very
restricted distribution, and these all occur in the two most
isolated regions, the Neotropical and the Australian. The
Neotropical region is by far the richest in varied forms of
Waders, having representatives of no less than 15 out of the 19
families, while 7 are altogether peculiar to it. The Australian
region has 11 families, with 1 peculiar. The other two tropical
regions each possess 11 families, but none are peculiar. The
Palæarctic region has 10, and the Nearctic 7 families. No less
than three families—Chionididæ, Thinocoridæ, and Cariamidæ—
are confined to the Temperate regions and highlands of South
America; while four others,—Aramidæ, Psophiidæ, Eurypygidæ
and Palamedeidæ,—are found in Tropical America only; and
these present such an array of peculiar and interesting forms as
no other part of the globe can furnish. The Phœnicopteridæ or
Flamingoes, common to the Tropical regions of Asia, Africa and
America, but absent from Australia, is the only other feature
of general interest presented by the distribution of the Waders.

The Order contains about 610 species, which gives about 32
species to each family, a smaller average than in the Gallinæ
or Accipitres, and only about one-fourth of the average number
in the Passeres. This is partly due to the unusual number
of very small families, and partly to the wide average range of
the species, which prevents that specialization of forms that
occurs in the more sedentary groups of birds.

Order IX.—ANSERES.

FAMILY 118.—ANATIDÆ.   (40 Genera, 180 Species.)

GENERAL DISTRIBUTION.					
NEOTROPICAL SUB-REGIONS.	NEARCTIC SUB-REGIONS.	PALÆARCTIC SUB-REGIONS.	ETHIOPIAN SUB-REGIONS.	ORIENTAL SUB-REGIONS.	AUSTRALIAN SUB-REGIONS.
1.2.3.4	1.2.3.4	1.2.3.4	1.2.3.4	1.2.3.4	1.2.3.4

The Anatidæ, comprehending the Ducks, Geese, and Swans with their allies, are of such universal distribution that there is probably no part of the globe where some of them are not occasionally found. They are, however, most abundant in temperate and cold regions ; and, contrary to what occurs in most other families, the most beautifully-coloured species are extratropical, and some even arctic. The distribution of the genera is as follows :—

Anseranas (1 sp.), Australia ; *Plectropterus* (2 sp.), Tropical Africa ; *Sarkidiornis* (1 sp.), South America, Africa, and India ; *Chenalopex* (1 sp.), Amazonia ; *Callochen* (1 sp.), South Europe, North, East, and South Africa ; *Cereopsis* (1 sp.), Australia ; *Anser* (13 sp.), Palæarctic and Nearctic regions to Central America and the Antilles ; *Bernicla* (12 sp.), Temperate regions of the Northern and Southern Hemispheres ; *Chloëphaga* (5 sp.), South Temperate America and Aleutian Islands ; *Nettapus* (4 sp.), Tropical Africa and Madagascar, India and Ceylon to Malaya and Australia ; *Cygnus* (10 sp.), Temperate regions of the Northern and Southern Hemispheres ; *Dendrocygna* (10 sp.), Tropical and sub-tropical regions ; *Tadorna* (3 sp.), Palæarctic and Australian regions ; *Casarca* (5 sp.), Palæarctic, Oriental, Ethiopian, and Australian regions, to New Zealand ; *Aix* (2 sp.), Temperate North America and Eastern Asia ; *Mareca* (4 sp.), Palæarctic region, North America, Temperate South America, and Australia ; *Dafila* (3 sp.), all America and the Palæarctic region ; *Anas* (16 sp.), cosmopolitan ; *Querquedula* (17 sp.), cosmopolitan; *Chaulelasmus* (2 sp.), Palæarctic region and North America; *Spatula* (5 sp.), all Temperate regions; *Malacorhynchus* (1 sp.), Australia; *Cairina* (1 sp.), Tropical South America; *Branta* (1 sp.), Palæarctic region and India; *Fuligula* (5 sp.), North Temperate regions and New Zealand; *Æthya* (5 sp), Palæarctic and Nearctic regions, India, Australia, and South Africa; *Metopiana* (1 sp.), South Temperate America; *Bucephala* (4 sp.), Nearctic and Palæarctic regions; *Harelda* (2 sp.), Northern Palæartic and Nearctic regions; *Hymenolaimus* (1 sp.), New Zealand; *Camptolaimus* (1 sp.), North-east of North America; *Micropterus* (1 sp.), Temperate South America; *Somateria* (5 sp.), Arctic and sub-arctic regions; *Œdemia* (5 sp.), Nearctic and Palæarctic regions; *Biziura* (1 sp.), Australia; *Thalassornis* (1 sp.), South Africa; *Erismatura* (6 sp.), all America, South-east Europe and South Africa; *Nesonetta* (1 sp.), Auckland Islands; *Merganetta* (3 sp.), Andes of Columbia to Chili; *Mergus* (6 sp.), Palæarctic and Nearctic regions, Brazil, and the Auckland Islands.

FAMILY 119.—LARIDÆ. (13 Genera, 132 Species)

GENERAL DISTRIBUTION.					
NEOTROPICAL SUB-REGIONS.	NEARCTIC SUB-REGIONS.	PALÆARCTIC SUB-REGIONS.	ETHIOPIAN SUB-REGIONS.	ORIENTAL SUB-REGIONS.	AUSTRALIAN SUB-REGIONS.
1.2.3.4	1.2.3.4	1.2.3.4	1.2.3.4	1.2.3.4	1.2.3.4

The Laridæ, or Gulls and Terns, are true cosmopolites, inhabiting the shores and islands of every zone; and most of the genera have also a wide range. They are therefore of little use in the study of geographical distribution. The genera are as follows :—

Stercorarius (6 sp.), cosmopolitan, most abundant in cold and temperate zones; *Rhodostethia* (1 sp), North America; *Larus* (60 sp.). cosmopolitan; *Xema* (1 sp.), North Temperate zone; *Creagrus* (1 sp.), North Pacific; *Pagophila* (1 sp.), Arctic seas; *Rissa* (3 sp.), Arctic and Northern seas; *Sterna* (36 sp.), cosmopolitan; *Hydrochelidon* (12 sp.), Tropical and Temperate zones;

Gygis (1 sp.), Indian Ocean and Tropical Pacific Islands; *Anous* (6 sp.), Tropical and Temperate zones; *Nœnia* (1 sp.), South Temperate America; *Rhynchops* (3 sp.), Tropical America, Africa, and India.

FAMILY 120.—PROCELLARIIDÆ.   (6 Genera, 96 Species.)

GENERAL DISTRIBUTION.					
NEOTROPICAL SUB-REGIONS.	NEARCTIC SUB-REGIONS.	PALÆARCTIC SUB-REGIONS.	ETHIOPIAN SUB-REGIONS.	ORIENTAL SUB-REGIONS.	AUSTRALIAN SUB-REGIONS.
1.2.3.4	1.2.3.4	1.2.3.4	1.2.3.4	1.2.3.4	1.2.3 4

The Procellariidæ, comprising the Shearwaters, Petrels, and Albatrosses, are universally distributed, but some of the genera are local.

Puffinus (20 sp.), *Procellaria* (18 sp.), and *Fulmarus* (40 sp.), are cosmopolitan; *Prion* (5 sp.) and *Pelecanoides* (3 sp.), belong to the South Temperate and Antarctic regions; *Diomedia* (10 sp.), comprises the Albatrosses, which are tropical, occasionally wandering into temperate seas.

FAMILY 121.—PELECANIDÆ.   (6 Genera, 61 Species.)

GENERAL DISTRIBUTION.					
NEOTROPICAL SUB-REGIONS.	NEARCTIC SUB-REGIONS.	PALÆARCTIC SUB-REGIONS.	ETHIOPIAN SUB-REGIONS.	ORIENTAL SUB-REGIONS.	AUSTRALIAN SUB-REGIONS.
1.2.3.4	1.2.3.4	1.2.3.4	1.2.3.4	1.2.3.4	1.2.3.4

The Pelecanidæ, comprising the Gannets, Pelicans, Darters, and Frigate-Birds, although universally distributed, are more abundant in tropical and temperate regions.

Sula (8 sp.) and *Phalacrocorax* (35 sp.), are cosmopolitan; *Pelecanus* (9 sp.) is tropical and temperate; *Fregetta* (2 sp.) and *Phaeton* (3 sp.) are confined to Tropical seas; *Ptotus* (4 sp.) to Tropical and warm Temperate zones.

FAMILY 122.—SPHENISCIDÆ.　(3 Genera, 18 Species.)

GENERAL DISTRIBUTION.					
NEOTROPICAL SUB-REGIONS.	NEARCTIC SUB-REGIONS.	PALÆARCTIC SUB-REGIONS.	ETHIOPIAN SUB-REGIONS.	ORIENTAL SUB-REGIONS.	AUSTRALIAN SUB-REGIONS.
1 . 2 — —	.. — — —	— — — —	— — 3 —	— .- — —	— 2 — 4

The Penguins are entirely confined to the Antarctic and South Temperate regions, except two species which are found on the coast of Peru and the Galapagos. They are most plentiful in the southern parts of South America, Australia, New Zealand, and most of the Antarctic islands, and one or two species are found at the Cape of Good Hope. The genera as given in the *Hand List* are :—

Spheniscus (1 sp.), South Africa and Cape Horn ; *Eudyptes* (15 sp.), with the range of the family ; *Aptenodytes* (2 sp.), Antarctic Islands.

FAMILY 123.—COLYMBIDÆ.　(1 Genus, 4 Species.)

GENERAL DISTRIBUTION.					
NEOTROPICAL SUB-REGIONS.	NEARCTIC SUB-REGIONS.	PALÆARCTIC SUB-REGIONS.	ETHIOPIAN SUB-REGIONS.	ORIENTAL SUB-REGIONS.	AUSTRALIAN SUB-REGIONS.
— — — —	— — — 4	1 — 3 . 4	— — — —	— — — —	— — — —

The Northern Divers are confined to the Arctic and North Temperate Seas. The only genus, *Colymbus*, has one species confined to the West Coast of North America, the others being common to the two northern continents.

FAMILY 124.—PODICIPIDÆ.　(2 Genera, 33 Species.)

GENERAL DISTRIBUTION.					
NEOTROPICAL SUB-REGIONS.	NEARCTIC SUB-REGIONS.	PALÆARCTIC SUB-REGIONS.	ETHIOPIAN SUB-REGIONS.	ORIENTAL SUB-REGIONS.	AUSTRALIAN SUB-REGIONS.
1 . 2 . 3 . 4	1 . 2 . 3 . 4	1 . 2 . 3 . 4	1 . 2 . 3 . 4	1 . 2 . 3 . 4	1 . 2 . 3 . 4

The Grebes are universally distributed. The genera are *Podiceps* (26 sp.), cosmopolitan ; and *Podilymbus* (2 sp.), confined to North and South America. Some ornithologists group these birds with the Colymbidæ.

FAMILY 125.—ALCIDÆ.   (7 Genera, 28 Species.)

GENERAL DISTRIBUTION.

NEOTROPICAL SUB-REGIONS.	NEARCTIC SUB-REGIONS.	PALÆARCTIC SUB-REGIONS.	ETHIOPIAN SUB-REGIONS.	ORIENTAL SUB-REGIONS.	AUSTRALIAN SUB-REGIONS.
— — — —	1 — — 4	1 – 3.4	— — — —	— — — — ··	— — — —

The Alcidæ, comprising the Auks, Guillemots, and Puffins, are confined to the North Temperate and Arctic regions, where they represent the Penguins of the Antarctic lands.   One of the most remarkable of these birds, the Great Auk, formerly abundant in the North Atlantic, is now extinct.   The genera are as follows :—

Alca (2 sp.), North Atlantic and Arctic seas ; *Fratercula* (4 sp.), Arctic and North Temperate zones ; *Ceratorhina* (2 sp.), North Pacific ; *Simorhynchus* (8 sp.), North Pacific ; *Brachyrhamphus* (3 sp.), North Pacific to Japan and Lower California ; *Uria* (8 sp.), Arctic and North Temperate zones ; *Mergulus* (1 sp.), North Atlantic and Arctic Seas.   The last three genera constitute the family Uriidæ, of some ornithologists.

General Remarks on the Distribution of the Anseres.

The Anseres, or Swimmers, being truly aquatic birds, possess, as might be expected, a large number of cosmopolitan families and genera.   No less than 5 out of the 8 families have a world-wide distribution, and the others are characteristic either of the North or the South Temperate zones.   Hence arises a peculiarity of distribution to be found in no other order of birds ; the Temperate being richer than the Tropical regions.   The Nearctic and Palæarctic regions each have seven families of Anseres, two of which, the Colymbidæ and Alcidæ, are peculiar to them. The Ethiopian, Australian, and Neotropical regions, which all extend into the South Temperate zone, have six families, with one peculiar to them; while the Oriental region, which is wholly tropical, possesses the five cosmopolitan families only.

There are about 78 genera and 552 species of Anseres, giving 69 species to a family, a high number compared with the Waders, and due to there being only one very small family, the Colymbidæ. The distribution of the Anseres, being more determined by temperature than by barriers, the great regions which are so well indicated by the genera and families of most other orders of birds, hardly limit these, except in the case of the genera of Anatidæ.

Order X.—STRUTHIONES.

FAMILY 126.—STRUTHIONIDÆ.   (2 Genera, 4 Species.)

GENERAL DISTRIBUTION.					
NEOTROPICAL SUB-REGIONS.	NEARCTIC SUB-REGIONS.	PALÆARCTIC SUB-REGIONS.	ETHIOPIAN SUB-REGIONS.	ORIENTAL SUB-REGIONS.	AUSTRALIAN SUB-REGIONS.
1 — — —	— — — —	— 2 — —	1 — 3 —	— — — —	- — — —

The Ostriches consist of two genera, sometimes formed into distinct families. *Struthio* (2 sp.) inhabits the desert regions of North, East, and South Africa, as well as Arabia and Syria. It therefore just enters the Palæarctic region. *Rhea* (3 sp.) inhabits Temperate South America, from Patagonia to the confines of Brazil.

FAMILY 127.—CASUARIIDÆ.   (2 Genera, 11 Species.)

GENERAL DISTRIBUTION.					
NEOTROPICAL SUB-REGIONS.	NEARCTIC SUB-REGIONS.	PALÆARCTIC SUB-REGIONS.	ETHIOPIAN SUB-REGIONS.	ORIENTAL SUB-REGIONS.	AUSTRALIAN SUB-REGIONS.
— — — —	— — — —	— — — —	— — — —	— — — —	1 . 2 — —

The Cassowaries and Emeus are confined to the Australian region. The Emeus, *Dromœus* (2 sp.), are found only on the main-land of Australia (Plate XII. Vol. I. p. 441). *Casuarius* (9 sp.) inhabits the islands from Ceram to New Britain, with one species in North Australia; it is most abundant in the Papuan Islands.

FAMILY 128.—APTERYGIDÆ.   (1 Genus, 4 Species.)

GENERAL DISTRIBUTION.					
NEOTROPICAL SUB-REGIONS.	NEARCTIC SUB-REGIONS.	PALÆARCTIC SUB-REGIONS.	ETHIOPIAN SUB-REGIONS.	ORIENTAL SUB-REGIONS.	AUSTRALIAN SUB-REGIONS.
— — — —	— — — —	— — — —	— — — —	— — — —	— — — 4

The species of *Apteryx* are entirely confined to the two larger islands of New Zealand. They are supposed to have some remote affinity with *Ocydromus*, a genus of Rails peculiar to Australia and New Zealand; but they undoubtedly form one of the most remarkable groups of living birds (Plate XIII. Vol. I. p. 445).

Struthious Birds recently extinct.

A number of sub-fossil remains of birds, mostly large and some of gigantic size, having affinities to the *Apteryx* and, less closely, to the Cassowaries, have been discovered in New Zealand. These are all classed by Professor Owen in the genus *Dinornis* and family *Dinornithidæ;* but Dr. Haast, from the study of the rich collections in the Canterbury (New Zealand) Museum, is convinced that they belong to two distinct families and several genera. His arrangement is as follows. (See *Ibis*, 1874, p. 209).

FAMILY 129.—DINORNITHIDÆ.   (2 Genera, 7 Species.)

Dinornis (5 sp.); *Meionornis* (2 sp.).

These had no hind toe, and include the largest species. Professor Newton thinks that they were absolutely wingless, being the only birds in which the fore limbs are entirely wanting.

FAMILY 130.—PALAPTERYGIDÆ.   (2 Genera, 4 Species.)

Palapteryx (2 sp.); *Euryapteryx* (2 sp.).
These had a well-developed hind toe, and rudimentary wings.

FAMILY 131.—ÆPYORNITHIDÆ.   (1 Genus, 3 Species.)

A gigantic Struthious bird (*Æpyornis*), belonging to a distinct family, inhabited Madagascar.

It was first made known by its enormous eggs, eight times the bulk of those of the ostrich, which were found in a sub-fossil condition. Considerable portions of skeletons have since been discovered, showing that these huge birds formed an altogether peculiar family of the order.

General Remarks on the Distribution of the Struthiones.

With the exception of the Ostrich, which has spread north-ward into the Palæarctic region, the Struthious birds, living and extinct, are confined to the Southern hemisphere, each continent having its peculiar forms. It is a remarkable fact that the two most nearly allied genera, *Struthio* and *Rhea*, should be found in Africa and South Temperate America respectively. Equally re-markable is the development of these large forms of wingless birds in Australia and the adjacent islands, and especially in New Zealand, where we have evidence which renders it probable that about 20 species recently coexisted. This points to the conclusion that New Zealand must, not long since, have formed a much more extensive land, and that the diminution of its area by subsidence has been one of the causes—and perhaps the main one—in bringing about the extinction of many of the larger species of these wingless birds.

The wide distribution of the Struthiones may, as we have already suggested (Vol. I., p. 287.), be best explained, by sup-posing them to represent a very ancient type of bird, developed at a time when the more specialized carnivorous mammalia had not come into existence, and preserved only in those areas which were long free from the incursions of such dangerous enemies. The discovery of Struthious remains in Europe in the Lower Eocene only, supports this view ; for at this time carnivora were few and of generalized type, and had probably not acquired sufficient speed and activity to enable them to exterminate powerful and quick-running terrestrial birds. It is, however, at a much more remote epoch that we may expect to find the remains of the earlier forms of this group ; while these Eocene birds may perhaps represent that ancestral wide-spread type which, when isolated in remoter continents and islands, became modified into the American and African ostriches, the Emeus and Cassowaries of Australia, the *Dinornis* and *Æpyornis* of New Zealand.

CHAPTER XIX.

THE DISTRIBUTION OF THE FAMILIES AND GENERA OF REPTILES AND AMPHIBIA.

REPTILIA.

Order I.—OPHIDIA.

FAMILY 1.—TYPHLOPIDÆ.—(4 Genera, 70 Species.)

GENERAL DISTRIBUTION.					
NEOTROPICAL SUB-REGIONS.	NEARCTIC SUB-REGIONS.	PALÆARCTIC SUB-REGIONS.	ETHIOPIAN SUB-REGIONS.	ORIENTAL SUB-REGIONS.	AUSTRALIAN SUB-REGIONS.
— 2 . 3 . 4	— — — —	— 2 — 4	1 . 2 . 3 . 4	1 . 2 . 3 . 4	1 . 2 — —

The Typhlopidæ, or Blind Burrowing Snakes, are widely scattered over the warmer regions of the earth, but are most abundant in the Oriental and Australian regions, and least so in the Neotropical. They are absent from the Nearctic region; and in the Palæarctic are found only in South-eastern Europe and Japan.

The most extensive genus is *Typhlops*, comprising over 60 species, and having a range almost as extensive as the entire family. The other well characterised genera are :—

Typhlina (1 sp.), ranging from Penang to Java and Hong Kong; *Typhline* (1 sp.), the Cape of Good Hope ; *Dibamus* (1 sp.), New Guinea.

FAMILY 2.—TORTRICIDÆ.   (3 Genera 5 Species.)

GENERAL DISTRIBUTION.					
NEOTROPICAL SUB-REGIONS.	NEARCTIC SUB-REGIONS.	PALÆARCTIC SUB-REGIONS.	ETHIOPIAN SUB-REGIONS.	ORIENTAL SUB-REGIONS.	AUSTRALIAN SUB-REGIONS.
— 2.3 —	1 — — —	— — — —	— — — —	1.2.3.4	1 — — —

The Tortricidæ, or Short-tailed Burrowing Snakes, are a small family, one portion of which ranges from India to Cambodja, and through the Malay islands as far as Celebes and Timor; these form the genus *Cylindrophis*. Another portion inhabits America, and consists of:—

Charina (1 sp.), found in California and British Columbia; and *Tortrix* (1 sp.), in Tropical America.

We have here a case of discontinuous distribution, indicating, either very imperfect knowledge of the group, or that it is the remnant of a once extensive family, on the road to extinction.

FAMILY 3.—XENOPELTIDÆ.   (1 Genus, 1 Species.)

GENERAL DISTRIBUTION.					
NEOTROPICAL SUB-REGIONS.	NEARCTIC SUB-REGIONS.	PALÆARCTIC SUB-REGIONS.	ETHIOPIAN SUB-REGIONS.	ORIENTAL SUB-REGIONS.	AUSTRALIAN SUB-REGIONS.
— — — —	— — — —	— — — —	— — — —	— — 3.4	1 — — —

The curious nocturnal carnivorous Snake, forming the genus *Xenopeltis*, and the sole representative of this family, ranges from Penang to Cambodja, and through the Malay Islands to Celebes.

FAMILY 4.—UROPELTIDÆ.   (5 Genera, 18 Species.)

GENERAL DISTRIBUTION.					
NEOTROPICAL SUB-REGIONS.	NEARCTIC SUB-REGIONS.	PALÆARCTIC SUB-REGIONS.	ETHIOPIAN SUB-REGIONS.	ORIENTAL SUB-REGIONS.	AUSTRALIAN SUB-REGIONS.
— — — —	— — — —	— — — —	— — — —	— 2 — —	— — — —

The Uropeltidæ, or Rough-tailed Burrowing Snakes, are strictly confined to Ceylon and the adjacent parts of Southern India, and would almost alone serve to mark out our second Oriental sub-region. The genera are :—
Rhinophis (7 sp.), Ceylon; *Uropeltis* (1 sp.), Ceylon; *Silybura* (8 sp.), Anamally Hills and Neilgherries; *Plecturus* (3 sp.), Neilgherries and Madras; and *Melancphidium* (1 sp.), the Wynand.

FAMILY 5.—CALAMARIIDÆ.   (32 Genera, 75 Species.)

GENERAL DISTRIBUTION.					
NEOTROPICAL SUB-REGIONS.	NEARCTIC SUB-REGIONS.	PALÆARCTIC SUB-REGIONS.	ETHIOPIAN SUB-REGIONS.	ORIENTAL SUB-REGIONS.	AUSTRALIAN SUB-REGIONS.
1.2.3.4	1.2.3 -	- 2 - -	1.2.3 -	1.2.3.4	1.2 - -

The Calamariidæ, or Dwarf Ground Snakes, are found in all warm parts of the globe, extending north into the United States as far as British Columbia and Lake Superior; but they are absent from the Palæarctic region, with the exception of a species found in Persia. The species are in a very confused state. The best characterised genera are the following :—
Calamaria (20 sp.), Persia, India to Java and the Philippine Islands, Celebes, and New Guinea; *Rhabdosoma* (18 sp.), Mexico and South America, and also the Malay Islands as far east as Amboyna, Timor, and New Guinea; *Typhlocalamus* (1 sp.), Borneo ; *Macrocalamus* (1 sp.), India; *Aspidura* (3 sp.), India and Ceylon; *Haplocerus* (1 sp.), Ceylon; *Streptophorus* (3 sp.), Central and South America ;—with a host of others of less importance or ill-defined.

FAMILY 6. — OLIGODONTIDÆ.   (4 Genera, 40 Species.)

GENERAL DISTRIBUTION.					
NEOTROPICAL SUB-REGIONS.	NEARCTIC SUB-REGIONS.	PALÆARCTIC SUB-REGIONS.	ETHIOPIAN SUB-REGIONS.	ORIENTAL SUB-REGIONS.	AUSTRALIAN SUB-REGIONS.
- 2.3 -	- - - 3	- - - 4	- - - -	1.2.3.4	- - - -

The Oligodontidæ are a small family of Ground Snakes which have been separated from the Calamariidæ, and, with the exception of a few species, are confined to the Oriental region. The best characterised genera are :—

Oligodon (12 sp.), India, Ceylon, and Philippines ; and, *Simotes* (24 sp.), India to China and Borneo. In addition to these, *Achalinus* is founded on a single species from Japan; and *Teleolepis* consists of three species from North and South America.

FAMILY 7.—COLUBRIDÆ.   (50 Genera, 270 Species.)

GENERAL DISTRIBUTION.					
NEOTROPICAL SUB-REGIONS.	NEARCTIC SUB-REGIONS.	PALÆARCTIC SUB-REGIONS.	ETHIOPIAN SUB-REGIONS.	ORIENTAL SUB-REGIONS.	AUSTRALIAN SUB-REGIONS.
1 . 2 . 3 . 4	1 . 2 . 3 . 4	1 . 2 . 3 . 4	1 . 2 . 3 . 4	1 . 2 . 3 . 4	1 . 2 — —

The Colubrine Snakes are universally distributed over the globe, and they reach the extreme northern limits of the order. They are, however, almost absent from Australia, being there represented only by a few species of *Tropidonotus* and *Coronella* in the northern and eastern districts.   This great family consists of four divisions or sub-families : the Coronellinæ (20 genera, 100 species), the Colubrinæ (16 genera, 70 species), the Dryadinæ (7 genera, 50 species), and the Natricinæ (7 genera, 50 species).   The more important genera of Colubridæ are the following:—

Ablabes, Coronella, Ptyas, Coluber, and *Tropidonotus*—all have a very wide distribution, but the two last are absent from South America, although *Tropidonotus* reaches Guatemala ; *Tomodon, Xenodon, Liopis, Stenorhina, Erythrolampus, Elapochrus, Callirhinus, Enophrys,* and *Dromicus*—are confined to the Neotropical region ; *Hypsirhynchus, Cryptodacus, Jaltris,* and *Coloragia,* are confined to the West Indian Islands; *Chilomeniscus, Conophis, Pituophis,* and *Ischcognathus,* to North America, the latter going as far south as Guatemala ; *Compsosoma, Zamenis, Zaocys, Atretium, Xenochrophys,* and *Herpetoreas,* are peculiarly Oriental, but *Zamenis* extends into South Europe ;

Lytorhynchus, *Rhamnophis*, *Herpetethiops* and *Grayia*, are Ethiopian; *Rhinechis* is peculiar to Europe; *Megablabes* to Celebes, and *Styporhynchus* to Gilolo; *Cyclophis*, is found in the Oriental region, Japan, and North America; *Spilotes*, in the Nearctic and Neotropical regions; *Xenelaphis* in the Oriental, Ethiopian, and Palæarctic regions; *Philodryas*, *Heterodon* and *Herpetodryas* in America and Madagascar, the latter genus being also found in China.

FAMILY 8.—HOMALOPSIDÆ.   (24 Genera, 50 Species.)

GENERAL DISTRIBUTION.					
NEOTROPICAL SUB-REGIONS.	NEARCTIC SUB-REGIONS.	PALÆARCTIC SUB-REGIONS.	ETHIOPIAN SUB-REGIONS.	ORIENTAL SUB-REGIONS.	AUSTRALIAN SUB-REGIONS.
1 – 3 –	– – 3 –	– 2 . 3 . 4	– 2 – –	1 . 2 . 3 . 4	1 . 2 – –

The Homalopsidæ, or Fresh-water Snakes, have been separated from the Hydridæ by Dr. Günther, and they include some groups which have been usually classed with the Natricinæ.   They are especially characteristic of the Oriental region, where considerably more than half the genera and species are found; next comes the Neotropical region which has 6 species; while none of the other regions have more than 4 or 5.   It is to be observed that the Ethiopian species occur in West Africa only, and mostly constitute peculiar genera, so that in this family the separation of the Ethiopian and Oriental regions is very well marked.   The best characterised genera of the family are the following:—

Cantoria (10 sp.), ranging from Europe to Japan, the Philippines, and Timor, with one species in Guinea; *Hypsirhina* (6 sp.), Bengal, China, and Borneo; *Fordonia* (3 sp.), Rangoon to Borneo and Timor; *Homalopsis* (2 sp,), Cambodja to Java; *Cerberus* (2 sp.), Ceylon and Siam, the Malay Islands, New Guinea, and North Australia; *Herpeton* (1 sp.), Siam; *Ferania* (1 sp.), Bengal to Penang; *Pythonopsis* (1 sp.), Borneo; *Myron* (2 sp.), India and North Australia; *Homalophis* (1 sp.), Borneo; *Hipistes* (1 sp.), Penang; *Xenodermus* (1 sp.), Java; *Neusterophis* and *Limnophis*, with one species each, are peculiar to West

Africa; *Helicops* (2 sp.), North and South America; *Farancia* and *Dimodes*, with one species each, are from New Orleans; and a few others imperfectly known from Tropical America.

FAMILY 9.—PSAMMOPHIDÆ.   (5 Genera, 20 Species.)

GENERAL DISTRIBUTION.					
NEOTROPICAL SUB-REGIONS.	NEARCTIC SUB-REGIONS.	PALÆARCTIC SUB-REGIONS.	ETHIOPIAN SUB-REGIONS.	ORIENTAL SUB-REGIONS.	AUSTRALIAN SUB-REGIONS.
— — — —	— — — —	— 2 — —	1.2.3.4	1 — 3.4	— — — —

The Psammophidæ, or Desert Snakes, are a small group characteristic of the Ethiopian and Oriental regions, but more abundant in the former. The distribution of the genera is as follows :—

Psammophis (16 sp.), ranges from West Africa to Persia and Calcutta; *Cœlopeltis* (1 sp.), North and West Africa; *Mimophis* (1 sp.), Madagascar; *Psammodynastes* (2 sp.), Sikhim to Cochin China, Borneo and the Philippine Islands; and *Dromophis* (1 sp.), Tropical Africa.

FAMILY 10.—RACHIODONTIDÆ.   (1 Genus, 2 Species.)

GENERAL DISTRIBUTION.					
NEOTROPICAL SUB-REGIONS.	NEARCTIC SUB-REGIONS.	PALÆARCTIC SUB-REGIONS.	ETHIOPIAN SUB-REGIONS.	ORIENTAL SUB-REGIONS.	AUSTRALIAN SUB-REGIONS.
— — — —	— — — —	— — — —	— 2.3 —	— — — —	— — — —

The Rachiodontidæ are a small and very isolated group of snakes of doubtful affinities. The only genus, *Dasypeltis* (2 sp.), is confined to West and South Africa.

FAMILY 11.—DENDROPHIDÆ.   (7 Genera, 35 Species.)

GENERAL DISTRIBUTION.					
NEOTROPICAL SUB-REGIONS.	NEARCTIC SUB-REGIONS.	PALÆARCTIC SUB-REGIONS.	ETHIOPIAN SUB-REGIONS.	ORIENTAL SUB-REGIONS.	AUSTRALIAN SUB-REGIONS.
1.2.3.4	— — — —	— — — —	1.2.3.4	1.2.3.4	1.2 — —

The Dendrophidæ, or Tree Snakes, are found in all the Tropical regions, but are most abundant in the Oriental. The genera are distributed as follows:—

Dendrophis ranges from India and Ceylon to the Pelew Islands and North Australia, and has one species in West Africa; *Ahætulla* is almost equally divided between Tropical Africa and Tropical America; *Gonyosoma* ranges from Persia to Java and the Philippines; *Chrysopelea* is found in India, Borneo, the Philippines, Amboyna, and Mysol; *Hapsidrophis* and *Bucephalus* are confined to Tropical Africa; and *Ithycyphus* (1 sp.), is peculiar to Madagascar.

FAMILY 12.—DRYIOPHIDÆ.   (5 Genera, 15 Species.)

GENERAL DISTRIBUTION.					
NEOTROPICAL SUB-REGIONS.	NEARCTIC SUB-REGIONS.	PALÆARCTIC SUB-REGIONS.	ETHIOPIAN SUB-REGIONS.	ORIENTAL SUB-REGIONS.	AUSTRALIAN SUB-REGIONS.
— 2.3 —	— — — —	— — — —	— 2 — 4	1.2.3.4	1 — — —

The Dryiophidæ, or Whip Snakes, are a very well characterised family of slender, green-coloured, arboreal serpents, found in the three tropical regions but absent from Australia, although they just enter the Australian region in the island of Celebes. In Africa they are confined to the West Coast and Madagascar. The genera are:—

Dryiophis (4 sp.), Tropical America and West Africa: *Tropidococcyx* (1 sp.), Central India; *Tragops* (4 sp.), Bengal to China, the Philippines, Java, and Celebes; *Passerita* (2 sp.), Ceylon and the Indian Peninsula; and *Langaha* (2 sp.), confined to Madagascar.

FAMILY 13.—DIPSADIDÆ. (11 Genera, 45 Species.)

GENERAL DISTRIBUTION.					
NEOTROPICAL SUB-REGIONS.	NEARCTIC SUB-REGIONS.	PALÆARCTIC SUB-REGIONS.	ETHIOPIAN SUB-REGIONS.	ORIENTAL SUB-REGIONS.	AUSTRALIAN SUB-REGIONS.
— 2 . 3 —	— — — —	— 2 — —	1 . 2 . 3 —	1 . 2 . 3 . 4	1 . 2 — —

The Dipsadidæ, or Nocturnal Tree Snakes, are distinguished from the last family by their dark colours and nocturnal habits. They are about equally abundant in the Oriental and Neotropical regions, less so in the Ethiopian, while only a single species extends to North Australia. The following are the best known genera:—

Dipsas, comprising all the Oriental species with one in Asia-Minor, and a few from the Moluccas, New Guinea, North Australia, West Africa, and Tropical America; *Thamnodyastes*, *Tropidodipsas*, and several others, from Tropical America; *Dipsadoboa*, from West Africa and Tropical America; *Leptodeira*, from Tropical and South Africa, South America, and Mexico; and *Pythonodipsas*, from Central Africa.

FAMILY 14.—SCYTALIDÆ. (3 Genera, 10 Species.)

GENERAL DISTRIBUTION.					
NEOTROPICAL SUB-REGIONS.	NEARCTIC SUB-REGIONS.	PALÆARCTIC SUB-REGIONS.	ETHIOPIAN SUB-REGIONS.	ORIENTAL SUB-REGIONS.	AUSTRALIAN SUB-REGIONS.
— 2 . 3 —	— — — —	— — — —	— — — —	— — — 4	— — — —

It is doubtful how far the three genera which constitute this family form a natural assemblage. We can therefore draw no safe conclusions from the peculiarity of their distribution— *Scytale* and *Oxyrhopus* being confined to Tropical America; while *Hologerrhum* inhabits the Philippine Islands.

FAMILY 15.—LYCODONTIDÆ.   (11 Genera, 35 Species.)

			GENERAL DISTRIBUTION.		
NEOTROPICAL SUB-REGIONS.	NEARCTIC SUB-REGIONS.	PALÆARCTIC SUB-REGIONS.	ETHIOPIAN SUB-REGIONS.	ORIENTAL SUB-REGIONS.	AUSTRALIAN SUB-REGIONS.
— — — —	— — — —	— — — —	1 . 2 . 3 —	1 . 2 . 3 . 4	1 — — —

The Lycodontidæ, or Fanged Ground Snakes, are confined to the Ethiopian and Oriental regions, over the whole of which they range, except that they are absent from Madagascar and extend eastward to New Guinea.   The genera have often a limited distribution :—

Lycodon ranges from India and Ceylon to China, the Philippines, and New Guinea; *Tetragonosoma*, the Malay Peninsula and Islands; *Leptorhytaon* and *Ophites*, India ; *Cercaspis*, Ceylon ; and *Cyclocorus*, the Philippines.   The African genera are *Boædon*, *Lycophidion*, *Holuropholis*, *Simocephalus*, and *Lamprophis*, the latter being found only in South Africa.   The species are nearly equally abundant in both regions, but no genus is common to the two.

FAMILY 16.—AMBLYCEPHALIDÆ.   (5 Genera, 12 Species.)

			GENERAL DISTRIBUTION.		
NEOTROPICAL SUB-REGIONS.	NEARCTIC SUB-REGIONS.	PALÆARCTIC SUB-REGIONS.	ETHIOPIAN SUB-REGIONS.	ORIENTAL SUB-REGIONS.	AUSTRALIAN SUB-REGIONS.
— 2 . 3 —	— — — —	— — — —	— — — —	— — 3 . 4	— — 3 ? —

The Amblycephalidæ, or Blunt Heads, are very singularly distributed, being nearly equally divided between Tropical America and the eastern half of the Oriental region, as will be seen by the following statement of the distribution of the genera :—

Amblycephalus (1 sp.), Malay Peninsula to Borneo and the Philippines ; *Pareas* (3 sp.), Assam, China, Java, and Borneo ;

Asthenodipsas (1 sp.), Malacca; *Leptognathus* (6 sp.), Central and South America; and *Anoplodipsas* (1 sp.), supposed to come from New Caledonia, and, if so, furnishing a link, though a very imperfect one, between the disconnected halves of the family.

FAMILY 17.—PYTHONIDÆ.　(21 Genera, 46 Species.)

GENERAL DISTRIBUTION.					
NEOTROPICAL SUB-REGIONS.	NEARCTIC SUB-REGIONS.	PALÆARCTIC SUB-REGIONS.	ETHIOPIAN SUB-REGIONS.	ORIENTAL SUB-REGIONS.	AUSTRALIAN SUB-REGIONS.
1.2.3.4	1 — — —	— — — —	1.2.3.4	1.2.3.4	1.2.3 —

The Pythonidæ, comprising the Rock Snakes, Pythons, and Boas, are confined to the tropics, with the exception of one species in California. They are very abundant in the Neotropical region, where nearly half the known species occur; the Australian region comes next, while the Oriental is the least prolific in these large serpents. The genera which have been described are very numerous, but they are by no means well defined. The following are the most important:—

Python is confined to the Oriental region; *Morelia, Liasis,* and *Nardoa* are Australian and Papuan; *Enygrus* is found in the Moluccas, New Guinea and the Fiji Islands; *Hortulia* is African; *Sanzinia* is peculiar to Madagascar; *Boa, Epicrates, Corallus, Ungalia,* and *Eunectes* are Tropical American; *Chilabothrus* is peculiar to Jamaica and Mexico; and *Lichanotus* to California.

An extinct species belonging to this family has been found in the Brown-coal formation of Germany, of Miocene age.

FAMILY 18.—ERYCIDÆ.　(3 Genera, 6 Species.)

GENERAL DISTRIBUTION.					
NEOTROPICAL SUB-REGIONS.	NEARCTIC SUB-REGIONS.	PALÆARCTIC SUB-REGIONS.	ETHIOPIAN SUB-REGIONS.	ORIENTAL SUB-REGIONS.	AUSTRALIAN SUB-REGIONS.
— — — —	— — — —	— 2 — —	— 2 — —	1 — 3 —	— — — —

The Erycidæ, or Land Snakes, form a small but natural family, chiefly found in the desert zone on the confines of the Palæarctic, Oriental, and Ethiopian regions. They range from South Europe to West Africa and to Sikhim. The three genera are distributed as follows :—

Cursoria (1 sp.), Afghanistan ; *Gongylophis* (1 sp.), India and Sikhim ; *Eryx* (4 sp.), has the range of the entire family.

FAMILY 19.—ACROCHORDIDÆ.   (2 Genera, 3 Species)

GENERAL DISTRIBUTION.

NEOTROPICAL SUB-REGIONS.	NEARCTIC SUB-REGIONS.	PALÆARCTIC SUB-REGIONS.	ETHIOPIAN SUB-REGIONS.	ORIENTAL SUB-REGIONS.	AUSTRALIAN SUB-REGIONS.
— — — —	— — — —	— — — —	— — — —	2 — 4	1 — — —

The Acrochordidæ, or Wart Snakes, form a small and isolated group, found only in two sub-divisions of the Oriental region— the South Indian and the Malayan, and in New Guinea.

Acrochordus, inhabits Penang, Singapore, and Borneo ; *Chersydrus*, Southern India and the Malay Peninsula, with a species recently discovered in New Guinea.

FAMILY 20.—ELAPIDÆ.   (23 Genera, 100 Species.)

GENERAL DISTRIBUTION.

NEOTROPICAL SUB-REGIONS.	NEARCTIC SUB-REGIONS.	PALÆARCTIC SUB-REGIONS.	ETHIOPIAN SUB-REGIONS.	ORIENTAL SUB-REGIONS.	AUSTRALIAN SUB-REGIONS.
1.2.3—	— — 3 —	— — — 4	1.2.3—	1.2.3.4	1.2.3—

The Elapidæ, or Terrestrial venomous Colubrine Snakes, are an extensive group, spread over the tropics of the whole world, but especially abundant in Australia, where half the known species occur, some of them being the most deadly of venomous serpents. In the Oriental region they are also abundant, containing amongst other forms, the well-known Cobras. The American species are almost equally numerous, but they all belong to one genus, and they are annulated with rings of various colours in a manner quite distinct from any other members of this family. The genera, which are all very distinct, are distributed as follows :—

Diemenia, Acanthophis, Hoplocephalus, Brachiurophis, Tropidechis, Pseudechis, Cacophis, Pseudonaje, Denisonia, and *Vermicella,* are Australian, the first two ranging to the Moluccas and New Guinea ; *Ogmodon* occurs in the Fiji Islands ; *Naja, Bungarus, Ophiophagus, Pseudonaje, Xenurelaps, Doliophis, Megærophis,* and *Callophis* are Oriental, one species of the latter genus being found in Japan, while an *Ophiophagus* has been discovered in New Guinea; *Cyrtophis, Elapsoidea,* and *Pœcilophis* are African : *Elaps* is American, ranging as far north as South Carolina, but not to the West Indian Islands.

FAMILY 21.—DENDRASPIDIDÆ.   (1 Genus, 5 Species.)

GENERAL DISTRIBUTION.					
NEOTROPICAL SUB-REGIONS.	NEARCTIC SUB-REGIONS.	PALÆARCTIC SUB-REGIONS.	ETHIOPIAN SUB-REGIONS.	ORIENTAL SUB-REGIONS.	AUSTRALIAN SUB-REGIONS.
— — — —	— — — —	— — — —	1 . 2 — —	— — — —	— — — —

The single genus *Dendraspis*, constituting the family, is confined to Tropical Africa.

FAMILY 22.—ATRACTASPIDIDÆ.   (1 Genus, 4 Species.)

GENERAL DISTRIBUTION.					
NEOTROPICAL SUB-REGIONS.	NEARCTIC SUB-REGIONS.	PALÆARCTIC SUB-REGIONS.	ETHIOPIAN SUB-REGIONS.	ORIENTAL SUB-REGIONS.	AUSTRALIAN SUB-REGIONS.
— — — —	— — — —	— — — —	— 2 . 3 —	— — — —	— — — —

This small family, consisting of the genus *Atractaspis*, is also confined to Africa, but has hitherto only been found in the West and South.

FAMILY 23.—HYDROPHIDÆ.   (8 Genera, 50 Species.)

GENERAL DISTRIBUTION.

NEOTROPICAL SUB-REGIONS.	NEARCTIC SUB-REGIONS.	PALÆARCTIC SUB-REGIONS.	ETHIOPIAN SUB-REGIONS.	ORIENTAL SUB-REGIONS.	AUSTRALIAN SUB-REGIONS.
— — — 3	— — — —	— — — —	— — — 4	1.2.3.4	1.2.3.4

The Hydrophidæ, or Sea Snakes, are a group of small-sized marine serpents, abundant in the Indian and Australian seas, and extending as far west as Madagascar, and as far east as Panama.   They are very poisonous, and it is probable that many species remain to be discovered.   The genera are distributed as follows :—

Hydrophis (37 sp.), ranging from India to Formosa and Australia; *Platurus* (2 sp.), from the Bay of Bengal to New Guinea and New Zealand; *Aipysurus* (3 sp.), Java to New Guinea and Australia; *Disteira* (1 sp.), unknown locality; *Acalyptus* (1 sp.), South-west Pacific; *Enhydrina* (1 sp.), Bay of Bengal to New Guinea; *Pelamis* (1 sp.), Madagascar to New Guinea, New Zealand, and Panama; *Emydocephalus* (1 sp.), Australian Seas.

FAMILY 24.—CROTALIDÆ.   (11 Genera, 40 Species.)

GENERAL DISTRIBUTION.

NEOTROPICAL SUB-REGIONS.	NEARCTIC SUB-REGIONS.	PALÆARCTIC SUB-REGIONS.	ETHIOPIAN SUB-REGIONS.	ORIENTAL SUB-REGIONS.	AUSTRALIAN SUB-REGIONS.
1.2.3.4	1.2.3.4	— — 3.4	— — — —	1.2.3.4	— — — —

The Crotalidæ, or Pit Vipers, including the deadly Rattlesnakes, form a well-marked family of fanged serpents, whose distribution is very interesting.   They abound most in the Oriental region, at least 5 of the genera and 20 species being found within its limits, yet they are quite unknown in the Ethiopian region —a parallel case to that of the Bears and Deer.   A few species are peculiar to the eastern portion of the Palæarctic region, while the Nearctic is actually richer than the Neotropical region both in genera and species. This would point to the conclusion, that the group originated in the Indo-Chinese sub-region and spread thence north-east to North America, and so onward to South America, which, having been the last to receive the group, has not had time to develop it largely, notwithstanding its extreme adaptability to Reptilian life. The genera are divided among the several regions as follows :—

Craspedocephalus (7 sp.), Tropical America and the West Indian Islands; *Cenchris, Crotalophorus, Uropsophorus,* and *Crotalus,* inhabiting North America from Canada and British Columbia to Texas, one species (*Crotalus horridus*) extending into South America; *Trimeresurus* (16 sp.), all India from Ceylon to Assam, Formosa, the Philippines and Celebes; *Peltopelor* and *Hypnale* (1 sp. each), peculiar to India ; *Calloselasma* (1 sp.), Siam ; *Atropos* (1 sp.), Java and Borneo ; *Halys* (3 sp.), peculiar to Tartary, Thibet, Japan, North China, and Formosa.

FAMILY 25.—VIPERIDÆ.   (3 Genera, 22 Species.)

GENERAL DISTRIBUTION.					
NEOTROPICAL SUB-REGIONS.	NEARCTIC SUB-REGIONS.	PALÆARCTIC SUB-REGIONS.	ETHIOPIAN SUB-REGIONS.	ORIENTAL SUB-REGIONS.	AUSTRALIAN SUB-REGIONS.
— — — —	— — — —	1 . 2 . 3 . 4	1 . 2 . 3 . 4	1 . 2 . 3 . 4	— — — —

The Viperidæ, or True Vipers, are especially characteristic of the Palæarctic and Ethiopian regions, only one species being found over a large part of the Oriental region, and another reaching Central India. They are especially abundant in Africa, and the Palæarctic confines in South-western Asia. The common Viper ranges across the whole Palæarctic region from Portugal to Saghalien Island, reaching to 67° North Latitude, in Scandinavia, and to 58° in Central Siberia. The genera, according to Dr. Strauch's synopsis, are distributed as follows :—

Vipera (17 sp.), which has the range of the family, extending over the whole of the Palæarctic and Ethiopian regions, except Madagascar, and as far as Ceylon, Siam, and Java, in the Oriental region; *Echis* (2 sp.), inhabiting North Africa to Persia and to Continental India; and *Atheris* (3 sp.), confined to West Africa.

Remarks on the General Distribution of Ophidia.

The Ophidia, being preeminently a Tropical order—rapidly diminishing in numbers as we go north in the Temperate Zone, and wholly ceasing long before we reach the Arctic Circle—we cannot expect the two Northern regions to. exhibit any great variety or peculiarity. Yet in their warmer portions they are tolerably rich; for, of the 25 families of snakes, 6 are found in the Nearctic region, 10 in the Palæarctic, 13 in the Australian, 16 in the Neotropical, 17 in the Ethiopian, and no less than 22 in the Oriental, which last is thus seen to be by far the richest of the great regions in the variety of its forms of Ophidian life. The only regions that possess altogether peculiar families of this order, are the Ethiopian (3), and the Oriental (2); the usually rich and peculiar Neotropical region not possessing exclusively, any family of snakes; and what is still more remarkable, the Neotropical and Australian regions together, do not possess a family peculiar to them. Every family inhabiting these two regions is found also in the Oriental; and this fact, taken in connection with the superior richness of the latter region both in families and genera, would indicate that the Ophidia had their origin in the northern hemisphere of the Old World (the ancient Palæarctic region) whence they spread on all sides, in successive waves of migration, to the other regions. The distribution of the genera peculiar to, or highly characteristic of, the several regions is as follows :—

The Nearctic possesses 9 ; four of these belong to the Colubridæ, one to the Pythonidæ, and four to the Crotalidæ. The Palæarctic region has only 2 peculiar genera, belonging to the Colubridæ and Crotalidæ. The Ethiopian has 25, belonging to 11 families ; four to Colubridæ, five to Lycodontidæ, and three to Elapidæ. The Oriental has no less than 50, belonging to 15 families ; five are Colubridæ, five Uropeltidæ, twelve Homalopsidæ, six Lycodontidæ, three Amblycephalidæ, eight Elapidæ, and four Crotalidæ.   The Australian has 16, belonging to three families only ; eleven being Elapidæ, and four Pythonidæ.   The Neotropical has about 24, belonging to eight families; ten are Colubridæ, six Pythonidæ, and the rest Dipsadidæ, Scytalidæ, Amblycephalidæ, Elapidæ, and Crotalidæ.

We find then, that in the Ophidia, the regions adopted in this work are remarkably distinct ; and that, in the case of the Oriental and Ethiopian, the difference is strongly marked, a very large number of the genera being confined to each region.   It is interesting to observe, that in many cases the affinity seems to be rather between the West Coast of Africa and the Oriental region, than between the East Coast and the plains of India ; thus the Homalopsidæ—a highly characteristic Oriental family— occur on the West Coast of Africa only ; the Dryiophidæ, which range over the whole Oriental region, only occur in Madagascar and West Africa in the Ethiopian ; the genus *Dipsas* is found over all the Oriental region and again in West Africa.   A cause for this peculiarity has been suggested in our sketch of the past history of the Ethiopian region, Vol. I. p. 288.   In the Lycodontidæ, which are strictly confined to these two regions, the genera are all distinct, and the same is the case with the more widely distributed Elapidæ ; and although a few desert forms, such as *Echis* and the Erycidæ, are common to Africa and the dry plains of India, this is evidently due to favourable climatic conditions, and cannot neutralise the striking differences in the great mass of the family and generic forms which inhabit the two regions. The union of Madagascar with the South-western part of the Oriental region under the appellation Lemuria, finds no support in the distribution of Ophidia ; which, however, strikingly accords with the views developed in the Third Part of this work, as to the great importance and high antiquity of the Euro-Asiatic continent, as the chief land-centre from which the higher organisms have spread over the globe.

Fossil Ophidia.—The oldest known remains of Ophidia occur in the Eocene formation in the Isle of Sheppey ; others are found in the Miocene (Brown Coal) of Germany, and in some Tertiary beds in the United States.   Most of these appear to have been large species belonging to the Pythonidæ, so that we are evidently still very far from knowing anything of the earliest forms of this order.　In some of the later Tertiary deposits the poison fangs of venomous species have been found; also a Colubrine snake from the Upper Miocene of the South of France.

Order II.—LACERTILIA.

FAMILY 26.—TROGONOPHIDÆ.　(1 Genus, 1 Species.)

GENERAL DISTRIBUTION.					
NEOTROPICAL SUB-REGIONS.	NEARCTIC SUB-REGIONS.	PALÆARCTIC SUB-REGIONS.	ETHIOPIAN SUB-REGIONS.	ORIENTAL SUB-REGIONS.	AUSTRALIAN SUB-REGIONS.
— — — —	— — — —	— 2 — —	— — — —	— — — —	— — — —

The single species of *Trogonophis*, forming this family, is found only in North Africa.

FAMILY 27.—CHIROTIDÆ.　(1 Genus, 1 Species.)

GENERAL DISTRIBUTION.					
NEOTROPICAL SUB-REGIONS.	NEARCTIC SUB-REGIONS.	PALÆARCTIC SUB-REGIONS.	ETHIOPIAN SUB-REGIONS.	ORIENTAL SUB-REGIONS.	AUSTRALIAN SUB-REGIONS.
— — 3 —	— — 3 —	— — — —	— — — —	— — — —	— — — —

Chirotes, the genus which constitutes this family, inhabits Mexico, and has also been found in Missouri, one of the Southern United States.

FAMILY 28.—AMPHISBÆNIDÆ.　(1 Genus, 13 Species.)

GENERAL DISTRIBUTION.					
NEOTROPICAL SUB-REGIONS.	NEARCTIC SUB-REGIONS.	PALÆARCTIC SUB-REGIONS.	ETHIOPIAN SUB-REGIONS.	ORIENTAL SUB-REGIONS.	AUSTRALIAN SUB-REGIONS.
1 . 2 — 4	— — — —	— 2 — —	1 . 2 — —	— — — —	— — — —

The Amphisbænidæ, which, in the opinion of Dr. Günther, are all comprised in the genus *Amphisbæna*, inhabit Spain and Asia Minor, North and Tropical Africa, South America as far as Buenos-Ayres and the West Indian Islands.

FAMILY 29.—LEPIDOSTERNIDÆ.   (3 Genera, 6 Species.)

GENERAL DISTRIBUTION.					
NEOTROPICAL SUB-REGIONS.	NEARCTIC SUB-REGIONS.	PALÆARCTIC SUB-REGIONS.	ETHIOPIAN SUB-REGIONS.	ORIENTAL SUB-REGIONS.	AUSTRALIAN SUB-REGIONS.
1 . 2 — —	— — — —	— — — —	— 2 . 3 —	— — — —	— — — —

The small family of Lepidosternidæ has nearly the same distribution as the last, indicating a curious relationship between the Tropical parts of Africa and America. *Lepidosternon* and *Cephalopeltis* are American genera, while *Monotrophis* is African.

FAMILY 30.—VARANIDÆ.   (3 Genera, 30 Species.)

GENERAL DISTRIBUTION.					
NEOTROPICAL SUB-REGIONS.	NEARCTIC SUB-REGIONS.	PALÆARCTIC SUB-REGIONS.	ETHIOPIAN SUB-REGIONS.	ORIENTAL SUB-REGIONS.	AUSTRALIAN SUB-REGIONS.
— — — —	— — — —	— 2 — —	1 . 2 . 3 —	1 . 2 . 3 . 4	1 . 2 — —

The Varanidæ, or Water Lizards, are most abundant in the Oriental region, whence they extend into the Austro-Malay Islands as far as New Guinea, and into Australia. Several species are found in Africa. *Psammosaurus* (1 sp), is found in North Africa and North-western India ; *Monitor* (18 sp.), has the range of the family ; while *Hydrosaurus* (8 sp) ranges from Siam to the Philippines, New Guinea, and Australia.

FAMILY 31.—HELODERMIDÆ.   (1 Genus, 1 Species.)

GENERAL DISTRIBUTION.

NEOTROPICAL SUB-REGIONS.	NEARCTIC SUB-REGIONS.	PALÆARCTIC SUB-REGIONS.	ETHIOPIAN SUB-REGIONS.	ORIENTAL SUB-REGIONS.	AUSTRALIAN SUB-REGIONS.
— — 3 —	— — — —	— — — —	— — — —	— — — —	— — — —

The genus *Heloderma*, which constitutes this family, is found in Mexico.

FAMILY 32.—TEIDÆ.   (12 Genera, 74 Species.)

GENERAL DISTRIBUTION.

NEOTROPICAL SUB-REGIONS.	NEARCTIC SUB-REGIONS.	PALÆARCTIC SUB-REGIONS.	ETHIOPIAN SUB-REGIONS.	ORIENTAL SUB-REGIONS.	AUSTRALIAN SUB-REGIONS.
1 . 2 . 3 . 4	1 . 2 . 3 —	— — — —	— — — —	— — — —	— — — —

The Teidæ, or Teguexins—a group of Lizards allied to the European Lacertidæ, but with differently formed superciliary scales—are highly characteristic of the Neotropical region, abounding almost everywhere from Patagonia to the Antilles and Mexico, and extending northwards to California on the west and to Pennsylvania on the east. The most extensive genus is *Ameiva*, containing nearly 60 species and having the range of the entire family; *Teius* (3 sp.), inhabits Brazil and Mendoza; *Callopistes* (2 sp.), Chili; *Centropyx* (3 sp.), Paraguay to Alabama; *Dicrodon* (Peru); *Monoplocus* (Western Ecuador); with *Acrantus, Acanthopyga, Emminia, Crocodilurus, Custa*, and *Ada*, which each consist of a single species, and all inhabit Tropical America.

FAMILY 33.—LACERTIDÆ.   (18 Genera, 80 Species.)

GENERAL DISTRIBUTION.

NEOTROPICAL SUB-REGIONS.	NEARCTIC SUB-REGIONS.	PALÆARCTIC SUB-REGIONS.	ETHIOPIAN SUB-REGIONS.	ORIENTAL SUB-REGIONS.	AUSTRALIAN SUB-REGIONS.
— — — —	— — — —	1 . 2 . 3 . 4	1 . 2 . 3 —	1 . 2 . 3 . 4	— 2 — —

The Lacertidæ, or Land Lizards, are small-sized, terrestrial, non-burrowing lizards, very characteristic of the Palæarctic region, which contains more than half the known species, and of the adjacent parts of the Oriental and Ethiopian regions, but extending also to South Africa, to Java, and even to Australia. The best-defined genera are the following :—

Lacerta (10 sp.), ranging over all Central and South Europe to Poland, and farther north in Russia and Siberia, eastward to Persia, and southward to North and West Africa; *Zootoca* (8 sp.), has nearly the same range in Europe as the last genus, but has representatives in Madeira, South Africa, and Australia; *Tachydromus* (7 sp.) is widely scattered in Chinese Asia, Japan, Borneo, and West Africa; *Acanthodactylus* (10 sp.) is most abundant in North Africa, but has a species in South Africa, and two in Central India; *Eremias* (18 sp.) is found all over Africa, and also in the Crimea, Persia, Tartary and China; *Psammodromus* (2 sp.), is confined to Spain, France, and Italy; *Ophiops* (6 sp.), inhabits India, Persia, and Asia Minor to South Russia. Less strongly marked and perhaps less natural genera are the following :—

Thetia (1 sp.), Algiers; *Teira* (1 sp.), Madeira; *Nucras* (4 sp.), Caucasus and South Africa; *Notopholis* (4 sp.), South Europe and South Africa; *Algira* (3 sp.), North and South Africa; *Scrapteira* (1 sp.), Nubia; *Aspidorhinus* (1 sp.), Caspian district; *Messalina* (4 sp.), North Africa, Persia, and North-west India *Cabrita*(1 sp.), Central India; *Pachyrhynchus* (1 sp.), Benguela.

FAMILY 34.—ZONURIDÆ.   (15 Genera, 52 Species.)

GENERAL DISTRIBUTION.					
NEOTROPICAL SUB-REGIONS.	NEARCTIC SUB-REGIONS.	PALÆARCTIC SUB-REGIONS.	ETHIOPIAN SUB-REGIONS.	ORIENTAL SUB-REGIONS.	AUSTRALIAN SUB-REGIONS.
— 2 . 3 . 4	1 . 2 . 3 . 4	— 2 — —	1 . 2 . 3 . 4	— — 3 —	— 2 — —

The Zonuridæ, or Land Lizards, characterised by a longitudinal fold of skin on each side of the body, have a very remarkable distribution.   Their head-quarters is the Ethiopian region, which contains more than half the known genera and species, most of which are found in South Africa and several in Madagascar.   Next to Africa the largest number of genera and species are found in Mexico and Central America, with a few in the Antilles, South America, and California, and even as far north as British Columbia.   Three of the genera form a distinct sub-group—the Glass Snakes,—the four species composing it being located in North Africa, North America, South-eastern Europe, and the Khasya Hills.

The prominent fact in the distribution of this family is, that the mass of the genera and species form two groups, one in South Africa, the other in Mexico,—countries between which it would be difficult to imagine any means of communication.   We have here, probably, an example of a once much more extensive group, widely distributed over the globe, and which has continued to maintain itself only in those districts especially adapted to its peculiar type of organization.   This must undoubtedly have been the case with the genus *Pseudopus*, whose two species now inhabit South-eastern Europe and the Khasya Hills in Assam respectively.

The genera   are,—*Cordylus*, *Pseudocordylus*, *Platysaurus*, *Cordylosaurus*, *Pleurostrichus*, and *Saurophis*, confined to South Africa ; *Zonurus*, South and East Africa and Madagascar; *Gerrhosaurus*, ranges over the whole Ethiopian region ; *Cicigna* is confined to Madagascar; *Gerrhonotus* (22 sp.), ranges from British Columbia, California, and Texas, to Cuba and South America, but is most abundant in Mexico and Central America ; *Abronia* and *Barissia*, are two genera of doubtful distinctness, peculiar to Mexico; *Ophisaurus* (the Glass Snake) is found in the Southern United States as far as Virginia ; the allied genus *Hyalosaurus* in North Africa; and *Pseudopus*, as above stated, in South-east Europe and the Khasya Hills.

FAMILY 35.—CHALCIDÆ.   (3 Genera, 8 Species.)

GENERAL DISTRIBUTION.					
NEOTROPICAL SUB-REGIONS.	NEARCTIC SUB-REGIONS.	PALÆARCTIC SUB-REGIONS.	ETHIOPIAN SUB-REGIONS.	ORIENTAL SUB-REGIONS.	AUSTRALIAN SUB-REGIONS.
1.2.3 —	— —?3 —	— — — —	— — — --	— — — —	— — — —

The Chalcidæ are a small group of Lizards characteristic of Tropical America, one species extending into the United States.

The genera are *Chalcis* (6 sp.), ranging from Central America to Chili; two other species, which have been placed in distinct genera, inhabit North America and Peru.

FAMILY 36.—ANADIADÆ.   (1 Genus, 1 Species.)

GENERAL DISTRIBUTION.					
NEOTROPICAL SUB-REGIONS.	NEARCTIC SUB-REGIONS.	PALÆARCTIC SUB-REGIONS.	ETHIOPIAN SUB-REGIONS.	ORIENTAL SUB-REGIONS.	AUSTRALIAN SUB-REGIONS.
— 2 — —	— — — —	— — — —	— — — —	— — — —	— — — —

The single species of *Anadia*, constituting this family, inhabits Tropical America.

FAMILY 37.—CHIROCOLIDÆ.   (1 Genus, 2 Species.)

GENERAL DISTRIBUTION.					
NEOTROPICAL SUB-REGIONS.	NEARCTIC SUB-REGIONS.	PALÆARCTIC SUB-REGIONS.	ETHIOPIAN SUB-REGIONS.	ORIENTAL SUB-REGIONS.	AUSTRALIAN SUB-REGIONS.
-- 2 — —	— — — —	— — — —	— — — —	— — — —	— — — —

The genus *Heterodactylus*, which constitutes this family, inhabits Brazil.

FAMILY 38.—IPHISADÆ.   (1 Genus, 1 Species.)

GENERAL DISTRIBUTION.					
NEOTROPICAL SUB-REGIONS.	NEARCTIC SUB-REGIONS.	PALÆARCTIC SUB-REGIONS.	ETHIOPIAN SUB-REGIONS.	ORIENTAL SUB-REGIONS.	AUSTRALIAN SUB-REGIONS.
— 2 — —	— — — —	— — — —	— — — —	— — . —	— — — —

The single species of *Iphisa,* has been found only at Para in Equatorial America.

FAMILY 39.—CERCOSAURIDÆ.   (1 Genus, 5 Species.)

GENERAL DISTRIBUTION.					
NEOTROPICAL SUB-REGIONS.	NEARCTIC SUB-REGIONS.	PALÆARCTIC SUB-REGIONS.	ETHIOPIAN SUB-REGIONS.	ORIENTAL SUB-REGIONS.	AUSTRALIAN SUB-REGIONS.
— 2 — —	— — — —	— — — —	— — — —	— — — —	— — — —

The genus *Cercosaura,* is known only from Brazil and Ecuador.

FAMILY 40.—CHAMÆSAURIDÆ.   (1 Genus, 1 Species.)

GENERAL DISTRIBUTION.					
NEOTROPICAL SUB-REGIONS.	NEARCTIC SUB-REGIONS.	PALÆARCTIC SUB-REGIONS.	ETHIOPIAN SUB-REGIONS.	ORIENTAL SUB-REGIONS.	AUSTRALIAN SUB-REGIONS.
— — — —	— — — —	— — — —	— — 3 —	— — — —	— — — —

This family, consisting of a single species of the genus *Chamæsaura,* is confined to South Africa.

FAMILY 41.—GYMNOPTHALMIDÆ.   (5 Genera, 14 Species.)

GENERAL DISTRIBUTION.					
NEOTROPICAL SUB-REGIONS.	NEARCTIC SUB-REGIONS.	PALÆARCTIC SUB-REGIONS.	ETHIOPIAN SUB-REGIONS.	ORIENTAL SUB-REGIONS.	AUSTRALIAN SUB-REGIONS.
— 2 - 4	— — — —	1 . 2 . 3 —	— 2 — 4	— — — —	1 . 2 . 3 —

The Gymnopthalmidæ, or Gape-eyed Scinks, so called from
their rudimentary eyelids, form a small group, which is widely
and somewhat erratically distributed, as will be seen by the
following account of the distribution of the genera :—

Lerista (1 sp.) and three other species for which Dr. Gray has
established the genera—*Morethria* (1 sp.), and *Menetia* (2 sp.),
are confined to Australia ; *Cryptoblepharus* (4 sp.), is found in
West Australia, Timor, New Guinea, the Fiji Islands, and
Mauritius ; *Ablepharus* (4 sp.), inhabits Eastern and South-
eastern Europe, Persia, Siberia, West Africa, and the Bonin
Islands ; and *Gymnopthalmus* (3 sp.), is found in Brazil and the
West Indies.

FAMILY 42.—PYGOPODIDÆ.   (2 Genera, 3 Species.)

GENERAL DISTRIBUTION.					
NEOTROPICAL SUB-REGIONS.	NEARCTIC SUB-REGIONS.	PALÆARCTIC SUB-REGIONS.	ETHIOPIAN SUB-REGIONS.	ORIENTAL SUB-REGIONS.	AUSTRALIAN SUB-REGIONS.
— — —	— — ·· —	— ·· — —	— — · —	— — — —	— 2 — —

This small family of two-legged Lizards, comprising the
genera *Pygopus* and *Delma,* is found only in Australia proper
and Tasmania.

FAMILY 43.—APRASIADÆ.   (1 Genus, 2 Species.)

GENERAL DISTRIBUTION.

NEOTROPICAL SUB-REGIONS.	NEARCTIC SUB-REGIONS.	PALÆARCTIC SUB-REGIONS.	ETHIOPIAN SUB-REGIONS.	ORIENTAL SUB-REGIONS.	AUSTRALIAN SUB-REGIONS.
— — — —	— — — —	— — — —	— — — —	— — — —	— . 2 — . .

The genus *Aprasia*, constituting this family, is found in West and South Australia.

FAMILY 44.—LIALIDÆ.   (1 Genus, 3 Species.)

GENERAL DISTRIBUTION.

NEOTROPICAL SUB-REGIONS.	NEARCTIC SUB-REGIONS.	PALÆARCTIC SUB-REGIONS.	ETHIOPIAN SUB-REGIONS.	ORIENTAL SUB-REGIONS.	AUSTRALIAN SUB-REGIONS.
— — — —	— — — —	— — — —	— — — —	— — — —	— 2 — —

This family is also confined to Australia, the single genus, *Lialis*, inhabiting the Western and Northern districts.

FAMILY 45.—SCINCIDÆ.   (60 Genera, 300 Species.)

GENERAL DISTRIBUTION.

NEOTROPICAL SUB-REGIONS.	NEARCTIC SUB-REGIONS.	PALÆARCTIC SUB-REGIONS.	ETHIOPIAN SUB-REGIONS.	ORIENTAL SUB-REGIONS.	AUSTRALIAN SUB-REGIONS.
1.2.3.4	1.2.3 —	1.2.3.4	1.2.3.4	1.2.3.4	1.2.3.4

The Scincidæ, or Scinks, are an extensive family of smooth-scaled lizards, frequenting dry and stony places, and almost universally distributed over the globe, being only absent from the cold northern and southern zones. The family itself is a very natural one, and it contains many natural genera; but a large number have been established which probably require careful revision. The following include the more important and the best established groups :—

Scincus (2 sp.), North Africa and Arabia; *Hinulia* (20 sp.), most of the Australian and Oriental regions; *Cyclodina* (1 sp.), *Hombronia* (1 sp.), and *Lygosomella* (1 sp.), all from New Zealand; *Keneuxia* (1 sp.), Philippines, Moluccas, and Papuan Islands; *Elania* (1 sp.) New Guinea; *Carlia* (2 sp.), North Australia and New Guinea; *Mocoa* (16 sp.), Australia and New Zealand, with species in Borneo, West Africa, and Central America; *Lipinia* (3 sp.), Philippine Islands and New Guinea; *Lygosoma* (12 sp.), Australia, New Caledonia, Pelew and Philippine Islands; *Tetradactylus* (1 sp.), *Hemierges* (2 sp.), *Chelomeles* (2 sp.), *Omolepida* (1 sp.), *Lissolepis* (1 sp.), *Siaphos* (1 sp.), *Rhodona* (3 sp.) *Anomalpus* (1 sp.), *Soridia* (2 sp.), and *Ophioscincus* (1 sp.) all confined to Australia; *Cophoscincus* (3 sp.), Philippine Islands, Celebes, and Queensland; *Plestiodon* (18 sp.), China and Japan, Africa, and America as far north as Pennsylvania and Nebraska; *Eumeces* (30 sp.), South Palæarctic, Oriental and Australian regions, to New Ireland and North Australia; *Mabouya* (20 sp.), Oriental region, Austro-Malaya, North Australia, the Neotropical region, and to Lat. 42° 30' in North America; *Amphixestus* (1 sp.), Borneo; *Hagria* 1 sp.), and *Chiamela* (1 sp.), India; *Senira* (1 sp.), Philippine Islands; *Brachymeles* (2 sp.). Philippine Islands and Australia; *Ophiodes* (1 sp.), Brazil; *Anguis* (3 sp.), West Palæarctic region and South Africa; *Tribolonotus* (1 sp.), New Guinea; *Tropidophorus* (2 sp.), Cochin-China and Philippine Islands; *Norbea* (2 sp.), Borneo and Australia; *Trachydosaurus* (1 sp.), Australia; *Cyclodus* (8 sp.), Australia, Aru Islands, and Ceram; *Silubosaurus* (2 sp.), *Egerina* (2 sp.), and *Tropidolepisma* (6 sp.), all peculiar to Australia; *Heteropus* (7 sp.), Australia, Austro-Malaya, and Bourbon; *Pygomeles* (1 sp.), Madagascar; *Dasia* (1 sp.), Malaya; *Euprepes* (70 sp.), Ethiopian and Oriental regions, Austro-Malaya, South America (?); *Celestus* (9 sp.), peculiar to the Antilles, except a species in Costa Rica; *Diploglossus* (7 sp.), the Neotropical region;—with a number of other genera founded on single species from various parts of the world.

FAMILY 46.—OPHIOMORIDÆ.   (2 Genera, 2 Species.)

GENERAL DISTRIBUTION.					
NEOTROPICAL SUB-REGIONS.	NEARCTIC SUB-REGIONS.	PALÆARCTIC SUB-REGIONS.	ETHIOPIAN SUB-REGIONS.	ORIENTAL SUB-REGIONS.	AUSTRALIAN SUB-REGIONS.
— — — —	— — — —	— 2 — —	— — — —	— — — —	— — — —

The snake-like Lizard constituting the genus *Ophiomorus*, is found in Southern Russia, Greece, and Algeria; while *Zygnopsis* having four weak limbs, has been recently discovered by Mr. Blanford in South Persia. The family is therefore confined to our Mediterranean sub-region.

FAMILY 47.—SEPIDÆ.   (7 Genera, 22 species.)

GENERAL DISTRIBUTION.					
NEOTROPICAL SUB-REGIONS.	NEARCTIC SUB-REGIONS.	PALÆARCTIC SUB-REGIONS.	ETHIOPIAN SUB-REGIONS.	ORIENTAL SUB-REGIONS.	AUSTRALIAN SUB-REGIONS.
— — — —	— — — —	— 2 — —	1.2.3.4	— — — —	— — — —

The Sepidæ, or Sand-Lizards, are a very natural group, almost confined to the Ethiopian region, but extending into the desert country on the borders of the Oriental region, and into the south of the Palæarctic region as far as Palestine, Madeira, Spain, Italy, and even the South of France. The genera are:—
Seps (10 sp.), South Europe, Madeira, Teneriffe, Palestine, North Africa, South Africa and Madagascar; *Sphenops* (2 sp.), North Africa, Syria, West Africa; *Scelotes* (3 sp.), Angola to South Africa, Madagascar; *Thyrus* (1 sp.), Bourbon and Mauritius; *Amphiglossus* (1 sp.), Madagascar; *Sphenocephalus* (1 sp.), Afghanistan; and *Sepsina* (4 sp.), South-west Africa.

FAMILY 48.—ACONTIADÆ.   (3 Genera, 7 Species.)

GENERAL DISTRIBUTION.

NEOTROPICAL SUB-REGIONS.	NEARCTIC SUB-REGIONS.	PALÆARCTIC SUB-REGIONS.	ETHIOPIAN SUB-REGIONS.	ORIENTAL SUB-REGIONS.	AUSTRALIAN SUB-REGIONS.
— — — —	— — — —	— — — —	— 2 . 3 . 4	— 2 — —	1 — — —

This small family of snake-like Lizards has a very curious distribution, being found in South and West Africa, Madagascar, Ceylon, and Ternate in the Moluccas. *Acontias* (4 sp.), is found in the four first-named localities; *Nessia* (2 sp.), is confined to Ceylon; *Typhloscincus* (1 sp.), to Ternate.

FAMILY 49.—GECKOTIDÆ.   (50 Genera, 200 Species.)

GENERAL DISTRIBUTION.

NEOTROPICAL SUB-REGIONS.	NEARCTIC SUB-REGIONS.	PALÆARCTIC SUB-REGIONS.	ETHIOPIAN SUB-REGIONS.	ORIENTAL SUB-REGIONS.	AUSTRALIAN SUB-REGIONS.
1 . 2 . 3 . 4	1 . 2 . 3 —	1 . 2 . 3 . 4	1 . 2 . 3 . 4	1 . 2 . 3 . 4	1 . 2 . 3 . 4

The Geckoes, or Wall-Lizards, form an extensive family, of almost universal distribution in the warmer parts of the globe; and they must have some exceptional means of dispersal, since they are found in many of the most remote islands of the great oceans,—as the Galapagos, the Sandwich Islands, Tahiti, New Zealand, the Loo-Choo and the Seychelle Islands, the Nicobar Islands, Mauritius, Ascension, Madeira, and many others. The following are the larger and more important genera :—

Oëdura (3 sp.), Australia; *Diplodactylus* (8 sp.), Australia, South Africa, and California; *Phyllodactylus* (8 sp.), widely scattered in Tropical America, California, Madagascar, and Queensland; *Hemidactylus* (40 sp.), all tropical and warm countries; *Peropus* (12 sp.), the Oriental region, Papuan Islands, Mauritius, and Brazil; *Pentadactylus* (7 sp.), Oriental region and Australia; *Gecko* (12 sp.), Oriental region to New Guinea and

North Australia; *Gehyra* (5 sp.), Australia, New Guinea and Fiji Islands; *Tarentola* (7 sp.), North Africa, North America, Madeira, Borneo, South Africa; *Phelsuma* (6 sp.), Madagascar, Bourbon, and Andaman Islands; *Pachydactylus* (5 sp.), South and West Africa, and Ascension Island; *Sphærodactylus* (5 sp.), the Neotropical region; *Naultinus*, (6 sp.), New Zealand; *Goniodactylus* (5 sp.), Australia, Timor, South America and Algiers; *Heteronota* (4 sp.), Australia, Fiji Islands, New Guinea and Borneo; *Cubina* (4 sp.), the Neotropical region; *Gymnodactylus* (16 sp.), all warm countries except Australia; *Phyllurus* (3 sp.), Australia; *Stenodactylus* (4 sp.), North and West Africa, and Rio Grande in North America.

The remaining genera mostly consist of single species, and are pretty equally distributed over the various parts of the world indicated in the preceding list. Madagascar, the Seychelle Islands, Chili, the Sandwich Islands, South Africa, Tahiti, the Philippine Islands, New Caledonia, and Australia—all have peculiar genera, while two new ones have recently been described from Persia.

FAMILY 50.—IGUANIDÆ. (56 Genera, 236 Species.)

GENERAL DISTRIBUTION.					
NEOTROPICAL SUB-REGIONS.	NEARCTIC SUB-REGIONS.	PALÆARCTIC SUB-REGIONS.	ETHIOPIAN SUB-REGIONS.	ORIENTAL SUB-REGIONS.	AUSTRALIAN SUB-REGIONS.
1 . 2 . 3 . 4	1 . 2 . 3 —	— — — —	— — — —	— — — —	— — 3 —

The extensive family of the Iguanas is highly characteristic of the Neotropical region, in every part of which the species abound, even as far as nearly 50° South Latitude in Patagonia. They also extend northwards into the warmer parts of the Nearctic region, as far as California, British Columbia, and Kansas on the west, and to 43° North Latitude in the Eastern States. A distinct genus occurs in the Fiji Islands, and one has been described as from Australia, and another from Madagascar, but there is some doubt about these. The most extensive genera are :—

Anolius (84 sp.), found in most parts of Tropical America and

north to California; *Tropidolepis* (15 sp.), which has nearly the
same range; *Leiocephalus* (14 sp.), Antilles, Guayaquil, and
Galapagos Islands; *Leiolœmus* (14 sp.), Peru to Patagonia;
Sceloporus (9 sp.), from Brazil to California and British Columbia,
and on the east to Florida; *Proctotretus* (6 sp.), Chili and Patagonia; *Phrynosoma* (8 sp.), New Mexico, California, Oregon
and British Columbia, Arkansas and Florida; *Iguana* (5 sp.),
Antilles and South America; *Cyclusa* (4 sp.), Antilles, Honduras, and Mexico.

Among the host of smaller genera may be noted:—

Brachylophus, found in the Fiji Islands; *Trachycephalus* and
Oreocephalus, peculiar to the Galapagos; *Oreodeira*, said to be from
Australia; *Diplolœmus* and *Phymaturus*, found only in Chili and
Patagonia; and *Callisaurus, Uta, Euphryne, Uma*, and *Hol-
brookia*, from New Mexico and California. All the other genera
are from various parts of Tropical America.

FAMILY 51.—AGAMIDÆ.   (42 Genera, 156 Species.)

GENERAL DISTRIBUTION.					
NEOTROPICAL SUB-REGIONS.	NEARCTIC SUB-REGIONS.	PALÆARCTIC SUB-REGIONS.	ETHIOPIAN SUB-REGIONS.	ORIENTAL SUB-REGIONS.	AUSTRALIAN SUB-REGIONS.
— — — —	— — — —	— 2.3.4	1.2.3.4	1.2.3.4	1.2.3 —

The extensive family Agamidæ—the Eastern representative
of the Iguanas—is highly characteristic of the Oriental region,
which possesses about half the known genera and species. Of the
remainder, the greater part inhabit the Australian region; others
range over the deserts of Central and Western Asia and Northern
Africa, as far as Greece and South Russia. One genus extends
through Africa to the Cape of Good Hope, and there are three
peculiar genera in Madagascar, but the family is very poorly
represented in the Ethiopian region. Many of these creatures
are adorned with beautifully varied and vivid colours, and the
little "dragons" or flying-lizards are among the most interesting
forms in the entire order. The larger genera are distributed as
follows:—

Draco (18 sp.), the Oriental region, excluding Ceylon; *Otocryptis* (4 sp.), Ceylon, North India, Malaya; *Ceratophora* (3 sp.), Ceylon; *Gonyocephalus* (8 sp.), Papuan Islands, Java, Borneo, Pelew Islands; *Dilophyrus* (7 sp.), Indo-Malaya and Siam; *Japalura* (6 sp.), Himalayas, Borneo, Formosa, and Loo Choo Islands; *Sitana* (2 sp.), Central and South India and Ceylon; *Bronchocela* (3 sp.), Indo-Malaya, Cambodja, and Celebes; *Calotes* (12 sp.), Continental India to China, Philippine Islands; *Oriocalotes* (2 sp.), Himalayas; *Acanthosaura* (5 sp.), Malacca and Siam; *Tiaris* (3 sp.), Andaman Islands, Borneo, Philippine and Papuan Islands; *Physignathus* (3 sp.), Cochin-China and Australia; *Uromastix* (5 sp.), South Russia, North Africa, Central India; *Stellio* (5 sp.), Caucasus and Greece to Arabia, High Himalayas and Central India; *Trapelus* (5 sp.), Tartary, Egypt, and Afghanistan; *Phrynocephalus* (10 sp.), Tartary and Mongolia, Persia and Afghanistan; *Lophura* (2 sp.), Amboyna and Pelew Islands; *Grammatophorus* (14 sp.), Australia and Tasmania; *Agama* (14 sp.), North Africa to the Punjaub, South Africa. The remaining genera each consist of a single species. Eight are peculiar to Australia, one to the Fiji Islands, one to the Aru Islands, three to Ceylon, five to other parts of the Oriental region, one to Persia, and one to South Russia.

FAMILY 52.—CHAMÆLEONIDÆ.   (1 Genus, 30 Species.)

GENERAL DISTRIBUTION.

NEOTROPICAL SUB-REGIONS.	NEARCTIC SUB-REGIONS.	PALÆARCTIC SUB-REGIONS.	ETHIOPIAN SUB-REGIONS.	ORIENTAL SUB-REGIONS.	AUSTRALIAN SUB-REGIONS.
— — — —	— — — —	— 2 — ··	1 . 2 . 3 . 4	1 . 2 — —	— — — —

The Chamæleons are an almost exclusively Ethiopian group, only one species, the common Chamæleon, inhabiting North Africa and Western Asia as far as Central India and Ceylon. They abound all over Africa, and peculiar species are found in Madagascar and Bourbon, as well as in the Island of Fernando Po.

General Remarks on the Distribution of the Lacertilia.

The distribution of the Lacertilia is, in many particulars, strikingly opposed to that of the Ophidia. The Oriental, instead of being the richest is one of the poorest regions, both in the number of families and in the number of peculiar genera it contains ; while in both these respects the Neotropical is by far the richest. The distribution of the families is as follows :—

The Nearctic region has 7 families, none of which are peculiar to it; but it has 3 peculiar genera—*Chirotes, Ophisaurus,* and *Phrynosoma.*

The Palæarctic region has 12 families, with two (Ophiomoridæ and Trogonophidæ, each consisting of a single species) peculiar ; while it has 6 peculiar or very characteristic genera, *Trogonophis* in North Africa, *Psammodromus* in South Europe, *Hyalosaurus* in North Africa, *Scincus* in North Africa and Arabia, *Ophiomorus* in East Europe and North Africa, and *Phrynocephalus* in Siberia, Tartary, and Afghanistan. We have here a striking amount of diversity between the Nearctic and Palæarctic regions with hardly a single point of resemblance.

The Ethiopian region has 13 families, only one of which (the Chamæsauridæ, consisting of a single species) is altogether peculiar; but it possesses 21 peculiar or characteristic genera, 9 belonging to the Zonuridæ, 2 to the Sepidæ, 7 to the Geckotidæ, and 3 to the Agamidæ.

The Oriental region has only 8 families, none of which are peculiar ; but there are 28 peculiar genera, 6 belonging to the Scincidæ, 1 to the Acontiadæ, 5 to the Geckotidæ, and 16 to the Agamidæ. Many lizards being sand and desert-haunters, it is not surprising that a number of forms are common to the borderlands of the Oriental and Ethiopian regions ; yet the Sepidæ, so abundant in all Africa, do not range to the peninsula of India ; and the equally Ethiopian Zonuridæ have only one Oriental species, found, not in the peninsula but in the Khasya Hills. The Acontiadæ alone offer some analogy to the distribution of the Lemurs, being found in Africa, Madagascar, Ceylon, and the Moluccas.

The Australian region has 11 families, 3 of which are peculiar; and it has about 40 peculiar genera in ten families, about half of these genera belonging to the Scincidæ. Only 3 families of almost universal distribution are common to the Australian and Neotropical regions, with one species of the American Iguanidæ in the Fiji Islands, so that, as far as this order is concerned, these two regions have little resemblance.

The Neotropical region has 15 families, 6 of which are peculiar to it, and it possesses more than 50 peculiar genera. These are distributed among 12 families, but more than half belong to the Iguanidæ, and half the remainder to the Teidæ,—the two families especially characteristic of the Neotropical region. All the Nearctic families which are not of almost universal distribution are peculiarly Neotropical, showing that the Lacertilia of the former region have probably been derived almost exclusively from the latter.

On the whole the distribution of the Lacertilia shows a remarkable amount of specialization in each of the great tropical regions, whence we may infer that Southern Asia, Tropical Africa, Australia, and South America, each obtained their original stock of this order at very remote periods, and that there has since been little intercommunication between them. The peculiar affinities indicated by such cases as the Lepidosternidæ, found only in the tropics of Africa and South America, and *Tachydromus* in Eastern Asia and West Africa, may be the results either of once widely distributed families surviving only in isolated localities where the conditions are favourable,—or of some partial and temporary geographical connection, allowing of a limited degree of intermixture of faunas. The former appears to be the more probable and generally efficient cause, but the latter may have operated in exceptional cases.

Fossil Lacertilia.

These date back to the Triassic period, and they are found in most succeeding formations, but it is not till the Tertiary period that forms allied to existing genera occur. These are at present too rare and too ill-defined to throw much light on the geographical distribution of the order.

Order III.—RHYNCOCEPHALINA.

FAMILY 53.—RHYNCOCEPHALIDÆ.   (1 Genus, 1 Species.)

GENERAL DISTRIBUTION.

NEOTROPICAL SUB-REGIONS.	NEARCTIC SUB-REGIONS.	PALÆARCTIC SUB-REGIONS.	ETHIOPIAN SUB-REGIONS.	ORIENTAL SUB-REGIONS.	AUSTRALIAN SUB-REGIONS.
— — — —	— — — —	— — — —	— — — —	— — — —	— — — 4

The singular and isolated genus *Hatteria*—the "Tuatara" or
fringed lizard—which alone constitutes this family, has peculiarities of structure which separate it from both lizards and crocodiles,
and mark it out as an ancestral type, as distinct from other living
reptiles as the Marsupials are from other Mammalia.   It is confined to New Zealand, and is chiefly found on small islands near
the north-east coast, being very rare, if not extinct, on the main
land.   A fossil reptile named *Hyperodapedon*, of Triassic age, has
been found in Scotland and India, and is supposed by Professor
Huxley to be more nearly allied to *Hatteria* than to any other
living animal.

Order IV.—CROCODILIA.

FAMILY 54.—GAVIALIDÆ.   (2 Genera, 3 Species.)

GENERAL DISTRIBUTION.

NEOTROPICAL SUB-REGIONS.	NEARCTIC SUB-REGIONS.	PALÆARCTIC SUB-REGIONS.	ETHIOPIAN SUB-REGIONS.	ORIENTAL SUB-REGIONS.	AUSTRALIAN SUB-REGIONS.
— — — —	— — — —	— — — —	— — — —	1 — — 4	1 — — —

The Gavials are long-snouted Crocodiles with large front teeth,
and canines fitting in notches of the upper jaw.   They consist
of two genera, *Gavialis* (1 sp.), inhabiting the Ganges ; *Tomistoma*
(2 sp.), found in the rivers of Borneo and North Australia.

FAMILY 55.—CROCODILIDÆ.　(1 Genus, 12 Species.)

GENERAL DISTRIBUTION.

NEOTROPICAL SUB-REGIONS.	NEARCTIC SUB-REGIONS.	PALÆARCTIC SUB-REGIONS.	ETHIOPIAN SUB-REGIONS.	ORIENTAL SUB-REGIONS.	AUSTRALIAN SUB-REGIONS.
– 2 . 3 . 4	– – – –	– – – –	1 . 2 . 3 . 4	1 . 2 . 3 . 4	1 – – –

The true Crocodiles, which have the canines in notches, and the large front teeth in pits in the upper jaw, are widely distributed over the tropical regions of the globe, inhabiting all the rivers of Africa, the shores and estuaries of India, Siam, and eastward to North Australia.　Other forms inhabit Cuba, Yucatan, and Guatemala, to Ecuador and the Orinooko.　Four species are Asiatic, one exclusively Australian, three African, and four American.　These have been placed in distinct groups, but Dr. Günther considers them all to form one genus, *Crocodilus.*

FAMILY 56.—ALLIGATORIDÆ.　(1 Genus, 10 Species.)

GENERAL DISTRIBUTION.

NEOTROPICAL SUB-REGIONS.	NEARCTIC SUB-REGIONS.	PALÆARCTIC SUB-REGIONS.	ETHIOPIAN SUB-REGIONS.	ORIENTAL SUB-REGIONS.	AUSTRALIAN SUB-REGIONS.
2 . 3 – –	– – 3 –	– – – –	– – – –	– – – –	– – – –

The Alligators, which are distinguished by having both the large front teeth and the canines fitting into pits of the upper jaw, are confined to the Neotropical, and the southern part of the Nearctic regions, from the lower Mississippi and Texas through all Tropical America, but they appear to be absent from the Antilles.　They are all placed by Dr. Günther in the single genus, *Alligator.*

General Remarks on the Distribution of Crocodilia.

These animals, being few in number and wholly confined to the tropical and sub-tropical regions, are of comparatively little interest as regards geographical distribution. America possesses both Crocodiles and Alligators; India, Crocodiles and Gavials; while Africa has Crocodiles only. Both Crocodiles and Gavials are found in the northern part of the Australian region, so that neither of the three families are restricted to a single region.

Fossil Crocodilia.

The existing families of the order date back to the Eocene period in Europe, and the Cretaceous in North America. In the south of England, Alligators, Gavials and Crocodiles, all occur in Eocene beds, indicating that the present distribution of these families is the result of partial extinction, and a gradual restriction of their range—a most instructive fact, suggesting the true explanation of a large number of cases of discontinuous distribution which are sometimes held to prove the former union of lands now divided by the deepest oceans. In more ancient formations, a number of Crocodilian remains have been discovered which cannot be classed in any existing families, and which, therefore, throw no light on the existing distribution of the group.

Order V.—CHELONIA.

FAMILY 57.—TESTUDINIDÆ.   (14 Genera, 126 Species.)

GENERAL DISTRIBUTION.					
NEOTROPICAL SUB-REGIONS.	NEARCTIC SUB-REGIONS.	PALÆARCTIC SUB-REGIONS.	ETHIOPIAN SUB-REGIONS.	ORIENTAL SUB-REGIONS.	AUSTRALIAN SUB-REGIONS.
1.2.3.4	1.2.3.4	1.2 — 4	1.2.3.4	1.2.3.4	— — —

The Testudinidæ, including the land and many fresh-water tortoises, are very widely distributed over the Old and New worlds, but are entirely absent from Australia. They are especially abundant in the Nearctic region, as far north as Canada and British Columbia, and almost equally so in the

Neotropical and Oriental regions; in the Ethiopian there is a considerable diminution in the number of species, and in the Palæarctic they are still less numerous, being confined to the warmer parts of it, except one species which extends as far north as Hungary and Prussia. The genera are:—

Testudo (25 sp.), most abundant in the Ethiopian region, but also extending over the Oriental region, into South Europe, and the Eastern States of North America; *Emys* (64 sp.), abundant in North America and over the whole Oriental region, less so in the Neotropical and the Palæarctic regions; *Cinosternon* (13 sp.), United States and California, and Tropical America; *Aromochelys* (4 sp.), confined to the Eastern States of North America; *Staurotypus* (2 sp.), Guatemala and Mexico; *Chelydra* (1 sp.), Canada to Louisiana; *Claudius* (1 sp.), Mexico; *Dermatemys* (3 sp.), South America, Guatemala, and Yucatan; *Terrapene* (4 sp.), Maine to Mexico, Sumatra to New Guinea, Shanghae and Formosa—a doubtfully natural group; *Cinyxis* (3 sp.), *Pyxis* (1 sp.), *Chersina* (4 sp.), are all Ethiopian; *Dumerilia* (1 sp.), is from Madagascar only.

FAMILY 58.—CHELYDIDÆ.   (10 Genera, 44 Species.)

GENERAL DISTRIBUTION.					
NEOTROPICAL SUB-REGIONS.	NEARCTIC SUB-REGIONS.	PALÆARCTIC SUB-REGIONS.	ETHIOPIAN SUB-REGIONS.	ORIENTAL SUB-REGIONS.	AUSTRALIAN SUB-REGIONS.
— 2 — —	— — — —	— — — —	1 . 2 . 3 . 4	— — — —	— 2 — —

The Chelydidæ, or fresh-water tortoises with imperfectly retractile heads, have a remarkable distribution in the three great southern continents of Africa, Australia, and South America; the largest number of species being found in the latter country. The genera are:—

Peltocephalus (1 sp.), *Podocnemis* (6 sp.), *Hydromedusa* (4 sp.), *Chelys* (1 sp.), and *Platemys* (16 sp.), inhabiting South America from the Orinooko to the La Plata, the latter genus occurring also in Australia and New Guinea; *Chelodina* (5 sp.), *Chelemys* (1 sp.), and *Elseya* (2 sp.) from Australia; while *Sternotheres*

(6 sp.), and *Pelomedusa* (3 sp.), inhabit Tropical and South Africa and Madagascar.

FAMILY 59.—TRIONYCHIDÆ.   (3 Genera, 25 Species.)

GENERAL DISTRIBUTION.					
NEOTROPICAL SUB-REGIONS.	NEARCTIC SUB-REGIONS.	PALÆARCTIC SUB-REGIONS.	ETHIOPIAN SUB-REGIONS.	ORIENTAL SUB-REGIONS.	AUSTRALIAN SUB-REGIONS.
— — — —	— — 3 —	— — — 4	1 . 2 . 3 —	1 . 2 . 3 . 4	— — — —

The distribution of the Trionychidæ, or Soft Tortoises, is very different from that of the Chelydidæ, yet is equally interesting. They abound most in the Oriental region, extending beyond it to Northern China and Japan. In the Nearctic region they are only found in the Eastern States, corresponding curiously to the distribution of plants, in which the affinity of Japan to the Eastern States is greater than to California. The Triony-chidæ are also found over the Ethiopian region, but not in Madagascar.

The genera are,—*Trionyx* (17 sp.), which extends over the whole area of the family as above indicated; *Cycloderma* (5 sp.), peculiar to Africa; *Emyda* (3 sp.), the peninsula of India, Ceylon, and Africa.

FAMILY 60.—CHELONIIDÆ.   (2 Genera, 5 Species.)

GENERAL DISTRIBUTION.—All the warm and tropical Seas.

The Marine Turtles are almost universally distributed. *Dermatochelys* (1 sp.), is found in the temperate seas of both the Northern and Southern Hemispheres; *Chelone* (4 sp.), ranges over all the tropical seas—*C. viridis*, the epicureans' species, inhabiting the Atlantic, while *C. imbricata* which produces the "tortoiseshell" of commerce is found in the Indian and Pacific oceans.

Remarks on the Distribution of the Chelonia.

The four families into which the Chelonia are classed have all of them a wide distribution, though none are universal. The Ethiopian region seems to be the richest, as it possesses 3 of the four families, while no other region has more than 2 ; and it also possesses 7 peculiar genera. Next comes the Neo-tropical region with 2 families and 6 peculiar genera; the Australian with 3, and the Nearctic with 2 peculiar genera; while the Oriental and Palæarctic regions possess none that are peculiar. There are about 30 genera and 200 species in the whole order.

Fossil Chelonia.—The earliest undoubted remains of this order occur in the Upper Oolite. These belong to the Cheloniidæ and Emydidæ, which are also found in the Chalk. In the Tertiary beds Chelonia are more abundant, and the Trionychidæ now appear. The Testudinidæ are first met with in the Miocene formation of Europe and the Eocene of North America, the most remarkable being the gigantic *Colossochelys Atlas* of the Siwalik Hills. It appears, therefore, that the families of the order Chelonia were already specialised in the Secondary period, a fact which, together with their more or less aquatic habits, sufficiently accounts for their generally wide distribution. Species of *Testudo, Emys,* and *Trionyx,* are found in the Upper Miocene of the south of France.

AMPHIBIA.

Order I.—PSEUDOPHIDIA.

FAMILY 1.—CÆCILIADÆ.   (4 Genera, 10 Species.)

GENERAL DISTRIBUTION.					
NEOTROPICAL SUB-REGIONS.	NEARCTIC SUB-REGIONS.	PALÆARCTIC SUB-REGIONS.	ETHIOPIAN SUB-REGIONS.	ORIENTAL SUB-REGIONS.	AUSTRALIAN SUB-REGIONS.
— 2 . 3 —	- - - -	- - - -	— 2 - -	1 . 2 . 3 —	- - - -

The Cæciliadæ are a curious group of worm-like Amphibia sparingly scattered over the three great tropical regions. The genera are,—*Cæcilia*, which inhabits West Africa, Malabar and South America; *Siphonopsis*, peculiar to Brazil and Mexico; *Ichthyopsis*, from Ceylon and the Khasya Mountains; and *Rhinatrema* from Cayenne.

Order II.—URODELA.

FAMILY 2.—SIRENIDÆ.   (1 Genus, 3 Species.)

GENERAL DISTRIBUTION.					
NEOTROPICAL SUB-REGIONS.	NEARCTIC SUB-REGIONS.	PALÆARCTIC SUB-REGIONS.	ETHIOPIAN SUB-REGIONS.	ORIENTAL SUB-REGIONS.	AUSTRALIAN SUB-REGIONS.
- - - -	- - 3 -	- - - -	- -- - -	- - - -	- - - -

The genus *Siren*, consisting of eel-like Batrachians with two anterior feet and permanent branchiæ, inhabits the South-Eastern States of North America from Texas to Carolina.

FAMILY 3.—PROTEIDÆ. (2 Genera, 4 Species.)

GENERAL DISTRIBUTION.					
NEOTROPICAL SUB-REGIONS.	NEARCTIC SUB-REGIONS.	PALÆARCTIC SUB-REGIONS.	ETHIOPIAN SUB-REGIONS.	ORIENTAL SUB-REGIONS.	AUSTRALIAN SUB-REGIONS.
— — — —	— — 3 —	1 — — —	— — — —	— — — —	— — — —

The Proteidæ have four feet and persistent external branchiæ. The two genera are,—*Proteus* (1 sp.), found only in caverns of Central Europe ; and *Menobranchus*, which are like newts in form, and inhabit the Eastern States of North America.

FAMILY 4.—AMPHIUMIDÆ. (1 Genus, 2 Species.)

GENERAL DISTRIBUTION.					
NEOTROPICAL SUB-REGIONS.	NEARCTIC SUB-REGIONS.	PALÆARCTIC SUB-REGIONS.	ETHIOPIAN SUB-REGIONS.	ORIENTAL SUB-REGIONS.	AUSTRALIAN SUB-REGIONS.
— — — —	— — 3 —	— — — —	— — — —	— — — —	— — — —

The genus *Amphiuma*, or *Murænopsis*, consists of slender eel-like creatures with four rudimentary feet, and no external branchiæ. The species inhabit the Southern United States from New Orleans to Carolina.

FAMILY 5.—MENOPOMIDÆ. (2 Genera, 4 Species.)

GENERAL DISTRIBUTION.					
NEOTROPICAL SUB-REGIONS.	NEARCTIC SUB-REGIONS.	PALÆARCTIC SUB-REGIONS.	ETHIOPIAN SUB-REGIONS.	ORIENTAL SUB-REGIONS.	AUSTRALIAN SUB-REGIONS.
— — — —	— — 3 —	— — — 4	— — — —	— — — —	— — — —

There are large Salamanders of repulsive appearance, found only in Eastern Asia and the Eastern United States. The genera are,—*Sieboldia* (2 sp.), Japan and north-west China ; *Menopoma* = *Protonopsis* (2 sp.), Ohio and Alleghany rivers.

FAMILY 6.—SALAMANDRIDÆ.   (20 Genera, 85 Species.)

GENERAL DISTRIBUTION.

NEOTROPICAL SUB-REGIONS.	NEARCTIC SUB-REGIONS.	PALÆARCTIC SUB-REGIONS.	ETHIOPIAN SUB-REGIONS.	ORIENTAL SUB-REGIONS.	AUSTRALIAN SUB-REGIONS.
— 2 . 3 —	1 . 2 . 3 . 4	1 . 2 . 3 . 4	— — — —	— — 3 —	— — — —

The Salamandridæ, of which our common Newts are characteristic examples, form an extensive family highly characteristic
of the North Temperate regions, a few species only extending
into the Neotropical region along the Andes to near Bogota, and
one into the Oriental region in Western China. The genera, as
arranged by Dr. Strauch, are as follows :—

Salamandra (2 sp.), Central and South Europe and North
Africa ; *Pleurodeles* (1 sp.), Spain, Portugal, and Morocco ; *Brady-
bates* (1 sp.), Spain ; *Triton* (16 sp.), all Europe except the
extreme north, Algeria, North China and Japan, Eastern States
of North America, California and Oregon ; *Chioglossa* (2 sp.)
Portugal and South Europe ; *Salamandrina* (1 sp.), Italy to Dal-
matia ; *Ellipsoglossa* (2 sp.), Japan ; *Isodactylium* (2 sp.), East
Siberia ; *Onychodactylus* (1 sp.), Japan ; *Amblystoma* (21 sp.),
Nearctic region from Canada and Oregon to Mexico, most abundant
in Eastern States ; *Ranodon* (1 sp.), Tartary and North-east China ;
Dicamptodon (1 sp.), California ; *Plethodon* (5 sp.), Massachusetts
to Louisiana, and Vancouver's Island to California ; *Desmognathus*
(4 sp.), Eastern United States south of latitude 43° ; *Anaides* (1
sp.), Oregon and Northern California ; *Hemidactylium* (2 sp.),
South-eastern United States and Southern California ; *Heredia*
(1 sp.), Oregon and California ; *Spelerpes* (18 sp.), Eastern United
States from Massachusetts to Mexico, Guatemala, Costa Rica and
Andes of Bogota, with a species in South Europe ; *Batrachoseps*
(2 sp.), South-eastern United States and California ; *Tylotriton*
(1 sp.), Yunan in West China.

Order III.—ANURA.

FAMILY 7.—RHINOPHRYNIDÆ.   (1 Genus, 1 Species.)

GENERAL DISTRIBUTION.					
NEOTROPICAL SUB-REGIONS.	NEARCTIC SUB-REGIONS.	PALÆARCTIC SUB-REGIONS.	ETHIOPIAN SUB-REGIONS.	ORIENTAL SUB-REGIONS.	AUSTRALIAN SUB-REGIONS.
— — 3 —	— — — —	— — — —	— — — —	— — — —	— — — —

The Rhinophrynidæ are Toads with imperfect ears and a tongue which is free in front. The single species of *Rhinophrynus*, is a native of Mexico.

FAMILY 8.—PHRYNISCIDÆ.   (5 Genera, 13 Species.)

GENERAL DISTRIBUTION.					
NEOTROPICAL SUB-REGIONS.	NEARCTIC SUB-REGIONS.	PALÆARCTIC SUB-REGIONS.	ETHIOPIAN SUB-REGIONS.	ORIENTAL SUB-REGIONS.	AUSTRALIAN SUB-REGIONS.
1.2.3—	— — — —	— — — —	1.2 — —	— — — 4	— 2 — —

The Phryniscidæ, or Toads with imperfect ears and tongue fixed in front, are widely distributed over the warmer regions of the earth, but are most abundant in the Neotropical region and Australia, while only single species occur in the Old World. The genera are :—

Phryniscus (7 sp.), from Costa Rica to Chili and Monte Video ; *Brachycephalus* (1 sp.), Brazil ; *Pseudophryne* (3 sp.), Australia and Tasmania ; *Hemisus* (1 sp.), Tropical Africa ; *Micrhyla* (1 sp.), Java.

FAMILY 9.—HYLAPLESIDÆ.   (1 Genus, 5 Species.)

GENERAL DISTRIBUTION.					
NEOTROPICAL SUB-REGIONS.	NEARCTIC SUB-REGIONS.	PALÆARCTIC SUB-REGIONS.	ETHIOPIAN SUB-REGIONS.	ORIENTAL SUB-REGIONS.	AUSTRALIAN SUB-REGIONS.
1 . 2 — 4	— — — —	— — — —	— — — —	— — — —	— — — —

The Hylaplesidæ are Toads with perfect ears, and they seem to be confined to the Neotropical region. The only genus, *Hylaplesia* (5 sp.), inhabits Brazil, Chili, and the Island of Hayti.

FAMILY 10.—BUFONIDÆ.   (6 Genera, 64 Species.)

GENERAL DISTRIBUTION.					
NEOTROPICAL SUB-REGIONS.	NEARCTIC SUB-REGIONS.	PALÆARCTIC SUB-REGIONS.	ETHIOPIAN SUB-REGIONS.	ORIENTAL SUB-REGIONS.	AUSTRALIAN SUB-REGIONS.
1.2.3.4	1.2.3.4	1.2.3.4	1.2.3 —	1.2.3.4	1.2 — —

The rather extensive family of the Bufonidæ, which includes our common Toad, and is characterised by prominent neck glands and tongue fixed in front, is almost universally distributed, but is very rare in the Australian region; one species being found in Celebes and one in Australia.   The genera are:—

Kalophrynus (2 sp.), Borneo; *Bufo* (58 sp.), has the range of the entire family, except Australia; *Otilophus* (1 sp.), South America; *Peltaphryne* (1 sp.), Porto Rico; *Pseudobufo* (1 sp.), Malay Peninsula; *Schismaderma* (1 sp.), Natal; *Notaden* (1 sp.), East Central Australia.

FAMILY 11.—XENORHINIDÆ.   (1 Genus, 1 Species.)

GENERAL DISTRIBUTION.					
NEOTROPICAL SUB-REGIONS.	NEARCTIC SUB-REGIONS.	PALÆARCTIC SUB-REGIONS.	ETHIOPIAN SUB-REGIONS.	ORIENTAL SUB-REGIONS.	AUSTRALIAN SUB-REGIONS.
— — — —	— — — —	— — — —	— — — —	— — — —	1 — — —

The Xenorhinidæ may be characterised as Toads with perfect ears and tongue free in front.   The only species of *Xenorhina* is a native of New Guinea.

FAMILY 12.—ENGYSTOMIDÆ.   (15 Genera, 31 Species.)

GENERAL DISTRIBUTION.					
NEOTROPICAL SUB-REGIONS.	NEARCTIC SUB-REGIONS.	PALÆARCTIC SUB-REGIONS.	ETHIOPIAN SUB-REGIONS.	ORIENTAL SUB-REGIONS.	AUSTRALIAN SUB-REGIONS.
1 . 2 . 3 —	— — 3 —	— — — —	— 2 . 3 —	1 . 2 . 3 . 4	— 2 — —

The Engystomidæ are Toads without neck-glands and with the tongue tied in front. They are most abundant in the Oriental and Neotropical regions, especially in the latter, which contains about half the known species, with isolated species in Australia, Africa, and the Southern States of North America. They appear to be the remnant of a once extensive and universally distributed group, which has maintained itself in two remote regions, but is dying out everywhere else. The genera are :—

Engystoma (9 sp.), Carolina to La Plata, with one species in South China ; *Diplopelma* (3 sp.), South India to China and Java ; *Cacopus* (2 sp.), Central India ; *Glyphoglossus* (1 sp.), Pegu ; *Callula* (4 sp.), Sikhim, Ceylon, China, and Borneo ; *Brachymerus* 1 sp.), South Africa ; *Adenomera* (1 sp.), Brazil ; *Pachybatrachus* (1 sp.), Australia ; *Breviceps* (2 sp.), South and West Africa ; *Chelydobatrachus* (1 sp.), West Australia ; *Hypopachus* (1 sp.), Costa Rica ; *Rhinoderma* (1 sp.), Chili ; *Atelopus* (1 sp.), Cayenne and Peru ; *Copea* (1 sp.), South America ; *Paludicola* (1 sp.), New Granada.

FAMILY 13.—BOMBINATORIDÆ.   (8 Genera, 9 Species.)

GENERAL DISTRIBUTION.					
NEOTROPICAL SUB-REGIONS.	NEARCTIC SUB-REGIONS.	PALÆARCTIC SUB-REGIONS.	ETHIOPIAN SUB-REGIONS.	ORIENTAL SUB-REGIONS.	AUSTRALIAN SUB-REGIONS.
1 . 2 — —	— — — —	1 . 2 — —	— — — —	— — — —	— — — 4

The Bombinatoridæ are a family of Frogs which have imperfect ears and no neck-glands, and they have a very peculiar and interesting distribution, being confined to Central and South Europe, the southern part of South America, and New Zealand. They consist of many isolated groups forming five separate sub-families. The genera are :—

Bombinator, Central Europe and Italy; *Pelobates* and *Didocus*, Central Europe and Spain; *Telmatobius* (2 sp.), Peru and Brazil; *Alsodes*, Chonos Archipelago; *Cacotus*, Chili; *Liopelma*, New Zealand; *Nannophryne*, Straits of Magellan.

FAMILY 14.—PLECTROMANTIDÆ.   (1 Genus, 1 Species.)

GENERAL DISTRIBUTION.					
NEOTROPICAL SUB-REGIONS.	NEARCTIC SUB-REGIONS.	PALÆARCTIC SUB-REGIONS.	ETHIOPIAN SUB-REGIONS.	ORIENTAL SUB-REGIONS.	AUSTRALIAN SUB-REGIONS.
1 — — —	— — — —	— — — —	— — — —	— — — —	— — — —

The Plectromantidæ, which are Frogs with neck-glands, and the toes but not the fingers dilated, consists of a single species of the genus *Plectromantis*.   It inhabits the region west of the Andes, and south of the Equator.

FAMILY 15.—ALYTIDÆ.   (5 Genera, 37 Species.)

GENERAL DISTRIBUTION.					
NEOTROPICAL SUB-REGIONS.	NEARCTIC SUB-REGIONS.	PALÆARCTIC SUB-REGIONS.	ETHIOPIAN SUB-REGIONS.	ORIENTAL SUB-REGIONS.	AUSTRALIAN SUB-REGIONS.
— 2 — —	1 . 2 . 3 —	1 — — —	1 . 2 . 3 —	— — — —	1 . 2 — —

The Alytidæ are Frogs with neck-glands and undilated toes. They are most abundant in the Ethiopian region, with a few species in the Nearctic and Australian regions, and one in Europe and Brazil respectively.   The genera are :—

Alytes (1 sp.), Central Europe; *Scaphiopus* (5 sp.), California to Mexico and the Eastern States; *Hyperolius* (29 sp.), all Africa, and two in New Guinea and North Australia; *Helioporus* (1 sp.), in Australia; *Nattereria* (1 sp.), Brazil.

FAMILY 16.—PELODRYADÆ. (3 Genera, 7 Species.)

GENERAL DISTRIBUTION.

NEOTROPICAL SUB-REGIONS.	NEARCTIC SUB-REGIONS.	PALÆARCTIC SUB-REGIONS.	ETHIOPIAN SUB-REGIONS.	ORIENTAL SUB-REGIONS.	AUSTRALIAN SUB-REGIONS.
1 . 2 – –	– – – –	– – – –	– – – –	– – – –	1 . 2 – –

The Pelodryadæ are Tree Frogs with neck-glands, and are confined to the Australian and Neotropical regions. The genera are :—

Phyllomedusa (3 sp.), South America to Paraguay; *Chirodryas*, Australia; and *Pelodryas* (3 sp.), Moluccas, New Guinea and Australia.

FAMILY 17.—HYLIDÆ. (11 Genera, 94 Species.)

GENERAL DISTRIBUTION.

NEOTROPICAL SUB-REGIONS.	NEARCTIC SUB-REGIONS.	PALÆARCTIC SUB-REGIONS.	ETHIOPIAN SUB-REGIONS.	ORIENTAL SUB-REGIONS.	AUSTRALIAN SUB-REGIONS.
1 . 2 . 3 . 4	1 . 2 . 3 . 4	1 . 2 . 3 –	– – – –	– – 3 –	1 . 2 – –

The Hylidæ are glandless Tree Frogs with a broadened sacrum. They are most abundant in the Neotropical region, which contains more than two-thirds of the species; about twenty species are Australian; six or seven are Nearctic, reaching northward to Great Bear Lake; while one only is European, and one Oriental. The genera are :—

Hyla (62 sp.), having the range of the whole family; *Hylella* (1 sp.), *Ololygon* (1 sp.), *Pohlia* (2 sp.), *Triprion* (1 sp.), *Opisthodelphys* (1 sp.), and *Nototrema* (4 sp.), are South American; while *Trachycephalus* (8 sp.), is peculiar to the Antilles, except one South American species; *Pseudacris* (1 sp.), ranges from Georgia, United States, to Great Bear Lake; *Litoria* (7 sp.), is Australian and Papuan, except one species in Paraguay; *Ceratohyla* (4 sp.), is only known from Ecuador.

FAMILY 18.—POLYPEDATIDÆ.   (24 Genera, 124 Species.)

		GENERAL DISTRIBUTION.			
NEOTROPICAL SUB-REGIONS.	NEARCTIC SUB-REGIONS.	PALÆARCTIC SUB-REGIONS.	ETHIOPIAN SUB-REGIONS.	ORIENTAL SUB-REGIONS.	AUSTRALIAN SUB-REGIONS.
1.2.3.4	— — 3 —	— — 3.4	1.2.3.4	1.2.3.4	1.2.3 —

The Polypedatidæ, or glandless Tree Frogs with narrowed sacrum, are almost equally numerous in the Oriental and Neotropical regions, more than forty species inhabiting each, while in the Ethiopian there are about half this number, and the remainder are scattered over the other three regions, as shown in the enumeration of the genera :—

Ixalus (16 sp.), Oriental, except one in Japan, and one in Western Polynesia; *Rhacophorus* (7 sp.), and *Theloderma* (1 sp.), are Oriental; *Hylarana* (10 sp.), Oriental, to the Solomon Islands and Tartary, Nicobar Islands, West Africa, and Madagascar; *Megalixalus* (1 sp.), Seychelle Islands; *Leptomantis* (1 sp.), Philippines; *Platymantis* (5 sp.), New Guinea, Philippines, and Fiji Islands; *Cornufer* (2 sp), Java and New Guinea; *Polypedates* (19 sp.), mostly Oriental, but two species in West Africa, one Madagascar, two Japan, one Loo-Choo Islands, and one Hong Kong; *Hylambates* (3 sp.), *Hemimantis* (1 sp.), and *Chiromantis* (1 sp.), are Ethiopian; *Rappia* (13 sp.), is Ethiopian, and extends to Madagascar and the Seychelle Islands; *Acris* (2 sp.), is North American; *Elosia* (1 sp.), *Epirhixis* (1 sp.), *Phyllobates* (9 sp.), *Hylodes* (26 sp.), *Hyloxalus* (1 sp.), *Pristimantis* (1 sp.), *Crossodactylus* (1 sp.), *Calostethus* (1 sp.), *Strabomantis* (1 sp.), and *Leiyla* (1 sp.), are Neotropical, the last two being Central American, while species of *Hylodes* and *Phyllobates* are found in the West Indian Islands.

FAMILY 19.—RANIDÆ.   (26 Genera, 150 Species.)

GENERAL DISTRIBUTION.					
NEOTROPICAL SUB-REGIONS.	NEARCTIC SUB-REGIONS.	PALÆARCTIC SUB-REGIONS.	ETHIOPIAN SUB-REGIONS.	ORIENTAL SUB-REGIONS.	AUSTRALIAN SUB-REGIONS.
1.2.3.4	1.2.3.4	1.2.3.4	1.2.3.4	1.2.3.4	1.2 — —

The Ranidæ, or true Frogs, are characterised by having simple undilated toes, but neither neck-glands nor dilated sacrum. They are almost cosmopolitan, extending to the extreme north and south from the North Cape to Patagonia, and they are equally at home in the tropics.   They are perhaps most abundant in South America, where a large number of the genera and species are found; the Ethiopian region comes next, while they are rather less abundant in the Oriental and Australian regions; the Nearctic region has much less (about 12 species), while the Palæarctic has only five, and these two northern regions only possess the single genus *Rana*.   The genera are distributed as follows :—

Rana (60 sp.), ranges all over the world, except Australia and South America, although it extends into New Guinea and into Mexico and Central America;   it is most abundant in Africa. *Pyxicephalus* (7 sp.), extends over the whole Ethiopian region, Hindostan, the Himalayas, and Japan; *Cystignathus* (22 sp.), is mainly Neotropical, but has three species Ethiopian.   All the other genera are confined to single regions.   The Neotropical genera are :—*Odontophrynus* (1 sp.), *Pseudis* (1 sp.), *Pithecopsis* (1 sp.), *Ensophleus* (1 sp.), *Limnocharis* (1 sp.), *Hemiphractus* (1 sp.), all Tropical South American east of Andes ; *Ceratophrys* (5 sp.), Panama to La Plata ; *Cycloramphus* (1 sp.), West Ecuador and Chili ; *Pleurodema* (6 sp.), Venezuela to Patagonia ; *Leiuperus* (12 sp.), Mexico and St. Domingo to Patagonia ; *Hylorhina* (1 sp.), Chiloe.   The Australian genera are :—*Myxophyes* (1 sp.), Queensland ; *Platyplectrum* (2 sp.), Queensland and West Australia ; *Neobatrachus* (1 sp.), South Australia ; *Limnodynastes* 7 sp.), and *Crinia* (11 sp.), Australia and Tasmania.   The

Oriental genera are :—*Dicroglossus* (1 sp.), Western Himalayas ; *Oxyglossus* (2 sp.), Siam to Java, Philippines and China ; *Hoplobatrachus* (1 sp.), Ceylon ; *Phrynoglossus* (1 sp.), Siam. The Ethiopian genera are :—*Phrynobatrachus* (1 sp.), *Stenorhynchus* (1 sp.), both from Natal.

FAMILY 20.—DISCOGLOSSIDÆ.   (14 Genera, 18 Species.)

GENERAL DISTRIBUTION.					
NEOTROPICAL SUB-REGIONS.	NEARCTIC SUB-REGIONS.	PALÆARCTIC SUB-REGIONS.	ETHIOPIAN SUB-REGIONS.	ORIENTAL SUB-REGIONS.	AUSTRALIAN SUB-REGIONS.
1 . 2 — —	— — — —	1 . 2 . 3 . 4	— 2 . 3 —	— 2 . 3 . 4	1 . 2 — —

The Discoglossidæ, or Frogs with a dilated sacrum, are remarkable for the number of generic forms scattered over a large part of the globe, being only absent from the Nearctic and the northern half of the Neotropical regions, and also from Hindostan and East Africa. The genera are :—

Chiroleptes (4 sp.), Australia ; *Calyplocephalus* (1 sp.), allied to the preceding, from Chili ; *Cryptotis* (1 sp.), Australia ; *Asterophys* (2 sp.), New Guinea and Aru Islands ; *Xenophrys* (1 sp.), Eastern Himalayas ; *Megalophrys* (2 sp.), Ceylon and the Malay Islands ; *Nannophrys* (1 sp.), Ceylon ; *Pelodytes* (1 sp.), France only ; *Leptobrachium* (1 sp.), Java ; *Discoglossus* (1 sp.), Vienna to Algiers ; *Laprissa* (1 sp.), *Latonia* (1 sp.), Palæarctic region ; *Arthroleptis* (2 sp.), West Africa and the Cape ; *Grypiscus* (1 sp.), South Brazil.

FAMILY 21.—PIPIDÆ.   (1 Genus, 1 Species.)

GENERAL DISTRIBUTION.					
NEOTROPICAL SUB-REGIONS.	NEARCTIC SUB-REGIONS.	PALÆARCTIC SUB-REGIONS.	ETHIOPIAN SUB-REGIONS.	ORIENTAL SUB-REGIONS.	AUSTRALIAN SUB-REGIONS.
— 2 — —	— — — —	— — — —	— — — —	— — — ..	— — — —

The Pipidæ are toads without a tongue or maxillary teeth, and with enormously dilated sacrum. The only species of *Pipa* is a native of Guiana.

FAMILY 22.—DACTYLETHRIDÆ.   (1 Genus, 2 Species.)

GENERAL DISTRIBUTION.					
NEOTROPICAL SUB-REGIONS.	NEARCTIC SUB-REGIONS.	PALÆARCTIC SUB-REGIONS.	ETHIOPIAN SUB-REGIONS.	ORIENTAL SUB-REGIONS.	AUSTRALIAN SUB-REGIONS.
— — — —	— — — —	— — — —	1.2.3 —	— — — —	— — — —

The Dactylethridæ are Toads with maxillary teeth but no tongue, and with enormously dilated sacrum. The species of *Dactylethra* are natives of West, East, and South Africa.

General Remarks on the Distribution of the Amphibia.

The Amphibia, as here enumerated, consist of 22 families, 152 genera, and nearly 700 species. Many of the families have a very limited range, only two (Ranidæ and Polypedatidæ) being nearly universal; five more extend each into five regions, while no less than thirteen of the families are confined to one, two, or three regions each. By far the richest region is the Neotropical, possessing 16 families (four of them peculiar) and about 50 peculiar or very characteristic genera. Next comes the Australian, with 11 families (one of which is peculiar) and 16 peculiar genera. The Nearctic region has no less than 9 of the families (two of them peculiar to it) and 15 peculiar genera, 13 of which are tailed Batrachians which have here their metropolis. The other three regions have 9 families each; the Palæarctic has no peculiar family but no less than 15 peculiar genera; the Ethiopian 1 family and 12 genera peculiar to it; and the Oriental, 19 genera but no family confined to it.

It is evident, therefore, that each of the regions is well characterised by its peculiar forms of Amphibia, there being only a few genera, such as *Hyla*, *Rana*, and *Bufo* which have a wide range. The connection of the Australian and Neotropical regions is well shown in this group, by the Phryniscidæ, Hylidæ, and Discoglossidæ, which present allied forms in both; as well as by the genus *Liopelma* of New Zealand, allied to the Bombinatoridæ of South America, and the absence of the otherwise cosmopolitan genus *Rana* from both continents. The affinity of the Nearctic and Palæarctic regions is shown by the Proteidæ, which are confined to them, as well as by the genus *Triton* and almost the whole of the extensive family of the Salamandridæ. The other regions are also well differentiated, and there is no sign of a special Ethiopian Amphibian fauna extending over the peninsula of India, or of the Oriental and Palæarctic regions merging into each other, except by means of genera of universal distribution.

Fossil Amphibia.—The extinct Labyrinthodontia form a separate order, which existed from the Carboniferous to the Triassic period. No other remains of this class are found till we reach the Tertiary formation, when Newts and Salamanders as well as Frogs and Toads occur, most frequently in the Miocene deposits. The most remarkable is the *Andrias scheuchzeri* from the Miocene of Œningen, which is allied to *Sieboldia maxima* the great salamander of Japan.

CHAPTER XX.

SUB-CLASS I.—TELEOSTEI.

Order I.—ACANTHOPTERYGII.

FAMILY 1.—GASTEROSTEIDÆ. (1 Genus, 11 Species.)

" Fresh-water or marine scaleless fishes, with elongate compressed bodies and with isolated spines before the dorsal fin."

DISTRIBUTION.—Palæarctic and Nearctic regions.

The species of *Gasterosteus*, commonly called Sticklebacks, are found in rivers, lakes, estuaries, and seas, as far south as Italy and Ohio. Four species occur in Britain.

FAMILY 2.—BERYCIDÆ. (10 Genera, 55 Species.)

" Marine fishes, with elevated compressed bodies covered with toothed scales, and large eyes."

DISTRIBUTION.—Tropical and temperate seas of both hemispheres.

Their northern limit is the Mediterranean and Japan. Most abundant in the Malayan seas.

FAMILY 3.—PERCIDÆ.   (61 Genera, 476 Species.)

"Marine or fresh-water carnivorous fishes, with oblong bodies covered with toothed scales."

DISTRIBUTION.—Seas, rivers and lakes, of all regions.

The genera which inhabit fresh-waters are the following:—

Perca (3 sp.), inhabits the Nearctic and Palæarctic regions as far south as Ohio and Switzerland; one species, the common perch, is British. *Percichthys* (5 sp.), Chili and Patagonia, with one species in Java; *Paralabrax* (2 sp.), California; *Labrax* (8 sp.), six species are marine, inhabiting the shores of Europe and North America, one being British, two species inhabit the rivers of the northern United States; *Lates* (2 sp.), Nile and large rivers of India and China; *Acerina* (3 sp.), Europe, from England to Russia and Siberia; *Percarina* (1 sp.), River Dniester; *Lucioperca* (6 sp.), North America and Europe; *Pileoma* (2 sp.), North America, Texas to Lake Erie; *Boleosoma* (3 sp.), Texas to Lake Superior; *Aspro* (2 sp.), Central Europe; *Huro* (1 sp.), Lake Huron; *Percilia*, (1 sp.), Rio de Maypu in Chili; *Centrarchus* (10 sp.), North America and Cuba; *Bryttus* (8 sp.), South Carolina to Texas; *Pomotis* (8 sp.), North America, Lake Erie to Texas.

Of the exclusively marine genera a species of *Polyprion* and one of *Serranus* are British. The latter genus has nearly 150 species spread over the globe, but is most abundant in the Tropics. *Mesoprion* is another extensive genus confined to the Tropics. *Apogon* abounds from the Red Sea to the Pacific, but has one species in the Mediterranean and one in the coast of Brazil.

FAMILY 4.—APHREDODERIDÆ.   (1 Genus, 1 Species.)

"Fresh-water fish, with oblong body covered with toothed scales, and wide cleft mouth."

DISTRIBUTION.—Atlantic States of North America.

FAMILY 5.—PRISTIPOMATIDÆ. (25 Genera, 206 Species.)

"Marine carnivorous fishes, with compressed oblong bodies, and without molar or cutting teeth."

DISTRIBUTION.—Seas of temperate and tropical regions, a few only entering fresh water.

Of the more extensive genera, nine, comprising more than half the species, are confined to the Indian and Australian seas, while only one large genus (*Hæmulon*) is found in the Atlantic on the coast of Tropical America. The extensive Pacific genus, *Diagramma*, has one species in the Mediterranean. One genus is confined to the Macquarie River in Australia. A species of *Dentex* has occurred on the English coast, and this seems to be the extreme northern range of the family, which does not regularly extend beyond the coast of Portugal, and in the East to Japan. Australia seems to form the southern limit.

FAMILY 6.—MULLIDÆ. (5 Genera, 34 Species.)

"Marine fishes, with elongate slightly compressed bodies covered with large scales, and two dorsal fins at a distance from each other."

DISTRIBUTION.—All tropical seas, except the West Coast of America, extending into temperate regions as far as the Baltic, Japan, and New Zealand.

Two species of *Mullus* (Mullets) are British, and these are the only European fish belonging to the family.

FAMILY 7.—SPARIDÆ. (22 Genera, 117 Species.)

"Herbivorous or carnivorous marine fishes, with oblong compressed bodies covered with minutely serrated scales, and with one dorsal fin."

DISTRIBUTION.—Seas of temperate and tropical regions, a few entering rivers.

Cantharus, Pagellus, and *Chrysophrys,* have occurred on the English Coast. *Haplodactylus* is confined to the West Coast of South America, and Australia ; *Sargus* to the temperate and warm parts of the Atlantic and the shores of East Africa ; *Pagellus* to the western coasts of Europe and Africa. The other large genera have a wider distribution.

FAMILY 8.—SQUAMIPENNES.   (12 Genera, 124 Species.)

"Carnivorous marine fishes, with compressed and elevated bodies, and scaly vertical fins."

DISTRIBUTION.—The seas between the tropics, most abundant in the Oriental and Australian regions, a few entering rivers or extending beyond the tropics.

The extensive genus *Chœtodon* (67 sp.), ranges from the Red Sea to the Sandwich Islands, and from Japan to Western Australia, while two species are found in the West Indies. *Holacanthus* (36 sp.), has a similar distribution, one species only occurring in the West Indies and on the coast of South America. Only one genus (*Pomacanthus*), with a single species, is confined to the West Atlantic.

FAMILY 9.—CIRRHITIDÆ.   (8 Genera, 34 Species.)

"Carnivorous marine fishes, with a compressed oblong body, covered with cycloid scales."

DISTRIBUTION.—The tropical and south temperate waters of the Indian and Pacific oceans, from Eastern Africa to Western America.   Absent from the Atlantic.

FAMILY 10.—TRIGLIDÆ.   (50 Genera, 259 Species.)

"Carnivorous, mostly marine fishes, with oblong compressed or subcylindrical bodies, and wide cleft mouths. They live at the bottom of the water."

DISTRIBUTION.—All seas, some entering fresh water, and a few inhabiting exclusively the fresh waters of the Arctic regions.

They are divided by Dr. Günther into four groups. The Heterolepidina (comprising 4 genera and 12 species) are confined to the North Pacific. The Scorpænina (23 genera 113 species) have an almost universal distribution, but the genera are each restricted to one or other of the great oceans. *Sebastes* has occurred on the English coast. The Cottina (28 genera 110 species) have also a universal distribution; the numerous species of *Cottus* are found either in the seas or fresh waters of Europe and North America; four species are British, as well as seven species of the wide-spread genus *Trigla*. *Ptyonotus* (1 sp.) is confined to Lake Ontario. The Cataphracti (5 genera, 23 species) have also a wide range; one genus, *Agonus*, is found in the British seas, and also in Kamschatka and on the coast of Chili. *Peristethus* is also British.

FAMILY 11.—TRACHINIDÆ.　(24 Genera, 90 Species.)

" Carnivorous marine fishes, with elongate bodies, living at the bottom, near the shore."

DISTRIBUTION.—Almost or quite universal.

Trachinus is a British genus. A species of *Aphritis* inhabits the fresh waters of Tasmania, while its two allies are found on the coasts of Patagonia.

FAMILY 12.　SCIÆNIDÆ.　(13 Genera, 102 Species.)

" Marine or fresh-water fishes, with compressed and rather elongate bodies, covered with toothed scales."

DISTRIBUTION.—Temperate and tropical regions, but absent from Australia.

Larimus is found in the Atlantic, and in African and American rivers. *Corvina, Sciæna,* and *Otilothus* are also marine and freshwater, both in the Atlantic and Pacific. The other genera are of small extent and more restricted range. *Umbrina* and *Sciæna* have occurred in British seas.

FAMILY 13.—POLYNEMIDÆ. (3 Genera, 23 Species.)

" Marine or fresh-water fishes, with compressed oblong bodies and entire or ciliated scales."

DISTRIBUTION.—Tropical seas and rivers of both the great oceans, but most abundant in the Pacific.

FAMILY 14.—SPHYRENIDÆ. (1 Genus, 15 Species.)

"Carnivorous marine fishes, with elongate sub-cylindrical bodies covered with small cycloid scales."

DISTRIBUTION.—The warm and tropical seas of the globe.

FAMILY 15.—TRICHIURIDÆ. (7 Genera, 18 Species.)

" Marine fishes, with elongate compressed bodies covered with minute scales or naked."

DISTRIBUTION.—All the tropical and sub-tropical seas.

FAMILY 16.—SCOMBRIDÆ. (20 Genera, 108 Species.)

" Marine fishes, with elongate compressed bodies, scaled or naked."

DISTRIBUTION.—All the temperate and tropical oceans. Mostly inhabiting the open seas.

Scomber, (the Mackerel) *Thynnus, Naucrates, Zeus, Centrolophus, Brama,* and *Lampris,* are genera which have occurred in the British seas.

FAMILY 17.—CARANGIDÆ. (27 Genera, 171 Species.)

" Marine fishes, with compressed oblong or elevated bodies covered with small scales or naked."

DISTRIBUTION.—All temperate and tropical seas ; some species occur in both the great oceans, ranging from New York to Australia.

Trachurus and *Capros* are genera which occur in British seas.

FAMILY 18.—XIPHIIDÆ.   (2 Genera, 8 Species.)

" Marine fishes, with elongate compressed body and a produced sword-shaped upper jaw."

DISTRIBUTION.—Mediterranean, and open seas between or near the Tropics.

Xiphias (the Sword-fish) has occurred on the English coast.

FAMILY 19.—GOBIIDÆ.   (24 Genera, 294 Species.)

" Carnivorous fishes, with elongate low, naked, or scaly bodies, living at the bottom of the shallow seas or fresh waters of temperate or tropical regions.   Individuals of the same species often differ in inhabiting exclusively fresh or salt water.

DISTRIBUTION.—All temperate and tropical regions, from Scotland and Japan to New Zealand.   Species of *Gobius, Latrunculus*, and *Callionymus* occur in Britain.   Several genera are confined to the East Indian seas and rivers, but none seem peculiar to America.   The genus *Periopthalmus* consists of the curious, large-headed, projecting-eyed fishes, so abundant on the muddy shores of African and Eastern tidal rivers, and which seem to spend most of their time out of water, hunting after insects, &c.

FAMILY 20.—DISCOBOLI.   (2 Genera, 11 Species.)

" Carnivorous fishes, with oblong naked or tubercular bodies, living at the bottom of shallow seas, and attaching themselves to rocks by means of a ventral disc.

DISTRIBUTION.—All northern seas, as far south as Belgium, England, and San Francisco.

Species of both genera (*Cyclopterus* and *Liparis*) occur in British seas.

FAMILY 21.—OXUDERCIDÆ.   (1 Genus, 1 Species.)

" A marine fish, with an elongate sub-cylindrical body and no ventral fins."

DISTRIBUTION.—Macao, China.

FAMILY 22.—BATRACHIDÆ.   (3 Genera, 12 Species.)

" Marine fishes, with sub-cylindrical body and broad depressed head."

DISTRIBUTION.—The coasts of nearly all tropical and south temperate regions, ranging from New York and Portugal to Chili and Tasmania.

FAMILY 23.—PEDICULATI.   (8 Genera, 40 Species.)

" Marine carnivorous fishes, with very large heads and without scales."

DISTRIBUTION.—Seas of all temperate and tropical regions, extending south to New Zealand and north to Greenland.

A species of *Lophius* (the Fishing-frog or Sea-Devil) is found in British seas.   The genus *Antennarius*, comprising two-thirds of the species, is wholly tropical.

FAMILY 24.—BLENNIDÆ.   (33 Genera, 201 Species.)

" Carnivorous fishes, with long sub-cylindrical naked bodies, living at the bottom of shallow water in seas, or tidal rivers."

DISTRIBUTION.—All seas from the Arctic regions to New Zealand, Chili, and the Cape of Good Hope.

Species of *Anarrhichas, Blennius, Blenniops, Centronotus* and *Zoarces* occur in British seas.   *Chasmodes* (3 sp.) is confined to the Atlantic coasts of Temperate North America; *Petroscirtes* (26 sp.) to the tropical parts of the Indian and Pacific Oceans ; and *Stichœus* (9 sp.) to the Arctic Seas.

FAMILY 25.—ACANTHOCLINIDÆ.   (1 Genus, 1 Species.)

"A carnivorous marine fish, with long flat body and very long dorsal fin."

DISTRIBUTION.—Coasts of New Zealand.

FAMILY 26.—COMEPHORIDÆ.   (1 Genus, 1 Species.)

"An elongate, naked, large-headed fish, with two dorsal fins."

DISTRIBUTION.—Lake Baikal.

Dr. Günther remarks, that this fish approaches the Scombrina (Mackerel) in several characters.   These are exclusively marine fishes, while Lake Baikal is fresh-water, and is situated among mountains, at an elevation of nearly 2000 feet, and more than a thousand miles from the ocean!

FAMILY 27.—TRACHYPTERIDÆ.   (3 Genera, 16 Species.)

"Deep sea fishes, with elongate, much compressed, naked bodies."

DISTRIBUTION.—Europe, East Indies, West Coast of South America, New Zealand.   Dr. Günther remarks, that little is known of these fishes, from their being so seldom thrown on shore, and then rapidly decomposing.   The Ribbon-fish (*Regalecus banksii*) has occurred frequently on our shores.   They have soft bones and muscles, small mouths, and weak dentition.

FAMILY 28.—LOPHOTIDÆ.   (1 Genus, 1 Species.)

"A marine fish, with elongate compressed naked body, and high crested head."

DISTRIBUTION.—Mediterranean Sea and Japan.

FAMILY 29.—TEUTHIDIDÆ.   (1 Genus, 29 Species.)

"Marine, herbivorous fishes, with compressed, oblong, small-scaled bodies."

DISTRIBUTION.—Eastern tropical seas, from Bourbon and the Red Sea to the Marianne and Fiji Islands.

FAMILY 30.—ACRONURIDÆ.   (5 Genera, 64 Species.)

"Marine, herbivorous fishes, with compressed, minutely-scaled bodies."

DISTRIBUTION.—All tropical seas, but most abundant in the Malay region, and extending to Japan and New Zealand.

FAMILY 31.—HOPLEGNATHIDÆ.   (1 Genus, 3 Species.)

"Marine fishes, with compressed elevated bodies, covered with very small toothed scales."

DISTRIBUTION.—Seas of Australia, China, and Japan.

FAMILY 32.—MALACANTHIDÆ.   (1 Genus, 3 Species.)

"Marine fishes, with elongate bodies covered with very small scales, and with very long dorsal and anal fins."

DISTRIBUTION.—Atlantic coasts of Tropical America, Mauritius, and New Guinea.

FAMILY 33.—NANDIDÆ.   (6 Genera, 14 Species.)

"Marine or fresh-water carnivorous fishes, with oblong, compressed, scaly bodies."

DISTRIBUTION.—From the Red Sea to the coasts of China and Australia; and the fresh waters of the Neotropical and Oriental regions.   *Badis*, *Nandus*, and *Catopra* inhabit the rivers of India and the Malay Islands; *Acharnes* the rivers of British Guiana.

FAMILY 34.—POLYCENTRIDÆ. (2 Genera, 3 Species.)

"Fresh-water carnivorous fishes, with compressed elevated scaly bodies, and many-spined dorsal and anal fins."

DISTRIBUTION.—Rivers of Tropical America.

FAMILY 35.—LABYRINTHICI. (9 Genera, 25 Species.)

"Fresh-water fishes, with compressed oblong bodies, and capable of living for some time out of water or in dried mud."

DISTRIBUTION.—Freshwaters of South Africa and the East Indies from the Mauritius to China, the Philippines, Celebes, and Amboyna.

FAMILY 36.—LUCIOCEPHALIDÆ. (1 Genus, 1 Species.)

"Fresh-water fish, with elongate scaled body, and a dilated branchial membrane."

DISTRIBUTION.—Rivers of Borneo, Biliton, and Banca.

FAMILY 37.—ATHERINIDÆ. (3 Genera, 39 Species.)

"Marine or fresh-water carnivorous fishes, with subcylindrical scaled bodies, and feeble dentition."

DISTRIBUTION.—All temperate and tropical seas, from Scotland and New York to the Straits of Magellan and Tasmania.

Atherina presbyter occurs in British seas. Species of *Atherina* and *Atherinichthis* are found in fresh-water lakes and rivers in Europe, America, and Australia.

FAMILY 38.—MUGILIDÆ. (3 Genera, 78 Species.)

"Fresh-water and marine fishes, with oblong compressed bodies, cycloid scales, and small mouths, often without teeth."

DISTRIBUTION.—Coasts and fresh waters of all temperate and tropical regions.

Mugil (66 sp.) is mostly marine, and is very widely distributed; several species (Grey Mullets) occur on the British coasts. *Agonostoma* (9 sp.) is confined to the fresh waters of the West Indies, Central America, New Zealand, Australia, Celebes, and the Comoro Islands. *Myxus* (3 sp.) is marine, and occurs both in the Atlantic and Pacific.

FAMILY 39.—OPHIOCEPHALIDÆ. (2 Genera, 26 Species.)

"Fresh-water fishes, with elongate subcylindrical scaled bodies; often leaving the water for a considerable time."

DISTRIBUTION.—Rivers of the Oriental region:—India, Ceylon, China, Malay Islands to Philippines and Borneo.

FAMILY 40.—TRICHONOTIDÆ. (2 Genera, 2 Species.)

"Marine carnivorous fishes, with elongate subcylindrical bodies, cycloid scales, and eyes directed upwards."

DISTRIBUTION.—Coasts of Celebes, Ceram, and New Zealand.

FAMILY 41.—CEPOLIDÆ. (1 Genus, 7 Species.)

"Marine fishes, with very long, compressed, band-like bodies, covered with small cycloid scales."

DISTRIBUTION.—Temperate seas of Western Europe and Eastern Asia, and one species in the Malayan Seas.

Cepola rubescens (the Band fish) ranges from Scotland to the Mediterranean. All the other species but one are from Japan.

FAMILY 42.—GOBIESOCIDÆ. (9 Genera, 21 Species.)

"Carnivorous marine fishes, elongate, anteriorly depressed and scaleless, with dorsal fin on the tail."

DISTRIBUTION.—Temperate and tropical seas; Scandinavia to the Cape, California to Chili, West Indies, Red Sea, Australia, New Zealand, and Fiji Islands.

Three species of *Lepadogaster* have occurred in the English Channel.

FAMILY 43.—PSYCHROLUTIDÆ. (1 Genus, 1 Species.)

"A large-headed, elongate, naked marine fish, with small teeth, and dorsal fin on the tail."

DISTRIBUTION.—West Coast of North America (Vancouver's Island.)

FAMILY 44.—CENTRISCIDÆ. (2 Genera, 7 Species.)

"Marine fishes, with compressed, oblong or elevated bodies, elongate tubular mouth and no teeth."

DISTRIBUTION.—West Coast of Europe and Africa, Mediterranean, Indian Ocean to Java, Philippines, and Japan.

A species of *Centriscus* has occurred on the South Coast of England, and another species is found both at Madeira and Japan.

FAMILY 45.—FISTULARIDÆ. (2 Genera, 4 Species.)

"Marine fishes, very elongate, with long tubular mouth and small teeth."

DISTRIBUTION.—Tropical seas, both in the Atlantic and Indian Ocean, and as far east as the New Hebrides.

FAMILY 46.—MASTACEMBELIDÆ.   (2 Genera, 9 Species.)

" Fresh-water fishes, with eel-like bodies and very long dorsal fin."

DISTRIBUTION.—Rivers of the Oriental region, one species from Ceram (?).

FAMILY 47.—NOTACANTHI.   (1 Genus, 5 Species.)

" Marine fishes, with elongate bodies covered with very small scales, and snout protruding beyond the mouth."

DISTRIBUTION.—Greenland, Mediterranean, and West Australia.

Order II.—ACANTHOPTERYGII PHARYNGOGNATHI.

FAMILY 48.—POMACENTRIDÆ.   (8 Genera, 143 Species.)

" Marine fishes, with short compressed bodies covered with toothed scales, and with feeble dentition."

DISTRIBUTION.—Tropical parts of Pacific and Indian Ocean, less numerous in Tropical Atlantic, a few reaching the Mediterranean, Japan, and South Australia. *Pomacentrus, Glyphidodon,* and *Heliastes* are Atlantic genera.

FAMILY 49.—LABRIDÆ.   (46 Genera, 396 Species.)

" Herbivorous or carnivorous marine fishes, with elongate bodies covered with cycloid scales, and teeth adapted for crushing the shells of mollusca."

DISTRIBUTION.—Temperate and tropical regions of all parts of the globe.

The genera *Labrus, Crenilabrus, Ctenolabrus, Acantholabrus, Centrolabrus,* and *Coris,* have occurred in British seas, and all of these, except the last, are confined to the Mediterranean and the Atlantic as far as Madeira. Eight other genera are characteristic of the Atlantic, most of them being West Indian, but one from the coasts of North America. Seven genera are common to all the great oceans ; the remainder being confined to the Indian and Pacific Oceans, ranging from Japan to New Zealand, but being far more abundant between the Tropics.

FAMILY 50.—EMBROTOCIDÆ. (2 Genera, 17 Species.)

" Marine viviparous fishes, with compressed elevated bodies covered with cycloid scales, and with small teeth."

DISTRIBUTION.—Pacific Ocean from Japan and California northwards. One species enters the fresh waters of California.

FAMILY 51.—GERRIDÆ. (1 Genus, 28 Species.)

" Marine fishes, with compressed oblong bodies covered with minutely serrated scales, and with small teeth."

DISTRIBUTION.—Tropical seas; ranging south as far as the Cape of Good Hope and Australia, and north to Japan and (one species) to New Jersey, U.S.

FAMILY 52.—CHROMIDÆ. (19 Genera, 100 Species.)

" Fresh-water herbivorous or carnivorous fishes, with elevated or elongate scaly bodies, and small teeth."

DISTRIBUTION.—The Oriental, Ethiopian, and Neotropical regions.

Eutroplus (2 sp.) is from the rivers of Southern India and Ceylon; *Chromis* (15 sp.), *Sarotherodon* (2 sp.), and *Hemichromis* (4 sp.), are from the rivers and lakes of Africa, extending to the Sahara and Palestine. The remaining 15 genera are American, and several of them have a restricted distribution. *Acara* (17 sp.) inhabits Tropical South America and the Antilles; *Theraps* (1 sp.), Guatemala; *Heros* (26 sp.), Texas and

Mexico to La Plata; *Mesonauta* (1 sp.), Brazil; *Petenia* (1 sp.), Lake Peten, Guatemala; *Uaru* (2 sp.), Brazil; *Hygrogonus* (1 sp.), Brazil; *Cichla* (4 sp.), Equatorial America; *Crenicichla* (9 sp.), Brazil and Guiana; *Chætobranchus* (3 sp.), Brazil and Guiana; *Mesops* (2 sp.), Brazil; *Satanoperca* (7 sp.), Amazon Valley and Guiana; *Geophagus* (1 sp.), North Brazil and Guiana; *Symphysodon* (1 sp.), Lower Amazon; *Pterophyllum* (1 sp.), Lower Amazon.

Order III.—ANACANTHINI.

FAMILY 53.—GADOPSIDÆ.   (1 Genus, 1 Species.)

"Fresh-water fish, with rather elongate body covered with very small scales, the upper jaw overhanging the lower, forming an obtuse snout."

DISTRIBUTION.—Rivers of Australia and Tasmania.

FAMILY 53a.—LYCODIDÆ.   (3 Genera, 14 Species.)

"Marine fishes, with elongate bodies, and the dorsal united with the anal fin."

DISTRIBUTION.—Arctic seas of America and Greenland, and Antarctic seas about the Falkland Islands and Chiloe Island.

FAMILY 54.—GADIDÆ.   (21 Genera, 58 Species.)

"Marine fishes, with more or less elongate bodies covered with small smooth scales."

DISTRIBUTION.—Cold and temperate regions of both hemispheres; in the North extending as far south as the Mediterranean, Canary Islands, New York and Japan (and one species to the Philippines and Bay of Bengal), and in the South to Chili and New Zealand.

Gadus (Cod), *Merluccius* (Hake), *Phycis, Lota, Molva, Couchia, Motella*, and *Raniceps*, are British.   *Lota* inhabits fresh waters.

FAMILY 55.—OPHIDIIDÆ.   (16 Genera, 43 Species.)

"Marine fishes, with more or less elongate bodies, the dorsal and anal fins united, and the ventral fins rudimentary or absent."

DISTRIBUTION.—Almost universal; from Greenland to New Zealand, but most abundant in the Tropics.

Ophidium and *Ammodytes* occur in British seas; *Lucifuga* inhabits subterranean fresh waters in Cuba.

FAMILY 56.—MACROURIDÆ.   (3 Genera, 21 Species.)

"Marine fishes, with the body terminating in a long, compressed tapering tail, and covered with spiny, keeled or striated scales."

DISTRIBUTION.—North Atlantic from Greenland to Madeira and the Canary Islands, Mediterranean, Japanese and Australian seas.

None of these fishes have occurred in the British seas.

FAMILY 57.—ATELEOPODIDÆ.   (1 Genus, 1 Species.)

"Marine fishes, with the naked body terminating in a long compressed, tapering tail."

DISTRIBUTION.—Japan.

FAMILY 58.—PLEURONECTIDÆ.   (34 Genera, 185 Species.)

"Marine carnivorous fishes, with strongly compressed flat bodies, one side of which is colourless, and eyes unsymmetrically placed, both on the coloured side.   They inhabit the sandy bottoms of shallow seas, and often ascend rivers."

DISTRIBUTION.—Universal, on Arctic, Temperate, and Tropical coasts.

Seven genera occur in British seas, viz. : *Hippoglossus, Hippoglossoides, Rhombus, Phrynorhombus, Arnoglossus, Pleuronectes* (Turbot), and *Solea* (Sole). There are 13 genera in the Atlantic and 23 in the Pacific, 4 being common to both ; and 2 found only in the Mediterranean. A Pacific genus, *Synaptura*, has one species in the Mediterranean.

Order IV.—PHYSOSTOMI.

FAMILY 59.—SILURIDÆ. (114 Genera, 547 Species.)

" Fresh-water or marine, scaleless fishes, often with bony shields, and the head always furnished with barbels."

DISTRIBUTION.—The fresh waters of all the temperate and tropical regions, those which enter the salt water keeping near the coast.

This extensive family is divided by Dr. Günther into eight sub-families and seventeen groups, the distribution of which is as follows :—

Sub-family 1 (SILURIDÆ HOMALOPTERÆ) is confined to the Old World. It consists of three groups : Clarina (2 genera, *Clarias* and *Heterobranchus*) ranges over the whole area of the Ethiopian and Oriental regions, to which it appears to be strictly confined ; Plotosina (3 genera, *Plotosus, Copidoglanis*, and *Cnidoglanis*) ranges from the eastern coasts of Africa to Japan, Polynesia, and Australia, in seas and rivers ; Chacina (1 genus, *Chaca*) ranges from India to Borneo.

Sub-family 2 (SILURIDÆ HETEROPTERÆ) is also confined to the Old World ; it consists of one group,—Silurina, containing 19 genera, viz. :—*Saccobranchus* (4 sp.), India to Cochin China and Ceylon ; *Silurus* (5 sp.), Palæarctic region from Central Europe to Japan, China, and Afghanistan, and a species in Cochin China; *Silurichthys* (3 sp.), Cashmere, Java, and Borneo ; *Wallago* (2 sp.), Hindostan, Sumatra, and Borneo ; *Belodontichthys* (1 sp.), Sumatra and Borneo ; *Eutropiichthys* (1 sp.), Bengal; *Cryptopterus*

(15 sp.), Java, Sumatra, and Borneo, with a species in the
Ganges, in Siam, and (?) in Amboyna; *Callichrous* (10 sp.),
Afghanistan to Borneo and Java; *Schilbe* (5 sp.), Tropical Africa;
Eutropius (6 sp.), Tropical Africa and Central India; *Hemisilurus*
(2 sp.), Java and Sumatra; *Siluranodon* (1 sp.), Nile; *Ailia*
(2 sp.), Bengal; *Schilbichthys* (1 sp.), Bengal; *Laïs* (1 sp.), Java,
Sumatra, Borneo; *Pseudeutropius* (6 sp.), India and Sumatra;
Pangasius (7 sp.), Ganges, Sumatra, Java, Borneo; *Helicophagus*
(2 sp.), Sumatra; *Silondia* (1 sp.), Ganges.

Sub-family 3 (SILURIDÆ ANOMALOPTERÆ) is confined to
Equatorial America; it consists of the group Hypopthalmina,
containing 2 genera: *Helogenes* (1 sp.), *Hypopthalmus* (4 sp.),
from the country north of the Amazon, Surinam, and the Rio
Negro.

Sub-family 4 (SILURIDÆ PROTEROPTERÆ) ranges over all the
tropical and most of the temperate parts of the globe, except
Europe and Australia.   It consists of four groups: Bagrina
(16 genera), ranging over most of the Old World and North
America; Pimelodina (15 genera), confined to Tropical America,
except one genus which is African; Ariina (10 genera), all
Tropical regions; and Bagarina (3 genera), Oriental region. The
distribution of the genera is as follows:—

Bagrus (2 sp.), Nile; *Chrysichthys* (5 sp.), Tropical Africa;
Clarotes (1 sp.), Upper Nile; *Macrones* (19 sp.), India, Ceylon
to Borneo, and one species in Asia Minor; *Pseudobagrus* (4 sp.),
Japan, China, and Cochin China; *Liocassis* (5 sp.), Japan, China,
Java, Sumatra, and Borneo; *Bagroides* (3 sp.), Sumatra and
Borneo; *Bagrichthys* (1 sp.), Sumatra and Borneo; *Rita* (5 sp.),
Continental India and Manilla; *Acrochordonichthys* (6 sp.), Java
and Sumatra; *Akysis* (3 sp.), Java and Sumatra; *Olyra* (1 sp.),
Khasya; *Branchiosteus* (1 sp.), Khasya; *Amiurus* (13 sp.),
Nearctic region to Guatemala and China; *Hopladelus* (1 sp.),
North America; *Noturus* (4 sp.), North America; *Sorubim*
(1 sp.), Amazon; *Platystoma* (11 sp.), Tropical South America;
Hemisorubim (1 sp.) Rio Negro, Brazil; *Platistomatichthys*
(1 sp.), Rio Branco, Brazil; *Phractocephalus* (1 sp.), Amazon;
Piramutana (2 sp.), Equatorial America; *Platynematichthys*

(1 sp.), northern and southern tributaries of Amazon ; *Piratinga* (3 sp.), Amazon Valley; *Sciades* (2 sp.), Amazon ; *Pimelodus* (42 sp.), Mexico to La Plata, single aberrant species from West Africa, Java and the Sandwich Islands ; *Pirinampus* (1 sp.), Brazil; *Conorhynchus* (1 sp.), Brazil; *Notoglanis* (1 sp.), Madeira, Amazon Valley; *Callophysus* (3 sp.), Tropical South America; *Auchenaspis* (1 sp.), Tropical Africa ; *Arius* (68 sp.), all Tropical regions; *Galeichthys* (1 sp.), Cape of Good Hope ; *Genidens* (1 sp.), Brazil; *Hemipimelodus* (3 sp.), India, Sumatra, and Borneo ; *Ketingus* (1 sp.), Sunda Islands; *Ælurichthys* (4 sp.), Eastern United States to Guiana; *Paradiplomystax* (1 sp.), Brazil ; *Diplomystax* (1 sp.), Chili; *Osteogeniosus* (3 sp.), India to Java ; *Batrachocephalus* (1 sp.), Java and Sumatra ; *Bagarius* (1 sp.), India to Java; *Euclyptosternum* (1 sp.), India; *Glyptosternum* (8 sp.), Himalayas, Central India, Java, and Sumatra; *Hara* (3 sp.), Continental India; *Amblyceps* (3 sp.), Continental India.

Sub-family 5 (SILURIDÆ STENOBRANCHIÆ) is confined to South America and Africa, with one genus and species in the Ganges. It consists of three groups: Doradina (12 genera), South America and Africa; Rhinoglanina (3 genera), Central Africa and the Ganges ; Malapterurina (1 genus), Tropical Africa. The distribution of the genera is as follows :—

Ageniosus (4 sp.), Surinam to La Plata ; *Tetranematichthys* (1 sp.), Central Brazil, Rio Guaporé ; *Euanemus* (1 sp.), Surinam and Brazil ; *Auchenipterus* (9 sp.), Equatorial America; *Centromochlus* (2 sp.), Equatorial America ; *Trachelyopterus* (2 sp.), Equatorial America ; *Cetopsis* (3 sp.), Brazil; *Asterophysus* (1 sp.), Rio Negro, North Brazil; *Doras* (13 sp.), Tropical South America east of Andes; *Oxydoras* (7 sp.), Amazon Valley and Guiana; *Rhinodoras* (3 sp.), Tropical South America east of Andes; *Synodontis* (12 sp.), Tropical Africa ; *Rhinoglanis* (1 sp.), Upper Nile; *Mochocus* (1 sp.), Nile ; *Callomystax* (1 sp.), Nile ; *Malapterurus* (3 sp.), Tropical Africa.

Sub-family 6 (SILURIDÆ PROTEROPODES) inhabits Tropical America and Northern India as far as Tenasserim. It consists of two groups: the Hypostomatina (17 genera), with the same distribution as the sub-family, and the Aspredinina (3 genera), confined to Equatorial America. The distribution of the genera is as follows :—

Arges (2 sp.), Andes of Peru and Ecuador ; *Stygogenes* (2 sp.), Andes ; *Brontes* (1 sp.), Andes ; *Astroblepus* (1 sp.), Popayan ; *Callichthys* (11 sp.), Tropical South America east of Andes, and Trinidad ; *Plecostomus* (15 sp.), Tropical South America east of Andes, and Trinidad ; *Liposarcus* (3 sp.), Surinam and Brazil ; *Chætostomus* (25 sp.), Tropical America, Trinidad, and Porto Rico ; *Pterygoplichthys* (4 sp.), Brazil ; *Rhinelepis* (1 sp.), Brazil ; *Acanthicus* (2 sp.), Equatorial America ; *Loricaria* (17 sp.), Tropical South America east of Andes ; *Acestra* (4 sp.), Brazil and Guiana ; *Sisor* (1 sp.), Northern Bengal ; *Erethistes* (1 sp.), Assam ; *Pseudecheneis* (1 sp.), Khasya Hills ; *Exostoma* (2 sp.), Assam and Tenasserim ; *Bunocephalus* (2 sp.), Guiana ; *Bunocephalichthys* (1 sp.), Rio Branco, North Brazil ; *Aspredo* (6 sp.), Guiana.

Sub-family 7 (SILURIDÆ OPISTHOPTERÆ) consists of two groups: Nematogenyina (2 genera), and Trichomycterina (3 genera), and is confined to South America. The distribution of the genera is as follows :—

Heptapterus (2 sp.), South America ; *Nematogenys* (1 sp.), Chili ; *Trichomycterus* (7 sp.), South America to 15,000 feet elevation ; *Eremophilus* (1 sp.), Andes of Bogota ; *Pariodon* (1 sp.), Amazon.

Sub-family 8 (SILURIDÆ BRANCHICOLÆ) is confined to Tropical South America. It consists of one group, Stegophilina, and 2 genera : *Stegophilus* (1 sp.), Brazil ; and *Vandellia* (2 sp.), Amazon Valley.

FAMILY 60.   CHARACINIDÆ.   (47 Genera, 230 Species.)

" Fresh-water fishes, with scaly bodies and without barbels."

DISTRIBUTION.—The Neotropical and Ethiopian regions.

This extensive family is divided by Dr. Günther into 10 groups, viz. : Erythrinina (5 genera), South America ; Curumatina

(6 genera), South America; Citharinina (1 genus), Tropical Africa; Anostomatina (3 genera), South America; Tetragonopterina (16 genera), South America and Tropical Africa; Hydrocyonina (9 genera), Tropical America and Tropical Africa; Distichodontina (1 genus), Tropical Africa; Icthyborina (1 genus), Africa; Crenuchina (1 genus), Equatorial America; Serrasalmonina (4 genera), South America.

The following is the distribution of the genera:—

Macrodon (4 sp.), Tropical America; *Erythrinus* (5 sp.), Brazil and Guiana; *Lebiasina* (1 sp.), West Equatorial America; *Pyrrhulina* (1 sp.), Guiana; *Corynopoma* (4 sp.), Trinidad only; *Curimatus* (15 sp.), Tropical South America and Trinidad; *Prochilodus* (12 sp.), South America to the La Plata; *Cœntropus* (2 sp.), East Equatorial America; *Hemiodus* (8 sp.), Equatorial America east of Andes; *Saccodon* (1 sp.), Ecuador; *Parodon* (1 sp.), Brazil; *Citharinus* (2 sp.), Tropical Africa; *Anostomus* (8 sp.), Tropical America; *Rhytiodus* (2 sp.), Equatorial America; *Leporinus* (14 sp.), South America East of Andes; *Piabucina* (2 sp.), Guiana; *Alestes* (4 sp.), Tropical Africa: *Brachyalestes* (5 sp.), Tropical Africa; *Tetragonopterus* (32 sp.), Tropical America; *Scissor* (1 sp.), South America; *Pseudochalceus* (1 sp.), West Ecuador; *Chirodon* (2 sp.), Chili; *Chalceus* (1 sp.), Guiana; *Brycon* (10 sp.), South America east of Andes; *Chalcinopsis* (4 sp.), Central America and Ecuador; *Bryconops* (2 sp.), Tropical America; *Creagrutus* (1 sp.), Western Ecuador; *Chalcinus* (4 sp.), Tropical South America; *Gastropelecus* (8 sp.), Tropical South America; *Piabuca* (2 sp.), Equatorial America; *Agoniates* (1 sp.), Guiana; *Anacyrtus* (7 sp.), Central and South America; *Hystricodon* (1 sp.), Equatorial America; *Salminus* (3 sp.), South America; *Hydrocyon* (3 sp.), Tropical Africa; *Sarcodaces* (1 sp.), West Africa; *Oligosarcus* (1 sp.), Brazil; *Xiphoramphus* (7 sp.), South America east of Andes; *Xiphostoma* (5 sp.), Equatorial America east of Andes; *Cynodon* (3 sp.), Tropical America East of Andes; *Distichodus* (7 sp.), Tropical Africa; *Icthyborus* (3 sp.), Nile; *Crenuchus* (1 sp.), Guiana; *Mylesinus* (1 sp.), Equatorial America; *Serrasalmo* (13 sp.), Tropical South America east of Andes; *Myletes* (18 sp.),

Tropical South America east of Andes; *Catoprion* (1 sp.), Brazil and Guiana.

FAMILY 61.—HAPLOCHITONIDÆ.   (2 Genera, 3 Species.)

" Fresh-water fishes, with naked or scaly bodies and without barbels."

DISTRIBUTION.—Temperate South America and South Australia.

The genera are, *Haplochiton* (2 sp.), Tierra del Fuego and the Falkland Islands; *Prototroctes* (2 sp.), Southern Australia and New Zealand.

FAMILY 62.—STERNOPTYCHIDÆ.   (6 Genera, 12 Species.)

" Marine fishes, with very thin deciduous scales or none, and with a row of phosphorescent spots or organs on the under surface of the body."

DISTRIBUTION.—Mediterranean and Atlantic.

These are deep-sea fishes found in the Mediterranean sea, and in the deep Atlantic from the coasts of Norway to the Azores and the Tropics.

FAMILY 63.—SCOPELIDÆ.   (11 Genera, 47 Species.)

"Marine fishes, somewhat resembling the fresh-water Siluridæ."

DISTRIBUTION.—Almost universal, but most abundant in warm and tropical seas.

These are deep-sea fishes, abounding in the Mediterranean and the great oceans, a few extending north to near Greenland and south to Tasmania.

FAMILY 64.—STOMIATIDÆ.   (4 Genera, 8 Species.)

"Small marine fishes, naked or with very fine scales."

DISTRIBUTION.—The Mediterranean and Atlantic.

These are deep-sea fishes, ranging from Greenland to beyond the Equator.

FAMILY 65.—SALMONIDÆ.   (15 Genera, 157 Species.)

"Fresh-water fishes, many species periodically descending to the sea and a few altogether marine :—Salmon and Trout."

DISTRIBUTION.—The Palæarctic and Nearctic Regions, and one genus and species in New Zealand. A considerable number of species are confined to single lakes or rivers, others have a wide distribution.

The genera are distributed as follows :—

Salmo (83 sp.), rivers and lakes of the Palæarctic and Nearctic Regions, as far south as Algeria, Asia Minor, the Hindoo-Koosh and Kamschatka, and to about 38° North Latitude in North America, many of the species migratory ; *Onchorhynchus* (8 sp.), American and Asiatic rivers entering the Pacific, as far south as San Francisco and the Amur ; *Brachymystax* (1 sp.), Siberian rivers, from Lake Baikal and the Atlai Mountains northwards ; *Luciotrutta* (2 sp.), Caspian Sea and Volga ; *Plecoglossus* (1 sp.), Japan and Formosa ; *Osmerus* (3 sp.), rivers of temperate Europe and North America entering the Atlantic, and one species in California ; *Thaleichthys* (1 sp.), Columbia River, Vancouver's Island ; *Hypomesus* (1 sp.), coasts of California, Vancouver's Island, and North-eastern Asia ; *Mallotus* (1 sp.), coasts of Arctic America from Greenland to Kamschatka ; *Retropinna* (1 sp.), fresh waters of New Zealand ; *Coregonus* (41 sp.), fresh waters of northern parts of temperate Europe, Asia and North America, many of the species migratory : *Thymallus* (6 sp.), fresh waters of temperate parts of

Europe, Asia, and North America ; *Argentina* (4 sp.), Mediterranean and deep seas of Western Europe ; *Microstoma* (2 sp.), Mediterranean, and seas of Greenland ; *Salarix* (2 sp.), China and Japan, in seas and rivers. *Salmo, Osmerus, Coregonus,* and *Thymallus,* are British genera.

FAMILY 66.—PERCOPSIDÆ. (1 Genus, 1 Species.)

" A fresh-water fish covered with toothed scales."

DISTRIBUTION.—Lake Superior, North America.

FAMILY 67.—GALAXIDÆ. (1 Genus, 12 Species.)

" Fresh-water fishes, with neither scales nor barbels."

DISTRIBUTION.—The temperate zone of the Southern Hemisphere.

The only genus, *Galaxias,* is found in New Zealand, Tasmania, and Tierra del Fuego, ranging north as far as Queensland and Chili ; and one of the species is absolutely identical in the two regions.

FAMILY 68.—MORMYRIDÆ. (3 Genera, 25 Species.)

" Fresh-water fishes, with scales on the body and tail but not on the head, and no barbels."

DISTRIBUTION.—The Ethiopian Region.

Most abundant in the Nile, a few from the Gambia, the Congo, and Rovuma. The genera are :—
Mormyrus (1 sp.), Nile, Gambia, West Africa, Mozambique, Rovuma ; *Hyperopsius* (2 sp.), Nile and West Africa ; *Mormyrops* (4 sp.), Nile, West Africa and Mozambique.

FAMILY 69.—GYMNARCHIDÆ.   (1 Genus, 1 Species.)

" Fresh-water fishes, resembling the Mormyridæ, but with tapering finless tail, and neither anal nor ventral fins."

DISTRIBUTION.—Ethiopian region.

The only genus, *Gymnarchus*, inhabits the Nile and the rivers of West Africa.

FAMILY 70.—ESOCIDÆ.   (1 Genus, 7 Species.)

" Fresh-water fishes, with scaly bodies, no barbels, and dorsal fins situated towards the tail."

DISTRIBUTION.—The Nearctic and Palæarctic regions.

One species, the Pike (*Esox lucius*) ranges from Lapland to Turkey, and in America from the Arctic regions to the Albany river; the remainder are American species extending South as far as New Orleans.

FAMILY 71.—UMBRIDÆ.   (1 Genus, 2 Species.)

" Small fresh-water scaly fishes, without barbels or adipose fin."

DISTRIBUTION.—Central Europe and Temperate North America.

FAMILY 72.—SCOMBRESOCIDÆ.   (5 Genera, 136 Species.)

" Marine or fresh-water fishes, with scaly bodies and a series of keeled scales along each side of the belly."

DISTRIBUTION.—Temperate and tropical regions.

All the genera have a wide distribution.   A species of *Belone* and one of *Scombresox* are found on the British coast. The Flying fishes (*Exocetus*, 44 sp.), belong to this family.   They abound in all tropical seas and extend as far as the Mediterranean and Australia. None of the genera are exclusively fresh-water, but a few species of *Belone* and *Hemiramphus* are found in rivers in various parts of the world.

FAMILY 73.—CYPRINODONTIDÆ.   (20 Genera, 106 Species.)

"Fresh-water fishes, covered with scales, the sexes frequently differing, mostly viviparous."

DISTRIBUTION.—Southern Europe, Asia, Africa and North America, but most abundant in Tropical America.

The distribution of the genera is as follows:—

Cyprinodon (11 sp.), Italy, North Africa and Western Asia to Persia, also North America from Texas to New York; *Fitzroya* (1 sp.), Montevideo; *Characodon* (1 sp.), Central America; *Tellia* (1 sp.), Alpine pools of the Atlas: *Limnurgus* (1 sp.), Mexican plateau; *Lucania* (1 sp.), Texas; *Haplochilus* (18 sp.), India, Java, Japan, Tropical Africa, Madagascar, and the Seychelle Islands, Carolina to Brazil, Jamaica; *Fundulus* (17 sp.), North and Central America and Ecuador, Spain and East Africa; *Rivulus* (3 sp.), Tropical America, Cuba and Trinidad; *Orestias* (6 sp.), Lake Titacaca, Andes; *Jenynsia* (1 sp.), Rio Plata; *Pseudoxiphophorus* (2 sp.), Central America; *Belonesox* (1 sp.), Central America; *Gambusia* (8 sp.), Antilles, Central America and Texas; *Anableps* (3 sp.), Central and Equatorial America; *Pœcilia* (16 sp.), Antilles, Central and South America; *Mollienesia* (4 sp.), Louisiana to Mexico; *Platypœcilus* (1 sp.), Mexico; *Girardinus* (10 sp.), Antilles and South Carolina to Uruguay; *Lepistes* (1 sp.), Barbadoes.

FAMILY 74.—HETEROPYGII.   (2 Genera, 2 Species.)

"Fresh-water fishes, with posterior dorsal fin, and very small scales."

DISTRIBUTION.—Fresh waters of the United States.

Amblyopsis (1 sp.) is a blind fish found in the caverns of Kentucky; while *Chologastes* (1 sp.), which only differs from it in having perfect eyes, is found in ditches in South Carolina.

FAMILY 75.—CYPRINIDÆ. (109 Genera, 790 Species.)

"Fresh-water fishes, generally scaly, with no adipose fin, and pharyngeal teeth only, the mouth being toothless."

DISTRIBUTION.—Fresh waters of the Old World and North America, but absent from Australia and South America.

This enormous family is divided by Dr. Günther into fourteen groups, the distribution of which is as follows:—
Catostomina (4 genera), North America and North-east Asia ; Cyprinina (39 genera), same range as the family ; Rohteichthyina (1 genus), Malay Archipelago ; Leptobarbina (1 genus), Malay Archipelago ; Rasborina (5 genera), East Africa to China and Borneo ; Semiplotina (2 genera), Western Asia ; Xenocypridina (3 genera), Eastern Asia ; Leuciscina (10 genera), Palæarctic and Nearctic regions ; Rhodeina (3 genera), Palæarctic region ; Danionina (9 genera), India to China and Japan ; Hypophthal-michthyina (1 genus), China ; Abramidina (16 genera), same range as the family ; Homalopterina (2 genera), India to Java ; Cobitidina (10 genera), Palæarctic and Oriental regions.

The following is the distribution of the genera:—
Catostomus (16 sp.), Nearctic region and Eastern Siberia ; *Moxostoma* (2 sp.), Eastern United States ; *Sclerognathus* (5 sp.), Temperate North America to Guatemala, also Northern China ; *Carpiodes* (1 sp.), United States ; *Cyprinus* (2 sp.), Temperate parts of Palæarctic region (1 sp. British) ; *Carassius* (3 sp.), Temperate Palæarctic region (1 sp. British) ; *Catla* (1 sp.), Continental India ; *Cirrhina* (5 sp.), Continental India to China ; *Dangila* (6 sp.), Java, Sumatra, Borneo ; *Osteochilus* (14 sp.), Siam to Java and Sumatra ; *Labeo* (27 sp.), Tropical Africa and Oriental region ; *Tylognathus* (10 sp.), Syria, India to Java ; *Abrostomus* (2 sp.), South Africa ; *Discognathus* (4 sp.), Syria to India and Java, mostly in mountain streams ; *Crossochilus* (9 sp.), India to Sumatra and Java ; *Gymnostomus* (7 sp.), Continental India ; *Epalzeorhynchus* (1 sp.), Sumatra and Borneo ; *Capoeta* (13 sp.), Western Asia ; *Barbus* (163 sp.), Temperate or Tropical

parts of Europe, Asia, and Africa (1 sp. British); *Thynnichthys*
(2 sp.), Pegu, Borneo, and Sumatra ; *Barbichthys* (1 sp.), Java,
Sumatra, and Borneo ; *Amblyrhynchichthys* (1 sp.), Sumatra and
Borneo ; *Albulichthys* (1 sp.), Sumatra and Borneo ; *Oreinus* (3
sp.), Himalayan region ; *Schizothorax* (13 sp.), Himalayan region
and west to Afghanistan and Persia ; *Ptychobarbus* (1 sp.), Thibet;
Gymnocypris (1 sp.), loc. unknown ; *Schizopygopsis* (1 sp.), Thibet;
Diptychus (1 sp.), Himalayas and Thibet; *Aulopyge* (1 sp.),
Western Asia ; *Gobio* (2 sp.), Temperate Europe (1 sp. British) ;
Pseudogobio (4 sp.), China, Japan, and Formosa; *Ceratichthys* (9
sp.), Temperate North America ; *Bungia* (1 sp.), Western Asia,
Herat ; *Pimephales* (2 sp.), Eastern United States; *Hyborhynchus*
(3 sp.), Eastern United States; *Ericymba* (1 sp), United States;
Pseudorasbora (1 sp.), Japan, China ; *Cochlognathus* (1 sp.), Texas;
Exoglossum (2 sp.), United States ; *Rhinichthys* (6 sp.), Eastern
United States ; *Rohteichthys* (1 sp.), Borneo and Sumatra; *Lepto-
barbina* (1 sp.), Sumatra and Borneo ; *Rasbora* (12 sp.), East
Coast of Africa, India, to Java and Borneo; *Luciosma* (3 sp.),
Java, Sumatra, and Borneo ; *Nuria* (2 sp.), India, Tenasserim,
and Ceylon ; *Aphyocypris* (1 sp.), North China; *Amblypharyn-
godon* (3 sp.), India to Tenasserim ; *Cyprinion* (3 sp.), Syria and
Persia ; *Semiplotus* (1 sp.), Assam; *Xenocypris* (1 sp.), China;
Paracanthobrama (1 sp.), China ; *Mystacoleucus* (1 sp.), Sumatra;
Leuciscus (84 sp.), Nearctic and Palæarctic regions (5 sp. are
British); *Ctenopharyngodon* (1 sp.), China; *Mylopharodon* (1
sp.), California ; *Paraphoxinus* (2 sp.), South-eastern Europe;
Meda (1 sp.), River Gila ; *Tinca* (1 sp.), Europe (Britain to Con-
stantinople); *Leucosomus* (8 sp.), Nearctic region; *Chondrostoma*
(7 sp), Europe and Western Asia; *Orthodon* (1 sp.), California ;
Acrochilus (1 sp.), Columbia River ; *Achilognathus* (6 sp.), China,
Japan, and Formosa; *Rhodeus* (3 sp.), Central Europe and China;
Pseudoperilampus (1 sp.), Japan ; *Danio* (8 sp.), India and Ceylon ; *Pterosarion* (2 sp.), Central India and Assam ; *Aspidoparia*
(3 sp.), Continental India; *Barilius* (15 sp.), East Africa and Continental India ; *Bola* (1 sp.), Ganges to Bramahputra ; *Schacra*
(1 sp.), Bengal; *Opsariichthys* (5 sp.), Japan and Formosa ;
Squaliobarbus (1 sp.), China ; *Ochetobius* (1 sp.), North China;

Hypophthalmichthys (2 sp.), China; *Abramis* (16 sp.), North America, Central Europe, and Western Asia (1 sp. is British); *Aspius* (3 sp.), East Europe, Western Asia, China; *Alburnus* (15 sp.), Europe and Western Asia (1 British sp.); *Rasborichthys* (1 sp.), Borneo; *Elopichthys* (1 sp.), China; *Pelotrophus* (2 sp.), East Africa; *Acanthobrama* (3 sp.), Western Asia; *Osteobrama* (5 sp.), Continental India; *Chanodichthys* (6 sp.), China and Formosa; *Smiliogaster* (1 sp.), Bengal; *Culter* (2 sp.), China; *Pelecus* (1 sp.), Eastern Europe; *Eustira* (1 sp.), Ceylon; *Chela* (16 sp.), India to Siam, Java and Borneo; *Pseudolabuca* (1 sp.), China; *Cachius* (1 sp.), Continental India; *Homaloptera* (8 sp.), India to Cochin China, Java, and Sumatra; *Psilorhynchus* (2 sp.), North-eastern India; *Misgurnus* (5 sp.), Europe to India, China, and Japan; *Nemachilus* (37 sp.), Europe and Asia; *Cobitis* (3 sp.), Europe, India, Japan; *Lepidocepalichthys* (3 sp.), India, Ceylon, and Java; *Acanthopsis* (2 sp.), Tenasserim, Sumatra, Java, and Borneo; *Botia* (7 sp.), India to Japan and Sunda Isles; *Oreonectes* (1 sp.), China; *Lepidocephalus* (1 sp.), Java and Sumatra; *Acanthopthalmus* (2 sp.), Java and Sumatra; *Apua* (1 sp.), Tenasserim; *Kneria* (2 sp.), Tropical Africa.

FAMILY 76 —GONORHYNCHIDÆ. (1 Genus, 1 Species.)

"A marine fish with spiny scales, mouth with barbels, and with short dorsal fin opposite the ventrals."

DISTRIBUTION.—Temperate parts of Southern Oceans, and Japan.

FAMILY 77.—HYODONTIDÆ (1 Genus, 1 Species.)

"A fresh-water fish with cycloid scales and posterior dorsal fin."

DISTRIBUTION.—Fresh waters of North America.

FAMILY 78.—OSTEOGLOSSIDÆ. (3 Genera, 5 Species.)

"Fresh-water fishes, with large hard scales, and dorsal fin opposite and equal to the anal fin."

DISTRIBUTION.—Tropical rivers.

The genera are:—*Osteoglossum* (3 sp.), Eastern South America, Sunda Islands, and Queensland; *Arapaima* (1 sp.), Eastern South America—the "Pirarucú" of the Amazon; *Heterotis* (1 sp.), Tropical Africa.

FAMILY 79.—CLUPEIDÆ. (18 Genera, 161 Species.)

"Marine scaly fishes, without barbels, and with the abdomen often compressed and serrated."

DISTRIBUTION.—Seas of the whole globe, many species entering rivers. They are most abundant in the Indian seas, less so in America, scarce in Africa, while they are almost absent from Australia. The Herring, Sprat, Shad, and Pilchard, are British species of *Clupea*, a genus which contains 61 species and ranges all over the world.

FAMILY 80.—CHIROCENTRIDÆ. (1 Genus, 1 Species.)

"A marine fish, with thin deciduous scales, no barbels, and posterior dorsal fin."

DISTRIBUTION.—The Eastern seas from Africa to China.

FAMILY 81.—ALEPOCEPHALIDÆ. (1 Genus, 1 Species.)

"A marine fish, covered with thin cycloid scales, no barbels, and posterior dorsal fin."

DISTRIBUTION.—Deep waters of the Mediterranean.

FAMILY 82.—NOTOPTERIDÆ. (1 Genus, 5 Species.)

" Fresh-water fishes, without barbels, head and body scaly, long tapering tail, and short posterior dorsal fin."

DISTRIBUTION.—Rivers of India, Siam, the Sunda Islands, and West Africa.

FAMILY 83.—HALOSAURIDÆ. (1 Genus, 1 Species.)

" Marine fishes, with cycloid scales, a short median dorsal fin, and no barbels."

DISTRIBUTION.—Deep waters of the Atlantic, Madeira.

FAMILY 84.—GYMNOTIDÆ. (5 Genera, 20 Species.)

" Fresh-water fishes, with elongate bodies, pointed tail, and no dorsal fin."

DISTRIBUTION.—Tropical America from Trinidad to the River Parana.

The genera are distributed as follows :—
Sternarchus (8 sp.), Guiana and Brazil; *Rhamphichthys* (6 sp.), Guiana and Brazil; *Sternophygus* (4 sp.), Tropical America; *Carapus* (1 sp.), Trinidad to Brazil; *Gymnotus*, (1 sp. —the Electric eel), Tropical South America.

FAMILY 85.—SYMBRANCHIDÆ. (4 Genera, 6 Species.)

" Marine and fresh-water fishes, having elongate bodies without fins, and very minute scales or none."

DISTRIBUTION.—Fresh waters and coasts of Western Australia and Tasmania.

The genera are :—
Amphipnous (1 sp.), Bengal; *Monopterus* (1 sp.), Siam to Northern China and Sunda Islands; *Symbranchus* (3 sp.), Tropical

America, and India to Australia; *Chilobranchus* (1 sp.), Australia and Tasmania.

FAMILY 86.—MURÆNIDÆ.   (26 Genera, 230 Species.)

" Marine or fresh-water fishes, with cylindrical or band-like bodies and no ventral fins."

DISTRIBUTION.—The seas and fresh waters of temperate and tropical regions. This family is divided by Dr. Günther into two sub-families and nine sections. The genus *Anguilla*, comprising our common Eel and a number of species from all parts of the world, is the only one which is found in fresh water, though even here most of the species are marine. *Anguilla* and *Conger* are the only British genera.

FAMILY 87.—PEGASIDÆ.   (1 Genus, 4 Species.)

" Small marine fishes, covered with bony plates, and short opposite dorsal and anal fins."

DISTRIBUTION.—Indian Ocean and seas of China and Australia.

Order V—LOPHOBRANCHII.

" Fish with a segmented bony covering, long snout, and small toothless mouth."

FAMILY 88.—SOLENOSTOMIDÆ.   (1 Genus, 3 Species.)

" Marine Lophobranchii, with wide gill openings and two dorsal fins."

DISTRIBUTION.—Indian Ocean, from Zanzibar to China and the Moluccas.

FAMILY 89.—SYNGNATHIDÆ.   (15 Genera, 112 Species.)

" Marine Lophobranchii, with very small gill opening and one soft dorsal fin."

DISTRIBUTION.—All the tropical and temperate seas. Some species of *Syngnathus*, *Doryichthys*, and *Cœlonotus* enter fresh water, and a few live in it exclusively. *Siphonostoma*, *Syngnathus*, *Nerophis*, and *Hippocampus* are British genera. The *Hippocampina* (5 genera, 25 sp.), or Sea-horses, are peculiar to the Indian and Pacific Oceans, except three or four species of *Hippocampus* in the Atlantic and Mediterranean.

Order VI.—PLECTOGNATHI.

" Fishes covered with rough scales or shields, having a narrow mouth, and soft posterior dorsal fin."

FAMILY 90 —SCLERODERMI.   (7 Genera, 95 Species.)

" Marine Plectognathi, with toothed jaws."

DISTRIBUTION.—Temperate and Tropical seas, but much more abundant in the Tropics.

FAMILY 91.—GYMNODONTES.   (10 Genera, 82 Species.)

" Marine or fresh-water Plectognathi, with jaws modified into a beak."

DISTRIBUTION.—Temperate and tropical regions.

Some species of *Tetrodon* are found in the rivers of Tropical America, Africa, and Asia. Species of *Tetrodon* and *Orthagoriscus* have been found on the British coasts.

Sub-class II.—DIPNOI.

Family 92.—Sirenoidei. (3 Genera, 3 Species.)

"Eel-shaped fresh-water fishes, covered with cycloid scales; the vertical fins forming a continuous border to the compressed tapering tail."

Distribution.—Rivers of Tropical Africa, South America, and Australia.

The genera are:—*Protopterus* (1 sp.), Tropical Africa; *Lepidosiren* (1 sp.), Amazon Valley; *Ceratodus* (1 sp.), Queensland.

Sub-class III.—GANOIDEI.

Order I.—HOLOSTEI.

"Body covered with scales."

Family 93.—AMIIDÆ. (1 Genus, 1 Species.)

"A fresh-water fish, with cycloid scales and a long soft dorsal fin."

Distribution.—United States.

Family 94.—POLYPTERIDÆ. (2 Genera, 2 Species.)

"Fresh-water fishes, with ganoid scales and dorsal spines."

Distribution.—Central and Western Africa.

The genera are :—
Polypterus (1 sp.), the Nile and rivers of West Africa; *Calamoichthys* (1 sp.), Old Calabar.

FAMILY 95.—LEPIDOSTEIDÆ.   (1 Genus, 3 Species.)

" Fresh-water fishes, with ganoid scales, and dorsal and anal fins composed of articulated rays."

DISTRIBUTION.—The genus *Lepidosteus*, the Garfishes or Bony Pikes, inhabits North America to Mexico and Cuba.

Order II.—CHONDROSTEI.

" Sub-cartilaginous scaleless fishes with heterocercal tail, the skin with osseous bucklers or naked."

FAMILY 96.—ACCIPENSERIDÆ.   (2 Genera, 20 Species.)

" Marine or fresh-water fishes with osseous bucklers and inferior mouth."

DISTRIBUTION.—Temperate and Arctic regions of the northern hemisphere. *Accipenser* (19 sp.), comprising the Sturgeons, has the distribution of the family ; most of the species are marine, but some are confined to the Caspian and Black Seas and the great American lakes with the rivers flowing into them, while the Danube, Mississippi, and Columbia River have peculiar species.   The other genus, *Scaphirhynchus* (1 sp.), is confined to the Mississippi and its tributaries.

FAMILY 97.—POLYDONTIDÆ.   (1 Genus, 2 Species.)

" Fresh-water fishes, with wide lateral mouth and naked skin."

DISTRIBUTION.—The Mississippi and Yang-tse-kiang rivers.

SUB CLASS IV.—CHONDROPTERYGII. (SHARKS AND RAYS.)

Order I.—HOLOCEPHALA. (Chimœras.)

FAMILY 98.—CHIMÆRIDÆ. (2 Genera, 4 Species.)

" Shark-like marine fishes, snout of the male with a prehensile organ."

DISTRIBUTION.—Northern and Southern temperate seas. Chimœra is British.

Order II.—PLAGIOSTOMATA.

Sub-order.—SELACHOIDEA. (Sharks.)

FAMILY 99.—CARCHARIIDÆ. (11 Genera, 59 Species.)

" Sharks with two dorsals and a nictitating membrane."

DISTRIBUTION.—Seas of the Arctic, temperate, and tropical regions. Species of Galeus and Mustelus have occurred on our coasts.

FAMILY 100.—LAMNIDÆ. (5 Genera, 7 Species.)

" Sharks with two dorsals and no nictitating membrane."

DISTRIBUTION.—Temperate and tropical seas. Species of Lamna, Alopecias, and Selache have occurred in British seas.

FAMILY 101.—RHINODONTIDÆ. (1 Genus, 1 Species.)

"Sharks with two dorsal fins, the second small, and no nictitating membrane."

DISTRIBUTION.—South and East Africa.

FAMILY 102.—NOTIDANIDÆ. (1 Genus, 4 Species.)

"Sharks with one dorsal fin and no nictitating membrane."

DISTRIBUTION.—Temperate and tropical seas, from the North Atlantic to the Cape of Good Hope and California. One species has occurred on our southern coasts.

FAMILY 103.—SCYLLIIDÆ. (7 Genera, 25 Species.)

"Sharks with one dorsal fin and no nictitating membrane."

DISTRIBUTION.—All temperate and tropical seas. Species of *Scyllium* and *Pristiurus* are British.

FAMILY 104.—CESTRACIONTIDÆ. (1 Genus, 4 Species.)

"Sharks with two dorsal fins and no nictitating membrane."

DISTRIBUTION.—Pacific Ocean from Japan to New Zealand, Moluccan Sea.

FAMILY 105.—SPINACIDÆ. (10 Genera, 21 Species.)

"Sharks with two dorsal fins and no nictitating membrane, no anal fin."

DISTRIBUTION.—Arctic, temperate, and tropical seas. Species of *Acanthias*, *Læmargus*, and *Echinorhinus* have occurred on our coasts.

FAMILY 106.—RHINIDÆ.   (1 Genus, 1 Species.)

"Sharks with depressed flat body and large expanded pectoral fins."

DISTRIBUTION.—Temperate and tropical seas, from Britain to California and Australia.

FAMILY 107.—PRISTIOPHORIDÆ.   (1 Genus, 4 Species.)

"Sharks with produced flat snout, armed with teeth on each edge."

DISTRIBUTION.—Seas of Japan and Australia.

Sub-order BATOIDEI.   (Rays.)

FAMILY 108.—PRISTIDÆ.   (1 Genus, 5 Species.)

"Rays with produced snout and lateral saw-like teeth."

DISTRIBUTION.—Seas of tropical and sub-tropical regions.

FAMILY 109.—RHINOBATIDÆ.   (3 Genera, 15 Species.)

"Rays with long and strong tail, having a caudal and two dorsal fins."

DISTRIBUTION.—Tropical and sub-tropical seas.

FAMILY 110.—TORPEDINIDÆ.   (6 Genera, 15 Species.)

"Rays with broad smooth disc, and an electric organ."

DISTRIBUTION.—Tropical and temperate seas, from Britain to Tasmania.

FAMILY 111.—RAIIDÆ.   (4 Genera, 29 Species.)

"Rays with broad rhombic disc and no serrated caudal spine."

DISTRIBUTION.—All temperate and tropical seas.   Several species of *Raia* are found on our coasts.

FAMILY 112.—TRYGONIDÆ.　(6 Genera, 43 Species.)

" Rays with the pectoral fins extending to end of snout."

DISTRIBUTION.—Seas of all temperate and tropical regions, and rivers of Tropical America. A species of *Trygon* has occurred on our Southern coast. *Ellipesurus* and *Tæniura* are found in the fresh waters of the interior of South America, while the latter genus occurs also in the Indian seas, but not in the Atlantic.

FAMILY 113.—MYLOBATIDÆ.　(5 Genera, 22 Species.)

" Rays with very broad pectoral fins not extending to end of snout."

DISTRIBUTION.—Temperate and tropical seas. A species of *Myliobatis* is British, but most of the species and genera are confined to tropical seas. *Dicerobatis* and *Ceratoptera* are very large Rays, commonly called Sea-devils.

SUB-CLASS V.—CYCLOSTOMATA.

" Cartilaginous fishes, with suctorial mouths and without lateral fins."

FAMILY 114.—PETROMYZONTIDÆ.　(4 Genera, 12 Species.)

" Marine or fresh-water eel-like fishes, with suctorial mouths and without barbels."

DISTRIBUTION.—Coasts and fresh waters of temperate regions of both hemispheres. Three species of *Petromyzon* (Lampreys), are British.

FAMILY 115.—MYXINIDÆ.   (2 Genera, 5 Species.)

" Marine eel-like fishes, with four pairs of barbels."

DISTRIBUTION.—Seas of the temperate regions of both hemispheres.

SUB-CLASS VI.—LEPTOCARDII.

FAMILY 116 —CIRRHOSTOMI.   (1 Genus, 1 Species.)

" A small marine fish with no jaws or fins, and with rudimentary eyes."

DISTRIBUTION.—The only species, the Lancelet (*Amphioxus*), is the lowest form of living vertebrate.   It is found in the temperate regions of both hemispheres, and has occurred on our southern coast.

Remarks on the Distribution of Fishes.

Marine Fish.—There are about 80 families of marine fishes, and of these no less than 50 are universally, or almost universally, distributed over the seas and oceans of the globe.   Of the remainder many are widely distributed, some species even ranging from the North Atlantic to Australia.   Six families are confined to the Northern Seas, but four of these consist of single species only, the other two being the Discoboli (2 genera, 11 sp.), and the Accipenseridæ (2 genera and 20 sp.).   Only one family (Acanthoclinidæ) is confined to the Southern oceans, and that consists of but a single species.   Four families (Sternoptychidæ), Stomiatidæ, Alepocephalidæ and Halosauridæ) are confined to the Atlantic Ocean, while 13 are found only in the Pacific ; and of the remainder several are more abundant in the Pacific than the Atlantic.   Two families (Lycodidæ and Gadidæ) are found in the Arctic and Antarctic seas only, though the latter family has a single species in the Indian seas. Among the curiosities of distribution are,—the extensive genus *Diagramma*, confined to the Pacific with the exception of one species in the Mediterranean ; the single species constituting the family Lophotidæ, found only in the Mediterranean and Japan ; the small family of Notacanthi, confined to Greenland, the Mediterranean, and West Australia ; and the four families, Sternoptychidæ, Stomiatidæ, Alepocephalidæ, and Halosauridæ, which are believed to inhabit exclusively the depths of the ocean, and are therefore very rarely obtained.

Fresh-water Fish.—There are 36 families of fishes which inhabit fresh water exclusively, and 5 others, which are both marine and fresh-water. These present many interesting peculiarities of distribution. The Neotropical region is the richest in families, and probably also in genera and species. No less than 22 families inhabit it, and of these 6 are altogether peculiar. The Ethiopian and Nearctic regions each have 18 families, the former with 3, and the latter with 5 peculiar. Several isolated forms, requiring to be placed in distinct families, inhabit the great American lakes ; and, no doubt, when the African lakes are equally well known, they will be found also to possess many peculiar forms. The Oriental region comes next, with 17 families, of which 3 are peculiar. The Palæarctic has 12, and the Australian 11 families, each with only 1 altogether peculiar to it.

If we take those regions which are sometimes supposed to be so nearly related that they should be combined, we shall find the fresh-water fishes in most cases markedly distinct. The Nearctic and Palæarctic regions, for example, together contain 20 families, but only 11 of these occur in both, and only 5 are exclusive inhabitants of these two regions. This shows an amount of diversity that would not, perhaps, be exhibited by any other class of animals. The Ethiopian and Oriental regions together possess 24 families, only 11 of which are found in both, and only 1 exclusively characteristic of the two. The Australian and Neotropical regions possess together 27 families, of which 7 are found in both, and 3 are exclusively characteristic of the two. This last fact is very interesting : the marine family of

Trachinidæ possesses a fresh-water genus, *Aphritis*, one species of which inhabits Tasmania, and two others Patagonia; the Haplochitonidæ (2 genera, 3 sp.) are found only in Tierra del Fuego, the Falkland Islands, and South Australia; and the Galáxidæ (1 genus, 12 sp.) inhabit the same regions, but extend to Chili, to New Zealand and to Queensland. We have here an illustration of that connection between South America and Australia which is so strongly manifested in plants, but of which there are only scattered indications in most classes of animals. The dividing line across the Malay Archipelago, separating the Oriental from the Australian regions, and which is so strikingly marked in mammalia and birds, is equally so in fresh-water fishes. No less than six families have their eastern limits in Java and Borneo; while the extensive family of Cyprinidæ has no less than 23 genera in Java and Borneo, but not a single species has been found in Celebes or the Moluccas.

The distribution of fresh-water fishes lends no support to the view that the peninsula of India belongs to the Ethiopian region. A large proportion of the Oriental families are common to the whole region; while there is hardly a single example, of a characteristic Ethiopian family or genus extending into the peninsula of India and no further.

Among the special peculiarities of distribution, is the curious fish, forming the family Comephoridæ, which is confined to Lake Baikal, among the mountains of Central Asia, 2,000 feet above the sea, and a thousand miles distant from the ocean; yet having its nearest allies in the exclusively oceanic family of the mackerels (Scomberidæ). The Characinidæ are confined to Africa and South America, distinct genera inhabiting each region. The Salmonidæ are confined to the two northern regions, except a single species of a peculiar genus in New Zealand. The genus *Osteoglossum* has a species in South America, another in the Sunda Islands, and a third in Queensland; while the curious Sirenoidei are represented by single species of peculiar genera in Tropical America, Tropical Africa, and Tropical Australia.

Fossil Fishes.—Fishes have existed from a very remote era and it is remarkable that the first whose remains have been discovered belong to the Ganoidei, a highly developed group which has continued to exist down to our times, and of which the sturgeon is the best known example. We may therefore be sure that the Upper Silurian rocks in which these are found, although so very far back in geological history, do not by any means lead us to the time when the primitive fish-type appeared upon the earth. In the Carboniferous and Permian formations numerous remains of fishes are found, allied to the *Lepidosteus* or Gar-pike of North America. The next group in order of appearance, are the Plagiostomata, containing the existing Sharks and Rays. Traces of these are found in the highest Silurian beds, and become plentiful in the Devonian and Carboniferous formations and in all succeeding ages, being especially abundant in Cretaceous and Eocene strata. The Holocephali appear first in the Oolitic period, and are represented by the living Chimæridæ. The Dipnoi, to which belong the *Lepidosiren* and *Ceratodus*, are believed to have existed in the Triassic period, from the evidence of teeth almost identical with those of the existing Australian fish. All the ancient fossil fishes belong to the above-mentioned groups, and many of them have little resemblance to existing forms. The Teleostean fishes, which form the great bulk of those now living, cannot be traced back further than the Cretaceous period, while by far the larger number first appear in the Tertiary beds. The Salmonidæ, Scopelidæ, Percidæ, Clupeidæ, Scombresocidæ, Mugilidæ, and Siluridæ, or forms closely allied to them, are found in the Cretaceous formation. In the Eocene beds we first meet with Squammipennes, Cyprinidæ, Pleuronectidæ, Characinidæ, Murænidæ, Gadidæ, Pediculati, Syngnathidæ, and Hippocampidæ.

Most of these fossils represent marine fishes, those of freshwater origin being rare, and of little importance as an aid in determining the causes of the distribution of living forms. To understand this we must look to the various changes of the land surface which have led to the existing distribution of all the higher vertebrates, and to those special means of dispersal which Mr. Darwin has shown to be possessed by all fresh-water productions.

CHAPTER XXI.

ALTHOUGH insects are, for the most part, truly terrestrial animals, and illustrate in a very striking manner the characteristic phenomena of distribution, it is impossible here to treat of them in much detail. This arises chiefly from their excessive numbers, but also from the minuteness and obscurity of many of the groups, and our imperfect knowledge of all but the European species. The number of described species of insects is uncertain, as no complete enumeration of them has ever been made; but it probably exceeds 100,000, and these may belong to somewhere about 10,000 genera—many times more than all vertebrate animals together. Of the eight Orders into which Insects are usually divided, only two—the Coleoptera and Lepidoptera —have been so thoroughly collected in all parts of the globe that they can be used, with any safety, to compare their distribution with that of vertebrate animals; and even of these it is only certain favourite groups which have been so collected. Among Lepidoptera, for example, although the extensive group of Butterflies may be said, in a general sense, to be thoroughly well known—every spot visited by civilized man having furnished its quota to our collections—yet the minute Tineidæ, or even the larger but obscure Noctuidæ, have scarcely been collected at all in tropical countries, and any attempt to study their geographical distribution would certainly lead to erroneous results. The same thing occurs, though perhaps in a less degree, among the Coleoptera. While the Carabidæ, Buprestidæ, and

Longicorns of the Tropics, are almost as well known as those of
the Temperate Zones, the Staphylinidæ, the smaller Elateridæ,
and many other obscure and minute groups, are very imperfectly
represented from extra-European countries. I therefore propose
to examine with some care the distribution of the Butterflies,
and the Sphingina among Lepidoptera, and the following large
and well-known families of Coleoptera :—Cicindelidæ, Carabidæ,
Lucanidæ, Cetoniidæ, Buprestidæ, and the three families of Longicorns. These families together contain over 30,000 species,
classed in nearly 3,000 genera, and comprise a large proportion
of the best known and most carefully studied groups. We may
therefore consider, that a detailed examination of their distribution will lead us to results which cannot be invalidated by any
number of isolated facts drawn from the less known members of
the class.

Range of Insects in Time.—In considering how much weight
is to be given to facts in insect distribution, and what inter-
pretation is to be put upon the anomalies or exceptional cases
that may be met with, it is important to have some idea of the
antiquity of the existing groups, and of the rate at which the
forms of insect life have undergone modification. The geological record, if imperfect in the case of the higher animals,
is fragmentary in the extreme as regards indications of former
insect life; yet the positive facts that it does disclose are of
great interest, and have an important bearing on our subject.
These facts and the conclusions they lead to have been discussed
in our first volume (p. 166), and they must be carefully weighed
in all cases of apparent conflict or incongruity between the distribution of insects and that of the higher animals.

Order—LEPIDOPTERA.

Sub-order—LEPIDOPTERA RHOPALOCERA, or BUTTERFLIES.

FAMILY 1.—DANAIDÆ.   (24 Genera, 530 Species.)

GENERAL DISTRIBUTION.					
NEOTROPICAL SUB-REGIONS.	NEARCTIC SUB-REGIONS.	PALÆARCTIC SUB-REGIONS.	ETHIOPIAN SUB-REGIONS.	ORIENTAL SUB-REGIONS.	AUSTRALIAN SUB-REGIONS.
1.2.3.4	1.2.3.4	— 2 — —	1.2.3.4	1.2.3.4	1.2.3.4

The Danaidæ are now held to comprehend, not only the whole of the group so named by Doubleday, but a large portion of the Heliconidæ of that author.   Their range is thus extended over the whole of the tropical regions.   A few species spread northwards into the Palæarctic and Nearctic regions, but these are only stragglers, and hardly diminish the exclusively tropical character of the group.   The more remarkable genera are,—*Hestia* (10 sp.), and *Ideopsis* (6 sp.), confined to the Malayan and Moluccan districts ; *Danais* (50 sp.), which has the range of the whole family ; *Euplœa* (140 sp.), confined to the Oriental and Australian regions, but especially abundant in the Malayan and Moluccan districts ; *Hamadryas* (4 sp.), Australian region only. The remaining genera constitute the Danaioid Heliconidæ, and are strictly confined to Tropical America, except a few species which extend into the southern parts of the Nearctic region. The chief of these genera are :—

Ithomia (160 sp.), *Melinœa* (18 sp.), *Napeogenes* (20 sp.), *Mechanitis* (4 sp.), *Ceratina* (32 sp.), *Dircenna* (10 sp.), and *Lycorea* (4 sp.).   Florida, Louisiana, and Southern California, mark the northern extent of these insects.

FAMILY 2.—SATYRIDÆ.   (60 Genera, 835 Species.)

GENERAL DISTRIBUTION.					
NEOTROPICAL SUB-REGIONS.	NEARCTIC SUB-REGIONS.	PALÆARCTIC SUB-REGIONS.	ETHIOPIAN SUB-REGIONS.	ORIENTAL SUB-REGIONS.	AUSTRALIAN SUB-REGIONS.
1.2.3.4	1.2.3.4	1.2.3.4	1.2.3.4	1.2.3.4	1.2.3.4

This family has an absolutely universal distribution, extending even into the Arctic and Antarctic regions. Many of the genera are, however, restricted in their range.

Hœtera, Lymanopoda, Calisto, Corades, Taygetis, Pronophila, Euptychia, and some allied forms (25 genera in all) are Neotropical, the last named extending north to Canada; *Debis, Melanitis, Mycalesis* and *Ypthima,* are mostly Oriental, but extending also into the Australian and the Ethiopian regions; *Gnaphodes, Leptoneura,* and a few other small genera, are exclusively Ethiopian; *Xenica, Hypocista,* and *Heteronympha,* are Australian; *Erebia, Satyrus, Hipparchia, Cœnonympha,* and allies, are mostly Palæarctic, but some species are Ethiopian, and others Nearctic; *Chionabas,* is characteristic of the whole Arctic regions, but is also found in Chili and the Western Himalayas. The peculiar genera in each region are,—Neotropical, 25; Australian, 7; Oriental, 11; Ethiopian, 5; Palæarctic, 3; Nearctic, 0.

FAMILY 3.—ELYMNIIDÆ.   (1 Genus, 28 Species.)

GENERAL DISTRIBUTION.					
NEOTROPICAL SUB-REGIONS.	NEARCTIC SUB-REGIONS.	PALÆARCTIC SUB-REGIONS.	ETHIOPIAN SUB-REGIONS.	ORIENTAL SUB-REGIONS.	AUSTRALIAN SUB-REGIONS.
— — — —	— — — —	— — — —	— 2 — —	— ·· 3 . 4	1 — — —

The genus *Elymnias,* which constitutes this family, is characteristic of the Malayan and Moluccan districts, with some species in Northern India and one in Ashanti. It thus agrees with several groups of Vertebrata, in showing the resemblance of Malaya with West Africa independently of the Peninsula of India.

FAMILY 4. MORPHIDÆ. (10 Genera, 106 Species.)

GENERAL DISTRIBUTION.					
NEOTROPICAL SUB-REGIONS.	NEARCTIC SUB-REGIONS.	PALÆARCTIC SUB-REGIONS.	ETHIOPIAN SUB-REGIONS.	ORIENTAL SUB-REGIONS.	AUSTRALIAN SUB-REGIONS.
— 2.3 —	— — — —	— — — —	— — — —	— — 3.4	1 — 3 —

The Morphidæ are a group of generally large-sized butterflies, especially characteristic of the Malayan and Moluccan districts, and of Tropical America; with a few species extending to the Himalayas on the west, and to Polynesia on the east. The genera are:—

Amathusia (6 sp.), Northern India to Java; *Zeuxidia* (9 sp.), the Malay district; *Discophora* (7 sp.), Northern India to Philippines, Java and Timor; *Enispe* (3 sp.), Northern India; *Hyades* (15 sp.), Moluccan and Polynesian districts, except one species in Java; *Clerome* (11 sp.), Northern India to Philippines and Celebes; *Æmona* (1 sp.), Sikhim; *Hyantis* (1 sp.), Waigiou; *Thaumantis* (10 sp.), Indo-Chinese and Malayan districts; *Morpho* (40 sp.), Neotropical region, Brazilian and Central American sub-regions.

FAMILY 5. BRASSOLIDÆ. (7 Genera, 62 Species.)

GENERAL DISTRIBUTION.					
NEOTROPICAL SUB-REGIONS.	NEARCTIC SUB-REGIONS.	PALÆARCTIC SUB-REGIONS.	ETHIOPIAN SUB-REGIONS.	ORIENTAL SUB-REGIONS.	AUSTRALIAN SUB-REGIONS.
— 2.3—	— — — —	— — — —	— — — —	— — — —	— — — —

The Brassolidæ have the same distribution as the genus *Morpho*. The genera are:—

Brassolis (5 sp.); *Opsiphanes* (17 sp.); *Dynastor* (2 sp.); *Penetes* (1 sp.); *Caligo* (21 sp.); *Narope* (5 sp.); and *Dasyopthalma* (3 sp.)

FAMILY 6.—ACRÆIDÆ.   (1 Genus, 90 Species.)

GENERAL DISTRIBUTION.					
NEOTROPICAL SUB-REGIONS.	NEARCTIC SUB-REGIONS.	PALÆARCTIC SUB-REGIONS.	ETHIOPIAN SUB-REGIONS.	ORIENTAL SUB-REGIONS.	AUSTRALIAN SUB-REGIONS.
— — 2.3	— — — —	— — — —	1.2.3.4	1.2.3.4	1.2 — —

The genus *Acræa* is especially abundant in the Ethiopian region, which contains two-thirds of all the known species; 3 or 4 species only, range over the whole Oriental, and most of the Australian regions ; while all the rest inhabit the same districts of the Neotropical region as the Brassolidæ.

FAMILY 7.—HELICONIDÆ.   (2 Genera, 114 Species.)

GENERAL DISTRIBUTION.					
NEOTROPICAL SUB-REGIONS.	NEARCTIC SUB-REGIONS.	PALÆARCTIC SUB-REGIONS.	ETHIOPIAN SUB-REGIONS.	ORIENTAL SUB-REGIONS.	AUSTRALIAN SUB-REGIONS.
— 2.3.4	— — 3 —	— — .. — —	— — — —	— — — —	— — — —

The true Heliconidæ are very characteristic of the Neotropical region ; one species only extending into the Southern States of North America as far as Florida.   The genus *Heliconius* (83 sp.), has the range of the family ; while *Eueides* (19 sp.), is confined to the Brazilian and Central American sub-regions.

FAMILY 8.—NYMPHALIDÆ.   (113 Genera, 1490 Species.)

GENERAL DISTRIBUTION.					
NEOTROPICAL SUB-REGIONS.	NEARCTIC SUB-REGIONS.	PALÆARCTIC SUB-REGIONS.	ETHIOPIAN SUB-REGIONS.	ORIENTAL SUB-REGIONS.	AUSTRALIAN SUB-REGIONS.
1.2.3.4	1.2.3.4	1.2.3.4	1.2.3.4	1.2.3.4	1.2.3.4

This is the largest and most universally distributed family of butterflies, and is well illustrated by our common Fritillaries,

Tortoise-shell, Peacock, Painted Lady, and Purple Emperor butterflies. They are found wherever butterfly-life can exist, and some single species—like the Painted Lady (*Pyrameis cardui*)—range almost over the globe. A few of the more extensive and remarkable genera only, can be here noticed:—

Colænis, Agraulis, Eresia, Synchloe, Epicalia, Eunica, Eubagis, Catagramma, Callithea, Ageronia, Timetes, Heterochroa, Prepona, Hypna, Paphia, and *Siderone,* are wholly Neotropical, as well as many others which have a smaller number of species. *Euryphene, Romaleosoma, Aterica,* and *Harma,* are exclusively Ethiopian. *Terinos, Athyma, Adolias,* and *Tanæcia,* are Oriental, but they mostly extend into the Moluccan region; the last however is strictly Malayan, and *Adolias* only reaches Celebes. *Mynes* alone, is exclusively Australian, but *Prothoe* is almost so, having only one outlying species in Java. *Eurytela* and *Ergolis* are confined to the Oriental and Ethiopian regions, but the latter reaches the Moluccas. *Cethosia, Cirrhochroa, Messaras,* and *Symphædra,* are both Oriental and Australian; while *Junonia, Cyrestis, Diadema, Neptis,* and *Nymphalis,* are common to the three tropical regions of the Eastern Hemisphere, the latter extending into the Mediterranean district, while *Junonia* occurs also in South America and the Southern United States.

The most cosmopolitan genus is *Pyrameis,* which has representatives in every region and every district. *Apatura* is found in all but the Ethiopian and the Australian, although it just enters the confines of the latter region in Celebes; *Limenitis* is abundant in the Oriental region, but extends eastward to Celebes and westward into Europe, North America, and even into South America. *Argynnis, Melitæa,* and *Vanessa,* are almost confined to the Palæarctic and Nearctic regions; the former however occurs in the Himalayas and in the mountains of Java, and also in Chili and in Jamaica. Two genera—*Dicrorrhagia* and *Helcyra*—have both one species in North India and another in the island of Ceram. The number of genera peculiar to each region is as follows:—Neotropical, 50; Australian, 2; Oriental 15; Ethiopian, 14; Palæarctic, 1; Nearctic, 0.

FAMILY 9.—LIBYTHEIDÆ. (1 Genus, 10 Species.)

GENERAL DISTRIBUTION.

NEOTROPICAL SUB-REGIONS.	NEARCTIC SUB-REGIONS.	PALÆARCTIC SUB-REGIONS.	ETHIOPIAN SUB-REGIONS.	ORIENTAL SUB-REGIONS.	AUSTRALIAN SUB-REGIONS.
2 — — 4	— 2 . 3 —	. 1 2 — —	— 2 — 4	1 . 2 . 3 . 4	1 — — —

The genus *Libythea*, which constitutes this family, appears to have its head-quarters in the Oriental region, but extends on all sides in an erratic manner, into various remote and disconnected portions of the globe, as indicated above.

FAMILY 10.—NEMEOBIIDÆ. (12 Genera, 145 Species.)

GENERAL DISTRIBUTION.

NEOTROPICAL SUB-REGIONS.	NEARCTIC SUB-REGIONS.	PALÆARCTIC SUB-REGIONS.	ETHIOPIAN SUB-REGIONS.	ORIENTAL SUB-REGIONS.	AUSTRALIAN SUB-REGIONS.
— 2 . 3 —	— — — —	1 — — —	— 2 — 4	— — 3 . 4	1 — — —

This group has been separated from the Erycinidæ of the older authors, and contains all the non-American genera and species. Half the genera and nearly four-fifths of the species of this group are, however, Neotropical; one is European; two or three African; and twenty-six Oriental and Australian. The genera are :—

Nemeobius (1 sp.), Europe; *Dodona* (6 sp.), North India; *Zemeros* (2 sp.), North India and Malaya; *Abisara* (11 sp.), North India, Malayan and Moluccan districts, Madagascar and West Africa; *Taxila* (8 sp.), North India and Malaya; *Dicallaneura* (2 sp.), Moluccan district; *Alesa* (6 sp.), *Eunogyra* (2 sp.), *Cremna* (7 sp.), *Bœotis* (3 sp.), are all from the Brazilian sub-region; *Eurybia* (10 sp.), *Mesosemia* (80 sp.), inhabit both the Brazilian and Mexican sub-regions.

FAMILY 11.—EURYGONIDÆ.   (2 Genera, 78 Species.)

GENERAL DISTRIBUTION.

NEOTROPICAL SUB-REGIONS.	NEARCTIC SUB-REGIONS.	PALÆARCTIC SUB-REGIONS.	ETHIOPIAN SUB-REGIONS.	ORIENTAL SUB-REGIONS.	AUSTRALIAN SUB-REGIONS.
— 2.3 —	— — — —	— — — —	— — — —	— — — —	— — — —

This small family, separated from the true Erycinidæ by Mr. Bates, is confined to the tropical forest-districts of continental America.   The genera are:—

Eurygona (71 sp.); *Methonella* (1 sp.); the latter found in Equatorial South America.

FAMILY 12.—ERYCINIDÆ.   (59 Genera, 560 Species.)

GENERAL DISTRIBUTION.

NEOTROPICAL SUB-REGIONS.	NEARCTIC SUB-REGIONS.	PALÆARCTIC SUB-REGIONS.	ETHIOPIAN SUB-REGIONS.	ORIENTAL SUB-REGIONS.	AUSTRALIAN SUB-REGIONS.
— 2.3.4	1.2.3 —	— — — —	— — — —	— .— — —	— — — —

This extensive family of small, but exquisitely beautiful butterflies, is especially characteristic of the virgin forests of the Neotropical region, only a few species of three genera extending into the Nearctic region.   The more important genera, and those which have an exceptional distribution, can alone be here noticed.   *Charis* extends from Brazil to New York; *Apodemia* from Brazil to California, Utah, and Oregon; *Amarynthis* inhabits the Brazilian and Antillean sub-regions; *Lepricornis* and *Metapheles* are small genera found only in the Mexican sub-region; *Lymnas, Necyria, Ancyluris, Diorhina, Esthemopsis, Anteros, Emesis, Symmachia, Cricosoma, Calydna, Lemonias, Nymphidium, Theope,* and *Aricoris* are common to the Brazilian and Mexican sub-regions.   All the other genera (40 in number) are only known from the Brazilian sub-region, and of these a considerable proportion are confined to the damp equatorial forests of the Amazon Valley.

FAMILY 13.—LYCÆNIDÆ.   (39 Genera, 1,220 Species.)

GENERAL DISTRIBUTION.

NEOTROPICAL SUB-REGIONS.	NEARCTIC SUB-REGIONS.	PALÆARCTIC SUB-REGIONS.	ETHIOPIAN SUB-REGIONS.	ORIENTAL SUB-REGIONS.	AUSTRALIAN SUB-REGIONS.
1.2.3.4	1.2.3.4	1.2.3.4	1.2.3.4	1.2.3.4	1.2.3.4

The Lycænidæ—of the variety and beauty of which in tropical regions our own "Blues" and "Coppers" give but a faint idea —are a group of universal distribution. We shall therefore indicate those genera which are restricted to one or more regions, or are nearly cosmopolitan. The large genus *Polyommatus* (containing 325 species) has the same universal distribution as the entire family. Our common "Blues" well represent this genus. *Lycæna* (comprising the "Coppers") is more especially characteristic of the Palæarctic and Nearctic regions, but straggling species occur also in North India, South Africa, Chili, and New Zealand. *Thecla* is especially characteristic of the Neotropical region, where there are about 370 species; in the Nearctic region, 36; in the Palæarctic 13; and in the Ethiopian 3 *Miletus, Lucia, Hypolycæna, Myrina,* and *Deudorix* are common to the three tropical regions of the Eastern Hemisphere—the Ethiopian, Oriental, and Australian. *Aphneus* and *Iolaus* are common to the Ethiopian and Oriental regions, the latter extending to Celebes. *Ialmenus, Pseudodipsas, Curetis,* and *Amblypodia* are common to the Oriental and Australian regions, but the first-named is found also in Madagascar. *Zephyrus* is found only in the Nearctic and Palæarctic, *Eumæus* in the Nearctic and Neotropical regions. The Nearctic region has one peculiar genus (*Feniseca*); the Palæarctic has two—*Thestor* and *Læosopis;* the Ethiopian has nine—*Pentila, Liptana, D'Urbania, Axiocerces, Capys, Phytala, Epitola, Hewitsonia,* and *Deloneura;* the Oriental has five—*Allotinus, Ilerda, Poritia, Camena,* and *Liphyra;* the Australian has three—*Hypochrysops, Utica,* and *Ogyris;* and the Neotropical also three—*Lamprospilus, Theorema,* and *Trichonis.*

FAMILY 14.—PIERIDÆ.   (35 Genera, 817 Species.)

GENERAL DISTRIBUTION.					
NEOTROPICAL SUB-REGIONS.	NEARCTIC SUB-REGIONS.	PALÆARCTIC SUB-REGIONS.	ETHIOPIAN SUB-REGIONS.	ORIENTAL SUB-REGIONS.	AUSTRALIAN SUB-REGIONS.
1.2.3.4	1.2.3.4	1.2.3.4	1.2.3.4	1.2.3.4	1.2.3—

The Pieridæ are distributed almost, if not quite, as widely over the globe as the last family, and we shall group the genera in the same manner. *Pieris* (130 sp.) is cosmopolitan; *Terias* and *Callidryas* are found in all the four tropical regions, and as far north as Pennsylvania in the Nearctic region; *Pontia, Tachyris, Eronia,* and *Thestias* are common to the Ethiopian, Oriental, and Australian regions, the last-named, however, only extending as far as Timor; *Colias* is pre-eminently Palæarctic and Nearctic, with a few Ethiopian species, one Indian, two in Chili, and one in the Sandwich Islands; *Anthocharis* is wholly Palæarctic and Nearctic; *Midea* has two species Nearctic, and one in Japan; *Gonepteryx* is Palæarctic and Neotropical, extending into Texas; *Idmais* and *Callosune* are Ethiopian and Oriental; *Thyca* and *Iphias* are Oriental and Australian; *Meganostoma* is Nearctic and Neotropical; *Nathalis* and *Kricogonia* are Neotropical, ranging into Florida, Texas, and Colorado.

The peculiar genera are pretty equally distributed. The Neotropical region has ten, two being confined to Chili; *Euterpe* and *Leptalis* are the most remarkable, the latter containing a number of forms mimicking the Heliconidæ and Danaidæ. The Oriental region has two, *Prioneris* and *Dercas*; the Australian one, *Elodina;* the Ethiopian two, *Teracolus* and *Pseudopontia;* the Palæarctic two, *Leucophasia* and *Zegris;* the Nearctic one, *Neophasia.*

FAMILY 15.—PAPILIONIDÆ.   (13 Genera, 455 Species.)

GENERAL DISTRIBUTION.					
NEOTROPICAL SUB-REGIONS.	NEARCTIC SUB-REGIONS.	PALÆARCTIC SUB-REGIONS.	ETHIOPIAN SUB-REGIONS.	ORIENTAL SUB-REGIONS.	AUSTRALIAN SUB-REGIONS.
1.2.3.4	1.2.3.4	1.2.3.4	1.2.3.4	1‖.2.3.4	1.2.3.4

The Papilionidæ, comprising many of the noblest and richest-coloured butterflies, and long placed at the head of the group, are almost as universally distributed as the Pieridæ, but they do not extend to so many remote islands nor so far into the Arctic and Antarctic regions. Nine-tenths of the species belong to the genus *Papilio*, and these are especially abundant in tropical regions, although species occur in every region and every sub-region. Well-marked sub-divisions of this large genus are characteristic of each great region—as the " Æneas" group in the Neotropical, the "Paris" group in the Oriental, the "Ægeus" group in the Australian, the " Zenobius " group in the Ethiopian, and many others. The few species of the Palæarctic region belong, on the other hand, to a group of universal distribution, and the Nearctic has a good number of species allied to Neotropical forms.

The other genera have mostly a very restricted range. *Parnassius* is an Alpine genus, confined to the Palæarctic and Nearctic regions. The Palæarctic region further possesses 5 peculiar genera—*Mesapia, Hypermnestra, Doritis, Sericinus,* and *Thais;* the Oriental has 4, *Calinaga, Teinopalpus, Bhutanitis,* and *Leptocircus,* the latter going as far as Celebes; the Australian has 1, *Eurycus;* and the Neotropical 1, *Euryades,* confined to the Chilian sub-region. The Ethiopian and the Nearctic regions have no peculiar genera.

FAMILY 16.—HESPERIDÆ. (52 Genera (?) 1,200 Species.)

GENERAL DISTRIBUTION.					
NEOTROPICAL SUB-REGIONS.	NEARCTIC SUB-REGIONS.	PALÆARCTIC SUB-REGIONS.	ETHIOPIAN SUB-REGIONS.	ORIENTAL SUB-REGIONS.	AUSTRALIAN SUB-REGIONS.
1.2.3.4	1.2.3.4	1.2.3.4	1.2.3.4	1.2.3.4	1.2.3.4

The Hesperidæ, or Skippers, are an immense group of mostly small obscurely coloured butterflies, universally distributed, and of which hosts of species still remain to be discovered and described. As the grouping of these into genera is not yet satisfactorily accomplished, only the more extensive and best known groups will be here noticed. *Pamphila* and *Hesperia* are universally distributed; *Nisoniades* seems to be only absent from the Australian region. The Neotropical region is pre-eminently rich in Hesperidæ, 33 genera being found there, of which 20 are peculiar to it; the Australian region has 12 genera, only 1 (*Euschemon*) being peculiar; the Oriental has 18, with 3 peculiar; the Ethiopian, 13, with 3 peculiar; the Palæarctic 6, with 1 (*Erynnis*) almost peculiar, a species occurring in Mexico; the Nearctic 9, with none peculiar, 4 being found also in the Neotropical region, 2 in the Palæarctic, and the rest being of wide distribution. Many new genera have, however, been recently described in the United States, but it is impossible yet to determine how many, if any, of these are peculiar. More than 100 species of the family are included in Mr. Edwards' "Synopsis of North American Butterflies,"—a very large number considering that Europe possesses only about 30.

Sub-order—LEPIDOPTERA HETEROCERA, or MOTHS.

The Lepidoptera Heterocera, or Moths, are of such immense extent, and are, besides, so imperfectly known compared with the Butterflies, that it would serve no purpose to go into the details of their distribution; especially as most of the families and a considerable number of the genera are cosmopolitan. We propose therefore to notice only the Sphingina, which, being generally of large size and finely marked or coloured, and many of them day-fliers, have been extensively collected; and whose numbers are more manageable than the succeeding groups.

Group I.—SPHINGINA.

FAMILY 17.—ZYGÆNIDÆ (46 Genera, about 530 Species).

The Zygænidæ are universally distributed, but many of the genera are restricted in their range. *Zygœna* (85 sp.) is mainly Palæarctic, but 2 species are South African, and 1 North American; *Procris* (22 sp.) has a scattered distribution, from the Palæarctic region to South America, South Africa and North India; *Heterogynis* (3 sp.) and *Dysauxis* (3 sp.) are European; *Pollanisus* (3 sp.) is Australian; *Glaucopis* (120 sp.) is mainly Neotropical, with a few Oriental; *Syntomis* (94 sp.) is found in all the Old-World regions; and *Euchromia* (150 sp.) is found in all warm countries, though especially abundant in South America.

FAMILY 18.—CASTNIIDÆ (7 Genera, 63 Species).

The Castniidæ have an interesting distribution, being mainly Neotropical, with four genera in Australia and New Guinea. *Castnia, Coronis,* and *Gazera,* with 51 species, are Neotropical; *Synemon, Euschemon, Damias* and *Cocytia,* with 12 species, are Australian, the latter being found only in the Papuan Islands.

FAMILY 19.—AGARISTIDÆ (13 Genera, 76 Species).

The Agaristidæ are beautiful diurnal moths, allied to the Castniidæ, but almost confined to the Australian and Oriental regions, with a few in the Ethiopian. The most important genera are,—*Agarista* (21 sp.), Australia and New Guinea; *Eusemia* (31 sp.), *Ægocera* (7 sp.), Oriental and Ethiopian regions; the other genera being confined to the islands from Java to New Guinea.

FAMILY 20—URANIIDÆ (2 Genera, 12 Species).

These magnificent insects have a singular distribution. The gold-spangled *Urania* (6 sp.) is characteristic of Tropical America, but a single species of great magnificence occurs in Madagascar. The large but sober-tinted *Nyctalemon* (6 sp.) is found in the Neotropical, Oriental, and Australian regions.

FAMILY 21.—STYGIIDÆ. (3 Genera, 14 Species.)

These insects are confined to the Palæarctic and Neotropical regions, 2 genera in the former, 1 in the latter.

FAMILY 22.—ÆGERIIDÆ. (24 Genera, 215 Species.)

This family is found in all parts of the world except Australia. *Ægeria* is most abundant in Europe, but is found also in North and South America.

FAMILY 23.—SPHINGIDÆ. (40 Genera, 345 Species.)

The Sphinx Moths are cosmopolitan. The most important genera are,—*Macroglossa* (26 sp.), *Chœrocampa* (46 sp.), and *Macrosila* (21 sp.), all cosmopolitan; *Sesia* (12 sp.), Europe, Asia, and North America; *Deilephila* (19 sp.), Palæarctic and Oriental regions, Nearctic region, and Chili; *Sphinx* (21 sp.), Europe,

North and South America; *Smerinthus* (29 sp.), all regions except Australia. Our Death's Head Moth (*Acherontia atropos*) ranges to Sierra Leone and the Philippine Islands.

General Remarks on the Distribution of the Diurnal Lepidoptera and Sphingidea.

The Diurnal Lepidoptera or Butterflies, comprehend 431 genera and 7,740 species, arranged in 16 families, according to Mr. Kirby's Catalogue published in 1871. The Sphingidea consist of 135 genera and 1,255 species, arranged in 7 families, according to the British Museum Catalogue dated 1864; and as this includes all Mr. Bates' collections in America and my own in the East, it is probable that no very large additions have since been made.

The distribution of the families and genera of Butterflies corresponds generally with that of Birds—and more especially with that of the Passerine birds—in showing a primary division of the earth into Eastern and Western, rather than into Northern and Southern lands. The Neotropical region is by far the richest and most peculiar. It possesses 15 families of butterflies, whereas the other regions have only from 8, in the Palæarctic, to 12 in the Ethiopian and Oriental regions; and as none of the Old World regions possess any peculiar families, the New World has a very clear superiority. In genera the preponderance is still greater, since the Neotropical region possesses about 200 altogether peculiar to it, out of a total of 431 genera, many of which are cosmopolitan. Comparing, now, the Eastern regions with the Western, we have two peculiar families in the former to 4 in the latter; while the Southern regions (Australian and Neotropical) possess not a single peculiar family in common.

In the Sphingidea the same general features recur in a less marked degree, the Neotropical being the richest region; but here we have one family (Castniidæ) which appears to be confined to the two southern regions,—the Australian and Neotropical.

The distribution of the genera affords us some facts of special interest, which must be briefly noticed. There are several genera typically characteristic of the North Temperate regions which have a few species widely scattered on mountains, or in the temperate parts of the Southern Hemisphere. Chili possesses representatives of four of these genera—*Argynnis, Lycœna, Colias,* and *Deilephila;* and this has been thought by some naturalists to be of such importance as to outweigh the purely Neotropical character of a large portion of the Chilian fauna, and to render it advisable to join it on, as an outlying portion of a great North Temperate zoological region. But when we remember that *Argynnis* occurs also in Java, and *Lycœna* in New Zealand, while *Colias* ranges to Southern Africa, Malabar, and the Sandwich Islands, we can hardly admit the argument to be a sound one. For a fuller discussion of this question see Vol. II., pp. 43—47. The remarkable fact of the existence of the otherwise purely Neotropical genus, *Urania,* in Madagascar is even more striking, supported as it is by the Antillean, *Solenedon,* belonging to a family of Mammalia otherwise confined to Madagascar, and by one or two Coleopterous genera, to be noticed farther on as common to the two countries. Our view as to the true explanation of this and analogous phenomena will be found at Vol. I., p. 284.

The division of the Castniidæ (a family almost confined to the Tropics), between the Neotropical and Australian regions, is also a very curious and important phenomenon, because it seems to point to a more remote connection between the two countries than that indicated by the resemblance between the productions of South Temperate America with those of Australia and New Zealand; but we have already shown that the facts may be explained in another way. (See Vol. I., pp. 398 and 404).

The division of the Malay Archipelago between the Oriental and Australian regions is clearly marked in the Lepidoptera, and it is very curious that it should be so, for in this, if in any group of animals, we should expect an almost complete fusion to have been effected. Lepidoptera fly readily across wide tracts of sea, and there is absolutely no climatal difference to interfere with their free migration from island to island. Yet we find no less than 10 genera abundant in the Indo-Malayan sub-region which never cross the narrow seas to the east of them; 6 others which only pass to Celebes; and 2 more which have extended from Java along the closely connected line of islands eastwards to Timor. On the other side, we find 5 strictly Austro-Malayan genera, and 2 others which have a single representative in Java. The following is a list of these genera:—

INDO-MALAYAN GENERA :—*Amathusia, Thaumantis, Tanæcia, Eurytela, Ilerda, Zemeros, Taxila, Aphneus, Prioneris, Dercas, Clerome, Adolias, Apatura, Limenitis, Iolaus, Leptocircus,* (the last six reach Celebes); *Discophora, Thestias*; (the last two reach Timor.)

AUSTRO-MALAYAN GENERA :—*Hamadryas, Hypocista, Mynes, Dicallaneura, Elodina, Hyades, Prothoe* (the last two reach Java).

The most characteristic groups, which range over the whole Archipelago and give it a homogeneous character, are the various genera of Danaidæ, the genus *Elymnias*, and *Amblypodia* with a few other Lycænidæ. These are all abundant and conspicuous groups, but they are nevertheless exceptions to the general rule of limitation to one or other of the regions. The cause of this phenomenon is probably to be found in the limitation of the larvæ of many Lepidoptera to definite species, genera, and families of plants; and we shall perhaps find, when the subject is carefully investigated, that the groups which range over the whole Archipelago feed on genera of plants which have an equally wide range, while those which are limited to one region or the other, have food-plants belonging to genera which are similarly limited. It is known that the vegetation of the two regions differs largely in a botanical sense, although its general aspect is almost identical ; and this may be the reason why the proportion of wide-ranging genera is greater among such insects as feed upon dead wood, than among those which derive their support from the juices of the living foliage. This subject will be again discussed under the various families of Coleoptera, and it will be well to bear in mind the striking facts of generic limitation which have been here brought forward.

Fossil Butterflies, apparently of existing genera, occur in the Miocene and Eocene formations, and an extinct form in the Lower Oolite; but these cannot be held to give any adequate idea of the antiquity of so highly specialised a group, which, in all probability, dates back to Palæozoic times, since one of the Bombycidæ,—a group almost as highly-organised—has been discovered in the coal formation of Belgium. (See Vol. I. p. 168.)

Order—COLEOPTERA.

GEODEPHAGA, or CARNIVOROUS GROUND BEETLES.

The Geodephaga consist of two families, Cicindelidæ and Carabidæ, differing in their form and habits no less than in their numbers and distribution. The former, comprising about 800 species, are far more abundant and varied in Tropical regions; the latter, more than ten times as numerous, are highly characteristic of the North Temperate zone, where fully half of all the known species occur.

CICINDELIDÆ.   (35 Genera, 803 Species.)

The Cicindelidæ, or Tiger Beetles, are a moderately extensive group, spread over the whole globe, but much more abundant in tropical than in temperate or cold countries. More than half of the species (418) belong to the single genus *Cicindela*, the only one which is cosmopolitan. The other large genera are,— *Collyris* (81 sp.), wholly Oriental; *Odontochila* (57 sp.), South American, with species in Java and Celebes; *Tetracha* (46 sp.), mostly South American, but with species in South Europe, North America, and Australia; *Tricondyla* (31 sp.), characteristic of the Oriental region, but extending eastward to New Guinea; *Ctenostoma* (26 sp.), wholly Neotropical; *Dromica* (24 sp.), wholly African, south of Lake Ngami and Mozambique; *Therates* (18 sp.), wholly Malayan, from Singapore to New Guinea.

The genera are distributed in the several regions as follows:— the Nearctic region has 5 genera, 3 of which are peculiar to it; the

Palæarctic has 2, but none peculiar; the Ethiopian 13, with 11 peculiar; the Oriental 8, with 3 peculiar; the Australian 9, with 2 peculiar; and the Neotropical 15, with 10 peculiar. The connection between South America and Australia is shown by the latter country possessing 9 species of the characteristic South American genus *Tetracha*, as well as one of *Megacephala*. The small number of peculiar genera in the Oriental and Australian regions is partly owing to the circumstance that two otherwise peculiar Oriental genera have spread eastward to the Moluccas and New Guinea, a fact to be easily explained by the great facilities such creatures have for passing narrow straits, and by the almost identical physical conditions in the Malayan portion of the two regions. The insects of Indo-Malaya were better adapted to live in the Austro-Malay Islands than those of Australia itself, and the latter group of islands have thus acquired an Oriental aspect in their entomology, though not without indications of the presence of an aboriginal insect-fauna of a strictly Australian type. The relation of the Australian and Neotropical regions is exhibited by this family in an unusually distinct manner. *Tetracha*, a genus which ranges from Mexico to La Plata, has 9 species in Australia; while *Megacephala* has 2 American and 1 Australian species. Another curious, and more obscure relation, is that between the faunas of Tropical America and Tropical Africa. This is also illustrated by the genus *Megacephala*, which has 4 African species as well as 2 South American; and we have also the genus *Peridexia*, which has 2 species in South America and 2 in Madagascar.

Several of the sub-regions are also well characterised by peculiar genera; as *Amblychila* and *Omus* confined to California and the Rocky Mountains; *Manticora*, *Ophryodera*, *Platychile* and *Dromica*, characteristic of South Africa; *Megalomma* and *Pogonostoma* peculiar to the Mascarene Islands; and *Caledonica* to the islands east of New Guinea. The extensive and elegant genus *Collyris* is highly characteristic of the Oriental region, over the whole of which it extends, only just passing the limits into Celebes and Timor.

The Cicindelidæ, therefore, fully conform to those divisions of the earth which have been found best to represent the facts of distribution in the higher animals.

CARABIDÆ. (620 Genera, 8500 Species.)

The enormous extent of this family, necessitates a somewhat general treatment. It has been very extensively collected, while its classification has been most carefully worked out, and a detailed exposition of its geographical distribution by a competent entomologist would be of the greatest interest. A careful study of Gemminger and Harold's Catalogue, however, enables me to sketch out the main features of its distribution, and to detail many of its peculiarities with considerable accuracy.

The Carabidæ are remarkable among insects, and perhaps among all terrestrial animals, as being a wonderfully numerous, varied, conspicuous, and beautiful group, which is pre-eminently characteristic of the Palæarctic region. So strikingly and unmistakably is this the case, that it must be held completely to justify the keeping that region distinct from those to which it has at various times been proposed to join it. Although the Carabidæ are thoroughly well represented by hosts of peculiar genera and abundant species in every part of the world without exception, yet the Palæarctic region alone contains fully one-third, or perhaps nearer two-fifths, of the whole. It may also be said, that the group is a temperate as compared with a tropical one; so that probably half the species are to be found in the temperate and cold regions of the globe, leaving about an equal number in the much more extensive tropical and warm regions. But, among the cold regions, the Palæarctic is pre-eminent. North America is also rich, but it contains, by far, fewer genera and fewer species.

The magnificent genus *Carabus*, with its allies *Procerus* and *Procrustes*, containing about 300 species, all of large size, is almost wholly confined to the Palæarctic region, only 10 species inhabiting North America, and 11 Temperate South America, with one on the African mountain of Kilimandjaro. Twelve large genera, containing together more than 2000 species, are truly cosmopolitan, inhabiting both temperate and tropical countries all over the globe; but many of these are more abundant in the Palæarctic region than elsewhere. Such are *Scarites*, *Calosoma*, *Brachinus*, *Cymindis*, *Lebia*, *Chlænius*, *Platynus*, *Harpalus*, *Bembecidium*, *Pœcilus*, and *Argutor*. Of tropical cosmopolites, or genera found in all the tropical regions, but not in the temperate zones, there seem to be only four,—*Catascopus*, *Coptodera*, *Colopodes*, and *Caasnonia*. *Pheropsophus* is confined to the tropics of the Old World ; while *Drimostoma*, though widely scattered, is characteristic of the Southern Hemisphere.

The Palæarctic region has about 50 genera of Carabidæ which are strictly confined to it, the most important being,—*Leistus* (30 sp.), *Procerus* (5 sp.), *Procrustes*, (17 sp.), *Zabrus* (60 sp.), *Pristonychus* (42 sp.), and *Ophonus* (60 sp.) ; but it possesses a large number in common with the Nearctic region. The more remarkable of these are,—*Carabus*, *Nebria*, *Amara*, *Cyrtonotus*, *Bradycellus*, *Anopthalmus*, *Celia*, *Cychrus*, *Patrobus*, *Elaphrus*, *Notiophilus*, *Bradytus*, *Callisthenus*, *Blethisa*, and several others. Many too, though not strictly confined to the North Temperate regions, are very abundant there, with a few species isolated in remote countries, or widely scattered, often in an eccentric manner. Among these may be mentioned, *Trechus* (120 sp.), all North Temperate but 8, which are scattered in Java, New Caledonia and South America ; *Dyschirus* (127 sp.), North Temperate, with 3 or 4 species in Australia, China and La Plata ; *Omaseus*, (88 sp.), *Steropus* (90 sp.), *Platysoma* (114 sp.), and *Pterostichus* (138 sp.), are mostly North Temperate, but each has a few species in the South Temperate zone, New Zealand, Australia, Chili, and the Cape of Good Hope. *Dromius* (54 sp.), is about two-thirds Palæarctic, the rest of the species being scattered over the world, in Chili, North and South America, South Africa, Burmah, Ceylon, and New Zealand. The North Temperate genera *Calathus* and *Olisthopus*, have each one species in New Zealand ; *Percus* has most of its species in South Europe, but 3 in Australia ; *Abax* is confined to the north temperate zone, but with one species in Madagascar while *Lœmosthenes* is said to have a species identically the same in South Europe and Chili. Some of these apparent anomalies may be due to wrong determination of the genera, but there can be little doubt that most of them represent important facts in distribution.

The Nearctic region is comparatively poor in Carabidæ. Its more important peculiar genera are,—*Dicœlus* (22 sp.), *Pasimachus* (17 sp.), *Eurytrichus* (9 sp.), *Sphœroderus* (7 sp.), *Pinacodera* (6 sp.), and others of smaller extent, about 30 in all. It also possesses representatives of a considerable number of Palæarctic genera, as already indicated ; and a few of South American genera, of which *Helluomorpha* and *Galerita* are the most important.

The Neotropical region is very rich in peculiar forms of Carabidæ, as in almost all other great groups. It possesses more than 100 peculiar genera, but about 30 of these are confined to the South Temperate sub-region. The more important peculiar genera of Tropical America are,—*Agra* (144 sp.), *Ardistomus* (44 sp.), *Schizogenius* (25 sp.), *Pelecium*, (24 sp.), *Calophena* (22 sp.), *Ctenodactyla* (7 sp.). Among the Chilian and South Temperate peculiar forms are,—*Antarctia* (29 sp.), *Scelodontis* (10 sp.), *Tropidopterus* (4 sp.). Among the Neotropical genera with outlying species are,—*Pachyteles* (50 sp.), one of which is West African; *Selenophorus* (70 sp.), with 4 African, 4 Oriental, and 1 from New Caledonia ; *Ega* (11 sp.), with one in the East Indies, and one in New Caledonia ; *Galerita*, with 36 American species, 8 African, and 3 Indian ; *Callida* and *Tetragonoderus*, mostly American, but with a few African, Oriental and Australian species ; and *Pseudomorpha*, common to America and Oceania.

The Australian region is almost equally rich, possessing about 95 peculiar genera of Carabidæ, no less than 20 of which are confined to New Zealand. The most important are, *Carenum, Promecoderus, Scaraphites, Notonomus, Ænigma, Sphallomorpha, Silphomorpha*, and *Adelotopus*. The gigantic *Catadromus* has 4 Australian species and 1 in Java; *Homalosoma* has 31 species in Australia and New Zealand, and 1 in Madagascar. Celebes and New Guinea have each peculiar genera, and one is common to Australia and the Cape of Good Hope.

The Oriental region possesses 80 peculiar genera, 10 of which are confined to Ceylon. The more important are,—*Pericallus, Planetes*, and *Mormolyce. Distrigus* is also characteristic of this region, with one species in Madagascar; while it has *Orthogo-nius*, *Hexagonia*, *Macrochilus*, and *Thyreopterus* in common with the Ethiopian region, and is rich in the fine tropical genus, *Catascopus*.

The Ethiopian region has 75 peculiar genera, 8 of which are confined to Madagascar. The more important are,—*Polyhirma*, *Graphipterus*, and *Piezia*. *Anthia* is chiefly African, with a few species in India; *Abacetus* is wholly African, except a species in Java, and another in South Europe; and *Hypolithus* is typically African, but with 7 species in South America and 1 in Java.

The facts of distribution presented by this important family, looked at broadly, do not support any other division of the earth into primary regions than that deduced from a study of the higher animals. The amount of speciality in each of these regions is so great, that no two of them can be properly united; and in this respect the Carabidæ accord wonderfully with the Vertebrates. In the details of distribution there occur many singular anomalies; but these are not to be wondered at, if we take into consideration the immense antiquity of Coleopterous insects—which existed under specialised forms so far back as the Carboniferous epoch,—the ease with which they may be dispersed as compared with larger animals, and the facilities afforded by their small size, habits of concealment, and often nocturnal habits, for adaptation to the most varied conditions, and for surviving great changes of surface and of the surrounding organic forms. The wonder rather is, not that there are so many, but so few cases of exceptional and anomalous distribution; and the fact that these creatures, so widely different from Vertebrates in organi-sation and mode of life, are yet on the whole subject to the same limitations of range as were found to occur among the higher animals, affords a satisfactory proof that the principles on which our six primary regions are founded, are sound; and that they are well adapted to exhibit the most interesting facts of geo-graphical distribution, among all classes of animals.

Much stress has been laid on the fact of a few species of such typical European genera as *Carabus*, *Dromius*, and others, being

found in Chili and Temperate South America; and it has been
thought, that in a system of Entomological regions this part of
the world must be united to the Northern Hemisphere. But these
writers omit to take into account, either the large numbers of
isolated and peculiar forms characteristic of South Temperate
America, or the indications of affinity with Tropical America
and Australia, both of which are really more important than the
connection with Europe. The three important Chilian genera,
Cascelius, Barypus, and *Cardiopthalmus,* are closely allied to the
Australian *Promecoderus :* others, as *Omostenus* and *Plagiotelium,*
are quite isolated; while *Antarctia* and *Metius,* according to
Lacordaire, form a distinct division of the family. Chili, too, has
many species of *Pachyteles, Coptodera,* and other South American
genera; and this affinity is far stronger in many other families
than in the Carabidæ. The existence of representatives of
typical northern forms in Chili, is a fact of great interest, and
may be accounted for in a variety of ways; (see Vol. II. p. 44)
but it is not of such a magnitude as to be of primary importance in geographical distribution, and it can only be estimated
at its fair value, by taking into account the affinities of all the
groups inhabiting that part of the world.

LUCANIDÆ. (45 Genera, 529 Species.)

Passing over a number of obscure families, we come to the
remarkable group of the Lucanidæ, or Stag-beetles, which, being
almost all of large size, and many of them of the most striking
forms, have been very thoroughly collected and assiduously
studied.

The most curious feature of their general distribution, is
their scarcity in Tropical South America, and their complete
absence from Tropical North America and the West Indian
Islands, though they appear again in Temperate North America.
In the New World they may, in fact, be looked upon as a
temperate group characteristic of the extra-tropical regions and
the highlands; while in the Old World, where they are far more
abundant, they are distinctly tropical, being especially numerous

in the Oriental and Australian regions. No genus has the range of the whole family, *Dorcus* and *Lucanus* being absent from Africa, while *Cladognathus* is unknown in the New World and on the continent of Australia. The Oriental region is the richest in peculiar forms, possessing 16 genera, 7 of which are wholly confined to it, while 3 others only just range beyond it to North China on the one side, or to the Austro-Malayan islands on the other. The Australian region comes next, with 15 genera, of which 7 are wholly peculiar. South America has 12 genera, 10 of which are peculiar. The Ethiopian region has 10 genera, 7 of which are peculiar, and 2 of these are confined to the island of Bourbon. The Palæarctic region has 8 genera, and the Nearctic 5; one genus being peculiar to Europe, and two confined to Europe and North America. The Ethiopian and Oriental regions have 3 genera in common and peculiar to them; the Oriental and Australian 3 ; while the Australian and Neotropical have 1 in common, to which may be added *Streptocerus*, which represents in Chili the Australian *Lamprima*.

Among the special features presented by the distribution of the Lucanidæ, may be mentioned—the remarkable group of genera, *Pholidotus*, *Chiasognathus*, and *Sphenognathus*, confined to Temperate South America, the Andes, and mountains of Brazil; *Lucanus* (19 sp.), almost confined to the Oriental and Palæarctic regions, three species only inhabiting North America ; *Odontolabris* (29 sp.), wholly Oriental, with 2 sp. in Celebes ; *Nigidius* (11 sp.), Ethiopian, but with species in Formosa, the Philippines, and Malacca; *Syndesus* (11 sp.), common to Australia, New Caledonia, and South America ; *Figulus* (20 sp.), divided between Africa and Madagascar on the one hand, and Australia, with the Malay and Pacific Islands, on the other.

The facts of distribution here sketched out are in perfect accordance with those of many groups of Vertebrates. The regions are sharply contrasted by their peculiar and characteristic genera ; the several relations of those regions are truly indicated ; while there is a comparatively small proportion of cases of anomalous or eccentric distribution.

CETONIIDÆ. (120 Genera, 970 Species.)

As representative of the enormous group of the Lamellicorns, which, according to continental entomologists, forms a single family numbering nearly 7,000 species, we take the Cetoniidæ or Rose-Chafers. These comprise a number of the most brilliant and beautifully-coloured insects, including the gigantic *Goliathi*, which are among the largest of known beetles. They have been assiduously collected in every part of the world, and their classification has been elaborated by many of our most eminent entomologists.

The Cetoniidæ are especially abundant in tropical and warm countries, yet far more so in the Old World than in the New; and in the Old World, the Ethiopian region exhibits a marvellous richness in this family, no less than 76 genera being found there, while 64, or more than half the total number, are peculiar to it. Next in richness, though still very far behind, comes the Oriental region, with 29 genera, 17 of which are peculiar. The Neotropical has only 14 genera, but all except two are peculiar to it, and one of these is not found out of the New World. The Australian region has 11 genera, three only being peculiar. The Palæarctic region has 13, with 4 peculiar; the Nearctic 7, with 2 peculiar. The affinities of the regions for each other, as indicated by the genera confined to two adjacent regions, are in this family somewhat peculiar. The Ethiopian and Oriental show the most resemblance, 6 genera being common and peculiar to the two; the Oriental and the Australian are unusually well contrasted, having only one genus exclusively in common, while 8 genera are found in the Indo-Malay Islands which do not cross the boundary to the Austro-Malayan division, and several others only pass to the nearest adjacent islands ; on the other hand, the only large Australian genus, *Schizorhina*, is found in many parts of the Moluccas, but not further west. The Australian and Neotropical regions exhibit no direct affinity, the nearest ally to the South American Gymnetidæ being *Clinteria*, an African and Asiatic genus; while not a single genus is common to Australia and South America. The Nearctic and Palæarctic regions have 3 genera in common, which are found in no other part of the world.

Among the special features of interest connected with the distribution of this family, we must first notice the exceptional richness of Madagascar, which alone possesses 21 peculiar genera. South Africa is also very rich, having 8 peculiar genera. *Stethodesma* is very peculiar, being divided between South America and Mexico on the one hand, and West and South Africa on the other. *Stalagmosoma* is a desert genus, ranging from Persia to Dongola. No genus is cosmopolitan, or even makes any approach to being so, except *Valgus*, which occurs in all the regions except the Neotropical; and even the family seems to be not universally distributed, since no species are recorded either from New Zealand, the Pacific Islands, or the Antilles.

The facts here brought forward, lead us to the conclusion that the Cetoniidæ are an Old-World tropical family, which had been well developed in Africa and Asia before it spread to Australia and America; and that it is only capable of being freely dispersed in the warmer regions of the earth. This view will explain the absence of affinity between the Australian and Neotropical regions, the only closer connection between which, has almost certainly occurred in the colder portions of the Temperate zone.

BUPRESTIDÆ. (109 Genera, 2,686 Species.)

The next family suited to our purpose is that of the Buprestidæ, consisting as it does of many large and some gigantic species, generally adorned with brilliant metallic colours, and attracting attention in all warm countries. Although these insects attain their full development of size and beauty only in the Tropics, they are not much less abundant in the warmer parts of the Temperate zone. In the Catalogue of the Coleoptera of Europe and the Mediterranean Basin, by M. de Marseul (1863), we find 317 species of Buprestidæ enumerated, although the district in question only forms a part of the Palæarctic region, which would thus seem to possess its full proportion of the species of this family. Confining ourselves to the generic forms, we find far less difference than usual between the numbers possessed by the tropical and the temperate regions; the richest being the Australian, with 47 genera, 20 of which are peculiar; and the poorest the Nearctic, with 24 genera, of which 7 are peculiar. The Oriental has 41 genera, 14 of which are peculiar; the Neotropical 39, of which the large proportion of 18 are peculiar; the Ethiopian 27, of which 6 are peculiar; and the Palæarctic also 27, but with 9 peculiar.

A most interesting feature in the distribution of this family, is the strong affinity shown to exist between the Australian and Neotropical regions, which have 4 genera common to both and found nowhere else; but besides this, the extensive and highly characteristic Australian genus, *Stigmodera*, is closely related to a number of peculiar South American genera, such as *Conognatha, Hyperantha, Dactylozodes,*—the last altogether confined to Chili and Temperate South America. Here we have a striking contrast to the Cetoniidæ, and we can hardly help concluding, that, as the latter is typically a tropical group, so the present family, although now so largely tropical, had an early and perhaps original development in the temperate regions of Australia, spreading thence to Temperate South America as well as to the tropical regions of Asia and Africa. The Australian and Oriental regions have 4 genera exclusively in common, but they also each possess a number of peculiar or characteristic genera, such as the Indo-Malayan *Catoxantha* (which has only a single species in the Moluccas) and nine others of less importance; and the exclusively Austro-Malayan genus, *Sambus*, with five smaller groups, and *Cyphogastra*, with only 2 Indo-Malay species. The Oriental and Ethiopian regions are very distinct, only possessing the single genus, *Sternocera*, exclusively in common. The Nearctic and Palæarctic are also distinct, only one genus, *Dicerca*, being confined to America (North and South) and Europe, a fact which again points to a southern origin for this family, and its comparatively recent extension into the

North Temperate zone. It must be remembered, however, that in view of the immense geological antiquity of the existing families of Beetles, dating back certainly to the Secondary and probably to the Palæozoic epoch, "comparatively recent" may still be of considerable antiquity.

It is somewhat singular that North and South America have no genera exclusively in common. The connection between South America and Africa seems to be shown,—by the genus *Psiloptera*, the mass of the species being divided between these regions, with a few widely scattered over the globe; and the American genus *Actenodes*, which has one species in West Africa. Somewhat allied, is the extensive genus *Polybothris*, strictly confined to Madagascar. The genus *Agrilus* is perhaps cosmopolitan, although no species of the family is recorded from New Zealand. Among the peculiarities of distribution we may notice,—the genus *Sponsor*, with 8 species in the island of Mauritius, 1 in Celebes, and 1 in New Guinea; *Ptosima*, scattered between the United States, Mendoza in South Temperate America, South Europe, the Philippine Islands, and North China; *Polycesta*, which besides inhabiting South America, North America, and Europe, has a single species in Madagascar; and *Belionota*, which has 8 species African, 8 Indo-Malayan, 2 Austro-Malayan, and 1 in California. The extensive genus *Acmæodera*, is most abundant in the warm and dry portions of the Palæarctic, Ethiopian, and Nearctic regions, with some in the Andes and South Temperate America, a few in Brazil and the West Indies, and 1 said to be from the Philippines. About one-third of the genera (containing more than half the species) have a tolerably extensive range, while the genera confined to single regions contain only about one-fourth of the total number of species.

It will, I think, be admitted, after a careful study of the preceding facts, that the regions and sub-regions here adopted, serve to exhibit, with great clearness, the chief phenomena of distribution presented by this interesting family.

LONGICORNIA. (1,488 Genera, 7,576 Species).

The elegant and admired group of the Longicorn Beetles, is treated by continental authors as a single family, consisting of three sub-divisions—the Prionidæ, Cerambycidæ, and Lamiidæ of English entomologists. These are so closely related, and are so similar in form, habits, and general distribution, that it will be best to consider the whole as one group, noticing whatever peculiarities occur in the separate divisions. The endless structural differences among these insects, have led to their being classed in an unusual number of genera, which average little more than 5 species each; a number far below that in any of the other families we have been considering, and probably below that which obtains in any of the more extensive groups of animals or plants. This excessive subdivision of the genera, a large number of which consist of only one or two species, renders it difficult to determine with precision the relations of the several regions, since the affinities of these genera for each other are in many cases undetermined. A group of such enormous extent as this, can only be properly understood after years of laborious study; we must therefore content ourselves with such results as may be obtained from a general survey of the group, and from a comparison of the range of the several genera, by means of a careful tabulation of the mass of details given in the recent Catalogue of Messrs. Gemminger and Harold and the noble work of Lacordaire.

The proportionate extent of the three families of Longicorns is very unequal; the Prionidæ comprising about 7 per cent., the Cerambycidæ 44 per cent., and the Lamiidæ 49 per cent. of the total number of species; and the genera are nearly in the same proportions, being almost exactly 10, 40, and 50 per cent. of the whole, respectively; or, 135 Prionidæ, 609 Cerambycidæ, and 746 Lamiidæ. The several regions, however, present marked differences in their proportions of these families. In the two North Temperate regions, the Cerambycidæ are considerably more numerous than the Lamiidæ, in the proportion of about 12 to

9; and in this respect the Neotropical region agrees with them, though the superiority in the proportion of Cerambycidæ is somewhat less. In the Old World tropical regions, however, and in Australia, the Lamiidæ greatly preponderate—being nearly double in the Oriental and Ethiopian regions (or as 11 to 6), while in the Australian it is as 6 to 5. The Prionidæ show a similar difference, though in a less degree; being proportionately more numerous in the North Temperate and Neotropical regions. Now, as regards the North Temperate regions, this difference can be, to some extent explained, by a difference in the habits of the insects. The Lamiidæ, which both in the larva and perfect state have exceedingly powerful jaws, exclusively frequent timber trees, and almost always such as are dead; while the Cerambycidæ, are generally more delicate and have weaker mandibles, and many of the species live on shrubs, dead twigs, foliage, and even on flowers. The immense superiority of the Tropics in the number and variety of their timber trees, and the extent of their forests, sufficiently accounts for their superiority to the Temperate regions in the development of Lamiidæ; but the great excess of Cerambycidæ in South America as compared with the rest of the Tropics, is not to be so readily explained.

Bearing in mind the different proportions of the families, as above noted, we may now consider the distribution of the Longicorns as a whole. In number of generic forms, the Neotropical region, as in so many other groups, has a marked superiority. It possesses 516 genera, 489 of which (or about $\frac{18}{20}$ of the whole) are peculiar to it. The Australian and Oriental regions come next, and are exactly equal, both possessing 360 genera, and having almost exactly the same proportion (in each case a little less than $\frac{3}{4}$) peculiar. The Ethiopian region has 262 genera, with about $\frac{5}{6}$ peculiar; the Palæarctic 196, with 51 (rather more than $\frac{1}{4}$) peculiar; and the Nearctic 111, with 59 (a little more than half) peculiar. The more isolated of the sub-regions are also well characterised by peculiar genera. Thus, Chili with Temperate South America possesses 37, a large proportion being Cerambycidæ; the Malagasi group 26, with a preponderance of Lamiidæ; and New Zealand 12, of which the Cerambycidæ are only slightly in excess.

The relations between the Longicorn fauna of the several regions, are such as are in accordance with the dependence of the group on a warm climate and abundant vegetation; and indicate the efficiency of deserts and oceans as barriers to their migration. The Neotropical and Australian regions have only 4 genera in common, but these are sufficient to show, that there must probably once have been some means of communication between the two regions, better adapted to these insects than any they now possess. The Nearctic and Neotropical regions have 5, and the Nearctic and Palæarctic 13 genera in common and peculiar to them, the latter fact being the most remarkable, because no means of inter-communication now exists, except in high latitudes where the species of the Longicorns are very few. The Oriental and Australian regions, on the other hand, are closely connected, by having no less than 52 genera of Longicorns in common and peculiar to them. Most of these are specially characteristic of the Malay Archipelago, often extending over all the islands from Sumatra to New Guinea. This large number of wide-spread genera of course gives a character of uniformity to the entire area over which they extend; and, with analogous facts occurring in other families, has led many entomologists to reject that division of the Archipelago between the Australian and Oriental regions, which has been so overwhelmingly demonstrated to be the natural one in the case of the higher animals. The general considerations already advanced in Chapter II. enable us, however, to explain such anomalies as this, by the great facilities that exist for the transfer from island to island of such small animals, so closely connected with woody vegetation in every stage of their existence. That this is the true and sufficient explanation, is rendered clear by certain additional facts, which those who object to the sharp division of the Indo-Malay and Austro-Malay sub-regions have overlooked.

An analysis of all the Malay Longicorns proves, that besides the 52 genera characteristic of the Archipelago as a whole, there are 100 genera which are confined to one or other of its component sub-regions. Many of these, it is true, consist of single species confined to a single island, and we will not lay any stress on these; but there are also several important groups, which extend over the Indo-Malay or the Austro-Malay islands only, stopping abruptly at the dividing-line between them. For example, on the Indo-Malay side we have *Euryarthrum*, *Leprodera*, *Aristobia*, *Cœlosterna*, and *Entelopcs*, and what is perhaps even more satisfactory, the large genera *Agelasta* and *Astathes*, abundant in all the Indo-Malay islands, but having only one or two species just passing the boundary into Celebes. On the other side we have *Tethionea*, *Sphingnotus*, *Arrhenotus*, *Tmesisternus* (the last three genera abounding from New Guinea to Celebes, but totally unknown further west), *Hestima*, *Trigonoptera*, *Amblymora*, *Stesilea*, *Enes*, and the large genus *Micracautha*, with but a single species beyond the boundary,—30 Austro-Malayan genera in all, each found in more than one island, but none of them extending west of Celebes. Here we have clear proof that the boundary line between the two great regions exists for Longicorns, as well as for all other animals; but in this case an unusually large number have been able to get across it. This, however, does not abolish the barrier, but only proves that it is not absolutely effectual in all cases. Those who maintain that the Malay Archipelago forms a single Coleopterous region, must disprove or explain the instances of limited range here adduced.

Out of nearly 1500 known genera of these insects, only one genus, *Clytus*, appears to be cosmopolitan. *Saperda* and *Callichroma* are the only others that perhaps occur in every region; but these are both wanting over wide tracts of the earth's surface, *Saperda* being absent from Tropical Africa and the Malay Archipelago; and *Callichroma* from the Australian region, except one species in Polynesia. Many of the genera of Longicorns have a somewhat wide and scattered distribution, indicative of decadence or great antiquity. *Mallodon* and *Parandra* are mostly South American, but have species in Australia and Africa; *Oeme* is found in Brazil and the United States, with one species in West Africa; *Ceratophorus* has 2 species in West Africa and 1 in New Zealand. *Xystrocera* is mostly African, but has single species in

Borneo, Java, Amboyna and South Australia; *Phyton* has one species in North America and the other in Ceylon; *Philagetes* has 2 in South Africa, and 1 in Malacca: *Toxotus* abounds in North America and Europe, with one species away in Madagascar. *Leptura* is also North Temperate, but has a species at the Cape, one at Singapore and a third in Celebes. *Necydalis* has species in North and South America, Europe and Australia. *Hylotrupes* has 1 species in North America and Europe, and 1 in Australia; *Leptocera* prefers islands, being found only in Ceylon, Madagascar, Bourbon, Batchian, the New Hebrides, New Caledonia and North Australia; *Hathliodes* is Australian, with 1 species in Ceylon; *Schœnionta* has 3 Malayan species, and 1 in Natal. Many other cases equally curious could be quoted, but these are sufficient. They cannot be held to indicate any close relation between the distant countries in which species of the same genus are now found, but perhaps serve to remind us that groups of great antiquity, and probably of great extent, have dwindled away, leaving a few surviving relics scattered far and wide, the sole proofs of their former predominance.

General Observations on the Distribution of Coleoptera.

We have now passed in review six of the most important and best known groups of the Coleoptera or Beetles, comprising about 2,400 genera, and more than 21,000 species. Although presenting certain peculiarities and anomalies, we have found that, on the whole, their distribution is in very close accordance with that of the higher animals. We have seen reason to believe that these great and well-marked groups have a high geological antiquity, and by constantly bearing this fact in mind, we can account for many of the eccentricities of their distribution. They have probably survived changes of physical geography which have altogether extinguished many of the more highly organised animals, and we may perhaps gain some insight into the bearing of those changes, by considering the cross relations between the several regions indicated by them. On carefully tabulating the indications given by each of the groups here discussed, I arrive at the following approximate result. The

best marked affinities between the regions are those between
the Nearctic and Palæarctic,—the Oriental and Australian,
—the Australian and Neotropical,—which appear to be about
equal in each case. Next comes that between the Ethiopian
and Oriental on the one side, and the Ethiopian and Neotropical
on the other, which also appear about equal. Then follows that
between the Nearctic and Neotropical regions; and lastly, and far
the least marked, that between the North Temperate and South
Temperate regions. That the relation between the Ethiopian
and Neotropical region should be so comparatively well marked,
is unexpected; but we must consider that in such a comparison
as the present, we probably get the result, not of any recent
changes or intermigrations, but of all the long series of changes
and opportunities of migration that have occurred during many
geological epochs,—probably during the whole of the Tertiary
period, perhaps extending far back into the Secondary age.

It appears evident that Insects exhibit in a very marked
degree in their actual distribution, the influence both of very
ancient and very modern conditions of the earth's surface. The
effects of the ancient geographical features of the earth, are to be
traced, in the large number of cases of discontinuous and widely
scattered groups which we meet with in almost every family,
and which, to some extent, obscure the broader features of distribution due to the period during which the barriers which divide
the several primary regions have continued to exist. And this,
which we may consider as the normal distribution, is still
further obscured in those cases where the barriers between
existing regions are of such a nature as to admit of the free
passage of insects or their larva in a variety of ways, and (what
is perhaps of more importance) in which the physical features
on both sides of the barrier are so nearly identical, as to admit
of the ready establishment of such immigrants as may occasionally arrive. These conditions concur, for some families of insects,
in the case of the Oriental and Australian portions of the Malay
Archipelago; and it is there that the normal distribution has
been sometimes greatly obscured, but never, as we have sufficiently shown, by any means obliterated.

CHAPTER XXII.

THE Mollusca being for the most part marine, it does not enter into the plan of this work to go into much detail as to their distribution. The orders and families will, however, be passed briefly in review, and all terrestrial and fresh-water groups discussed in somewhat more detail; with the object of showing how far their distribution accords with that of the higher animals, and to what extent the anomalies they present can be explained by peculiarities of organisation and habits. If the views advocated in our fifth chapter are correct, the regions there marked out must apply to all classes of animals; and it will be the task of the students of each group, to work out in detail the causes which have led to any special features of distribution. All I can hope to do here, is to show, generally and tentatively, that such a mode of treatment is possible; and that it is not necessary, as it is certainly not convenient or instructive, to have a distinct set of "Regions" established for each class or order in the Animal and Vegetable Kingdoms.

For all the Marine groups I have merely summarised the information contained in Mr. Woodward's *Manual of the Mollusca*, but in the case of the Land Shells I have consulted the most recent general works, and endeavoured to give an accurate, though doubtless a very incomplete, account of the most interesting facts in their distribution. As their classification is very unsettled, I have followed that of the two latest great works, by Martens and Pfeiffer.

CLASS.—CEPHALOPODA.

Order I.—DIBRANCHIATA.

FAMILY 1.—ARGONAUTIDÆ. "Paper Nautilus." (1 Genus, 4 Species).

DISTRIBUTION.—Open seas of all warm regions. Two species fossil in Tertiary deposits.

FAMILY 2.—OCTOPODIDÆ. "Polypi." (7 Genera, 60 Species).

DISTRIBUTION.—Norway to New Zealand, all tropical and temperate seas and coasts.

FAMILY 3.—TEUTHIDÆ. "Squids or Sea-pens." (16 Genera, 102 Species.)

DISTRIBUTION.—Universal, to Greenland; 2 other genera are fossil, in the Lias and Oolite.

FAMILY 4.—SEPIADÆ. "Cuttle Fish." (1 Genus, 30 Species).

DISTRIBUTION.—All seas: 4 other genera are fossil, in Eocene and Miocene deposits.

FAMILY 5.—SPIRULIDÆ. (1 Genus, 3 Species).

DISTRIBUTION.—All the warmer seas.

FAMILY 6. — BELEMNITIDÆ. Fossil. (6 Genera, 100 Species).

DISTRIBUTION.—Lias to Chalk in Europe, India and North America.

Order II.—TETRABRANCHIATA.

FAMILY 7.—NAUTILIDÆ. (1 Genus, 3 Species, Living; 4 Genera, 300 Species, Fossil).

DISTRIBUTION.—Indian and Pacific Oceans; and the fossil species from the Silurian Period to the Tertiary, in all parts of the world.

FAMILY 8.—ORTHOCERATIDÆ. Fossil. (8 Genera, 400 Species).

DISTRIBUTION.—Lower Silurian to Lias.

FAMILY 9.—AMMONITIDÆ. Fossil. (14 Genera, 1100 Species).

DISTRIBUTION.—Upper Silurian to Chalk. Found at 16,000 feet elevation in the Himalayas.

CLASS.—GASTEROPODA.

Order I.—PROSOBRANCHIATA.

FAMILY 1.—STROMBIDÆ.   (4 Genera, 86 Species.)

DISTRIBUTION.—The Strombidæ, or Wing-shells, inhabit tropical and warm seas from the Mediterranean to New Zealand ; most abundant in the Indian and Pacific Oceans.   There are nearly 200 fossil species, from the Lias to Miocene and recent deposits.

FAMILY 2.—MURICIDÆ.   (12 Genera, 1000 Species.)

DISTRIBUTION.—All seas, most abundant in the Tropics. *Trichotropis* is confined to Northern seas; *Murex* and *Fusus* are cosmopolitan.   There are about 700 fossil species, ranging from the Oolite to the Miocene and recent formations.

FAMILY 3.—BUCCINIDÆ.   (24 Genera, 1100 Species.)

DISTRIBUTION.—The Buccinidæ, or " Whelks," range over the whole world, but some of the genera are restricted.   *Buccinum* inhabits the north and south temperate seas; *Monoceros* the West Coast of America ; *Cassidaria* the Mediterranean ; *Phos, Harpa, Eburna,* and *Ricinula,* are confined to the Pacific ; *Dolium* inhabits the Mediterranean as well as the Pacific.   There are about 350 fossil species, mostly from the Eocene and Miocene beds.

FAMILY 4.—CONIDÆ.   (3 Genera, 850 Species.)

DISTRIBUTION.—The Cones are universally distributed, but this applies only to the genus *Pleurotoma*. *Conus* is tropical and sub-tropical, and *Cithara* is confined to the Philippine Islands. There are about 460 fossil species, from the Chalk formation to the most recent deposits.

FAMILY 5.—VOLUTIDÆ.   (5 Genera, 670 Species.)

DISTRIBUTION.—The Volutes are mostly tropical; but a small species of *Mitra* is found at Greenland, and a *Marginella* in the Mediterranean. *Cymba* is confined to the West Coast of Africa and Portugal. *Voluta* extends south to Cape Horn. There are about 200 fossil species, from the Chalk and Eocene to recent formations.

FAMILY 6.—CYPRÆIDÆ.   (3 Genera, 200 Species.)

DISTRIBUTION.—The well-known Cowries are found all over the world, but they are much more abundant in warm regions. One small species extends to Greenland. There are nearly 100 fossil species, from the Chalk to the Miocene and recent formations.

FAMILY 7.—NATICIDÆ.   (5 Genera, 270 species.)

DISTRIBUTION.—The Naticidæ, or Sea-snails, though most abundant in the Tropics, are found also in temperate seas, and far into the Arctic regions. Two other genera are fossil; and there are about 300 extinct species, ranging from the Devonian to the Pliocene formations.

FAMILY 8.—PYRAMIDELLIDÆ. (10 Genera, 220 Species.)

DISTRIBUTION.—These turreted shells are very widely distributed both in temperate and tropical seas; and most of the genera have also a wide range. There are about 400 extinct species, from so far back as the Lower Silurian to the Pliocene formations.

FAMILY 9.—CERITHIADÆ. (5 Genera, 190 Species.)

DISTRIBUTION.—These are marine, estuary, or fresh-water shells, of an elongated spiral form; they have a world-wide distribution, but are most abundant in the Tropics. *Potamides* (41 sp.), is the only fresh-water genus, and is found in the rivers of Africa, India and China, to North Australia and California. Another genus is exclusively fossil, and there are about 800 extinct species, ranging from the Trias to the Eocene and recent formations.

FAMILY 10.—MELANIADÆ. (3 Genera, 410 Species.)

DISTRIBUTION.—Fresh-water only: lakes and rivers in warm countries, widely scattered. South Palæarctic and Australian regions, from Spain to New Zealand ; South Africa, West Africa, and Madagascar; United States. There are about 50 fossil species, from the Wealden and Eocene to recent formations.

FAMILY 11.—TURRITELLIDÆ. (5 Genera, 230 Species.)

DISTRIBUTION.—Universal. *Cæcum* is found in north temperate seas only. The other genera are mostly tropical, but some species reach Iceland and Greenland. There are near 300 species fossil, ranging from the Neocomian to the Pliocene formations.

FAMILY 12.—LITTORINIDÆ.   (9 Genera, 310 Species.)

DISTRIBUTION.—The Littorinidæ are mostly found on the coasts in shallow water; as the common Periwinkle (*Littorina littorea*). They are of world-wide distribution; but *Solarium* and *Phorus* are tropical; while *Lacuna*, *Skcnea*, and most species of *Rissoa* are Northern.   About 180 species are fossil, ranging from the Permian to the Pliocene formations.

FAMILY 13.—PALUDINIDÆ.   (4 Genera, 217 Species.)

DISTRIBUTION.—The Paludinidæ, or River-snails, are all fresh-water, and range over the whole world.   *Paludina* (60 sp.), is confined to the Northern Hemisphere; *Ampullaria* (136 sp.), is tropical; *Amphibola* (3 sp.), inhabits New Zealand and the Pacific Islands; *Valvata* (18 sp.), North America and Britain. There are 72 fossil species of *Paludina* and *Valvata*, in the Wealden formation and more recent fresh-water deposits.

FAMILY 14.—NERITIDÆ.   (10 Genera, 320 Species)

DISTRIBUTION.—All warm seas, ranging north to Norway and the Caspian Sea.   *Neritina* and *Navicella* inhabit fresh or brackish waters, the latter confined to the countries bordering the Indian Ocean and the islands of the Pacific.   There are 80 fossil species, from the Trias, Lias, and Eocene formations down to recent deposits.

FAMILY 15.—TURBINIDÆ.   (10 Genera, 425 Species).

DISTRIBUTION.—The genus *Trochus* (200 sp.) has a world-wide range, but the other genera are mostly tropical, and are most abundant in the Indian and Pacific Oceans.   There are more than 900 fossil species, found in all parts of the world, from the Lower Silurian to the Tertiary formations.

FAMILY 16.—HALIOTIDÆ. (6 Genera, 106 Species),

DISTRIBUTION.—The Ear-shells are most abundant in the Indian and Pacific Oceans; some are found on the east coasts of the Atlantic, but there are very few in the West Indies. *Ianthina* (10 sp.) consists of floating oceanic snails found in the warm parts of the Atlantic. Three other genera are fossil, and there are near 500 fossil species of this family ranging from the Lower Silurian to the Pliocene formations.

FAMILY 17.—FISSURELLIDÆ. (5 Genera, 200 Species).

DISTRIBUTION.—All seas. *Puncturella* (6 sp.) is confined to Northern and Antarctic seas; *Rimula* to the Philippines; and *Parmophorus* (15 sp.) from the Cape of Good Hope to the Philippines and New Zealand. There are about 80 fossil species, ranging from the Carboniferous formation to the deposits of the Glacial epoch.

FAMILY 18.—CALYPTRÆIDÆ. (4 Genera, 125 Species).

DISTRIBUTION. — The Calptræidæ, or Bonnet-Limpets, are found on the coasts of all seas from Norway to Chili and Australia; but are most abundant within the Tropics. The genera are all widely scattered. There are 75 fossil species, ranging from the Devonian to recent formations.

FAMILY 19.—PATELLIDÆ. (4 Genera, 254 Species).

DISTRIBUTION.—The Patellidæ, or Limpets, are universally distributed, and are as abundant in the temperate as in tropical seas. There are about 100 fossil species, ranging from the Silurian to the Tertiary formations.

FAMILY 20.—DENTALIADÆ.   (1 Genus, 50 Species).

DISTRIBUTION.—The genus *Dentalium* is found in the North Atlantic, Mediterranean, West Indies and India. There are 125 fossil species, found in various formations as far back as the Devonian in Europe and in Chili.

FAMILY 21.—CHITONIDÆ.   (1 Genus, 250 Species).

DISTRIBUTION.—On rocky shores in all parts of the world. There are 37 fossil species ranging back to the Silurian period.

Order II.—PULMONIFERA.   (" *Terrestrial Molluscs.*")

The Land and Fresh-water snails are so important and extensive a group, and their classification has been so carefully studied, that their geographical distribution is a subject of much interest. The range of the genera will therefore be given in some detail. For the Helicidæ I follow the classical work of Albers—*Die Helicien*, Von Martens' Edition (1860); and for the Operculate families, Pfeiffer's *Monographia Pneumonopomorum Viventium*, 2nd Supplement, 1865. The number of species is, of course, very considerably increased since these works were published (and the probable amount of the increase I have in most cases indicated), but this does not materially affect the great features of their geographical distribution.

FAMILY 22.—HELICIDÆ.   (33 Genera, 3,332 Species) (1860).

GENERAL DISTRIBUTION.—Universal.

The Helicidæ, or Snails, are a group of immense extent and absolutely cosmopolitan in their range, being found in the most barren deserts and on the smallest islands, all over the globe. They reach to near the line of perpetual snow on mountains, and to the limit of trees or even considerably beyond it, in the Arctic regions ; but they are comparatively very scarce in all cold countries. The Antilles, the Philippine Islands, Equatorial America, and the Mediterranean sub-region are especially rich in this family. Comparatively few of the genera, and those generally small ones, are restricted to single regions ; but on the other hand very few are generally distributed, only two—*Helix* and *Pupa*—occurring in all the six regions, while *Helix* alone is truly cosmopolitan, occurring in every sub-region, in every country, and perhaps in every island on the globe.

The Neotropical region is, on the whole, the richest in this family, the continental Equatorial districts producing an abundance of large and handsome species, while the Antilles are pre-eminent for the number of their peculiar forms. This region possesses 22 of the genera, and 6 of them are peculiar.

The Palæarctic region seems to come next in productiveness, but this may be partly owing to its having been so thoroughly explored. It possesses 16 of the genera, and 3 of them are confined to it. The great mass of the species are found in the warm and fertile countries surrounding the Mediterranean Sea.

The Ethiopian region has 13 genera, only one of which is peculiar.

The Australian region has 14 genera, 2 of which are confined to the Pacific Islands.

The Oriental has 15 genera and the Nearctic 12, but in neither case are there any peculiar generic types.

The following is the distribution of the several genera taken in the order of their magnitude :—

Helix (1,115 sp.), cosmopolitan. This genus is divided into 88 sub-genera, a number of which have a limited distribution. An immense quantity of species have been recently described, so that the number now exceeds 2,000.

Nanina (290 sp.) is characteristic of the Oriental and Australian regions, over the whole of which it extends, just entering the Palæarctic region as far as North China and Japan. Isolated from this area is a small group of 4 species occurring in West Africa. The number of species in this genus have now been increased to about 400.

Clausilia (272 sp.) is most abundant in Europe, with a few species widely scattered in India, Malaya, China, Japan, Equatorial America, and one in Porto Rico. The described species have been increased to nearly 500.

Bulimulus (210 sp.) is American, and almost exclusively Neotropical, ranging from Montevideo and Chili, to the West Indian Islands, California and Texas; with two sub-genera confined to the Galapagos Islands. About 100 new species have been described since the issue of the second edition of Dr. Woodward's Manual.

Pupa (210 sp.) abounds most in Europe and the Arctic regions, but has a very wide range, being scattered throughout Africa, continental India, Australia, the Pacific Islands, North America to Greenland, and the Antilles; but it is absent from South America, the Himalayan and Malayan sub-regions, China and Japan. An extinct species has occurred abundantly in the carboniferous strata of North America. About 160 additional species have been described.

Bulimus (172 sp.) abounds most in Tropical South America; it is also found from Burmah eastward through Malaya to the Solomon and Fiji Islands; there are also scattered species in Patagonia, St. Vincents, Texas, St. Helena, and New Zealand. More than 100 additional species have been described.

Buliminus (132 sp.) ranges from Central and South Europe over the whole Ethiopian and Oriental regions to North China, and through the Australian to New Zealand; there is also a single outlying species in the Galapagos Islands. About 50 more species have been described.

Cochlostyla (127 sp.) is almost peculiar to the Philippine Islands, beyond which, are a species in Borneo, one in Java, and two in Australia. Very few new species have been added to this genus.

Achatinella (95 sp.) is absolutely confined to the Sandwich Island group. Recent researches have more than tripled the number of described species.

Achatina (87 sp.) is most abundant and finest in the Ethiopian region, over the whole of which it ranges; but there are also species in Florida, the Antilles, the Sandwich Islands, Ceylon and India. The described species are now more than doubled.

Hyalina (84 sp.) inhabits all Tropical America and the Antilles, North America to Greenland, and Europe to the Arctic regions. Comparatively few new species have been described.

Cylindrella (83 sp.) inhabits the West Indian islands and Guatemala to Texas, with a sub-genus in the Philippine Islands. Species since described have more than trebled the number in this genus.

Cionella (67 sp.) is widely scattered; in India from Ceylon to the Khasia Mountains, Brazil, New Granada, the West Indian islands, Palæarctic, and northern part of Nearctic regions, Pacific Islands, New Zealand, and Juan Fernandez. About 20 new species have since been described.

Glandina (66 sp.), Peru to South Carolina and the Antilles, with three species in Central Africa and one in South Europe. About 40 species have been added to this genus.

Stenogyra (49 sp.), widely distributed: Tropical America and West Indies to Florida, South and West Africa, the Mediterranean region, India and the Philippines. About a dozen new species have been described.

Succinea (41 sp.), widely scattered in all the regions, and in St. Helena, Juan Fernandez, Tahiti, Chiloe, Greenland, West Africa, Himalayas and Australia. The described species are now more than 100.

Partula (39 sp.), Solomon Islands to Tahiti and Sandwich Islands. This genus has also been increased to near 100 species.

Streptaxis (34 sp.), most abundant in Tropical South America, but occurs in West Africa, the Seychelles and Rodriguez Islands, Ceylon and Burmah. It now contains over 100 described species.

Spiraxis (33 sp.), Yucatan to Mexico, and less abundant in the West Indian Islands. About 20 species have been added.

L L 2

Macroceramus (27 sp.), Antilles, Florida, and Peru. The species have been more than doubled.

Vitrina (26 sp.), widely scattered through North and Central Europe, North-west America and Greenland, Abyssinia, Madagascar and South Africa, Himalayas to Burmah and Australia. Species since described have more than doubled the number in this genus.

Orthalicus (23 sp.), Bolivia to Mexico and Antilles. This genus has been increased to about 40 species.

Sagda (19 sp.), Antilles only. Very few new species, if any, have been described.

Zonites (12 sp.), South Europe, with one species of a distinct type in Guatemala. The number of species in this genus has been since about tripled.

Leucochroa (11 sp.), Mediterranean region to Syria and Arabia Petrea.

Simpulopsis (7 sp.), Bahia, Antilles, and far away in the Solomon Islands. Two or three have been added.

Balea (6 sp.), Middle and North Europe, Brazil, and the Island of Tristan d'Acunha.

Daudebardia (6 sp.), Central and South Europe; and a species has since been discovered in New Zealand.

Macrocycles (4 sp.), Chili, California, Oregon and Central North America.

Columna (3 sp.), West Africa, Princes Islands and Madagascar.

Stenopus (2 sp.), Island of St. Vincent (West Indies.)

Pfeifferia (2 sp.), Philippines and Moluccas.

Testacella (2 sp.), West Europe and Teneriffe. About 8 species have been since described, including one from New Zealand.

Fossil species of *Helix, Bulimus, Achatina, Balea,* and *Clausilia,* are found in all the Tertiary formations; while a species of *Pupa* (as already stated) occurs in the carboniferous formation. For interesting details of the distribution of the subgenera and species of *Achatinella* in the Sandwich Islands, see a paper by Rev. J. T. Gulick in the *Journal of the Linnean Society.* (Zoology, vol. xi. p. 496.)

FAMILY 23.—LIMACIDÆ.—(12 Genera, 116 Species.)

GENERAL DISTRIBUTION.

NEOTROPICAL SUB-REGIONS.	NEARCTIC SUB-REGIONS.	PALÆARCTIC SUB-REGIONS.	ETHIOPIAN SUB-REGIONS.	ORIENTAL SUB-REGIONS.	AUSTRALIAN SUB-REGIONS.
— — — —	1.2.3.4	1.2.3.4	— — 3 —	1.2.3.4	1.2.3.4

The Limacidæ, or Slugs, are widely distributed, but they are absent from South America, where they are represented by the next family. They also seem to be absent from the greater part of Africa. The genera are distributed as follows :—

Limax (51 sp.), Palæarctic region, Australia and the Sandwich Islands; *Anadenus* (2 sp.), Himalayas ; *Philomychus* (9 sp.), North America, China and Java ; *Arion* (25 sp.), Norway to Spain and South Africa; *Parmacella* (7 sp.), South Europe, Canary Islands and North India ; *Janella* (1 sp.), New Zealand ; *Aneitea* (1 sp.), NewHebrides and New Caledonia; *Parmarion* (4 sp.), India; *Triboniophorus* (3 sp.), Australia ; *Testacella* (3 sp.), South Europe, Canary Islands, and New Zealand ; *Hyalimax* (2 sp.), Bourbon and Mauritius; *Krynickia* (8 sp.), Eastern Europe and North America. A few species of *Limax, Arion,* and *Testacella* have been found fossil in Tertiary deposits.

FAMILY 24.—ONCIDIADÆ. (2 Genera, 36 Species.)

GENERAL DISTRIBUTION.

NEOTROPICAL SUB-REGIONS.	NEARCTIC SUB-REGIONS.	PALÆARCTIC SUB-REGIONS.	ETHIOPIAN SUB-REGIONS.	ORIENTAL SUB-REGIONS.	AUSTRALIAN SUB-REGIONS.
1.2.3.4	— — — —	1.2 — —	— — — 4	1.2.3.3	— 2 — 4

The Oncidiadæ, or Slugs with a coriaceous mantle, inhabit the Oriental region, Mauritius, Australia, the Pacific Islands, South America and South Europe. The genera are :—

Oncidium (16 sp.), South Europe (1 sp. British), Mauritius, Australia and Pacific Islands; *Vaginulus* (20 sp.), Neotropical and Oriental regions.

FAMILY 25.—LIMNÆIDÆ.   (7 Genera, 332 Species.)

GENERAL DISTRIBUTION.

NEOTROPICAL SUB-REGIONS.	NEARCTIC SUB-REGIONS.	PALÆARCTIC SUB-REGIONS.	ETHIOPIAN SUB-REGIONS.	ORIENTAL SUB-REGIONS.	AUSTRALIAN SUB-REGIONS.
1.2.3.4	1.2.3.4	1.2.3.4	— — 3 -	1.2.3.4	— — — —

The Limnæidæ, or Fresh-water Snails, inhabit ponds and rivers in most parts of the world, but appear to be absent from the Australian region.   The genera are distributed as follows:— *Limnœa* (95 sp.), Nearctic, Palæarctic, and Oriental regions; *Choanomphalos* (2 sp.), Lake Baikal ; *Pompholyx* (2 sp.), Western America; *Chilinia* (18 sp.), South America ; *Physa* (20 sp.), Nearctic, Palæarctic, Ethiopian and Oriental regions, and extends to above 73° North Latitude in Siberia, being the most Arctic of land or fresh-water shells; *Ancylus* (49 sp.), Nearctic and Neotropical regions, Europe and New Zealand; *Planorbis* (145 sp.), Nearctic, Palæarctic and Oriental regions.   Several genera are found fossil, chiefly in the Wealden, Eocene, and Miocene formations.

FAMILY 26.—AURICULIDÆ.   (3 Genera, 210 Species.)

GENERAL DISTRIBUTION.

NEOTROPICAL SUB-REGIONS.	NEARCTIC SUB-REGIONS.	PALÆARCTIC SUB-REGIONS.	ETHIOPIAN SUB-REGIONS.	ORIENTAL SUB-REGIONS.	AUSTRALIAN SUB-REGIONS.
1 — — 4	1.2.3.4	1.2 — —	1.2.3 —	1.2.3.4	1.2 — 4

The Auriculidæ are chiefly found near the sea in hot countries, and are most abundant in the Eastern tropics.   They are absent from the East coast of South America. The genera have a somewhat restricted distribution as follows:—

Auricula (128 sp.), India, Pacific Islands, Peru and West Indies; *Melampus* (56 sp.), West Indies and Europe; *Carychium* (9 sp.), Europe and North America; *Plectrotrema* (14 sp.), Australia, Malay Islands, China, Cuba; *Blauneria* (2 sp.), West Indian and Sandwich Islands. There are many fossil species ranging back to the Eocene formation.

FAMILY 27.—ACICULIDÆ. (4 Genera, 65 Species.) (1865.)

GENERAL DISTRIBUTION.					
NEOTROPICAL SUB-REGIONS.	NEARCTIC SUB-REGIONS.	PALÆARCTIC SUB-REGIONS.	ETHIOPIAN SUB-REGIONS.	ORIENTAL SUB-REGIONS.	AUSTRALIAN SUB-REGIONS.
— 2 . 3 . 4	1 . 2 — —	1 . 2 — 4	— — — 4	— 2 — 4	1 2 . 3 —

The Aciculidæ are small cylindrical shells chiefly found in the West Indian Islands, but with representatives widely scattered over the globe.

Acicula (5 sp.) is European only; *Geomelania* (21 sp.), and *Chittya* (1 sp.), are confined to the Island of Jamaica; *Truncatella* (38 sp.), is most abundant in the Antilles, but is also found in some part of each of the six regions, as indicated by the diagram of the family. But few new species have been added to this group.

FAMILY 28.—DIPLOMMATINIDÆ. (3 Genera, 23 Species.) (1865.)

GENERAL DISTRIBUTION.					
NEOTROPICAL SUB-REGIONS.	NEARCTIC SUB-REGIONS.	PALÆARCTIC SUB-REGIONS.	ETHIOPIAN SUB-REGIONS.	ORIENTAL SUB-REGIONS.	AUSTRALIAN SUB-REGIONS.
— 2 — —	— — — —	— — — —	— — — —	1 — 3 . 4	1 . 2 . 3 4

The Diplommatinidæ are minute shells of the Oriental and Australian regions.

Diplommatina (18 sp.) inhabits India to Burmah, and the greater part of the Australian region; the number of species has now been doubled, and one has been discovered in the island of Trinidad; *Clostophis* (1 sp.), Moulmein; *Paxillus* (3 sp.), Borneo, Hong Kong, and Loo Choo Islands.

FAMILY 29.—CYCLOSTOMIDÆ.   (41 Genera, 1009 Species.)
(1865.)

GENERAL DISTRIBUTION.					
NEOTROPICAL SUB-REGIONS.	NEARCTIC SUB-REGIONS.	PALÆARCTIC SUB-REGIONS.	ETHIOPIAN SUB-REGIONS.	ORIENTAL SUB-REGIONS.	AUSTRALIAN SUB-REGIONS.
— 2 . 3 . 4	— — 3 —	— — — 4	— — 3 . 4	1 . 2 . 3 . 4	1 — — —

This extensive group, comprising the largest of the operculated land-shells, is especially characteristic of the Oriental region, which possesses 25 genera, no less than 12 of them being wholly confined to it. The Neotropical region comes next, with 15 genera, 9 of which are peculiar; but a large number of these are confined to the West Indian Islands, South America itself being very poor in this group. The Palæarctic region has 3 peculiar genera; the Ethiopian and Australian 1 each. The Nearctic region has but a single West Indian species in Florida. The distribution of the genera is as follows:—

Peculiar to or characteristic of the Oriental region are, *Opisthoporus* (11 sp.), *Rhiostoma* (6 sp.), *Alycaeus* (39 sp.), *Opisthostoma* (1 sp.), *Hybocistis* (3 sp.), *Pterocyclos* (19 sp.), extending to the Moluccas; *Aulopoma* (4 sp.), *Dermatocera* (4 sp.), *Leptopoma* (54 sp.), extending west to the Seychelles and east to the Moluccas and New Guinea; *Cyclophorus* (163 sp.), most abundant in the Oriental region, but ranges to Japan, to Chili, and all Tropical America, over the whole Australian region, and to Natal and Madagascar; *Cataulus* (15 sp.), confined to Ceylon, the Neilgherries and Nicobar Islands; *Rhaphaulus* (4 sp.), Penang to Ceram; *Streptaulus* (1 sp.), *Arinia* (3 sp.), *Pupinella* (2 sp.), *Pupina* (24 sp.), half in North India to Philippines and

Japan, the other half in Moluccas, New Guinea and Australia ;
Cyclotopsis (2 sp.), India and Malaya : *Registoma* (9 sp.), Philippines and Moluccas, New Caledonia and Pacific.

Characteristic of the Neotropical region are :—*Cyclotus* (111
sp.), half in the Antilles and Tropical America, the rest in the
Moluccas, China, Malaya, India, Natal, and the Seychelle
Islands ; *Megalomastoma* (27 sp.), abundant in Cuba, West
Indies and South America, others in India, Malaya, and
Mauritius ; *Jamaicia* (2 sp.), Jamaica ; *Licina* (5 sp.), Antilles ;
Choanopoma (49 sp.), Antilles ; *Ctenopoma* (25 sp.), Antilles ;
Diplopoma (1 sp.), Cuba ; *Adamsiella* (15 sp.), Jamaica, Cuba,
Guatemala ; *Cyclostomus* (113 sp.), abundant in Antilles, also
occurs in Madagascar, Arabia, Syria, Hungary and New Zealand ;
Tudora (34 sp.), Antilles, and one species in Algeria ; *Cistula*
(40 sp.), *Chondropoma* (94 sp.), *Bourcieria* (2 sp.), Tropical
America.

Peculiar to or characteristic of the Palæarctic region are :—
Craspedopoma (5 sp.), confined to Madeira, the Azores and
Canaries ; *Leonia* (1 sp.), Spain and Algeria ; *Pomatias* (22 sp.),
Europe and Canaries with a species in the Himalayas ; *Cecina*
(1 sp.), Manchuria.

The Ethiopian region has the peculiar genus *Lithodion* (5 sp.),
Madagascar, Socotra and Arabia ; and *Otopoma* (19 sp.), Mascarene
Islands and Socotra, with a species in Western India and another
in New Ireland.

The Australian region is characterised by *Callia* (3 sp.), in
Ceram, Australia, and the Philippines respectively ; *Realia* (7
sp.), New Zealand and the Marquesas Islands ; *Omphalotropis*
(38 sp.), the Australian region, with some species in India,
Malaya, and the Mauritius.

The remaining genus, *Hydrocena* (27 sp.), has a very
widely scattered distribution, being found in South Europe,
Japan, the Cape, China, Malaya, New Zealand, the Pacific
Islands and Chili. From 10 to 20 per cent. of new species have
been since described in most of the genera of this family.

FAMILY 30.—HELICINIDÆ. (7 Genera, 433 Species.) (1868.)

GENERAL DISTRIBUTION.

NEOTROPICAL SUB-REGIONS.	NEARCTIC SUB-REGIONS.	PALÆARCTIC SUB-REGIONS.	ETHIOPIAN SUB-REGIONS.	ORIENTAL SUB-REGIONS.	AUSTRALIAN SUB-REGIONS.
— 2 . 3 . 4	— — 3 —	— — — —	— — — —	— — 3 . 4	1 . 2 . 3 —

The Helicinidæ are very characteristic of the Antilles, comparatively few being found in any other part of the world except the Islands of the Pacific. The genera are :—
Trochatella (33 sp.), Antilles with a species in Venezuela, and another in Cambodja; *Lucidella* (5 sp.), Antilles; *Helicina* (274 sp.), Antilles, Pacific Islands, Tropical America, Southern United States, Moluccas, Australia, Philippines, Java, Andaman Islands and North China; *Schasicheila* (5 sp.), Mexico, Guatemala and Bahamas; *Alcadia* (28 sp.), Antilles; *Georissa* (5 sp.) Moulmein to Burmah. About 10 per cent. of new species appear to have been since described in the larger genera of this family.

General Observations on the Distribution of the Land Mollusca.

A consideration of the distribution of the families and genera of land-shells shows us, that although they possess some special features, yet they agree in many respects with the higher animals in their limitation by great natural barriers, such as oceans, deserts, mountain ranges, and climatal zones. A remarkable point in the distribution of these animals, is the number of genera which have a very limited range, and also the prevalence of genera having species scattered, as it were at random, all over the earth. No less than 14 genera (or about one-sixth of the whole number) are confined to the Antilles, while the greater part of the sub-genera of modern authors are restricted to limited areas.

If we first compare the New World with the Old, we find the difference as regards genera quite as great as in most of the vertebrates. In the Helicidæ, 10 genera are confined to the New, and 7 to the Old World, 16 being common to both. In the Operculata the number of genera of restricted range is greater,— the New World having 15, the Old World 32 genera, only 8 being common to both. Of the New World genera 12 out of the 15 do not occur at all in South America; and of those of the Old World, 22 out of the 32 occur in a single region only. If we take the northern and southern division proposed by Professor Huxley (the latter comprising the Australian and Neotropical regions), we find a much less well-marked diversity. Among the Helicidæ only 4 are exclusively northern, 8 southern; while among the Operculata 22 are northern, 16 southern. The best way to compare these two kinds of primary division will be to leave out all those genera confined to a single region each, and to take account only of those characteristic of two or more of the combined regions; which will evidently show which division is the most natural one for this group. The result is as follows :—

GENERA COMMON TO TWO OR MORE REGIONS IN, AND CONFINED TO, EACH PRIMARY DIVISION OF THE EARTH.

	Helicidæ	Operculata.	Totals.
Northern	0	0	0
Southern	0	0	0
Old World	1	12	13
New World	4	0	4

We find then that the northern and southern division of the globe is not at all supported by the distribution of the terrestrial molluscs. It is indeed very remarkable, that the connection so apparent in many groups between Australia and South America is so scantily indicated here. The only facts supporting it seem to be, the occurrence of *Geotrochus* (a sub-genus of Helix) in Brazil, as well as in the Austro-Malayan and West Pacific Islands and North Australia; and of *Bulimus* in the same two parts of the globe, but peculiar sub-genera in each. But in neither case is there any affinity shown between the temperate portions of the two regions, so that we must probably trace this resemblance to some more ancient diffusion of types than that which led to the similarity of plants and insects. Still more curious is the entire absence of genera confined to, and characteristic of Africa and India. One small sub-genus of *Helix*, (*Rachis*), and one of *Achatina*, (*Homorus*), appear to have this distribution,—a fact of but little significance when we find another sub-genus of *Helix*, (*Hapalus*), common and confined to Guinea and the Philippine Islands ; and when we consider the many other cases of scattered distribution which cannot be held to indicate any real connection between the countries implicated. No genus is confined to the Palæarctic and Nearctic regions as a whole. A large number of sub-genera, many of them of considerable extent, are peculiar to one or other of these regions, but only 3 sub-genera of *Helix* and 2 of *Pupa* are common and peculiar to the two combined, and these are always such as have an Arctic range and whose distribution therefore offers no difficulty.

We find, then, that each of our six regions and almost all of our sub-regions are distinctly confirmed by the distribution of the terrestrial mollusca ; while the different combinations of them which have at various times been suggested, receive little or no support whatever. Even those remarkably isolated sub-regions, New Zealand and Madagascar, have no strictly peculiar genera of land-shells, although they both possess several peculiar sub-genera; being thus inferior in isolation to some single West Indian Islands, to the Sandwich Islands, and even to the North Atlantic Islands (Canaries, Madeira, and Azores), each of which have peculiar genera. This of course, only indicates that the means by which land mollusca have been dispersed are somewhat special and peculiar. To determine in what this speciality consists we must consider some of the features of the specific distribution of this group.

The range of genera, and even of sub-genera is, as we have seen, often wide and erratic, but as a general rule the species have a very restricted area.

Hardly a small island on the globe but has some land-shells peculiar to it. Juan Fernandez has 20 species, all peculiar. Madeira and Porto Santo have 109 peculiar species out of a total of 134. Every little valley, plain, or hill-top, in the Sandwich Islands, though only a few square miles in extent, has its peculiar species of *Achatinella*. Another striking feature of the distribution of land molluscs, is the richness of islands as compared with continents. The Philippines contain more species than all India; and those of the Antilles according to Mr. Bland almost exactly equal the numbers found in the entire American continent from Greenland to Patagonia. Taking the whole world, it appears that many more species of land-shells are found in the islands than on the continents of the globe, a peculiarity that obtains in no other extensive group of animals.

Looking at these facts it seems probable, that the air-breathing molluscs have been chiefly distributed by air- or water-carriage, rather than by voluntary dispersal on the land. Even seas and oceans have not formed impassable barriers to their diffusion; whereas they only spread on dry land with excessive slowness and difficulty. The exact mode in which their diffusion is effected is not known, and it may depend on rare and exceptional circumstances; but it seems likely to occur in two ways. Snails frequently conceal themselves in crevices of trees or under bark, or attach themselves to stems or foliage, and either by their operculum or mucous diaphragm, are able to protect themselves from the injurious effects of salt water for long periods. They might therefore, under favourable conditions, be drifted across arms of the sea or from island to island; while wherever there are large rivers and occasional floods, they would by similar means be widely scattered over land areas. Another possible mode of distribution is by means of storms and hurricanes, which would carry the smaller species for long distances, and might occasionally transport the eggs of the larger forms. Aquatic birds might occasionally get both shells and eggs attached to their feet or their plumage, and convey them across a wide extent of sea. But whether these, or some other unknown agency has acted, the facts of distribution clearly imply that some means of transport over water is, and has been, the chief agent in the distribution of these animals; but that its action is very rare or intermittent, so that its effects are hardly perceptible in the distribution of single species.

Another important factor in enabling us to account for the

distribution of these animals is the geological antiquity of the
group, and the amount of change exhibited in time, by species
and genera. Now we find that most of the genera of land-shells
range back to the Eocene period, while those inhabiting fresh
water are found almost unchanged in the Wealden. In North
America a species of *Pupa* and one of *Zonites*, have been discovered in the coal measures, along with Labyrinthodonts; and
this fact seems to imply, that many more terrestrial molluscs
would be discovered, if fresh-water deposits, made under favourable conditions, were more frequently met with in the older
rocks. If then the existing groups of land-molluscs are of such
vast antiquity, and possess some means, however rarely occurring,
of crossing seas and oceans, we need not wonder at the wide and
erratic distribution now presented by so many of the groups;
and we must not expect them to conform very closely to those
regions which limit the range of animals of higher organization
and less antiquity.

The total number of species of pulmoniferous mollusca is about
7,000, according to the estimate of Mr. Woodward, brought down
to 1868 by Mr. Tate. But this number would be largely increased if the estimates of specialists were taken. Mr. Woodward
for example, gives 760 as the number of species in the West
Indian Islands; whereas Mr. Thomas Bland, who has made the
shells of these islands a special study, considers that there were
1,340 species in 1866. So, the land-shells of the Sandwich
Islands are given at 267; but Mr. Gulick has added 120 species
of Achatinellidæ, bringing the numbers up to nearly 400,—but
no doubt several of these are so closely related that many conchologists would class them as varieties. The land-shell fauna
of the Antilles is undoubtedly the most remarkable in the world,
and it has been made the subject of much interesting discussion
by Mr. Bland and others. This fauna differs from that of all
other parts of the globe in the proportions of the operculate to
the inoperculate shells. The Operculata of the globe are about
one-seventh, the Inoperculata about six-sevenths of the whole;
and some general approximation to this proportion (or a much
smaller one) exists in almost all the continents, islands, and

archipelagoes.   In the Philippines, for example, the proportion
of the Operculata is a little more than one-seventh; in the
Mauritius, between one-third and one-fourth; in Madeira, one-
fourteenth; in the whole American continent about one-eighth;
but when we come to the Antilles we find them to amount to
nearly five-sixths, about half the Operculata of the globe being
found there!

Mr. Bland endeavours to ascertain the source of some of the
chief genera found in the West Indian Islands, on the principle
that "each genus has had its origin where the greatest number
of species is found;" and then proceeds to determine that some
have had an African, some an Asiatic, and some an American
origin, while others are truly indigenous.   But we fear there is
no such simple way of arriving at so important a result ; and in
the case of groups of extreme antiquity like the genera of mollusca, it would seem quite as possible that the origin of a genus
is generally *not* where the greatest number of species are now
found.   For during the repeated changes of physical conditions
that have everywhere occurred since the Eocene period (to go
no further back) every genus must have made extensive migrations, and have often become largely developed in some other
district than that in which it first appeared.   As a proof of this,
we not unfrequently find fossil shells where the species and even
the genus now no longer exists; as *Auricula*, found fossil in
Europe, but only living in the Malay and Pacific Islands; *Anas-
toma* and *Megaspira*, now peculiar to Brazil, but fossil in the
Eocene of France ; and *Proserpina* of the West Indies, found in
the Eocene formation of the Isle of Wight.   The only means by
which the origin of a genus can satisfactorily be arrived at, is by
tracing back its fossil remains step by step to an earlier form ;
and this we have at present no means of doing in the case of
the land-shells.   Taking existing species as our guide we should
certainly have imagined that the genus *Equus* originated in
Africa or Central Asia ; but recent discoveries of numerous
extinct species and of less specialized forms of the same type,
seem to indicate that it originated in North America, and that
the whole tribe of " horses " may be, for anything we yet know

to the contrary, recent immigrants into the Old World! This example alone must convince us, that it is impossible to form any conclusion as to the origin of a genus, from the distribution of existing species only.

The general conclusion we arrive at, therefore, is, that the causes that have led to the existing distribution of the genera and higher groups of the terrestrial mollusca are so complex, and have acted through such long periods, that most of the barriers which limit the range of other terrestrial animals do not apply to them, although the species are, in most cases, strictly limited by them. Some means of diffusion—which, though probably acting very slowly and at long intervals, and more powerfully on continents than between islands, is yet highly efficient when we consider the long duration of genera—has, to a considerable extent, dispersed them across continents, seas, and oceans. On the other hand, those mountain barriers which separate many groups of the higher vertebrates, are generally less ancient than the genera of land-shells, which are thus often distributed independently of them. In order to compare the distribution of the terrestrial mollusca on equal terms with those of land animals generally, we must take genera of the former as equivalent to family groups of the latter; and we shall, I believe, then find that the distribution of the sub-genera and smaller groups of species do accord mainly with those divisions of the earth into regions and sub-regions which we have here indicated. Mr. Harper Pease, in a communication on Polynesian Land Shells in the *Proceedings of the Zoological Society* for 1871 (p. 449), marks out the limits of the Polynesian sub-region, so as exactly to agree with that arrived at here from a consideration of the distribution of vertebrata; and he says that this sub-region, (or region, as he terms it) is distinctly characterised by its land-shells from all the surrounding regions. The genera (or sub-genera) *Partula, Pitys, Achatinella, Palaina, Omphalotropis,* and many others, are either wholly confined to this sub-region or highly characteristic of it. Mr. Binney, in his *Catalogue of the Air-breathing Molluscs of North America,* marks out our Nearctic region (with almost identical limits) as most clearly characterised. He also arrives at a series of sub-divisions, which generally (though not exactly) agree with the sub-regions which I have here adopted. The Palæarctic, the Ethiopian, and the Oriental regions, are also generally admitted to be well. characterised by their terrestrial molluscs. There only remain the Australian and the Neotropical regions, in which some want of homogeneity is apparent, owing to the vast development and specialisation of certain groups in the islands which belong to these regions. The Antilles, on the one hand, and the Polynesian Islands, on the other, are so rich in land-shells and possess so many peculiar forms, that, judged by these alone, they must form primary instead of secondary divisions. We have, however, already pointed out the inconvenience of any such partial systems of zoological geography, and the causes have been sufficiently indicated which have, in the case of land-shells as of insects, produced certain special features of distribution.

We therefore venture to hope, that conchologists will give us the advantage of their more full and accurate knowledge both of the classification and distribution of this interesting group of animals, not to map out new sets of regions for themselves, but to show what kind of barriers have been most efficient in limiting the range of species, and how their distribution is actually effected, so as to be able to explain whatever discrepancies exist between the actual distribution of land-shells and that of the higher animals.

Order III.—OPISTHO-BRANCHIATA.

There are ten families in this order, all of which, as far as known, are widely or universally distributed. Some of them are found fossil, ranging back to the Carboniferous epoch. They are commonly termed Sea-slugs, and have either a thin small shell or none. We shall therefore simply enumerate the families, with the number of genera and species as given by Mr. Woodward.

FAMILY 31.—TORNATELLIDÆ. (7 Genera, 62 Species living, 166 fossil.)

FAMILY 32.—BULLIDÆ. (12 Genera, 168 Species living, 88 fossil.)

FAMILY 33.—APHYSIADÆ. (8 Genera, 84 Species living, 4 fossil.)

FAMILY 34.—PLEUROBRANCHIDÆ. (7 Genera, 28 Species living, 5 fossil.)

FAMILY 35.—PHYLLIDIADÆ. (4 Genera, 14 Species living, 0 fossil.)

FAMILY 36.—DORIDÆ. (23 Genera, 160 Species living, 0 fossil.)

FAMILY 37.—TRITONIADÆ. (9 Genera, 38 Species living, 0 fossil.)

FAMILY 38.—ÆOLIDÆ. (14 Genera, 101 Species living, 0 fossil.)

FAMILY 39.—PHYLLYRHOIDÆ. (1 Genus, 6 Species living, 0 fossil.)

FAMILY 40.—ELYSIADÆ (5 Genera, 13 Species living, 0 fossil.)

Order IV.—NUCLEO-BRANCHIATA.

These are oceanic, swimming molluscs, of a delicate texture. They are found in all warm seas, and range back to the Lower Silurian epoch. There are only two families.

FAMILY 41.—FIROLIDÆ. (2 Genera, 33 Species living, 1 fossil.)

FAMILY 42.—ATLANTIDÆ. (5 Genera, 22 Species living, 159 fossil.)

CLASS.—PTEROPODA.

These are swimming, oceanic mollusca, inhabiting both Arctic, Temperate, and Tropical seas. The three families have each a wide distribution in all the great oceans. They range back to the Silurian period.

FAMILY 1.—HYALEIDÆ. (9 Genera, 52 Species living, 95 fossil.)

FAMILY 2.—LIMACINIDÆ. (4 Genera, 19 Species living, 0 fossil.)

FAMILY 3.—CLIONIDÆ. (4 Genera, 14 Species living, 0 fossil.)

CLASS.—BRACHIOPODA.

These are sedentary, bivalve, marine mollusca, having laterally symmetrical shells, but with unequal valves. Both in space and time they are the most widely distributed molluscs. They are found in all seas, and at all depths; and when any of the families or genera have a restricted range, it seems to be due to our imperfect knowledge, rather than to any real geographical limitations. In time they range back to the Cambrian formation, and seem to have had their maximum development in the Silurian period. It is not, therefore, necessary for our purpose, to do more than give the names of the families with the numbers of the genera and species, as before.

FAMILY 1.—TEREBRATULIDÆ.   (5 Genera, 67 Species living, 340 fossil.)

FAMILY 2.—SPIRIFERIDÆ.   (4 Genera, 0 Species living, 380 fossil.)

FAMILY 3.—RHYNCHONELLIDÆ.   (3 Genera, 4 Species living, 422 fossil.)

FAMILY 4.—ORTHIDÆ.   (4 Genera, 0 Species living, 328 fossil.)

FAMILY 5.—PRODUCTIDÆ.   (3 Genera, 0 Species living, 146 fossil.)

FAMILY 6.—CRANIADÆ.   (1 Genus, 5 Species living, 37 fossil.)

FAMILY 7.—DISCINIDÆ.   (2 Genera, 10 Species living, 90 fossil.)

FAMILY 8.—LINGULIDÆ.   (2 Genera, 16 Species living, 99 fossil.)

CLASS.—CONCHIFERA.

The Conchifera, or ordinary Bivalve Molluscs, may be distinguished from the Brachiopoda by having their shells laterally unsymmetrical, while the valves are generally (but not always) equal. They are mostly marine, but a few inhabit fresh water. As the distribution of some of the families presents points of interest, we shall treat them in the same manner as the marine Gasteropoda.

FAMILY 1.—OSTREIDÆ.   (5 Genera, 426 Species.)

DISTRIBUTION.—The Ostreidæ, including the Oysters and Scallops, are found in all seas, Arctic as well as Tropical. There are nearly 1,400 species fossil, ranging back to the Carboniferous period.

FAMILY 2.—AVICULIDÆ.   (3 Genera, 94 Species.)

DISTRIBUTION,—The Aviculidæ, or Wing-shells and Pearl Oysters, are characteristic of Tropical and warm seas, a few only ranging into temperate regions. Nearly 700 fossil species are known from various formations ranging back to the Devonian, and Lower Silurian.

FAMILY 3.—MYTILIDÆ.   (3 Genera, 217 Species.)

DISTRIBUTION.—The Mytilidæ, or Mussels, have a world-wide distribution. There is one fresh-water species, which inhabits the Volga. There are about 350 fossil species, ranging back to the Carboniferous epoch.

FAMILY 4.—ARCADÆ.   (6 Genera, 360 Species.)

DISTRIBUTION.—The Arcadæ are universally distributed, and are most abundant in warm seas.   The genus *Leda* is, however, abundant in Arctic and Temperate regions, and *Solenella* is confined to the South Temperate zone.   There are near 1,200 fossil species, found in all strata as low as the Lower Silurian.

FAMILY 5.—TRIGONIADÆ.   (1 Genus, 3 Species.)

DISTRIBUTION.—The living *Trigoniæ* are confined to Australia, but there are 5 other genera fossil, containing about 150 species, and found in various formations from the Chalk to the Lower Silurian.

FAMILY 6.—UNIONIDÆ.   (7 Genera, 549 Species.)

DISTRIBUTION.—The Unionidæ, or Fresh-water Mussels, are found in all the fresh waters of the globe, but some of the genera are restricted.   *Castalia, Mycetopus* and *Mulleria* are confined to the rivers of South America ; *Anodon*, to the Nearctic and Palæarctic regions; *Iridina*, and *Etheria*, to the rivers of Africa; *Unio* has a universal distribution, but is especially abundant in North America.   About 60 fossil species are found in the Tertiary and Wealden formations.

FAMILY 7.—CHAMIDÆ.   (1 Genus, 50 Species.)

DISTRIBUTION.—The Chamidæ, or Giant Clams, are confined to Tropical seas, chiefly among coral reefs.   There are two other genera and 62 species fossil, ranging from the Chalk to the Oolite formations.

FAMILY 8.—HIPPURITIDÆ.   (5 Genera, 103 Species.)

Fossils of doubtful affinity, from the Chalk formation.

FAMILY 9.—TRIDACNIDÆ. (1 Genus, 8 Species.)

DISTRIBUTION.—The Tridacnidæ, or Clam-shells, are of very large size, and are confined to the Tropical regions of the Indian and Pacific Oceans. A few species have been found fossil in the Miocene formation.

FAMILY 10.—CARDIADÆ. (1 Genus, 200 Species.)

DISTRIBUTION.—The Cardiadæ, or Cockles, are of world-wide distribution. Another genus is fossil, and nearly 400 fossil species are known, ranging back to the Upper Silurian formation.

FAMILY 11.—LUCINIDÆ. (8 Genera, 178 Species.)

DISTRIBUTION.—The Lucinidæ inhabit the Tropical and Temperate seas of all parts of the world ; but the genus *Corbis* is confined to the Indian and Pacific Oceans, *Montacuta* and *Lepton*, to the Atlantic. There are nearly 500 extinct species, ranging from the Tertiary back to the Silurian formation.

FAMILY 12.—CYCLADIDÆ. (3 Genera, 176 Species.)

DISTRIBUTION.—The Cycladidæ are small fresh- or brackish-water shells found all over the globe. The genus *Cyclas* is most abundant in the North Temperate zone, while *Cyrena* inhabits the warmer shores of the Atlantic and Pacific, but is absent from the West Coast of America. There are about 150 species fossil, ranging back from the Pliocene to the Wealden formations.

FAMILY 13.—CYPRINIDÆ. (10 Genera, 176 Species).

DISTRIBUTION.—Universal. *Cyprina* and *Astarte* are Arctic and North Temperate ; *Cardita* is Tropical and South Temperate. There are several extinct genera and about 1,000 species found in all formations as far back as the Lower Silurian.

FAMILY 14.—VENERIDÆ, (10 Genera, 600 Species.)

DISTRIBUTION.—Universal. *Lucinopsis* is confined to the North Atlantic; *Glauconeza* to the mouths of rivers in the Oriental region; *Meroe* and *Trigona* to warm seas. There are about 350 fossil species, ranging back to the Oolitic period.

FAMILY 15.—MACTRIDÆ. (5 Genera, 147 Species.)

DISTRIBUTION.—All seas, but more abundant in the Tropics. *Gnathodon* is found in the Gulf of Mexico; *Anatinella* in the Oriental region. There are about 60 fossil species, ranging back to the Carboniferous period.

FAMILY 16.—TELLINIDÆ. (11 Genera, 560 Species.)

DISTRIBUTION.—All seas; most abundant in the Tropics. *Galatea* is confined to African rivers. There are about 60 fossil species, mostly Tertiary, but ranging back to the Carboniferous period.

FAMILY 17.—SOLENIDÆ. (3 Genera, 63 Species.)

DISTRIBUTION.—All Temperate and Tropical seas. There are 80 fossil species which range back to the Carboniferous epoch.

FAMILY 18.—MYACIDÆ. (6 Genera, 121 Species.)

DISTRIBUTION.—All seas. *Panopœa* inhabits both North and South Temperate seas; *Glycimeris*, Arctic seas. There are near 350 fossil species, ranging back to the Lower Oolite formation.

FAMILY 19.—ANATINIDÆ. (8 Genera, 246 Species.)

DISTRIBUTION.—All seas. *Pholadomya* is from Tropical Africa; *Myadora* from the Western Pacific; *Myochama* and *Chamostræa* are Australian. There are about 400 fossil species, ranging back to the Lower Silurian formation.

FAMILY 20.—GASTROCHÆNIDÆ.   (5 Genera, 40 Species.)

DISTRIBUTION.—Temperate and warm seas. *Aspergillum* ranges from the Red Sea to New Zealand. There are 35 fossil species, ranging back to the Lower Oolite.

FAMILY 21.—PHOLADIDÆ.   (4 Genera, 81 Species.)

DISTRIBUTION.—These burrowing molluscs inhabit all Temperate and warm seas from Norway to New Zealand. There are about 50 fossil species, ranging back to the epoch of the Lias.

General Remarks on the Distribution of the Marine Mollusca.

The marine Mollusca are remarkable for their usually wide distribution. About 48 of the families are cosmopolitan, ranging over both hemispheres, and in cold as well as warm seas. About 15 are restricted to the warmer seas of the globe; but several of these extend from Norway to New Zealand, a distribution which may be called universal, and only 2 or 3 are absolutely confined to Tropical seas. Two small families only, are confined to the Pacific and Indian Oceans. Marine fishes, on the other hand, have a much less cosmopolitan character, no less than 30 families having a limited distribution, while 50 are universal. Some of these 30 families are confined to the Northern seas, some to the Atlantic and Mediterranean, and a considerable number to the Indian Ocean and Western Pacific. Many of these families, it is true, are much smaller than those of the Mollusca, which seem to possess very few of those small isolated families of two or three species only, which abound in all the Vertebrate classes. These differences are no doubt connected with the higher organisation of fishes, which renders them more susceptible to changed conditions of life ; and this is indicated by the much less antiquity of existing families of fishes, the greater part of which do not date back beyond the Cretaceous epoch, and many of them only to the Eocene. In striking contrast we have the vast antiquity of most of the families of Mol-

lusca, as shown in the following table of their range taken from
Mr. Woodward's work, but re-arranged, and somewhat modified.

Range of Families of Mollusca in Time; arranged in their order of appearance and disappearance.	Lower Silurian.	Upper Silurian.	Devonian.	Carboniferous.	Permian.	Trias.	Lower Oolite.	Upper Oolite.	Lower Cretaceous.	Upper Cretaceous.	Eocene.	Miocene.	Pliocene.	Recent.
Productidæ	—	—	—	—	—									
Orthoceratidæ	—	—	—	—	—	—								
Spiriferidæ, Orthidæ	—	—	—	—	—	—								
Atlantidæ, Hyaleidæ	—	—	—	—	—	—	—	—	—	—	—	—	—	—
Pyramidellidæ, Turbinidæ	—	—	—	—	—	—	—	—	—	—	—	—	—	—
Ianthidæ, Chitonidæ	—	—	—	—	—	—	—	—	—	—	—	—	—	—
Lingulidæ	—	—	—	—	—	—	—	—	—	—	—	—	—	—
Aviculidæ, Mytilidæ	—	—	—	—	—	—	—	—	—	—	—	—	—	—
Arcadæ, Trigoniadæ	—	—	—	—	—	—	—	—	—	—	—	—	—	—
Cyprinidæ, Anatinidæ	—	—	—	—	—	—	—	—	—	—	—	—	—	—
Nautilidæ	—	—	—	—	—	—	—	—	—	—	—	—	—	—
Rhynchonellidæ, Craniadæ, Discinidæ	—	—	—	—	—	—	—	—	—	—	—	—	—	—
Cardiadæ, Lucinidæ	—	—	—	—	—	—	—	—	—	—	—	—	—	—
Ammonitidæ	—	—	—	—	—	—	—	—	—	—				
Naticidæ, Calyptræidæ.	—	—	—	—	—	—	—	—	—	—	—	—	—	—
Dentalidæ, Terebratulidæ	—	—	—	—	—	—	—	—	—	—	—	—	—	—
Helicidæ	—	—	—	—	—	—	—	—	—	—	—	—	—	—
Fissurellidæ, Tornatellidæ		—	—	—	—	—	—	—	—	—	—	—	—	—
Pectinidæ, Solenidæ		—	—	—	—	—	—	—	—	—	—	—	—	—
Cerithiadæ, Littorinidæ, Astartidæ			—	—	—	—	—	—	—	—	—	—	—	—
Belemnitidæ			—	—	—	—	—	—	—	—				
Teuthidæ, Sepiadæ			—	—	—	—	—	—	—	—	—	—	—	—
Neritidæ, Patellidæ, Bullidæ				—	—	—	—	—	—	—	—	—	—	—
Gastrochænidæ, Pholadidæ				—	—	—	—	—	—	—	—	—	—	—
Limnæidæ, Melaniadæ				—	—	—	—	—	—	—	—	—	—	—
Chamidæ, Myadæ				—	—	—	—	—	—	—	—	—	—	—
Cycladidæ, Veneridæ, Tellinidæ					—	—	—	—	—	—	—	—	—	—
Hippuritidæ										—	—			
Unionidæ										—	—	—	—	—
Strombidæ, Buccinidæ										—	—	—	—	—
Conidæ, Volutidæ										—	—	—	—	—
Auriculidæ, Cyclostomidæ										—	—	—	—	—
Mactridæ —										—	—	—	—	—
Limacidæ											—	—	—	—
Argonautidæ												—	—	—
Tridacnidæ													—	—

Nor is this enormous antiquity confined to family types alone.
Many genera are equally ancient. The genus *Lingula* has

existed from the earliest Palæozoic times down to the present day ; while *Terebratula, Rhynchonella, Discina, Nautilus, Natica, Pleurotomaria, Patella, Dentalium, Mytilus* and many other living forms, range back to the Palæozoic epoch. That groups of such immense antiquity, and having power to resist such vast changes of external conditions as they must have been subject to, should now be widely distributed, is no more than might reasonably be expected. It is only in the case of sub-genera and species, that we can expect the influence of recent geological or climatal changes to be manifest ; and it must be left to special students to work out the details of their distribution, with reference to the general principles found to obtain among the more highly organised animals.

CHAPTER XXIII.

HAVING already given summaries of the distribution of the several orders, and of some of the classes of land animals, we propose here to make a few general remarks on the special phenomena presented by the more important groups, and to indicate where possible, the general lines of migration by which they have become dispersed over wide areas.

MAMMALIA.

This class is very important, and its past history is much better known than that of most others. We shall therefore briefly summarise the results we have arrived at from our examination of the distribution of extinct and living forms of each order.

Primates.—This order, being pre-eminently a tropical one, became separated into two portions, inhabiting the Eastern and Western Hemispheres respectively, at a very early epoch. In consequence of this separation it has diverged more radically than most other orders, so that the two American families, Cebidæ and Hapalidæ, are widely differentiated from the Apes, Monkeys, and Lemurs of the Old World. The Lemurs were probably still more ancient, but being much lower in organisation, they became extinct in most of the areas where the higher forms of Primates became developed. Remains found in the Eocene formation indicate, that the North American and European

Primates had, even at that early epoch, diverged into distinct series, so that we must probably look back to the secondary period for the ancestral form from which the entire order was developed.

Chiroptera.—These are also undoubtedly very ancient. The most generalised forms—the Vespertilionidæ and Noctilionidæ—are the most widely distributed; while special types have arisen in America, and in the Eastern Hemisphere. Remains found in the Upper Eocene formation of Europe differ little from species still living in the same countries; so that we can form no conjecture as to the origin or migration of the group. Their power of flight would, however, enable them rapidly to spread over all the great continents of the globe.

Insectivora.—This very ancient group, now probably verging towards extinction, appears to have originated in the Northern continent, and never to have reached Australia or South America. It may, however, have become extinct in the latter country owing to the competition of the numerous Edentata. The Insectivora now often maintain themselves amidst more highly developed forms, by means of some special protection. Some burrow in the earth,—like the moles; others have a spiny covering,—as the hedgehogs and several of the Centetidæ; others are aquatic,—as the *Potamogale* and the desman; others have a nauseous odour,—as the shrews; while there are several which seem to be preserved by their resemblance to higher forms,—as the elephant-shrews to jerboas, and the tupaias to squirrels. The same need of protection is shown by the numerous Insectivora inhabiting Madagascar, where the competing forms are few; and by one lingering in the Antilles, where there are hardly any other mammalia.

Carnivora.—Although perhaps less ancient than the preceding, this form of mammal is far more highly organised, and from its earliest appearance appears to have become dominant in the world. It would therefore soon spread widely, and diverge into the various specialised types represented by existing families. Most of these appear to have originated in the Eastern Hemisphere, the only Carnivora occurring in North

American Miocene deposits being ancestral forms of Canidæ and Felidæ. It seems probable, therefore, that the order had attained a considerable development before it reached the Western Hemisphere. The Procyonidæ, now confined to America, are not very ancient; and the occurrence of a few allied forms in the Himalayas (*Ælurus* and *Æluropus*) render it probable that their common ancestors entered North America from the Palæarctic region during the Miocene period, but being a rather low type they have succumbed under the competition of higher forms in most parts of the Eastern Hemisphere. Bears and Weasels are probably still more recent emigrants to America. The aquatic carnivora (Seals, &c.) are, as might be expected, more widely and uniformly distributed, but there is little evidence to show at what period the type was first developed.

Ungulata.—These are the dominant vegetable-feeders of the great continents, and they have steadily increased in numbers and in specialisation from the oldest Tertiary times to the present day. Being generally of larger size and less active than the Carnivora, they have somewhat more restricted powers of dispersal. We have good evidence that their wide range over the globe is a comparatively recent phenomenon. Tapirs and Llamas have probably not long inhabited South America, while Rhinoceroses and Antelopes were once, perhaps, unknown in Africa, although abounding in Europe and Asia. Swine are one of the most ancient types in both hemispheres; and their great hardiness, their omnivorous diet, and their powers of swimming, have led to their wide distribution. The sheep and goats, on the other hand, are perhaps the most recent development of the Ungulata, and they seem to have arisen in the Palæarctic region at a time when its climate already approximated to that which now prevails. Hence they are pre-eminently a Temperate group, never found within the Tropics except upon a few mountain ranges.

Proboscidea.—These huge animals (the Elephants and Mastodons) appear to have originated in the warmer parts of the Palæarctic region, but they soon spread over all the great continents, even reaching the southern extremity of America. Their extinction has probably depended more on physical than on organic changes, and we can clearly trace their almost total disappearance to the effects of the Glacial epoch.

Rodentia.—Rodents are a very dominant group, and a very ancient one. Owing to their small size and rapid powers of increase, they soon spread over almost every part of the globe, whence has resulted a great specialisation of family types in the South American continent which remained so long isolated. They are capable of living wherever there is any kind of vegetable food, hence their range will be determined rather by organic than by physical conditions ; and the occupation of a country by enemies or by competing forms, is probably the chief cause which has prevented many of the families from acquiring a wide range. The occurrence of isolated species of the South American families, Octodontidæ and Echimyidæ in the Ethiopian and Palæarctic regions, is an indication that the range of many of the families has recently become less extensive.

Edentata.—These singular and lowly-organised animals appear to have become almost restricted to the two great Southern lands—South Africa and South America—at an early period ; and, being there free from the competition of higher forms, developed a number of remarkable types often of huge size, of which the Megatherium is one of the best known. The incursion of the highly-organised Ungulates and Carnivora into Africa during the Miocene epoch, probably exterminated most of them in that continent ; but in America they continued in full force down to the Post-Pliocene period ; and even now, the comparatively diminutive Sloths, Ant-eaters, and Armadillos, form a large and important portion of the fauna.

Marsupialia and Monotremata.—These are probably the representatives of the most ancient and lowly-organised types of mammal. They once existed in the northern continents, whence they spread into Australia; and being isolated, and preserved from the competition of the higher forms which soon arose in other parts of the world, they have developed into a variety of types, which, however, still preserve a general uniformity of organisation. One family, which continued to exist in Europe till the latter part of the Miocene period, reached America, and has there been preserved to our day.

Lines of Migration of the Mammalia.—The whole series of phenomena presented by the distribution of the Mammalia, looked at broadly, are in harmony with the view that the great continents and oceans of our own epoch have been in existence, with comparatively small changes, during all Tertiary times. Each one of them has, no doubt, undergone considerable modifications in its area, its altitude, and in its connection with other lands. Yet some considerable portion of each continent has, probably, long existed in its present position, while the great oceans seem to have occupied the same depressions of the earth's crust (varied, perhaps, by local elevations and subsidences) during all this vast period of time. Hence, allowing for the changes of which we have more or less satisfactory evidence, the migrations of the chief mammalian types can be pretty clearly traced. Some, owing to their small size and great vitality, have spread to almost all the chief land masses ; but the majority of the orders have a more restricted range. All the evidence at our command points to the Northern Hemisphere as the birth-place of the class, and probably of all the orders. At a very early period the land communication with Australia was cut off, and has never been renewed ; so that we have here preserved for us a sample of one or more of the most ancient forms of mammal. Somewhat later the union with South America and South Africa was severed ; and in both these countries we have samples of a somewhat more advanced stage of mammalian development. Later still, the union by a northern route between the Eastern and Western Hemispheres appears to have been broken, partly by a physical separation, but almost as effectually by a lowering of temperature. About the same period the separation of the Palæarctic region from the Oriental was effected, by the rise of the Himalayas and the increasing contrast of climate ; while the formation of the great desert-belts of the Sahara, Arabia, Persia, and Central Asia, helped to complete the separation of the Temperate and Tropical zones, and to render further intermigration almost impossible.

In a few cases—of which the Rodents in Australia and the pigs in Austro-Malaya are perhaps the most striking examples —the distribution of land-mammals has been effected by a sea-passage either by swimming or on floating vegetation ; but, as a rule, we may be sure that the migrations of mammalia have taken place over the land; and their presence on islands is, therefore, a clear indication that these have been once connected with a continent. The present class of animals thus affords the best evidence of the past history of the land surface of our globe; and we have chiefly relied upon it in sketching out (in Part III.) the probable changes which each of our great regions has undergone.

Birds.

Although birds are, of all land-vertebrates, the best able to cross seas and oceans, it is remarkable how closely the main features of their distribution correspond with those of the Mammalia. South America possesses the low Formicaroid type of Passeres,—which, compared with the more highly developed forms of the Eastern Hemisphere, is analogous to the Cebidæ and Hapalidæ as compared with the Old World Apes and Monkeys; while its Cracidæ as compared with the Pheasants and Grouse, may be considered parallel to the Edentata as compared with the Ungulates of the Old World. The Marsupials of America and Australia, are paralleled. among birds, in the Struthionidæ and Megapodiidæ ; the Lemurs and Insectivora preserved in Madagascar are represented by the Mascarene Dididæ; the absence of Deer and Bears from Africa is analogous to the absence of Wrens, Creepers, and Pheasants ; while the African Hyracidæ and Chrysochloridæ among mammals, may well be compared with the equally peculiar Coliidæ and Musophagidæ among birds.

From these and many other similarities of distribution, it is clear that birds have, as a rule, followed the same great lines of migration as mammalia ; and that oceans, seas, and deserts, have always to a great extent limited their range. Yet these barriers have not been absolute; and in the course of ages birds have been able to reach almost every habitable land upon the globe. Hence have arisen some of the most curious and interesting phenomena of distribution; and many islands, which are entirely destitute of mammalia, or possess a very few species, abound in birds, often of peculiar types and remarkable for some unusual character or habit. Striking examples of such interesting bird-fauuas are those of New Zealand, the Sandwich Islands, the Galapagos, the Mascarene Islands, the Moluccas, and the Antilles; while even small and remote islets,—such as Juan Fernandez and Norfolk Island, have more light thrown upon their past history by means of their birds, than by any other portion of their scanty fauna.

Another peculiar feature in the distribution of this class is the extraordinary manner in which certain groups and certain external characteristics, have become developed in islands, where the smaller and less powerful birds have been protected from the incursions of mammalian enemies, and where rapacious birds—which seem to some degree dependent on the abundance of mammalia—are also scarce. Thus, we have the Pigeons and the Parrots most wonderfully developed in the Australian region, which is pre-eminently insular; and both these groups here acquire conspicuous colours very unusual, or altogether absent, elsewhere. Similar colours (black and red) appear, in the same two groups, in the distant Mascarene islands; while in the Antilles the parrots have often white heads, a character not found in the allied species on the South American continent. Crests, too, are largely developed, in both these groups, in the Australian region only; and a crested parrot formerly lived in Mauritius,—a coincidence too much like that of the colours as above noted, to be considered accidental.

Again, birds exhibit to us a remarkable contrast as regards the oceanic islands of tropical and temperate latitudes; for while most of the former present hardly any cases of specific identity with the birds of adjacent continents, the latter often show hardly any differences. The Galapagos and Madagascar are examples of the first-named peculiarity ; the Azores and the Bermudas of the last; and the difference can be clearly traced to the frequency and violence of storms in the one case and to the calms or steady breezes in the other.

It appears then, that although birds do not afford us the same convincing proof of the former union of now disjoined lands as we obtain from mammals, yet they give us much curious and suggestive information as to the various and complex modes in which the existing peculiarities of the distribution of animals have been brought about. They also throw much light on the relation between distribution and the external characters of animals ; and, as they are often found where mammalia are quite absent, we must rank them as of equal value for the purposes of our present study.

Reptiles.

These hold a somewhat intermediate place, as regards their distribution, between mammals and birds, having on the whole rather a wider range than the former, and a more restricted one than the latter.

Snakes appear to have hardly more facilities for crossing the ocean than mammals ; hence they are generally absent from oceanic islands. They are more especially a tropical group, and have thus never been able to pass from one continent to another by those high northern and southern routes, which we have seen reason to believe were very effectual in the case of mammalia and some other animals. Hence we find no resemblance between the Australian and Neotropical regions, or between the Palæ-arctic and Nearctic; while the Western Hemisphere is comparatively poor as regards variety of types, although rich in genera and species. Deserts and high mountains are also very effectual barriers for this group, and their lines of migration have probably been along river valleys, and occasionally across narrow seas by means of floating vegetation.

Lizards, being somewhat less tropical than snakes, may have passed by the northern route during warm epochs. They are also more suited to traverse deserts, and they possess some unknown means of crossing the ocean, as they are not unfrequently found in remote oceanic islands. These various causes have modified their distribution. The Western Hemisphere is much richer in lizards than it is in snakes; and it is also very distinct from the Eastern Hemisphere. The lines of migration of lizards appear to have been along the mountains and deserts of tropical countries, and, under special conditions, across tropical seas from island to island.

Crocodiles are a declining group. They were once more generally distributed, all the three families being found in British Eocene deposits. Being aquatic and capable of living in the sea, they can readily pass along all the coasts and islands of the warmer parts of the globe. Tortoises are equally ancient, and the restriction of certain groups to definite areas seems to be also a recent phenomenon.

Amphibia.

The Amphibia differ widely from Reptiles in their power of enduring cold; one of their chief divisions, the Urodela or Tailed-Batrachia, being confined to the temperate parts of the Northern Hemisphere. To this class of animals the northern and southern routes of migration were open; and we accordingly find a considerable amount of resemblance between South America and Australia, and a still stronger affinity between North America and the Palæarctic continent. The other tropical regions are more distinct from each other; clearly indicating that, in this group, it is tropical deserts and tropical oceans which are the barriers to migration. The class however is very fragmentary, and probably very ancient; so that descendants of once widespread types are now found isolated in various parts of the globe, between which we may feel sure there has been no direct transmission of Batrachia. Remembering that their chief lines of migration have been by northern and southern land-routes, by floating ice, by fresh-water channels, and perhaps at rare intervals by ova being carried by aquatic birds or by violent storms,—we shall be able to comprehend most of the features of their actual distribution.

Fresh-water Fishes.

Although it would appear, at first sight, that the means of dispersal of these animals are very limited, yet they share to some extent the wide range of other fresh-water organisms. They are found in all climates; but the tropical regions are by far the most productive, and of these South America is perhaps the richest and most peculiar. There is a certain amount of identity between the two northern continents, and also between those of the South Temperate zone; yet all are radically distinct, even North America and Europe having but a small proportion of their forms in common. The occurrence of allied fresh-water species in remote lands—as the *Aphritis* of Tasmania and Patagonia, and the *Comephorus* of Lake Baikal, distantly allied to the mackerels of Northern seas— would imply that marine fishes are often modified for a life in fresh waters; while other facts no less plainly show that permanent fresh-water species are sometimes dispersed in various ways across the oceans, more especially by the northern and southern routes.

The families of fresh-water fishes are often of restricted range, although cases of very wide and scattered distribution also occur. The great zoological regions are, on the whole, very well characterized; showing that the same barriers are effectual here, as with most other vertebrates. We conclude, therefore, that the chief lines of migration of fresh-water fishes have been across the Arctic and Antarctic seas, probably by means of floating ice as well as by the help of the vast flocks of migratory aquatic birds that frequent those regions. On continents they are, usually, widely dispersed; but tropical seas, even when of small extent, appear to have offered an effectual barrier to their dispersal. The cases of affinity between Tropical America, Africa, Asia, and Australia, must therefore be imputed either to the survival of once widespread groups, or to analogous adaptation to a fresh-water life of wide-spread marine types; and these cases cannot be taken as evidence of any former land connection between such remote continents.

Insects.

It has already been shown (Vol. I. pp. 209-213 and Vol. II. pp. 44-48) that the peculiarities of distribution of the various groups of insects depend very much on their habits and general economy Their antiquity is so vast, and their more important modifications of structure have probably occurred so slowly, that modes of dispersal depending on such a combination of favourable conditions as to be of excessive rarity, may yet have had time to produce large cumulative effects. Their small specific gravity and their habits of flight render them liable to dispersal by winds to an extent unknown in other classes of animals; and thus, what are usually very effectual barriers have been overstepped, and sometimes almost obliterated, in the case of insects. A careful examination will, however, almost always show traces of an ancient fauna, agreeing in character with other classes of animals, intermixed with the more prominent and often more numerous forms whose presence is due to this unusual facility of dispersal.

The effectual migration of insects is, perhaps more than in any other class of animals, limited by organic and physical conditions. The vegetation, the soil, the temperature, and the supply of moisture, must all be suited to their habits and economy; while they require an immunity from enemies of various kinds, which immigrants to a new country seldom obtain. Few organisms have, in so many complex ways, become adapted to their special environment, as have insects. They are in each country more or less adapted to the plants which belong to it; while their colours, their habits, and the very nature of the juices of their system, are all modified so as to protect them from the special dangers which surround them in their native land. It follows, that while no animals are so well adapted to show us the various modes by which dispersal may be effected, none can so effectually teach us the true nature and vast influence of the organic barrier in limiting dispersal.

It is probable that insects have at one time or another taken advantage of every line of migration by which any terrestrial organisms have spread over the earth, but owing to their small size and rapid multiplication, they have made use of some which are exclusively their own. Such are the passage along mountain ranges from the Arctic to the Antarctic regions, and the dispersal of certain types over all temperate lands. It will perhaps be found that insects have spread over the land surface in directions dependent on our surface zones—forests, pastures, and deserts;—and a study of these, with a due consideration of the fact that narrow seas are scarcely a barrier to most of the groups, may assist us to understand many of the details of insect-distribution.

Terrestrial Mollusca.

The distribution of land-shells agrees, in some features, with that of insects, while in others the two are strongly contrasted. In both we see the effects of great antiquity, with some special means of dispersal ; but while in insects the general powers of motion, both voluntary and involuntary, are at a maximum, in land-molluscs they are almost at a minimum. Although to some extent dependent on vegetation and climate, the latter are more dependent on inorganic conditions, and also to a large extent on the general organic environment. The result of these various causes, acting through countless ages, has been to spread the main types of structure with considerable uniformity over the globe ; while generic and sub-generic forms are often wonderfully localized.

Land-shells, even more than insects, seem, at first sight, to require regions of their own; but we have already pointed out the disadvantages of such a method of study. It will be far more instructive to refer them to those regions and sub regions which are found to accord best with the distribution of the higher animals, and to consider the various anomalies they present as so many problems, to be solved by a careful study of their habits and economy, and especially by a search after the hidden causes which have enabled them to spread so widely over land and ocean.

The lines of migration which land-shells have followed, can

hardly be determined with any definiteness. On continents they seem to spread steadily, but slowly, in every direction, checked probably by organic and physical conditions rather than by the barriers which limit the higher groups. Over the ocean they are also slowly dispersed, by some means which act perhaps at very long intervals, but which, within the period of the duration of genera and families, are tolerably effective. It thus happens that, although the powers of dispersal of land-shells and insects are so very unequal, the resulting geographical distribution is almost the opposite of what might have been expected,—the former being, on the whole, less distinctly localized than the latter.

CONCLUSION.

The preceding remarks are all I now venture to offer, on the distinguishing features of the various groups of land-animals as regards their distribution and migrations. They are at best but indications of the various lines of research opened up to us by the study of animals from the geographical point of view, and by looking upon their range in space and time as an important portion of the earth's history. Much work has yet to be done before the materials will exist for a complete treatment of the subject in all its branches; and it is the author's hope that his volumes may lead to a more systematic collection and arrangement of the necessary facts. At present all public museums and private collections are arranged zoologically. All treatises, monographs, and catalogues, also follow, more or less completely, the zoological arrangement; and the greatest difficulty the student of geographical distribution has to contend against, is the total absence of geographical collections, and the almost total want of complete and comparable local catalogues. Till every well-marked district,—every archipelago, and every important island, has all its known species of the more important groups of animals catalogued on a uniform plan, and with a uniform nomenclature, a thoroughly satisfactory account of the Geographical Distribution of Animals will not be possible. But more than this is wanted. Many of the most curious relations between animal forms and their habitats, are entirely unnoticed, owing to the productions of the same locality *never* being associated in our museums and collections. A few such relations have been brought to light by modern scientific travellers, but many more remain to be discovered; and there is probably no fresher and more productive field still unexplored in Natural History. Most of these curious and suggestive relations are to be found in the productions of islands, as compared with each other, or with the continents of which they form appendages; but these can never be properly studied, or even discovered, unless they are visibly grouped together. When the birds, the more conspicuous families of insects, and the land-shells of islands, are kept together so as to be readily compared with similar associations from the adjacent continents or other islands, it is believed that in almost every case there will be found to be peculiarities of form or colour running through widely different groups, and strictly indicative of local or geographical influences. Some of these coincident variations have been alluded to in various parts of this work, but they have never been systematically investigated. They constitute an unworked mine of wealth for the enterprising explorer; and they may not improbably lead to the discovery of some of the hidden laws (supplementary to Natural Selection), which seem to be required, in order to account for many of the external characteristics of animals.

In concluding his task, the author ventures to suggest, that naturalists who are disposed to turn aside from the beaten track of research, may find in the line of study here suggested a new and interesting pursuit, not inferior in attractions to the lofty heights of transcendental anatomy, or the bewildering mazes of modern classification. And it is a study which will surely lead them to an increased appreciation of the beauty and the harmony of nature, and to a fuller comprehension of the complex relations and mutual interdependence, which link together every animal and vegetable form, with the ever-changing earth which supports them, into one grand organic whole.

GENERAL INDEX.

GENERAL INDEX.

ALL names in Italics refer, either to the genera and other groups of Extinct Animals in Part II. of the First Volume ;—or to the genera whose distribution is given under Geographical Zoology (Part IV.) in the Second Volume ; the Families and higher groups being in small capitals. All other references are in ordinary type.

The various matters discussed under Zoological Geography (Part III.), are indexed as much as possible by subjects and localities. None of the genera mentioned in this Part are indexed, as this would have more than doubled the extent of the Index, and would have served no useful purpose, because the general distribution of each genus is given in Part IV., and the separate details can always be found by referring to the region, sub-region, and class.

A.

QQ 2

THE END.

LONDON : R. CLAY, SONS, AND TAYLOR, PRINTERS.